TRENDS IN
ASTROPARTICLE
PHYSICS

TRENDS IN

ASTROPARTICLE PHYSICS

Santa Monica, California   Nov 28 – Dec 1, 1990

Edited by

D. Cline & R. Peccei

Physics Dept, UCLA

World Scientific

Singapore • New Jersey • London • Hong Kong

Published by

World Scientific Publishing Co. Pte. Ltd.
P O Box 128, Farrer Road, Singapore 9128
USA office: Suite 1B, 1060 Main Street, River Edge, NJ 07661
UK office: 73 Lynton Mead, Totteridge, London N20 8DH

Library of Congress Cataloging-in-Publication data is available.

TRENDS IN ASTROPARTICLE PHYSICS

ISBN 981-02-0825-1

Printed in Singapore by Utopia Press.

PREFACE

During the last week of November in 1990 we held two meetings in Santa Monica, California. These meetings were the "SuperNova Watch Workshop" and the "First International Trends in Astroparticle Physics Conference". Contributions from both meetings are included in this book.

The SuperNova Watch is a world wide search for neutrino bursts from a galactic supernova (SN) as well as the automated Berkeley search for the optical pulse from the SuperNova. Detection of SuperNova pulses would be the ultimate neutrino laboratory as well as advancing our understanding of nuclear physics and the search for new particles such as axions.

The Trends meeting covered the interface between elementary particles and the universe. Topics covered range from massive neutrinos, axions, WIMPS, large scale structure of the universe, COBE results to the early universe particle physics at the GUT and Planck scales. There was an entire session devoted to microwave background measurements, emphasizing the COBE results (organized by E. Wright).

Organizers who worked conscientiously on preparations for the meeting and deserve special thanks are G. Gelmini, I. McLean, Roger Ulrich, Mark Morris, R. Peccei and E. Wright. Representing the conference secretariat, both M. Laraneta and M. Barrett deserve our thanks. But most importantly we wish to thank the speakers for first sharing their results at the conference and then taking the time to produce the papers which have made this publication possible.

Since the meeting we have received many votes of thanks for providing the opportunity for interchange on the issues discussed in the papers that follow. Due to this enthusiastic response from the community in general, and taking into consideration the many requests received from participants, we are beginning to plan the Second UCLA International Conference on Trends in Astroparticle Physics. This second meeting will again be held in Santa Monica, California in either late 1993 or early 1994. We want to take this opportunity to invite you all to join us for what we hope will prove to be an equally productive experience.

David B. Cline
UCLA

CONTENTS

III. LARGE SCALE STRUCTURE OF THE UNIVERSE AND STRUCTURE FORMATION

IV. COBE AND THE ANISOTROPY IN THE MICROWAVE BACKGROUND

V. NEUTRINOS, COSMOLOGY AND NUCLEOSYNTHESIS

VI. NEW DETECTORS FOR DARK MATTER AND OTHER NEW PARTICLES

VII. THE SUPERNOVA WATCH

I. DARK MATTER IN THE UNIVERSE

MICHAEL S. TURNER

NASA/Fermilab Astrophysics Center
Fermi National Accelerator Laboratory
Batavia, IL 60510-0500
and
Departments of Physics and Astronomy & Astrophysics
Enrico Fermi Institute
The University of Chicago
Chicago, IL 60637-1433

What is the quantity and composition of material in the Universe? This is one of the most fundamental questions we can ask about the Universe, and its answer bears on a number of important issues including the formation of structure in the Universe, and the ultimate fate and the earliest history of the Universe. Moreover, answering this question could lead to the discovery of new particles, as well as shedding light on the nature of the fundamental interactions. At present, only a partial answer is at hand: Most of the material in the Universe does not give off detectable radiation, i.e., is "dark;" the dark matter associated with bright galaxies contributes somewhere between 10% and 30% of the critical density (by comparison luminous matter contributes less than 1%); baryonic matter contributes between 1.1% and 12% of critical. The case for the spatially-flat, Einstein–de Sitter model is supported by three compelling theoretical arguments—structure formation, the temporal Copernican principle, and inflation—and by some observational data. If Ω is indeed unity—or even just significantly greater than 0.1—then there is a strong case for a Universe comprised of nonbaryonic matter. There are three well motivated particle dark-matter candidates: an axion of mass 10^{-6} eV to 10^{-4} eV; a neutralino of mass $10\,\mathrm{GeV}$ to about $3\,\mathrm{TeV}$; or a neutrino of mass $20\,\mathrm{eV}$ to $90\,\mathrm{eV}$. All three possibilities can be tested by experiments that are either being planned or are underway.

I. Weighing the Universe: Dark Matter Dominates!

The Friedmann–Robertson–Walker cosmology, also known as the hot big bang model, provides a reliable and tested accounting of the Universe from about 10^{-2} sec after the bang until the present. It is so successful that it is known as the standard cosmology. In the context of this cosmology the critical density separates models that expand forever ($\rho < \rho_{\mathrm{CRIT}}$) from those that ultimately recollapse ($\rho > \rho_{\mathrm{CRIT}}$); $\rho_{\mathrm{CRIT}} \equiv 3H_0^2/8\pi G \simeq 1.88h^2 \times 10^{-29}\,\mathrm{g\,cm^{-3}} \simeq 1.05h^2 \times 10^4\,\mathrm{eV\,cm^{-3}}$, where the present value of the Hubble parameter $H_0 = 100h\,\mathrm{km\,sec^{-1}\,Mpc^{-1}} \simeq 1/3000h^{-1}\,\mathrm{Mpc}$. I will denote the ratio of the *total* energy density ρ (including a possible vacuum energy) to the critical density by $\Omega \equiv \rho/\rho_{\mathrm{CRIT}}$, and the fraction of critical density contributed by species i by, $\Omega_i \equiv \rho_i/\rho_{\mathrm{CRIT}}$. The flat Einstein–de Sitter model corresponds to $\Omega = 1$; the negatively curved model to $\Omega < 1$; and the positively curved model to $\Omega > 1$. The radius of curvature can be expressed in terms of H_0 and Ω: $R_{\mathrm{CURV}} = H_0^{-1}/|\Omega - 1|^{1/2}$.

There are a variety of methods for determining Ω.[1] Broadly speaking they can be divided into two qualitatively different categories. First, there are the dynamical methods where the mass density is inferred by its gravitational effects; these include measuring the "rotation curves" of spiral galaxies, the virial masses of clusters of galaxies, and the local peculiar-velocity field. Second, there are the kinematic methods, which are sensitive to both the space-time geometry and the time evolution of the cosmic scale factor $R(t)$. They include the classic Hubble diagram (red shift–luminosity relation), the red shift–galaxy count relation, red shift–angular size relation, and others.[2]

Dynamical Methods

One can use Kepler's third law to determine the mass of a galaxy: $GM = v^2 r$, where v is the orbital velocity of a "test particle," r is its orbital radius, and M is the mass interior to the orbit (valid for a spherical mass distribution); or its statistical analogue, the virial theorem, to determine the mass of a gravitationally bound cluster: $GM = \langle v^2 \rangle r$ where M is the cluster mass, $\langle v^2 \rangle^{1/2}$ is the velocity dispersion of the galaxies, and r is the core radius of the cluster (orbits are assumed to be distributed isotropically).

For simplicity, one can imagine that one uses these methods to determine the "average mass per galaxy" and then multiplies it by the number density of galaxies to determine the average mass density ρ. In reality, astronomers use these methods to determine the mass-to-light ratio for spiral galaxies and for clusters of galaxies; from the mass-to-light ratio they infer the average mass density

$$\rho = \langle M/L \rangle \, \mathcal{L}, \tag{1}$$

where $\langle M/L \rangle$ is the mass-to-light ratio, and $\mathcal{L}$ is the luminosity density, whose value is about $2.4h \times 10^8 \, L_{B\odot} \, \mathrm{Mpc}^{-3}$ in the B_T system. The critical mass-to-light ratio is $(M/L)_{\mathrm{CRIT}} \simeq 1200h \, M_\odot/L_\odot$, where subscript $\odot$ refers to solar units.

"Rotation curves"—that is orbital velocity as a function of orbital distance—have been determined for numerous spiral galaxies. They are obtained by measuring the Doppler shifts of stellar spectral features and of the 21 cm radiation from neutral gas clouds (HI regions)—the stars and clouds act as gravitational test particles. Rotation curves are all qualitatively similar; they rise rapidly from the galactic center and remain flat ($v = $ const) out to the furthest distances that can be probed—eventually, one "runs out" of test particles, i.e., stars and gas clouds. Since $v = $ const implies $M(r) \propto r$, this means that one "runs out" of stellar light and 21 cm radiation before the mass of the galaxy has "converged." In some cases the 21 cm rotation curves have been determined to a distance that is three times that where the light has fallen to 1% of its value at the center of the galaxy.

By restricting oneself to the bright central regions of a galaxy one can use the rotation velocity to infer the amount of mass associated with the "luminous" part of the galaxy; doing so one finds that luminous matter contributes

$$\Omega_{\mathrm{LUM}} \lesssim 0.01, \tag{2}$$

which is far from the critical density. A similarly small value is obtained by using the mass-to-light ratio determined for the local solar neighborhood, $\langle M/L \rangle_{\mathrm{local}} \sim 2 - 3$.

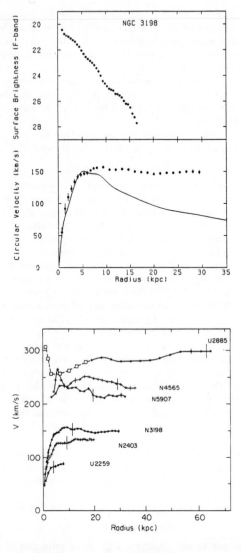

Fig. 1. Upper: F-band surface brightness of NGC 3198 in units mag arcsec^{-2} (F-band covers a "red" part of the spectrum from about 5000Å to about 7000Å). 21 cm rotation curve for NGC 3198 (dots with error flags) and rotation curve predicted from the luminous matter alone (assuming constant mass-to-light ratio $M/L_B = 4$). Lower: Rotation curves for a number of spiral galaxies determined from 21 cm observations. Vertical bars indicate the point beyond which the surface brightness is less than 25 (blue) mag arcsec^{-2} (less than about 1% of the central surface brightness). [From Sancisi and van Albada in Kormendy and Knapp, Ref. 1.]

Based upon the fact that many rotation curves stay flat out to distances far beyond where the surface luminosity of the galaxy is negligible, one can infer that there is much more matter associated with spiral galaxies that is dark (i.e., does not give off visible radiation) than is luminous. For our own galaxy the rotation velocity has been measured out to a distance of about 20 kpc, at which point the dark matter contributes about three times more mass than the luminous matter (for reference the solar system is about 8.5 kpc from the center of the galaxy). There is weaker evidence that this dark matter exists in a spherically-symmetric, extended halo with a density that varies as r^{-2} at large distances from the center of the galaxy.

Based upon the rotation curves, one can conclude that the dark halo material in spiral galaxies contributes *at least* three to ten times the mass density that luminous matter does,

$$\Omega_{\text{HALO}} \gtrsim 0.03 - 0.10. \tag{3}$$

Since there is no convincing evidence for a rotation curve that "turns over" and decreases as $r^{-1/2}$ indicating that the halo mass has converged, it is possible that the halos of spiral galaxies extend a factor of order ten further and thereby provide the critical density.[a]

The existence of dark matter halos in spiral galaxies provides the answer to one puzzle—the stability of galactic disks—and raises another—the apparent conspiracy of the luminous matter and dark matter to produce smooth rotation curves. A disk-like structure is subject to many instabilities, and a massive halo stabilizes a disk-like structure against these instabilities thereby resolving a longstanding puzzle. However, the existence of dark matter halos raises another question: Why do the inner and outer parts of the rotation curve join so smoothly, in light of the fact that the inner part of the rotation curve is supported by luminous matter and the outer part by dark matter? (The rotation curves of most spiral galaxies are very similar, with the rotation velocity rising rapidly from zero at the center to a nearly constant value. The rotation curve for our own galaxy is quite flat and smooth at our position, in spite of the fact that the gravitational support for rotation velocities at our position are about equally split between luminous disk material and dark halo material.) Some (e.g., Peebles) have argued that this is evidence that the halo and disk have a similar composition—baryons—while others put their faith in numerical simulations of the formation of galactic halos and disks that indicate that this occurs quite naturally when the ratio of nondissipative dark matter and dissipative luminous matter is of order ten.

There is some evidence that individual elliptical galaxies contain significant amounts of dark matter, although the case is not as well established as that for spirals. Most cluster galaxies are ellipticals, and as I will now discuss there is strong evidence for dark matter in clusters.

Estimates of the mass density based upon the virial masses of clusters lead to

$$\Omega_{\text{CLUSTER}} \simeq 0.1 - 0.3, \tag{4}$$

---

[a] There are arguments to the contrary; e.g., mass estimates of the Milky Way and Andromeda based upon their velocity of approach seem to indicate that their halos could not be this large, although such arguments assume that the Milky Way and Andromeda are on a radial orbit and are approaching each other for the first time. Likewise, mass estimates of the Milky Way based upon the orbits of its satellite galaxies indicate the same, although it is assumed that the orbits are isotropically distributed.[3]

again indicating substantially more mass than that required to account for the light. Several points should be noted: (1) X-ray emission from hot intracluster gas indicates the presence of comparable or greater amounts of baryonic mass than that associated with the visible light (dark is a relative term!), but no where near enough to account for the cluster's virial mass. (2) Since only about one in ten galaxies resides in a large cluster, one can question whether or not the mass-to-light ratio—and value of Ω—deduced from clusters is indicative. However, there seems to be no question that clusters contain significant amounts of dark matter. (3) These determinations are based upon the assumption that the clusters are well virialized, single objects and that the galaxy orbits are distributed isotropically; moreover, the cluster core radius is inferred from the distribution of the visible galaxies. If galaxies have sunk deep into the cluster potential, e.g., due to dynamical friction, then the actual core radius of the cluster—and cluster mass—could be much larger[3a] (just as with galactic halos).[b]

Peculiar Velocities

The velocity of a galaxy can be split into two pieces: the velocity due to the general expansion of the Universe (or Hubble velocity) which is radial and proportional to galaxy's distance from us; and the peculiar velocity, the velocity the galaxy has in addition to its Hubble velocity.[c] Any peculiar velocity that is not "supported" by a gravitational field will decay with time, inversely with the cosmic scale factor $R(t)$. Put another way, peculiar velocities arise due to the lumpy distribution of matter—and thereby offer a probe of the density field. In contrast, the distribution of bright galaxies only probes the distribution of light—and the two distributions need not be the same.

In the linear perturbation regime, i.e., $\delta\rho/\rho \lesssim 1$, the Fourier expansion of the velocity field, $\mathbf{v}_k$, is related to that of the density field, δ_k, $\mathbf{v}_k = -i\mathbf{k}R(t)\dot{\delta}_k(t)/|\mathbf{k}|^2$, and to a good approximation $|\mathbf{v}_k| \simeq \Omega^{0.6} H_0 |\delta_k|/k$. Suppose the peculiar velocity of an object is primarily due to linear perturbations on the scale λ, then

$$\frac{\delta v}{c} \sim \Omega^{0.6} \left(\frac{\lambda}{H_0^{-1}}\right) \left(\frac{\delta\rho}{\rho}\right)_\lambda. \tag{5}$$

Even if the contribution from one Fourier component does not dominate, Eq. (5) still illustrates the correct dependence of the peculiar velocity upon Ω.[d] One can exploit this relationship in different ways: (i) input Ω and $\delta\mathbf{v}$ to infer $\delta\rho(\mathbf{r})/\rho$; (ii) input Ω and $\delta\rho(\mathbf{r})/\rho$ to infer $\delta\mathbf{v}$; or (iii) input $\delta\mathbf{v}$ and $\delta\rho(\mathbf{r})/\rho$ to infer Ω. The last of these alternatives is the one we are interested in here; however, what one can directly measure is $\delta n_G(\mathbf{r})/n_G$, and so one must relate $\delta n_G/n_G$ to $\delta\rho/\rho$ (n_G is the number density of "bright" galaxies). The simplest *ansatz* is to take them to be equal: "light traces mass." A slightly more general

---

[b] It should be mentioned that in 1933 the astrophysicist Fritz Zwicky pointed out that the mass associated with the light in several clusters was much less than the mass required to bind the cluster—and thus was the first to identify the dark matter problem.

[c] Of course, we can only measure the component of the peculiar velocity that is parallel to the line of sight.

[d] More precisely, the peculiar velocity at position $\mathbf{r}$ is $\delta\mathbf{v}(\mathbf{r}) = -\Omega^{0.6}(H_0/4\pi) \int \delta\rho(\mathbf{r}')(\mathbf{r} - \mathbf{r}')d^3r'/|\mathbf{r} - \mathbf{r}'|^3\rho$.

approach is to assume that "light is a biased tracer of mass:" $\delta\rho/\rho = b^{-1}(\delta n_G/n_G)$, where $1 \lesssim b \lesssim 3$ is the biasing factor.

Using the IRAS catalogue of infrared-selected galaxies to determine the mass distribution (i.e., $\delta n_G/n_G$), several groups have used measurements of the local peculiar-velocity field[4] to infer $\Omega^{0.6}/b \simeq 1.$, with an estimated uncertainty of about 0.3 or so.[5] With some delight, I note that this technique seems to suggest that Ω is indeed close to unity. Although I caution the reader that these results are still preliminary, if they hold up, they will provide the strongest evidence to date for a large value of Ω![e]

Before going on to the kinematic methods, I mention that there are other dynamical methods for determining Ω, including the use of gravitational lens systems to measure cluster and galaxy masses, Virgo infall (which is similar to the peculiar-velocity method mentioned above), cosmic virial theorems, and pair-wise velocities of galaxies.[6]

In addition, there may or may not be another, more local dark matter problem. The mass density of the disk in our neighborhood can be determined by studying the motions of stars perpendicular to the plane of the disk, and by a "direct inventory" of the material in the local neighborhood (stars, white dwarfs, gas, dust, etc.). In principle the two results should agree. The local mass density inferred from dynamics,[f] $1.3 \times 10^{-23}\,\mathrm{g\,cm^{-3}}$, is about a factor of two larger than can be accounted for by the local inventory.[7] This discrepancy of a factor of two may or may not be significant. In any case, it has little bearing on the "big" dark matter problem. Since the mass density of the local neighborhood is dominated by luminous matter, this additional dark matter—if it exists—makes a contribution to Ω that is at most comparable to that of luminous matter.

Moreover, this local dark matter cannot be due to halo material: Based upon the rotation curve of our galaxy and detailed models for the distribution of matter in our galaxy, the local halo density is estimated to be[8]

$$\rho_{\mathrm{HALO}} \simeq 5 \times 10^{-25}\,\mathrm{g\,cm^{-3}} \simeq 0.3\,\mathrm{GeV\,cm^{-3}}, \qquad (6)$$

with an uncertainty of about a factor of two. The local halo density is about a factor of ten smaller than the local disk dark-matter density; put another way, if the halo material accounted for the disk dark-matter density, the local rotation velocity would be about a factor of three larger than its measured value!

Kinematic Determinations

There are a number of classic kinematic tests—luminosity–red shift (or Hubble diagram), angle–red shift, galaxy-number count–red shift—that can in principle be used to determine our cosmological model.[2] These tests depend upon the global space-time geometry and the time evolution of the scale factor. For example, the luminosity distance to

---

[e] The infrared bright galaxies tend to be spiral galaxies in the field, and so clusters are under-represented. The authors have tried to correct for this by including some important clusters, and find that their results do change significantly. One must also worry about convergence; that is, has one reached the point where the contribution of galaxies at still larger distances has become insignificant.

[f] This density is known as the Oort limit, in honor of the first astronomer to address this problem.

a galaxy at red shift z, $d_L^2 \equiv \mathcal{L}/4\pi\mathcal{F}$, is related to the coordinate distance to the galaxy, $r(z)$, by

$$d_L^2 = r(z)^2(1+z)^2,$$

$$\int_0^{r(z)} \frac{dr}{\sqrt{1-kr^2}} = \int_{t(z)}^{t_0} \frac{dt}{R(t)}, \tag{7}$$

where the present value of the scale factor $R(t_0)$ is taken to be one and $(1+z) = R(t)^{-1}$. Since the evolution of the scale factor depends on the equation of state, e.g., $p = 0$, matter-dominated, $R \propto t^{2/3}$; $p = \rho/3$, radiation-dominated, $R \propto t^{1/2}$; $p = -\rho$, vacuum-dominated, $R \propto \exp(Ht)$, the functional dependence of $r(z)$ does too. Thus, the red shift–luminosity distance relation depends upon both the curvature of space and the composition of the Universe. For a matter-dominated model

$$H_0 d_L = q_0^{-2}\left[zq_0 + (q_0 - 1)\left(\sqrt{2q_0 z + 1} - 1\right)\right] = z\left[1 + (1 - q_0)z/2 + \cdots\right]. \tag{8}$$

where $q_0 \equiv -\ddot{R}_0/H_0^2 = \Omega(1 + 3p/\rho)/2 = \Omega/2$, and the second expression is an expansion in z.

The success or failure of this technique depends upon obtaining accurate luminosity distances for objects out to red shifts of order unity. Accurate luminosity distances requires the existence of objects of known luminosity (standard candles). Here lies the problem; evolutionary effects are likely to be important, especially at high red shifts, and it is difficult to determine even the sign of the evolutionary effects let alone reliably estimate the magnitude! Nevertheless, there are some who believe that the K-band ($2.2\,\mu$m) version of the Hubble diagram will prove useful,[9] as evolutionary effects are lessened.[g]

A kinematic test with great cosmological leverage and promise is the galaxy count–red shift relation. The number of galaxies seen in the red shift interval dz and solid angle $d\omega$ depends upon the number density of galaxies $n_G(z)$ and the spatial volume element, $dV = r^2 dr d\omega/\sqrt{1-kr^2}$. This relationship too depends upon both the spatial curvature and the time evolution of the scale factor. For a matter-dominated model,

$$\frac{dN_{\text{GAL}}}{d\omega dz} = \frac{n_{\text{GAL}}(z)[zq_0 + (q_0 - 1)(\sqrt{2q_0 z + 1} - 1)]^2}{H_0^3(1+z)^3 q_0^4[1 - 2q_0 + 2q_0(1+z)]^{1/2}},$$

$$\simeq z^2 n_{\text{GAL}}(z)[1 - 2(q_0 + 1)z + \cdots]/H_0^3. \tag{9}$$

For fixed (comoving) number density of galaxies, the galaxy count increases with decreasing Ω (or q_0) because of the increase in spatial volume. Loh and Spillar[10] have used the galaxy count–red shift test with a sample of about 1000 field galaxies—red shifts out to 0.75— to infer $\Omega = 0.9^{+0.7}_{-0.5}$ (95% confidence). Their result has drawn much criticism; in part because their red shifts are not spectroscopically determined (they are determined by six-band photometry) and because their results are sensitive to the assumptions made about galactic evolution.[11]

---

[g] When one observes a galaxy of moderate red shift in the visible, the light one sees comes from the blue or UV part of the spectrum and is produced by massive stars that evolve rapidly. By contrast, observing in K-band, the light one sees was emitted in the red part of the spectrum and is produced by lower mass stars that evolve much more slowly.

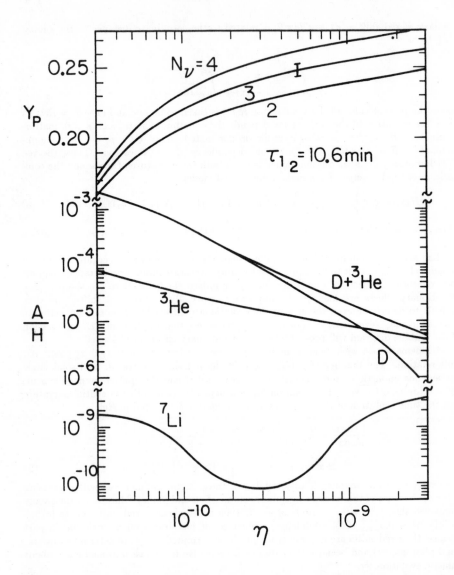

Fig. 2. Predicted light-element abundances as a function of the baryon-to-photon ratio η in the standard scenario of big-bang nucleosynthesis; error flag indicates the change in ^{4}He that arises for $\Delta\tau_{1/2}(n) = \pm 0.2\,$min. The inferred primordial abundances are: $Y_P = 0.24 \pm 0.01$; D/H $\gtrsim 10^{-5}$; (D+^{3}He)/H $\lesssim 1.1 \times 10^{-4}$; and ^{7}Li/H $\simeq 1.2 \pm 0.3 \times 10^{-10}$. Concordance between the predicted and measured abundances requires: $3 \times 10^{-10} \lesssim \eta \lesssim 5 \times 10^{-10}$; or $0.011 \lesssim 0.011h^{-2} \lesssim \Omega_B \lesssim 0.019h^{-2} \lesssim 0.12$.

In principle, this test is less sensitive to evolution, provided that the number of galaxies remains constant and their luminosities do not evolve so drastically that they cannot be seen. Recent deep galaxies counts indicate an excess of galaxies at higher red shifts—indicative of a low value of Ω.[12] (If galaxy mergers are very important—as they may well be in cold dark matter scenarios—the number density of galaxies at higher red shifts would be expected to be larger.) At the moment, determinations of Ω based upon the galaxy number count test are not conclusive. However, many believe that this method has great potential because a large sample of objects can be used and it is less sensitive to evolution.

Primordial Nucleosynthesis and Ω_B

Primordial nucleosynthesis provides the most stringent and earliest test of the standard cosmology, probing it back to the epoch when $T \sim$ MeV and $t \sim$ sec. The primordial abundances of D, ^{3}He, ^{4}He, and ^{7}Li predicted in the standard (and simplest) model of primordial nucleosynthesis agree with the inferred primordial abundances of these light elements.[13] Moreover, this agreement can be used to constrain one cosmological parameter—the baryon-to-photon ratio η—and one parameter of the standard model—the number of light neutrino species N_ν.[14] Concordance between theory and observation requires:

$$3 \times 10^{-10} \lesssim \eta \lesssim 5 \times 10^{-10} \qquad \text{and} \qquad N_\nu \leq 3.4$$

The constraint to the number of light neutrino species has recently been confirmed by precise measurements of the properties of the Z^0 boson,[15] which imply $N_\nu = 3.0 \pm 0.1$. This is an impressive confirmation of the standard cosmology at this very early epoch.

Primordial nucleosynthesis provides the most precise determination of the baryon density. In converting the baryon-to-photon ratio to the fraction of critical density contributed by baryons two other parameters are needed: (i) the temperature of the cosmic microwave background (CMBR), which is now accurately determined to be 2.736 ± 0.01 K;[16] and (ii) the not so well known value of the present Hubble parameter, $0.4 \lesssim h \lesssim 1.0$.[h] The nucleosynthesis constraint can be written as

$$0.011 \lesssim 0.011 h^{-2} \lesssim \Omega_B \lesssim 0.019 h^{-2} \lesssim 0.12. \tag{10}$$

Summary of Our Knowledge of Ω

What then is the present state of our knowledge concerning the mass density of the Universe? Let me try to summarize:

- Luminous matter contributes only a small fraction of the critical density: $\Omega_{LUM} \lesssim 0.01$.
- Based upon primordial nucleosynthesis baryonic matter contributes: $0.011 \lesssim \Omega_B \lesssim 0.12$.[i]
- Based upon dynamical methods, the mass density associated with bright galaxies is $\Omega_{ABG} \simeq 0.2 \pm 0.1$ (the ± 0.1 is not meant to be a formal uncertainty estimate).

---

[h] It also assumed that the only change in the baryon-to-photon ratio since the start of nucleosynthesis is the factor of 4/11 decrease caused by the transfer of the entropy in $e^{\pm}$ pairs to photons when $T \sim m_e/3$.

[i] If Ω is close to unity and the cosmological constant is zero, then h must be close to 0.5 to insure a sufficiently elderly Universe; in this case: $0.04 \lesssim \Omega_B \lesssim 0.12$.

• There is some evidence that Ω might be close to unity; e.g., analyses of the local peculiar-velocity field based upon the IRAS catalogue of galaxies, and the result of Loh and Spillar.

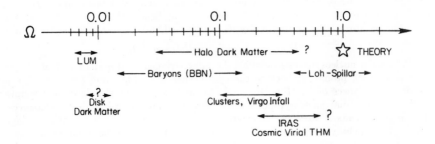

From this I would make the following inferences:

★ The dark component of the mass density dominates the luminous component by at least a factor of ten, and closer to a factor of 100 if $\Omega = 1$, and is more diffuse than the luminous component, e.g., the halos of spiral galaxies.

★ There is strong evidence for the existence of a dark component of baryons. This should not be too surprising since baryons can exist in a variety of low luminosity objects—white dwarfs, neutron stars, black holes, brown dwarfs, jupiters, etc.

★ At present there is no irrefutable case for a universal mass density that is larger than that permitted for baryons.

★ If Ω is significantly greater than 0.1—which is already suggested by mass-to-light ratios determined for clusters and the local peculiar velocity field—then there is a strong case for nonbaryonic dark matter. As I will discuss, there are three attractive particle dark-matter candidates whose relic abundance is expected to be close to critical: the axion, the neutralino, and a light neutrino.

★ If Ω is one, a discrepancy must be explained: why the estimates for the amount of material associated with bright galaxies is a factor of about five smaller. There are two possibilities. The first, as previously mentioned, the halos of spiral galaxies could extend far enough to account for $\Omega = 1$ (and likewise for clusters). Second, there could be a component of the mass density that is more smoothly distributed, contributes $\Omega_{SM} \simeq 0.8$, and is not associated with bright galaxies; e.g., a population of low-luminosity galaxies that is more smoothly distributed than the bright galaxies—so-called biased galaxy formation—or a relic cosmological constant (more later).

★ There may be several dark matter problems—and with different solutions. While the most economical approach is to assume that all dark matter has the same composition, that need not be the case. As mentioned above there is already evidence that some of the baryonic matter is dark. Moreover, if there is indeed a local dark matter problem, its solution must involve "particles" that can dissipate energy and condense into the disk; it is very unlikely that axions, neutralinos, or neutrinos can do so. Taken at face value the observations seem to indicate that there is more dark matter in clusters (per galaxy) than in the halos of spiral galaxies—and if $\Omega = 1$—even more dark matter that is not associated with clusters.

To give a concrete example, consider an $\Omega = 1$, neutrino-dominated Universe ($m_\nu \simeq 92h^2$ eV). Because of their high speeds, neutrinos would be unlikely to find their way into potential wells as shallow as those of galaxies or perhaps even clusters. They would likely remain smooth on scales up to the neutrino free-streaming length, $\lambda_{FS} \simeq 40\,\mathrm{Mpc}/(m_\nu/30\,\mathrm{eV})$. The dark matter in galaxies would be baryons—perhaps white dwarfs that formed relatively recently in the local neighborhood and brown dwarfs that formed when the galaxy did in the halo—and the dark matter in clusters would be the neutrinos that eventually made their way into clusters.[j]

A Theoretical Prejudice

While the hard observational evidence for the flat, Einstein–de Sitter model is less than overwhelming, there are several compelling theoretical arguments: (i) the temporal Copernican principle—if $\Omega \neq 1$ the deviation of Ω from unity grows as a power of the scale factor, begging one to ask why Ω is just now beginning to differ from unity; (ii) structure formation—in $\Omega < 1$ models there is less time for the growth of density perturbations and larger initial perturbations are required; in fact, $\Omega < 0.3$ models with adiabatic density perturbations are inconsistent with the isotropy of the CMBR (see Bond's contribution to these proceedings); and (iii) the flat, Einstein–de Sitter model is an inescapable prediction of inflation. To be sure, these arguments are not rooted in hard facts; however, the are sufficiently compelling to create a strong theoretical prejudice for $\Omega = 1$. From this point forward I will adopt this prejudice!

Dark Matter: New Physics or New Particles

Finally, there are some who have suggested another explanation for the dark matter problem: A deviation from Newtonian (Einsteinian) gravity at large distances.[18] Newtonian gravity (i.e., the weak field, slow velocity limit of general relativity) is well tested at distances from order 10^2 cm to the size of the solar system, order 10^{14} cm. However, the dark matter problem involves distance scales of order 10^{23} cm and greater. If gravity were for some reason stronger on these scales there would perhaps be no need for additional "unseen" matter to explain flat rotation curves. For example, if G were a function of distance, say $G(r) \propto r$, then flat rotation curves would be consistent with constant mass interior to r—eliminating the need for unseen matter.

I opt for unseen matter. First, it seems unlikely that the same functional dependence for the strength of gravity could fit all the observations: While all spiral galaxies have flat rotation curves, the size of the luminous part of the galaxy can vary by almost a factor of ten, and clusters are even larger. Perhaps a more important reason is that of aesthetics: Not only is there no theoretical motivation for such a theory, but it seems difficult, if not impossible, to construct a relativistic theory of gravity in which G increases with distance.

[j] In a neutrino-dominated Universe it is probably necessary for the dark matter in galaxies to be baryonic, as there seems to be evidence for dark matter in several dwarf galaxies in which there is not enough phase space to contain the necessary numbers of neutrinos.[17]

The one such theory I am aware is extremely complicated and leads to an unsatisfactory cosmology.[18] Were it the other way around—lack of compelling dark matter candidates and an attractive alternative theory of gravity—I would opt for new physics in the gravitational sector.

II. Why Not Baryons?

Given the existing observational evidence one has to be bold to insist that $\Omega = 1$. Moreover, this assumption seems to require one to go still further and postulate that most of the matter in the Universe is comprised of particles whose existence is still hypothetical! Before taking the big leap, I will comment on the possibility that baryons could contribute the critical density. There are two obstacles to this possibility: the nucleosynthesis constraint, $\Omega_B \lesssim 0.12$; and finding a place to hide the more than 99 invisible baryons for every visible baryon.

A number of different schemes have been suggested to evade the nucleosynthesis bound, for example, massive relic particles that decay into hadrons shortly after nucleosynthesis and initiate a second epoch of nucleosynthesis.[19] This scenario requires an unstable particle species with very special properties, and seems to lead to the overproduction of ^{6}Li and the underproduction of ^{7}Li. Perhaps the most clever idea is the scenario where the baryon-to-photon ratio is reduced after nucleosynthesis because photons suddenly come into thermal contact with "shadow particles" at a lower temperature, which leads to entropy transfer from the photons to the shadow world.[20]

Inhomogeneous Nucleosynthesis

The alternative to the standard scenario that has attracted the most attention is inhomogeneous nucleosynthesis.[21] If the quark/hadron transition is strongly first order and occurs at a relatively low temperature ($\lesssim 125\,\text{MeV}$), baryon number can become concentrated in regions where the quark–gluon plasma persisted the longest. Moreover, due to the difference in the mean free paths of the proton and neutron around the time of nucleosynthesis, the high baryon density regions will become proton rich. Clearly, nucleosynthesis proceeds very differently, and two new parameters arise: the density contrast between the high and low baryon density regions and the separation of the high density regions.

While early calculations, done with two independent "zones" of differing baryon number density and proton fraction, suggested that $\Omega_B \sim 1$ could be made consistent with the observed light element abundances by an appropriate choice of these two parameters, more detailed calculations that allow for diffusion between the zones indicate that the predicted abundances for *all four* light elements conflict with observations if $\Omega_B \sim 1$—for all values of the two parameters.[22] While this appears to be a sad end to an interesting idea, it does serve to emphasize the brilliant success of standard nucleosynthesis: The simplest model with no extra dials or knobs correctly predicts the primordial abundances of D, ^{3}He, ^{4}He, and ^{7}Li.

Where Is It?

Should one be able to evade the nucleosynthesis bound the next problem that one faces is where to put all those dark baryons. Ordinary stars, dust, and gas would all be "visible" in one way or another. Black holes and neutron stars do not necessarily provide an easy way out either. If, as seems likely, black holes and neutron stars evolve from massive stars, where are the heavy elements these stars produced? And remember, one is trying to hide 99 baryons for every baryon that is in a star. Perhaps massive black holes can form without overproducing heavy elements; however, there are other worries. If these black holes are too massive they will puff up the disk of the galaxy and disrupt binary stars by their gravitational effects, and lead to the (unobserved) lensing of distant QSOs. These considerations restrict the mass of black holes in the halo to be less than about $10^5 M_\odot$.[23]

White dwarfs, brown dwarfs (stars less massive than about $0.08 M_\odot$ which do not get hot enough for to burn hydrogen), or jupiters are better candidates.[24] All could have escaped detection thus far and *might* be detectable in planned experiments to look for microlensing of stars in the LMC by such objects in the halo of our galaxy. However, there is the issue of the large number of these objects needed. When one smoothly extrapolates the observed IMF (initial mass function of the most recent generation of stars) to these very small masses, one concludes that are far too few of these objects to account for the dark matter in the halo. It should be noted that the IMF is an empirical, rather than fundamental, relation, and some have suggested that when the galaxy formed most of its mass could have fragmented into small objects.

To summarize, it is not impossible to evade the nucleosynthesis bound, and there is no devastating argument to preclude astrophysical objects comprised of baryons from contributing critical density. However, the elegance of the nucleosynthesis argument and the difficulty of hiding so many baryons seem to suggest that nonbaryonic dark matter is a more promising option to pursue! .

III. Particle Dark Matter

According to the standard cosmology,[25] at times earlier than the epoch of matter–radiation equality, $t \lesssim t_{EQ} = 4.4 \times 10^{10} (\Omega h^2)^{-2}$ sec and $T \gtrsim T_{EQ} = 5.5 (\Omega h^2)$ eV, the energy density of the Universe was dominated by a thermal bath of particles at temperature T. For reference, for $t \lesssim t_{EQ}$, $T \sim \text{GeV}/\sqrt{t/10^{-6} \text{ sec}}$.

While the extrapolation of the standard cosmology to very early times ($t \ll 1$ sec) is a bold step, there are several reasons to expect that such an extrapolation is at least self consistent, if not correct: (1) The splendid success of big bang nucleosynthesis, which tests the standard cosmology well into its radiation-dominated phase; (2) The fact that according to the standard model of particle physics the fundamental degrees of freedom are pointlike quarks and leptons, gauge bosons, and Higgs (scalar) bosons[k] whose interactions are expected to remain perturbatively weak at very high energies; and (3) Quantum corrections to general relativity should be very small for times $t \gg 10^{-43}$ sec and temperatures $T \ll 10^{19}$ GeV.

The implications of this hot, early epoch for cosmology, and dark matter in particular, are manifold: At temperature T all particles of mass less than T should be present in numbers comparable to that of the photons; several phase transitions should take place (e.g., quark/hadron transition, chiral symmetry restoration, and electroweak symmetry restoration); and in the symmetry restored phase, the strength of all interactions—including "very weak" interactions that have yet to be discovered—should be comparable.

---

[k] Of course, the existence of the Higgs sector has yet to be confirmed, and there could well be some surprises at energies greater than $1/\sqrt{G_F} \sim 300$ GeV, corresponding to times earlier than 10^{-11} sec.

While the standard $SU(3)_C \otimes SU(2)_L \otimes U(1)_Y$ gauge theory of the strong and electroweak interactions does not offer any dark matter candidates—beyond the now dim hope that inhomogeneous nucleosynthesis could resurrect $\Omega_B \sim 1$—the speculations about fundamental physics beyond the standard model do. These well founded speculations include Peccei–Quinn (PQ) symmetry, technicolor, supersymmetry, grand unification, and superstrings. In the context of the hot big bang model these speculations lead to the prediction of various cosmological relics, including particles, topological defects (cosmic strings, domain walls, monopoles, and textures), and the baryon asymmetry of the Universe. The discovery—or nondiscovery—of an expected relic provides an important cosmological window on fundamental physics beyond the standard model. Since terrestrial experiments are hard pressed to probe the physics beyond the standard model, the Heavenly Laboratory has become an indispensable testing ground for fundamental physics.

In the context of the dark matter problem, the implications of theories that go beyond the standard model have great significance. Many of these theories predict particle relics whose contribution to the present mass density is comparable to the critical density! This is no mean feat, and for many of us is a strong hint that the idea of nonbaryonic particle relics as the dark matter is on the right track.

I have organized my discussion of particle dark-matter candidates into six broad categories: thermal relics; "skew" relics; axions; nonthermal relics; "significant-other" relics; and exotic relics. I have given the axion is own category not just because it is my favorite candidate, but also because the story of relic axions is a very rich one and spans three categories!

Thermal Relics

Because the Universe was in thermal equilibrium at early times essentially all the known particles—and perhaps many particles that are yet to be discovered—were present in great abundance: When the temperature T was greater than the mass m of a species, a number comparable to that of the photons If thermal equilibrium were the whole story, it would be a very uninteresting one indeed: At low temperatures the equilibrium abundance of a species is exponentially negligible, a factor of order $(m/T)^{3/2} \exp(-m/T)$ less than that of the photons.

A massive particle species can only maintain its equilibrium abundance so long as the rate for interactions that regulate its abundance is greater than the expansion rate of the Universe: $\Gamma \gtrsim H$, where the expansion rate of the Universe $H = 1.67 g_*^{1/2} T^2 / m_{\rm Pl}$ (g_* counts the total number of degrees of freedom of all relativistic species and $m_{\rm Pl} = 1.22 \times 10^{19}$ GeV). The expansion rate enters because it sets the rate at which the temperature is decreasing, $H = |\dot{T}|/T$, and therefore the rate at which phase-space distribution functions must change.

If we specialize to the case of interest for particle dark matter, a stable (or very long lived) particle, the reactions that control the abundance are pair production and annihilation, and their rates are related by detailed balance. The problem now reduces to a textbook example! The particle's number density n is governed by the Boltzmann equation, which takes the form[26]

$$\frac{dn}{dt} + 3Hn = -\langle \sigma|v|\rangle_{\rm ANN} \left(n^2 - n_{\rm EQ}^2 \right),$$ (11)

where $\langle \sigma|v|\rangle_{\rm ANN}$ is the thermally averaged annihilation cross section times relative velocity and $n_{\rm EQ}$ is the equilibrium number density. It is more convenient to recast Eq. (11) in terms of the number of particles per comoving volume,[l] $Y = n/s$, where $s = 2\pi^2 g_* T^3/45$ is the entropy density, and the dimensionless evolution variable $x = m/T$:

$$\frac{dY}{dx} = -\frac{x\,s\,\langle\sigma|v|\rangle_{\text{ANN}}}{H(T = m)}\left(Y^2 - Y_{\text{EQ}}^2\right),\tag{12}$$

where $Y_{\text{EQ}} = 0.278 g_{\text{eff}}/g_*$ (for $x \ll 3$) and $0.145(g/g_*)x^{3/2}\exp(-x)$ (for $x \gg 3$), g is the species' number of internal degrees of freedom, and $g_{\text{eff}} = g$ (for bosons) or $0.75g$ (for fermions). Eq. (12) is a particular form of the Ricatti equation that has no closed form solutions; it can be solved easily by approximation or numerical integration. I will highlight the evolution of a species' abundance.

Roughly speaking, the abundance tracks equilibrium until "freeze out," which occurs at temperature T_F, defined by $\Gamma = H$, where $\Gamma = n_{\text{EQ}}\langle\sigma|v|\rangle_{\text{ANN}}$ is the annihilation rate per particle. After that, annihilations cannot keep pace with the decreasing equilibrium abundance ("they freeze out"), and thereafter the number of particles per comoving volume remains roughly constant, at approximately its equilibrium value at freeze out: $Y_\infty \simeq Y(T_F)$. The mass density contributed by the relic particles today is

$$\rho = mY_\infty s_0 \qquad \text{or} \qquad \Omega h^2 = 0.28 Y_\infty(m/\text{eV}),\tag{13}$$

where $s_0 \simeq 7.1 n_\gamma \simeq 2970\,\text{cm}^{-3}$ is the present entropy density.

Hot and cold relics

There are two limiting cases: *hot* relics—species whose annihilations freeze out while they are still relativistic ($x_F \lesssim 3$); and *cold* relics—species whose annihilations freeze out while they are nonrelativistic ($x_F \gtrsim 3$). For a hot relic the present abundance is comparable to that of the photons, i.e., Y is of order unity. The weak interactions keep ordinary neutrinos in thermal equilibrium until a temperature of a few MeV; thus a neutrino species lighter than a few MeV is a hot relic, and

$$Y_\infty = \frac{0.278 g_{\text{eff}}}{g_*(T_F)} \simeq 3.9 \times 10^{-2} \qquad \Omega_\nu = \frac{m_\nu}{92 h^2\,\text{eV}}.\tag{14}$$

(There is an intermediate regime, referred to as *warm* relics; in this case the freeze out temperature is sufficiently high so that $g_*(T_F) \gg 1$ and Y_∞ is significantly less than order unity. For example, if $T_F \gtrsim 300\,\text{GeV}$, g_* is at least 106.75, which is the total number of degrees of freedom in the standard model, and for a fermion with two degrees of freedom, e.g., a light axino or gravitino, $\Omega = m/910 h^2\,\text{eV}$.)

Freeze out for a cold relic occurs when the species is very nonrelativistic and the species' present abundance is significantly less than that of photons ($Y_\infty \ll 1$). In this very interesting case the relic abundance is inversely proportional to the annihilation cross section,

$$Y_\infty \sim \frac{4 x_F/\sqrt{g_*}}{m m_{\text{Pl}}\langle\sigma|v|\rangle_{\text{ANN}}},\tag{15}$$

---

[l] In the absence of appreciable entropy production, the entropy per comoving volume $S \equiv R^3 s$ is conserved, implying that $s \propto R^{-3}$; thus the number of particles per comoving volume $N \equiv R^3 n \propto n/s$.

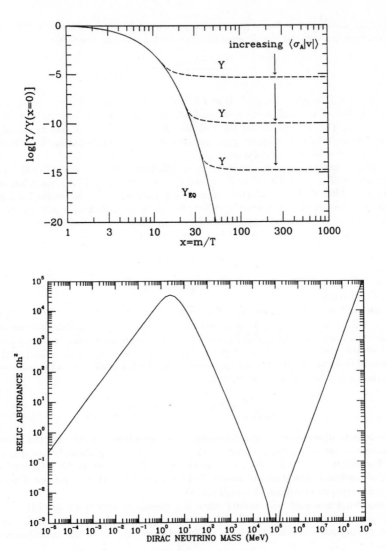

Fig. 3. Upper: Freeze out of a stable, massive particle species. Solid curves indicate equilibrium abundance, and broken curves indicate actual abundance (for different values of the annihilation cross section). Lower: Contribution of a massive, stable Dirac neutrino species to the present mass density as a function of mass. As explained in the text, Ωh^2 increases as m_ν for $m_\nu \lesssim$ MeV; decreases as m_ν^{-2} for $m_\nu \gtrsim$ MeV; and increases as m_ν^2 for $m_\nu \gtrsim 10\,\mathrm{GeV}$, thereby achieving $\Omega h^2 = 1$ for three values of m_ν. The general behaviour of Ωh^2 vs. mass for any stable particle species is similar (e.g., the neutralino).

where freeze out occurs for $x = x_F \simeq \ln[0.04 m_{\rm Pl} m \langle \sigma|v| \rangle_{\rm ANN} g/\sqrt{g_*}]$. In most cases of interest freeze occurs at $x_F \sim 20 - 30$, corresponding to $T_F \sim m/20 - m/30$ (in any case x_F only varies logarithmically).

This is a rather remarkable result: The relic abundance varies inversely with the strength of the species' interactions—implying that the weak shall prevail! Moreover, specifying that the species provides the critical density determines the annihilation cross section: $\langle \sigma|v| \rangle_{\rm ANN} \sim 10^{-37} \, {\rm cm}^2$—roughly that of the weak interactions!

Massive neutrinos

The simplest example of a cold relic is a "heavy" neutrino (mass greater than a few MeV). Provided its mass is less than that of the Z^0 boson, $\langle \sigma|v| \rangle_{\rm ANN} \sim G_F^2 m^2$ and

$$Y_\infty \simeq 6 \times 10^{-9} \left(\frac{m}{\rm GeV} \right)^{-3} \qquad \Omega_\nu h^2 \simeq 3 \left(\frac{m}{\rm GeV} \right)^{-2}. \tag{16}$$

That is, the relic abundance of stable neutrino whose mass is a few GeV would provide closure density. Since none of the three known neutrino species can be this massive and the SLC/LEP results rule out a fourth neutrino (unless it is heavier than about 45 GeV), this result, first discussed by Lee and Weinberg,[27] is only an interesting example.

For a neutrino more massive than about 100 GeV the annihilation cross section begins to decrease as m^{-2}, due to the momentum dependence of the Z^0 propagator. In this regime $Y_\infty \propto m$, and Ωh^2 varies as m^2, increasing to order unity for a mass of order a few TeV.[28]

Bringing everything together, the relic mass density of a stable neutrino species increases as m up to a mass of a few MeV; it then decreases as m^2 up to a mass of order 100 GeV; and finally it increases as m^2 for larger masses. A stable neutrino species can contribute critical density for three values of its mass: $\mathcal{O}(100 \, {\rm eV})$; $\mathcal{O}(1 \, {\rm GeV})$; and $\mathcal{O}({\rm TeV})$. This behavior is generic for a particle whose annihilations proceed through a massive boson (here the Z^0).

Griest and Kamionkowski[29] have generalized this result. Unitarity provides a bound on the annihilation cross section of any pointlike species: $\langle \sigma|v| \rangle_{\rm ANN} \lesssim 8\pi/m^2$. This implies a lower bound to Ωh^2 that increases as m^2; requiring that Ωh^2 be no larger than one (based upon the age of the Universe[30]) results in an upper bound of 340 TeV to the mass of any stable, pointlike species.

Neutralinos

A more viable cold relic is the lightest supersymmetric partner or LSP. In supersymmetric extensions of the standard model a discrete symmetry, R-parity, is usually imposed (to ensure the longevity of the proton); it also guarantees the stability of the LSP. In most supersymmetric extensions of the standard model the LSP is the (lightest) neutralino (it could in principle be the sneutrino or gluino). The neutralino(s) are the four mass eigenstates that are linear combinations of the Bino, Wino, and two Higgsinos. In many models discussed early on, especially ones where the LSP was relatively light, the neutralino (by which I mean the lightest neutralino) was almost a pure photino state, and thus was referred to as the photino.

The minimal supersymmetric extension of the standard model has a number of parameters that must be specified: μ and M, two soft supersymmetry breaking mass parameters which are expected to be 100 GeV to few TeV; $\tan\beta = v_2/v_1$, the ratio of the two Higgs vacuum expectation values; the top quark mass; and the scalar quark and scalar lepton masses.[31] These parameters determine the composition of the neutralino, its mass, and its interactions. The parameter space of supersymmetric models is multidimensional and cumbersome to deal with.

To determine the relic neutralino abundance *all* one has to do is calculate the cross section for neutralino annihilation (the neutralino is a Majorana fermion). For a neutralino that is lighter than the $W^{\pm}$ boson, the final states are fermion–antifermion pairs and light Higgs bosons. For the most general neutralino this task has been done by Griest.[32] For neutralinos that are heavier than the $W^{\pm}$ boson, many additional final states open up: $W^{+}W^{-}$, $Z^{0}Z^{0}$, HH, HW, and HZ. This complicated cross section has been calculated by Kamionkowski and his collaborators.[33] Let me summarize the salient points.

- Because the scale of supersymmetry breaking is roughly of order the weak scale, "spartner" masses are of order the weak scale; since the interactions of the neutralino with ordinary matter involve the exchange of spartners, $W^{\pm}$ bosons, Higgs bosons, or Z^{0} bosons, the neutralino's interactions are roughly weak in strength. Many of the qualitative features of the relic neutralino abundance are the same as for a neutrino.

- Over almost the entirety of the parameter space of the minimal supersymmetric extension of the standard model the relic neutralino abundance $\Omega_{\chi}h^{2}$ is greater than 10^{-3}; and in large regions of parameter space $\Omega_{\chi}h^{2}$ is of order unity. This of course traces to the fact that the neutralino's interactions with ordinary matter are roughly weak, and makes the neutralino a rather compelling dark matter candidate.

- Neutralinos can provide the critical density for masses from order $10\,\text{GeV}$ to order $3\,\text{TeV}$ (depending upon the model parameters). Fixing some of the parameters and examining $\Omega_{\chi}h^{2}$ as function of m_{χ} reveals a similar behavior as for neutrinos: $\Omega_{\chi}h^{2}\sim 1$ for a mass in the GeV range and for a mass in the TeV range.

- Just as with a heavy neutrino, for large neutralino masses the annihilation cross section decreases as $1/m_{\chi}^{2}$; this results in a maximum neutralino mass that is cosmologically acceptable: $3.5\,\text{TeV}$. For $m_{\chi}\geq 3.5\,\text{TeV}$, $\Omega_{\chi}h^{2}$ is greater unity for all models.

- Finally, the parameter space of models is constrained by unsuccessful accelerator-based searches for evidence of supersymmetry. Broadly speaking, the failure to find any evidence for supersymmetry has slowly pushed the expected mass of the neutralino upward.[34]

Axinos—A Dark Horse LSP[53]

In low-energy supersymmetric models that also incorporate Peccei–Quinn symmetry (see *Axions* below) the axion has a supersymmetric fermionic partner called the axino. There are two possibilities for the mass of the axino: (i) of order $\alpha_{\text{SM}}m_{\text{SUSY}}\sim 10\,\text{GeV}-100\,\text{GeV}$; or (ii) of order $m_{\text{SUSY}}^{2}/(f_{a}/N)$ which is $\mathcal{O}(\text{few keV})$ for $f_{a}/N\sim 10^{12}\,\text{GeV}$ ($m_{\text{SUSY}}\sim 100\,\text{GeV}-1\,\text{TeV}$ is the scale of supersymmetry breaking). This makes the axino a serious candidate for the LSP. In case (i), if the axino is the LSP its relic abundance is far too large; even if it isn't the LSP its decays lead to cosmological havoc, including overproduction of the LSP and disruption of primordial nucleosynthesis. Case (i) appears to be cosmologically excluded.

Case (ii) is very intriguing. The axino has a mass in the keV range and is clearly the LSP. Such axinos would be brought into thermal equilibrium in the early Universe (gluon + gluino → gluon + axino) and decouple at a temperature of order $10^{10}\,\text{GeV}$ when

$g_*(T_F) \gtrsim 230$. Their relic abundance is $Y_\infty = 0.278 g_{\text{eff}}/g_*(T_F) \lesssim 2 \times 10^{-3}$ leading to $\Omega_{\text{axino}} h^2 \lesssim m_{\text{axino}}/2\,\text{keV}$. That is, for interesting values of m_{SUSY} and f_a/N axinos could provide closure density as a warm relic as well as rendering the neutralino impotent.

How accurately are relic abundances known?

Calculating the relic abundance of a species that was once in thermal equilibrium has become a routine chore for the particle cosmologist. Because of the importance of this calculation, it is prudent to consider the inherent uncertainties. They are easy to identify.[35] Recall that freeze out involves the competition between the expansion rate and the annihilation rate. The annihilation rate as a function of temperature is determined by the properties of the species—and is thus a given. In calculating the expansion rate we have assumed that the Universe was radiation dominated at freeze out; further we assumed that there was no entropy production since freeze out, so that Y_∞ remains constant.

- If there the entropy per comoving volume increased by a factor of γ after freeze out, then the relic abundance Y_∞ is *decreased* by the same factor γ. Entropy release could occur in a first-order phase transition, or through the out-of-equilibrium decay of a massive particle species.

- Additional forms of energy density in the early Universe (e.g., scalar fields, or shear) serve to increase the expansion rate at fixed temperature. This in turn leads to an earlier freeze out, at a larger abundance. Increasing $H(T)$ then can *increase* Y_∞. While we can be confident that the Universe was radiation dominated by the epoch of nucleosynthesis, freeze out for most dark matter candidates occurs earlier, at a time when we cannot exclude the possibility that there were additional contributions to the energy density.

Skew Relics[36]

In discussing thermal relics I tacitly assumed that the abundance of the particle and its antiparticle were equal. For a Majorana fermion (like the neutralino) this is necessarily so; a Dirac fermion (or a scalar species) can carry a conserved (or at least approximately conserved) quantum number, and if the net particle number is sufficiently large it will determine the relic abundance of the species. Baryon number provides a simple example; if there were no net baryon number, baryons and antibaryons would annihilate down to a relic abundance $n_b/s = n_{\bar{b}}/s \simeq 10^{-19}$, which is significantly smaller than that observed, $n_b/s \simeq \eta/7 \sim 10^{-10}$. As is well appreciated the relic baryon abundance is determined by the net baryon number: $n_b/s = n_B/s$ (the net baryon number density $n_B = n_b - n_{\bar{b}}$).

The same can occur for any species whose net particle number is conserved, e.g., a heavy Dirac neutrino whose net particle number is conserved because of conservation of family lepton number. Denote the net particle number per comoving volume by n_L/s (L for lepton number). Since the relic abundance cannot be less than the net particle number, it follows roughly that: If the net particle number is greater than the would-be freeze out abundance, the relic abundance is determined by it, $Y_\infty = n_L/s$; on the other hand, if the net particle number is smaller than the would-be freeze out abundance, the net particle number plays no important role and the relic abundance is given by the usual freeze out abundance, $Y_\infty \simeq Y(x_F)$.

In the case that the relic abundance is determined by the net particle number

$$\Omega h^2 = \left[\frac{n_L/s}{10^{-10}}\right] \left(\frac{m}{35\,\text{GeV}}\right);$$ (17)

that is, a particle species of mass 35 GeV with a net particle number comparable to the baryon asymmetry would contribute the critical density.

Axions

Peccei-Quinn (PQ) symmetry with its attendant pseudo-Nambu-Goldstone boson—the axion—remains the most attractive and promising solution to the strong-CP problem.[37] Moreover, the axion arises naturally in supersymmetric and superstring models. One might call PQ symmetry and the axion the simplest and most compelling extension to the standard model!

The axion mass and PQ symmetry breaking scale are related by

$$m_a \simeq \frac{\sqrt{z}}{1+z} \frac{f_\pi m_\pi}{(f_a/N)} \simeq \frac{0.62\,\text{eV}}{(f_a/N)/10^7\,\text{GeV}},$$ (18)

where f_a is the PQ symmetry breaking scale, $z \simeq 0.56$ is the ratio of the up to down quark masses, f_π and m_π are pion decay constant and mass, and N is the color anomaly of PQ symmetry. At present there is little theoretical guidance as to the key parameter: the axion mass, although a variety of astrophysical and cosmological arguments leave open only two "windows" for the axion mass:[38] 10^{-6} eV to 10^{-3} eV and 3 eV to 8 eV (hadronic axions only).

Relic axions arise due to three distinct mechanisms: thermal production[39]—for an axion of mass greater than about 10^{-4} eV axions thermalize shortly after the QCD transition and, today, like neutrinos, should have a relic abundance of order 30 cm^{-3}; and two coherent processes, the "misalignment" mechanism[40] (see below) and axionic string decay[41]—since PQ symmetry breaking involves the spontaneous breakdown of a global $U(1)$ symmetry, strings are produced; they decay by radiating (among other things) axions. While the thermal population of axions dominates for axion masses greater than about 10^{-2} eV, there are strong astrophysical constraints in this mass range which preclude an axion more massive than about 8 eV. Thus, thermal axions can contribute at most 10% of critical density (more later on thermal axions).

For axion masses greater than about 10^{-2} eV misalignment and axionic string decay are the dominant production processes, and sufficient numbers of axions can be produced to provide closure density. The importance of axionic string decay is still a matter of intense debate. It seems to be agreed that axion production through this mechanism is somewhere between being comparable to and about 100 times more important than the misalignment mechanism,[41] further that if the Universe inflated either before or during PQ symmetry breaking, the number of axions produced by axionic strings is negligible. In the "no inflation" case, if axionic string decay is as potent as is claimed by some authors, axions provide the critical density for an axion mass of about 10^{-3} eV.

Let me briefly describe the misalignment mechanism. The free energy of the vacuum depends upon the axion field because this field modulates the phase of the instanton amplitude. At low temperatures the free energy has a maximum value of about Λ_{QCD}^4, is periodic in the "axion angle" $\theta \equiv a/(f_a/N)$, and is minimized at a value of $\theta = 0$. The mass of the axion is determined by the curvature of the free energy at $\theta = 0$ and is given approximately by Eq. (18). At high temperatures instanton effects are strongly suppressed, and for $T \gg \Lambda_{\text{QCD}}$ the free energy is essentially independent of the axion field. Thus, when PQ symmetry breaking occurs $(T \sim f_a)$, no value of the axion angle is singled out dynamically, and one expects that the value of the axion angle in different causally distinct regions will be randomly distributed between $-\pi$ and π. Thus the primeval energy density associated with the misalignment of the axion field should be of order Λ_{QCD}^4. Around a temperature of order Λ_{QCD} instanton effects become potent, and the axion mass starts to "turn on." When the axion mass exceeds $3H$ the axion field will begin to relax toward $\theta = 0$. Because it has no efficient way to shed energy, the field is left oscillating. The energy density in oscillations of the axion field behaves as nonrelativistic matter during the subsequent evolution of the Universe, and may be interpreted in particle language as a gas of zero-momentum axions.

The contribution of these axions to the present mass density of the Universe is estimated to be[40]

$$\Omega_a h^2 \simeq 0.13 \times 10^{\pm 0.4} \Lambda_{200}^{-0.7} f(\theta_1^2) \theta_1^2 (m_a/10^{-5}\,\text{eV})^{-1.18}. \tag{19}$$

where $\Lambda_{\text{QCD}} = \Lambda_{200} 200\,\text{MeV}$, and θ_1 is the initial misalignment angle. The function $f(\theta_1^2)$ accounts for anharmonic effects, and is of order unity (and specifically $f \to 1$ for $\theta_1 \ll 1$). The $10^{\pm 0.4}$ factor is an estimate of theoretical uncertainties—e.g., in the temperature dependence of the axion mass. Provided that $\theta_1 \sim \mathcal{O}(1)$ closure density in axions is achieved for a mass somewhere between $10^{-6}\,\text{eV}$ and $10^{-4}\,\text{eV}$, and for a mass less than about $10^{-6}\,\text{eV}$ axions "overclose" the Universe.[m]

The unusual dependence of the axion energy density upon the axion mass is easily understood. Regardless of the value of the axion mass, the energy density associated with the initial misalignment of the axion field is of order Λ_{QCD}^4; once the axion field starts to oscillate that energy density red shifts as R^{-3}. The axion field begins to oscillate when the axion mass $m_a(T) \simeq 3H$: For smaller masses the axion oscillations begin later, and the energy density trapped in the misalignment of the axion field is diminished less.

Since the initial misalignment angle θ_1 is a random variable, at the time of PQ symmetry breaking the value of θ_1 will be different and uncorrelated in different causally distinct regions of the Universe. In the absence of inflation, these different regions are very small, and today the Universe is comprised of a very large number of regions that each had a different value of θ_1. To obtain the average axion energy density, one uses the rms average of θ_1, which is just $\pi/3$, in Eq. (19). In this circumstance axions provide closure density for a mass in the range of $10^{-6}\,\text{eV}$ to $10^{-4}\,\text{eV}$.

If the Universe inflated before or during PQ symmetry breaking the fluctuations in the axion field take an entirely different form. While the average of θ_1^2 over many causally-separate volumes is still $\pi/3$, the practical relevance of this fact is nil, because the entire

---

[m] Overclose is not completely accurate; if the Universe is open, the production of axions— or any other particle—cannot change the geometry and close it. More precisely, a larger value of Ωh^2 leads to an earlier epoch of matter–radiation equality and ultimately to a more youthful Universe. Requiring that the Universe be at least 10 Gyr old and $h \gtrsim 0.4$ constrains $\Omega h^2 \lesssim 1$.[30]

24

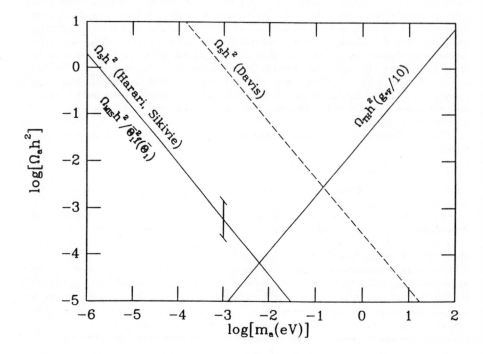

Fig. 4. Contribution of relic axions to the present mass density as a function of axion mass. Subscript "TH" indicates the contribution of thermal relic axions; "MIS" the contribution of axions produced by the misalignment process; "S" the contribution of axions produced by the decay of axionic strings. Note, in the case that the Universe inflated after or during PQ SSB breaking $\Omega_S = 0$, and Ω_{MIS} is proportional to the misalignment angle squared, whose value is unknown.

presently observable Universe lies within one causal region where θ_1 is constant. A number of authors[42] have pointed out that an axion of mass smaller than 10^{-6} eV could lead to $\Omega_a \sim 1$, provided that θ_1 was sufficiently small:

$$\theta_1 \simeq h(m_a/10^{-6}\,\text{eV})^{0.59}. \tag{20}$$

In this case, then, we would be living in a rare, axion-poor region of the Universe. If the Universe did indeed undergo inflation, the fundamental laws of physics do not determine θ_1. Despite its cosmic import the local value of this parameter is an "historical accident," and can only be determined through direct measurement of $\Omega_a h^2$ and m_a.[n]

Nonthermal Relics

The axion provides *two* examples of how a relic can be produced coherently rather than thermally: the misalignment mechanism and axionic string decay. For both of these processes the number of axions produced is highly superthermal, as is clear since theses productions mechanism dominate thermal production for $m_a \lesssim 10^{-2}$.

There are other examples of nonthermal relics. The most familiar is the superheavy magnetic monopole. The monopole is a topologically nontrivial configuration of gauge and Higgs fields. Monopoles are produced as topological defects in a symmetry breaking phase transition where a semi-simple group $\mathcal{G}$ is broken down to a smaller group $\mathcal{H}$ that contains a $U(1)$ factor; e.g., $SU(5) \rightarrow SU(3) \otimes SU(2) \otimes U(1)$. Because of the finite size of the particle horizon in the standard cosmology, after symmetry breaking the Higgs field can only be correlated on distance scales less than $H^{-1} \sim ct$—and thus must be uncorrelated on larger scales. Because of this fact of order one monopole per horizon volume will be produced. Monopole annihilation is ineffective, the monopoles produced should be with us today. This production process, which relieves on the fact that the Higgs field cannot be correlated on scales larger than the horizon, implies quite generally that order one topological defect per horizon volume should arise in a phase transition. It is known as the "Kibble mechanism."

For the simplest symmetry breaking patterns GUT monopoles are so copiously produced by the Kibble mechanism that they overclose the Universe by a factor of about 10^{10}! Moreover, there are other stringent astrophysical bounds to their relic abundance. Inflation solves the monopole problem by expanding the horizon to a size that is larger than our present Hubble volume, and thus predicts less than one monopole in the Universe due to the Kibble mechanism.

More complicated symmetry breaking schemes can reduce the relic monopole abundance to an acceptable level; and it is possible that significant numbers of monopoles can be produced as thermal pairs after inflation. It is very difficult to make a sensible prediction for the relic abundances of monopoles; however, magnetic monopoles of mass 10^{19} GeV could provide closure density and have a flux that is consistent with all the astrophysical constraints.[44]

---

[n] One might then be left with the impression that if the Universe underwent inflation, any axion mass can provide closure density provided that θ_1 is appropriately small. Additional, very important constraints emerge when fluctuations in the axion field that arise during inflation are taken into account.[43]

There are other examples of nonthermal relics, including soliton stars.[45] Soliton stars are regions of false vacuum that are stabilized by dynamics rather than topology. (By contrast, magnetic monopoles, domain walls, and cosmic string are regions of false vacuum that are stable for topological reasons.) For example, imagine a closed region of false vacuum associated with a scalar field ϕ. Such a region is unstable and should collapse. However, if there are particles inside this region whose mass when they are in the false vacuum is *less* than when they are in the true vacuum, they can exert pressure and stabilize the region. Whether soliton stars are an interesting dark matter candidate remains to be seen.

"Significant-Other" Relics

Up to this point I have focused on relics that contribute the critical density. A relic from the early Universe can be interesting and significant even if it contributes only a fraction of the critical density; e.g., most cosmologists consider baryons ($\Omega_B \sim 0.1$) and microwave photons ($\Omega_\gamma \sim 10^{-4}$) to be interesting relics, in spite of their small contributions to Ω! I will use the term "significant-other" for such relics.

I will mention two possible significant-other relics: a neutralino and an axion of mass $3\,\mathrm{eV}$ to $8\,\mathrm{eV}$. While it is possible that the neutralino contributes the critical density, it need not be the case. However, in the minimal supersymmetric extension of the standard model, the neutralino contributes at least 0.1% of the critical density; thus, if Nature exhibits low-energy supersymmetry, the neutralino is at the very least a significant-other relic! Moreover, efforts to directly detect relic neutralinos could still be successful even if they are only a significant-other relic. Needless to say the implications of their discovery for cosmology and particle physics would be almost as profound.

Axions of mass $3\,\mathrm{eV}$ to $8\,\mathrm{eV}$ arise as thermal relics and would contribute only about 1% of the critical density. Such an abundance is sufficient to permit their detection through their decay to two photons.[46] The axion mean lifetime

$$\tau(a \to 2\gamma) \simeq 6.8 \times 10^{24}\, \xi^{-2}\, (m_a/\,\mathrm{eV})^{-5}\,\mathrm{sec}, \tag{21}$$

where $\xi \equiv [E/N - 2(z+4)/3(z+1)]/0.72 \simeq (E/N - 1.95)/0.72$ and E is the electromagnetic anomaly of PQ symmetry. In the simplest axion models, $E/N = 8/3$ and $\xi = 1$.

Relic thermal axions will fall into the various potential wells that develop in the Universe as structure formation proceeds. Today they will be found in extended structures such as the halos of galaxies and clusters of galaxies, as they cannot dissipate energy and collapse further. They will decay and produce photons of wavelength $\lambda_a \simeq 24800\,\text{Å}/(m_a/\,\mathrm{eV})$. This radiation will be Doppler-broadened due to the velocities that axions have in these objects—for galaxies $\Delta\lambda/\lambda \simeq v/c \sim 10^{-3}$ and for clusters $\Delta\lambda/\lambda \simeq v/c \sim 10^{-2}$—and for distant objects the line will also be red shifted. The most favorable case for their detection is to search for the radiation from decaying axions in clusters. The intensity of the axion line is approximately[47]

$$I_{\mathrm{cluster}} \sim 10^{-17}\, \xi^2\, (m_a/3\,\mathrm{eV})^7\,\mathrm{erg\,cm}^{-2}\,\mathrm{arcsec}^{-2}\,\text{Å}^{-1}\,\mathrm{s}^{-1}/(1 + z_c)^4,$$

where z_c is the red shift of the cluster.

The background against with which this line must compete is the "night sky," which at a ground-based observatory is dominated by the glow of the atmosphere and includes many

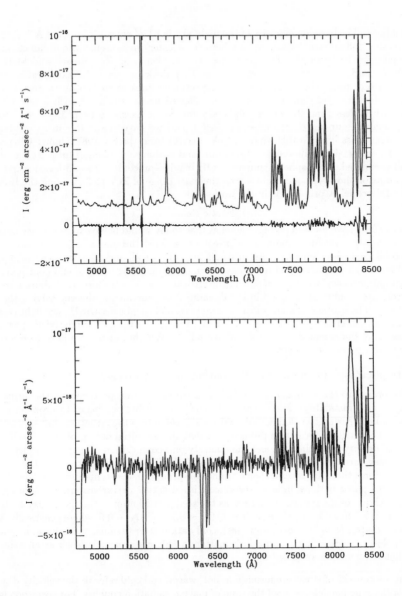

Fig. 5. Upper: Spectrum of A2218 close to the cluster core (top curve). The spectrum is dominated by the "night sky" (atmospheric emission). Spectrum of A2218 near the cluster core minus a spectrum at five cluster-core radii from the core (bottom curve). The narrow features at 5035Å and 5350Å are cosmic-ray hits. Lower: "On–off" spectrum for A2256 with the line expected for a 3.2 eV axion artificially introduced. Note the factor of ten change in scale from the upper Figure to the lower Figure.

strong lines. The baseline intensity of the night sky is $10^{-17}\,\mathrm{erg\,cm^{-2}\,arcsec^{-2}\,\AA^{-1}\,s^{-1}}$. By subtracting "off-cluster" measurements from "on-cluster" measurements one can eliminate the night-sky background. This past May, two students, M. Ted Ressell and Matthew Bershady, and I used the 2.1 m telescope at Kitt Peak to search for axion radiation in three clusters using this technique. The spectra we took span $3600\,\AA$ to $8600\,\AA$ with $10\,\AA$ resolution. Our "on-off" subtractions allowed us to search for such a line with a sensitivity of less than 3% of the night sky for the mass range from 3.1 eV to 7.9 eV. Unfortunately, our search proved unsuccessful, and we have closed this mass window.[48] (The lower mass limit to this window, 3 eV, derives from the SN 1987A limit. Obviously there are uncertainties inherent in this limit, and perhaps an axion of mass 2 eV to 3 eV is still permitted. Atmospheric emission—OH bands—preclude a ground-based search for such an axion; however, one could search the mass range of 2 eV to 3 eV using the Hubble Space Telescope! A proposal is in the works.)

Exotic Relics

Thus far I have focused on particle relics that today would behave like ordinary nonrelativistic matter. There are more exotic possibilities. Since the amount of matter associated with bright galaxies seems to contribute only 20% of critical, and a strong theoretical prejudice for $\Omega = 1$ exists, several relics have been suggested that today would contribute an almost uniform energy density of 80% of the critical density. A uniform contribution to the mass density would not show up in the dynamical measurements, thereby solving the "Ω problem." The exotic candidates include a relic cosmological constant,[49] very light cosmic strings that are either fast moving or exist in a tangled network,[50] or relativistic particles produced by the recent decays of a massive relic.[51] Whether or not we have to resort to such exotics to savage our strong prejudice remains to be seen.

IV. Implications for Structure Formation in the Universe

According to the standard cosmology, structure formation proceeds via the Jeans (or gravitational) instability: Small primeval density perturbations begin to grow once the Universe becomes matter dominated, and then develop into the structure that we observe today. The structure-formation problem is essentially an initial data problem: Specify the primeval density perturbations and the quantity and composition of matter, and let it go!

We now have well motivated suggestions for both pieces of initial data.[52] For the density perturbations, there are several choices: inflation-produced, constant-curvature (Harrison–Zel'dovich) perturbations; inflation-produced, isocurvature perturbations; and topological relics, such as cosmic strings or texture, as the seed perturbations. For the matter content, there are the following suggestions: $\Omega = 1$, $\Omega_B \sim 0.1$, and $\Omega_X \sim 0.9$, where generically X is hot dark matter (a light neutrino species), cold dark matter (axions, neutralinos, magnetic monopoles, ...), or perhaps warm dark matter (a keV mass particle, such as an axino,[53] righthanded neutrino, or gravitino).[o]

---

[o] In the context of structure formation, hot, warm, and cold refer to the velocity dispersion of the relic particles around the time of matter–radiation equality; hot corresponds to relativistic and cold to very nonrelativistic. For thermal relics this matches the previous nomenclature; in general nonthermal relics have very small velocity dispersions and behave like cold dark matter.

The suggestion that weakly interacting relic particles comprise the bulk of the mass density of the Universe and contribute $\Omega \sim 1$ has been a particularly important one, and virtually all scenarios of structure formation now include nonbaryonic dark matter. For good reason; in a "particle dark-matter" Universe density perturbations can begin growing as soon as the Universe becomes matter dominated, while in a baryon-dominated Universe density perturbations cannot begin to grow until decoupling. Further, linear perturbations in a low-Ω model cease growing at a red shift $z \sim \Omega^{-1}$. Thus, in a low-Ω model larger amplitude perturbations are required. Low-Ω models with curvature-perturbations conflict with the observed isotropy of the CMBR if $\Omega \lesssim 0.3$.

Two "stories" of structure formation have been studied in some detail: hot dark matter and cold dark matter (both with inflation-produced, constant-curvature perturbations). Hot dark matter seems to be ruled out, as galaxies form too late.[54] Cold dark matter is the most successful paradigm for structure formation yet proposed.[55] Other scenarios involving cosmic strings and texture are presently less well developed. In any case, the "hints from the early Universe" as to the initial data for structure formation have served well to bring this problem into sharper focus. Next, I will digress briefly to discuss my candidate for the "best-fit model" of the Universe.

The Best-fit Universe

Cold dark matter does a remarkably good job of describing the Universe on scales less than about $20h^{-1}$ Mpc. However, it appears to have a number of shortcomings: deficient large-scale structure, deficient galaxy counts, the age problems, and the Ω problem. No one of these problems is sufficiently troublesome to falsify the cold dark matter paradigm— yet—but taken together they are worrisome. As we shall see, the addition of a cosmological constant simultaneously addresses all of these problems.

As a reference point, the conventional cold dark matter scenario is: a flat Universe whose composition is $\Omega_B \sim 0.1 \ll \Omega_{CDM} \sim 0.9$, with $h \sim 0.5$ (to have a sufficiently old Universe) and inflation-produced Harrison–Zel'dovich curvature perturbations whose spectrum after the epoch of matter–radiation equality is[55]

$$|\delta_k|^2 = \frac{A\,k}{(1 + \beta k + \omega k^{1.5} + \gamma k^2)^2}. \tag{22}$$

Here δ_k is the amplitude of the Fourier component of comoving wavenumber k ($\equiv 2\pi/\lambda$), A is an overall normalization constant, $\beta = 1.7(\Omega h^2)^{-1}$ Mpc, $\omega = 9.0(\Omega h^2)^{-1.5}$ Mpc$^{1.5}$, and $\gamma = 1.0(\Omega h^2)^{-2}$ Mpc^2.

The basic idea of the best-fit model is simple; retain the flatness, but add a cosmological constant.[49,56] The model I discuss here is: (i) Hubble constant of around $70\,\mathrm{km\,s^{-1}\,Mpc^{-1}}$ ($h = 0.7$)—a nice compromise value; (ii) $\Omega_B \sim 0.03$—near the central value implied by nucleosynthesis; (iii) $\Omega_{CDM} \sim 0.17$—sufficiently greater than the baryonic component so that the mass density is dominated by that of the cold dark matter; (iv) Ω_Λ—cosmological constant corresponding to an energy density $\rho_\Lambda \equiv \Omega_\Lambda \rho_{CRIT} \simeq 3.2 \times 10^{-47}\,\mathrm{GeV^4} = (2.4 \times 10^{-3}\,\mathrm{eV})^4$. I am not wed to these particular values and I simply use this set for definiteness. (If the ratio of the mass densities of CDM and baryons is somewhat smaller, then the decoupling of matter and radiation can have an effect on the spectrum of density perturbations, which is to boost power on large scales.[57] If the "best-fit model" is still deficient in large-scale power, this effect could improve the situation.)

For this model the total matter contribution $\Omega_{NR} = 0.2$, and today the vacuum energy density dominates the matter energy density by a factor of four. In general the ratio $\rho_{NR}/\rho_\Lambda = 0.25(1 + z)^3$. At red shifts greater than about $z_\Lambda \simeq 0.59$ the matter energy density dominates, and the model behaves just a flat, CDM model. To determine when this model becomes matter dominated one simply sets $\Omega h^2 = \Omega_{NR}h^2 \simeq 0.098$: $T_{EQ} = 0.54\,\text{eV}$; $t_{EQ} \simeq 4.5 \times 10^{12}\,\text{sec}$; and $z_{EQ} \simeq 2300$. Once the radiation energy density is negligible ($z \ll z_{EQ}$), the scale factor evolves as

$$R(t) = \left(\frac{\Omega_{NR}}{\Omega_\Lambda}\right)^{1/3} \sinh^{2/3}\left(3\sqrt{\Omega_\Lambda}H_0 t/2\right), \qquad (23)$$

where the value of the scale factor today is taken to be one.

The Ω problem

A cosmological constant behaves just like a uniform mass density (with equation of state $p = -\rho$). As such, it would not affect determinations of Ω based upon dynamics (galactic halos and cluster virial masses). These measurements of the masses of tightly bound systems are insensitive to the contribution of a uniform background energy density because the average density in these objects is much greater than the average density of the Universe. Likewise, determinations of Ω based upon the peculiar velocities induced by the clumpy matter distribution would only reveal the clumpy, matter component. Thus, all current dynamical determinations that indicate $\Omega \simeq 0.1 - 0.3$, would be consistent with a flat Universe ($\Omega = 1$) with $\Omega_{NR} = 0.2$.

The age problems

As is well appreciated the addition of a cosmological constant *increases* the age of a flat Universe. The age of a Λ model is

$$t(z) = \frac{2H_0^{-1}}{3\sqrt{\Omega_\Lambda}} \sinh^{-1}\left[\sqrt{\Omega_\Lambda/\Omega_{NR}}/(1 + z)^{3/2}\right]; \qquad (24a)$$

$$t_0 \equiv t(z = 0) = \frac{2H_0^{-1}}{3\sqrt{\Omega_\Lambda}} \sinh^{-1}\left[\sqrt{\Omega_\Lambda/\Omega_{NR}}\right] = \frac{2H_0^{-1}}{3\sqrt{\Omega_\Lambda}} \ln\left[\frac{1 + \sqrt{\Omega_\Lambda}}{\sqrt{\Omega_{NR}}}\right]. \qquad (24b)$$

The present age of a Λ-model is always greater than $2H_0^{-1}/3$ and for $\Omega_\Lambda = 0.8$, $t_0 = 1.1H_0^{-1} \simeq 15.5\,\text{Gyr}$, an age which is comfortably consistent with the age as determined from the radioactive elements, from the oldest globular clusters, and from white dwarf cooling (e.g., see Ch. 1 of Ref. 52 and references therein). Moreover, a Λ model is older than its matter-dominated counterpart at any given epoch, so that objects at a given red shift have had more time to evolve. For $z \gg z_\Lambda$, $t(z) \to 2H_0^{-1}/3\sqrt{\Omega_{NR}}(1 + z)^{3/2}$, which is a factor of $\Omega_{NR}^{-1/2}$ older than a flat, matter-dominated model; at these early epochs the "best-fit model" is a factor of 1.6 older than the conventional CDM model.

Large-scale structure

The spectrum of density perturbations at matter-radiation equality, $(\delta M/M) \propto k^{3/2}|\delta_k|$, decreases monotonically with λ and its wavelength scale is determined by the value of Ωh^2. The spectrum "shifts" to larger length scales as Ωh^2 is decreased. Supposing that the spectrum is normalized on the scale $\lambda = 8h^{-1}\,\text{Mpc}$ (a common normalization

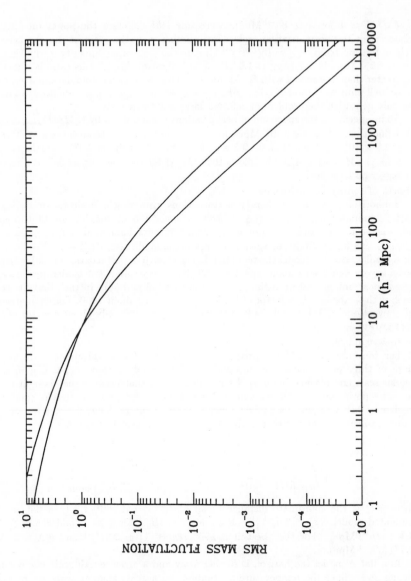

Fig. 6. The spectra of *rms* mass fluctuations for the best-motivated (conventional CDM) and the best-fit models. The *rms* mass fluctuation is computed using the "top-hat" window function (sharp sphere of radius R) and normalized to $\delta M/M = 1$ on the scale $R \simeq 8h^{-1}$ Mpc. The spectrum for the best-fit model is shifted to the right (relative to conventional CDM) because $\Omega_{\rm NR} h^2$ is smaller; for this reason it has more power on large scales.

is: $\delta M/M \simeq 1$ for $\lambda \simeq 8h^{-1}\,\mathrm{Mpc}$), decreasing Ωh^2 increases the power on all scales greater than the normalization scale. Put another way, the ratio of the characteristic scale in the spectrum, $\lambda_{\mathrm{EQ}} = 13(\Omega h^2)^{-1}\,\mathrm{Mpc}$, to the scale of nonlinearity in the Universe, $\lambda_{\mathrm{NL}} \simeq 8h^{-1}\,\mathrm{Mpc}$, is $\lambda_{\mathrm{EQ}}/\lambda_{\mathrm{NL}} \simeq 1.6/\Omega h$; in the "best-fit model" this ratio is a factor of 3.5 greater than in a model with $\Omega = 1$ and $h = 0.5$ (conventional cold dark matter, or the "most well motivated model"), implying more power on large scales. Needless to say, this can only help with the problem of deficient large-scale structure.

To be specific, if the spectrum of perturbations is normalized by $(\delta M/M)_{\lambda=8h^{-1}\,\mathrm{Mpc}} = 1$,[p] I find that: $A = 4.4 \times 10^6\,\mathrm{Mpc}^4$ for $\Omega = 1$ and $h = 0.5$ (conventional CDM) and $A = 2.5 \times 10^7\,\mathrm{Mpc}^4$ for $\Omega_{\mathrm{NR}} = 0.2$ and $h = 0.7$ ("best-fit model"). On large scales $(\lambda \gg \lambda_{\mathrm{EQ}})\ \delta M/M \propto \sqrt{A}/\lambda^2$; it follows that $\delta M/M$ for the "best-fit model" is a factor of 4.7 bigger on large scales.

Growth of density perturbations

Subhorizon-sized, linear density perturbations grow as the scale factor during the matter-dominated regime ($z \lesssim z_{\mathrm{EQ}} \simeq 23000\Omega h^2$), and remain roughly constant in amplitude when the Universe is radiation dominated, curvature dominated ($z \lesssim z_{\mathrm{CURV}} \simeq \Omega^{-1} - 2$; $z_{\mathrm{CURV}} \simeq 3$ for $\Omega = 0.2$), or vacuum-energy dominated ($z_\Lambda \simeq [\Omega_\Lambda^{-1} - 1]^{1/3} - 1 \simeq 0.59$). For a nonflat, $\Omega = 0.2$ model the reduction in the growth of perturbations relative to a flat model is very significant: about a factor of 20. By contrast, in flat-Λ models perturbations grow almost unhindered until the present (see Refs. 49 and 58). In the "best-fit model" the growth factor is only a factor of 0.8 less than z_{EQ}, or about 1800. For comparison, in the conventional CDM model the growth factor $z_{\mathrm{EQ}} \simeq 5800$, only about a factor of three more growth.

Microwave anisotropies

For conventional CDM the predicted CMBR temperature anisotropies are about a factor of three or so below the current level of observed isotropy (depending upon the angular scale and biasing factor b).[59] One might worry that because the "best-fit model" has more power on large scales and the growth factor for perturbations is smaller the predicted CMBR anisotropies might violate current bounds. That is not the case. The reason involves the angular size on the sky θ of a given scale λ at epoch z:

$$\theta(\lambda, z) = \lambda/r(z); \tag{25a}$$

$$r(z) = \int_{t(z)}^{t_0} \frac{dt}{R(t)} = \frac{2H_0^{-1}}{3\Omega_\Lambda^{1/6}\Omega_{\mathrm{NR}}^{1/3}} \int_{\sinh^{-1}\left[\sqrt{\Omega_\Lambda/(1+z)^3\Omega_{\mathrm{NR}}}\right]}^{\sinh^{-1}\left[\sqrt{\Omega_\Lambda/\Omega_{\mathrm{NR}}}\right]} \frac{du}{\sinh^{2/3} u}, \tag{25b}$$

where $r(z)$ is the coordinate distance to an object at red shift z. In a flat, matter-dominated model $r(z) = 2H_0^{-1}\left[1 - 1/\sqrt{1+z}\right] \to 2H_0^{-1}$ for $z \gg 1$, and $\theta(\lambda, z \gg 1) \simeq 34.4''\,(\lambda/h^{-1}\,\mathrm{Mpc})$. For the "best-fit model" $r(z \gg 1) \simeq 3.9H_0^{-1}$ and $\theta(\lambda, z \gg 1) \simeq 17.7''\,(\lambda/h^{-1}\,\mathrm{Mpc})$.

In a flat Λ-model the horizon is further away and a given length scale has a smaller angular size. Since the temperature fluctuations on a given angular scale are related to

[p] I have used the "top hat" window function $[W(r) = 1$ for $r \le r_0$ and $= 0$ for $r \ge r_0]$ to define M, so that $(\delta M/M)^2 = (9/2\pi^2) \int_0^\infty k^2 |\delta_k|^2 [\sin(kr_0)/k^3 r_0^3 - \cos(kr_0)/k^2 r_0^2]^2\, dk$, where $r_0 = 8h^{-1}\,\mathrm{Mpc}$.

the density perturbations on the length scale that subtends that angle at decoupling, in the "best-fit model" temperature fluctuations on a given angular scale are related to density perturbations on a *larger* scale λ. While the "best-fit model" has more power on a *fixed* (large) length scale, a fixed angle θ corresponds to a *larger* length scale, where the amplitude of perturbations is smaller because $\delta M/M$ decreases with λ.

Consider the temperature fluctuations on large-angular scales ($\theta \gg 1°$); they arise due to the Sachs–Wolfe effect and $(\delta T/T)_\theta \simeq (\delta \rho/\rho)_{HOR}/2$ on the scale $\lambda(\theta)$ when that scale crossed inside the horizon. For the Harrison–Zel'dovich spectrum the horizon-crossing amplitude is constant, so that $\delta T/T$ is independent of angular scale (for $\theta \gg 1°$). The CMBR quadrupole anisotropy is related to the amplitude of the perturbation that is just now crossing inside the horizon: $\lambda_{HOR} \sim 2H_0^{-1} \sim 12000\,\mathrm{Mpc}$ (conventional CDM) and $\lambda_{HOR} \sim 3.9H_0^{-1} \sim 16700\,\mathrm{Mpc}$ ("best-fit model"). Evaluating the normalized spectra on these scales it follows that the large-angle temperature fluctuations in the "best-fit model" are only a factor of 1.2 larger than for conventional CDM, in spite of the fact that the "best-fit model" has significantly more power on large scales.

The amplitude of the temperature fluctuations on small angular scales ($\theta \ll 1°$) is proportional to the amplitude of the density perturbations at the time of decoupling ($z_{DEC} \sim 1000$), on the scale $\lambda(\theta)$. In the "best-fit model" perturbations have grow by a factor of about $0.8z_{DEC}$ since decoupling, while those in the "most well motivated model" have grown by a factor of z_{EQ}. On the other hand the length scale corresponding to the angular scale θ is larger for the "best-fit model." The net result is that the temperature fluctuations on an angular scale of 1° are also only about a factor of 1.2 larger.

Galaxy counts

Because the coordinate distance to an object of given red shift is greater in a flat Λ model, there is greater volume per red shift interval per solid angle, which increases the number of galaxies in $dzd\omega$. To see roughly how this goes, consider the deceleration parameter

$$q_0 = \Omega(1 + 3p/\rho)/2 = (1 - 3\Omega_\Lambda)/2 \simeq -1.2, \tag{26}$$

where Ω is the total energy density ρ divided by the critical energy density and p is the total pressure. From Eq. (2) one can see that the galaxy-number count is significantly increased by the addition of a cosmological constant, $dN_{GAL}/dz = z^2 n_{GAL}[1 - 3z + \cdots]$ compared to $z^2 n_{GAL}[1 + 0.4z + \cdots]$.

Large-scale motions

The *rms* peculiar velocity of a volume defined by the "window function" $W(r)$, averaged over all such volumes in the Universe, is

$$\langle v^2 \rangle = \frac{1}{2\pi^2} \int_0^\infty k^2 |\mathbf{v}_k|^2 |W(k)|^2 dk, \tag{27}$$

Using a gaussian window function $[W_{r_0}(r) = \exp(-r^2/2r_0^2)]$ and normalizing the spectrum as above, the *rms* peculiar velocity expected on the scale $r_0 = 50h^{-1}$ Mpc is[60]

$$v_{50} \simeq 83h^{-0.9}\,\mathrm{km\,s^{-1}} \simeq 160\,\mathrm{km\,s^{-1}} \qquad (\Omega = 1,\ h = 0.5);$$

$$v_{50} \simeq 83\Omega_{NR}^{-0.33}h^{-0.9}\,\mathrm{km\,s^{-1}} \simeq 200\,\mathrm{km\,s^{-1}} \qquad (\Omega_{NR},\ h = 0.7).$$

While the *rms* peculiar velocity on the scale of 50 Mpc is still far short of $700 \, \text{km s}^{-1}$, it is larger, owing to fact that there is more power on large scales.[q]

Motivation

As its name suggests, it is a model motivated by observations and not aesthetics: Conventional cold dark matter is clearly better motivated. In this regard one should keep in mind the words of Francis Crick: "Any theory that agrees with *all* the data at a given time must be wrong!" While the conventional CDM model has one question to answer—why the ratio of the baryon density to that of cold dark matter is of order unity (see below)—in the "best-fit model" one must also address "why now?"—why is the cosmological constant just now becoming dynamical important? (This problem is similar to the flatness problem, where the question is, why is the curvature radius just now becoming comparable to the Hubble radius?) Moreover, there is the issue of the cosmological constant itself: At present there is every reason to expect a cosmological constant $\rho_\Lambda = \Lambda/8\pi G \sim m_{\text{Pl}}^4$ that is some 122 orders of magnitude larger than observations permit[r] (Supersymmetry *might* be able to help in this regard, reducing the estimate to $\rho_\Lambda \sim G_F^{-2}$, which is only 56 orders of magnitude too large!) The strongest statement that one can make in defense of a relic cosmological constant of the desired size is that no good argument exists for *excluding* it!

V. A New Dimensionless Cosmic Ratio[63]

Dimensionless numbers play a crucial role in physics and in cosmology, and attempts to understand their origin often lead to important insights. There are a number of dimensionless ratios in cosmology: the baryon-to-photon ratio, the fractional abundances of the light elements, the amplitude of the primeval density perturbations, and the ratio of the neutrino and photon temperatures. If there is a significant amount of nonbaryonic matter in the Universe, we have a new dimensionless ratio to understand

$$r \equiv \frac{\Omega_B}{\Omega_X} \sim 0.1. \tag{28}$$

In particular we can ask why r is order unity, and not say 10^{-20} or 10^{20}?

We can try to express r in terms of fundamental quantities. To begin, write

$$r = \frac{m_B}{m_X} \frac{n_B/s}{n_X/s}. \tag{28'}$$

One of the great successes of particle cosmology is the dynamical explanation of the baryon asymmetry, or baryogenesis.[64] While the specific details of baryogenesis are still lacking, generally one expects that $n_B/s \sim \epsilon/g_*$, where $g_* \sim 100 - 1000$ counts the number of degrees of freedom at the epoch of baryogenesis (10^{14} GeV?) and $\epsilon \sim 10^{-8} - 10^{-7}$ is a measure of the C, CP violation in the baryon number violating sector and—on general

---

[q] The comparison of theoretical expectations to the peculiar-velocity data is far more complicated than just computing $\langle v^2 \rangle$ for a gaussian window function.[61] The point I wish to make here is that adding a cosmological constant increases peculiar velocities.

[r] There is one interesting explanation of why the cosmological constant is "probably" zero: Coleman and others[62] have argued that due to wormhole effects the wavefunction of the Universe is very sharply peaked at zero cosmological constant.

grounds—is order $(\alpha/\pi)^N$ ($\alpha = g^2/4\pi$, g is a Higgs coupling). The quantity n_X/s is the relic abundance of X particles per comoving volume, a quantity that can be calculated as we seen above. Now consider the implications of $r \sim 0.1$ for the various relics previously discussed.

Heavy neutrino/neutralino

For a thermal relic like a heavy neutrino or neutralino whose interactions are weak, $n_X/s \sim 1/m_X^3 m_{\rm Pl} G_F^2$ (see *Thermal relics* above). The condition that r be of order unity implies

$$\frac{G_N}{G_F} \sim \sqrt{\frac{\epsilon m_B m_X^2}{g_* m_{\rm Pl}^3}} \ll 1, \tag{29}$$

and thus is related to the fact that the weak scale is much smaller than the Planck scale.

Axion

The relic abundance of axions can be expressed as $n_a/s \sim f_a^2/\Lambda_{\rm QCD} m_{\rm Pl}$. The condition that r be of order unity implies

$$\frac{f_a}{m_{\rm Pl}} \sim \epsilon \left(\frac{m_B}{\Lambda_{\rm QCD}}\right)^2 \sim \epsilon, \tag{30}$$

and thus is related to the fact that the PQ symmetry breaking scale is somewhat less than the Planck scale.

Light neutrino

The relic abundance of a light neutrino species, n_ν/s, is of order unity. If we assume that light neutrino masses arise through the see-saw mechanism, then $m_\nu \sim m_f^2/M$, where m_f is a typical fermion mass and M is the large energy that characterizes lepton number violation. The condition that r be of order unity implies

$$m_f^2/m_B \sim \epsilon M/g_*, \tag{31}$$

and thus is related to the fact that fermion masses are much smaller than the scale of lepton number violation.

Skew relic

Consider a skew relic whose net particle number per comoving volume is comparable to that of baryon number (perhaps its net particle number was produced at the same time as the baryon number, e.g., a heavy neutrino). In this case the fact that r is of order unity is related to the fact that the mass of the skew relic is comparable to that of a nucleon.

In a sense, all of these relations only tell us what we already knew and put in. However, this exercise does illustrate the fact that r can be related to fundamental quantities in physics, and raises the hope that this very important dimensionless cosmological ratio may some day have a more fundamental explanation. Apparently, that explanation will have to wait until we have a better understanding of the various energy scales that arise in particle physics.

VI. Summary

What do we know about the quantity and composition of the matter in the Universe? Most of the matter in the Universe is dark, with luminous matter contributing less than

DARK MATTER CANDIDATES

NB: VERY EXOTIC CANDIDATES NOT LISTED!

FOR REFERENCE:
$$\rho_{crit} = 10^{-29} g\,cm^{-3} \simeq 10^4 eV\,cm^{-3}$$
$$n_\gamma \simeq 422\,cm^{-3}$$
$$n_B \simeq 2\times10^{-7}\,cm^{-3} \quad \& \quad 1\,yr = 7.000\,dog\text{-}yr$$

'BEWARE OF THE DARKSIDE'

SUSPECT	MASS	'ABUNDANCE'	BIRTH SITE	
INVISIBLE AXION	$10^{-5}eV$	$10^9\,cm^{-3}$	$10^{-30}sec$	$10^{12}GeV$
LIGHT NEUTRINO[†]	$30eV$	$109\,cm^{-3}$	$1\,sec$	$1\,MeV$
PHOTINO/GRAVITINO/ "MIRROR" PARTICLES	keV	$10\,cm^{-3}$	$10^{-5}sec$	$300\,MeV$
HVY NEUTRINO[†]/PHOTINO/HIGGSINO SNEUTRINO/AXINU/ GRAVITINO/GLUINO/... SHADOW MATTER!	GeV	$10^{-5}\,cm^{-3}$	$10^{-4}sec$	$100\,MeV$
CHAMPs ➚ CHUMPS?	TeV	$10^{-8}\,cm^{-3}$	$10^{-10}sec$	$100\,GeV$
KRYPTO BARYONS ↗	$10^{12}GeV$	$10^{-17}\,cm^{-3}$	$10^{-30}sec$	$10^{12}GeV$
SUPERHEAVY MAGNETIC MONOPOLES	$10^{16}GeV$ $(\simeq 10^{-8}g)$	$10^{-21}\,cm^{-3}$	$10^{-34}sec$	$10^{14}GeV$
PYRGONS/MAXIMONS/ PERRY POLES/SCHWARZ-SCHILDS	$\gtrsim 10^{19}GeV$ $(\gtrsim 10^{-5}g)$	$\lesssim 10^{-26}\,cm^{-3}$	$10^{-43}sec$	$10^{19}GeV$
QUARK NUGGETS[†]	$\sim 10^{15}g$	$\sim 10^{-44}\,cm^{-3}$ Flux $\sim$ 1 Earth⁻¹ 30 MI.y⁻¹	$10^{-5}sec$	$300\,MeV$
PRIMORDIAL BLACK HOLES[†] SOLITON STARS	$\gtrsim 10^{15}g$	$\lesssim 10^{-44}cm^{-3}$	$\gtrsim 10^{-12}sec$	$\lesssim 10^3 GeV$

[†] PARTICLE ACTUALLY KNOWN TO EXIST !!

1% of the critical density. The best estimates of the amount of matter associated with bright galaxies is $\Omega_{ABG} \simeq 0.1 - 0.3$; however, there are some observations that suggest that Ω might be larger, perhaps even equal to one. Based upon primordial nucleosynthesis, we can be confident that baryons contribute between 1.1% and 12% of the critical density— more than that of luminous matter, but far less than the critical density. It is no means impossible that baryons account for the entire mass density of the Universe.

While there may already be evidence for nonbaryonic matter—if Ω is indeed 0.2—if our strong theoretical prejudice for $\Omega = 1$ is correct, nonbaryonic matter *must* account for the bulk of the mass density in the Universe. In any case, it is certainly a hypothesis worthy of careful consideration.

Theories of fundamental physics that go beyond the standard model have profound implications for the earliest moments of the Universe; indeed, many of us believe that the "blueprint" for the Universe traces to events that took place during that epoch. Theories that unify the particles and interactions predict the existence of new, stable particles (or additional properties for known particles, e.g., neutrino masses), and remarkably enough, the relic abundances calculated for a number of these new particles is comparable to that required to close the Universe. For many, this is what makes the particle dark-matter hypothesis so compelling. Needless to say, the discovery of such a relic would not only solve a cosmological puzzle, but would also shed light on the theory that unifies the forces and particles.

By now there is a virtual zoo of particle dark-matter candidates.[65] However, three candidates are particularly well motivated and attractive. They are an axion of mass 10^{-6} eV to 10^{-4} eV, a neutrino of mass $92h^2$ eV, and a neutralino of mass 10 GeV to 3 TeV. Peccei–Quinn symmetry and its axion resolve a nagging and serious difficulty of the standard model: the strong-CP problem. The neutralino is a very robust prediction of theories that incorporate low-energy supersymmetry. Low-energy supersymmetry provides some understanding of the hierarchy problem (the large disparity between the weak scale and the Planck scale), and is further motivated by superstring theories. Neutrinos actually exist—and come in three flavors!—and in many extensions of the standard model small neutrino masses are predicted. Moreover, the first results of the SAGE experiment,[66] together with the results of the Homestake and Kamiokande II solar neutrino experiments, suggest that nonadiabatic MSW neutrino oscillations may be *the solution* to the solar neutrino problem.[67] If this is so, it implies a mass for the μ or τ neutrino in the range 10^{-4} eV to 10^{-2} eV. Speculating (upon supposition to be sure) that this is the mass for the μ neutrino, a simple see-saw scaling estimate for the τ neutrino mass might just put it in the cosmologically interesting range. While cold dark matter provides a far more promising paradigm for structure formation than does hot dark matter, I am certain that cosmology could learn how to live with a neutrino-dominated Universe.[s]

Particle dark matter is an attractive and compelling hypothesis, and the next step is to test it. A variety of experiments are underway, and more are planned.[68] The experimental efforts encompass a diversity of approaches, involving conventional laboratory and accelerator experiments, large-underground detectors, and experiments built expressly to

---

[s] Dennis Sciama has recently touted the multitude of astrophysical virtues of an unstable neutrino, and has gone so far as to precisely "predict" its mass, 28 eV to 30 eV, and lifetime, $\tau = 2 \pm 1 \times 10^{21}$ sec.[67a]

detect the dark matter particles in our local neighborhood. The search for evidence of supersymmetry is going on at accelerator laboratories all over the world. Indirect evidence for the existence of particle dark matter in our own halo could come from the annihilation products of particle dark matter in the halo or from particle dark matter that has accumulated in the sun or earth. The GALLEX and SAGE experiments may well provide information about neutrino masses, and a nearby supernova or a long-baseline neutrino oscillation experiment could provide definite evidence for neutrino masses. The MACRO experiment in the Gran Sasso Laboratory is operating and can search for both relic magnetic monopoles and high-energy neutrinos from particle dark-matter annihilations in the sun or earth. First-generation Sikivie-type detectors to search for cosmic axions have been built and successfully operated;[69] a second generation detector with sufficient sensitivity to detect halo axions in the our neighborhood has been proposed.[70] Low-background, cryogenic detectors designed to detect the keV energies deposited by halo neutralinos that elastically scatter within the detector are under development in laboratories all over the world, and low-background ionization detectors have already been used to search for heavy neutrinos and cosmions.[71]

The answer to the simple question—What is the Universe made of?—may well be answered soon. If the bulk of the matter in the Universe is nonbaryonic, this discovery will rank as one of the most important of the century, and will have profound implications for both cosmology and particle physics.

Acknowledgments

I would like to thank our hosts at UCLA for putting together a very stimulating meeting. This work was supported in part by the DOE (at Chicago).

References

1. For a review of the dark-matter problem and determinations of Ω see e.g., V. Trimble, *Ann. Rev. Astron. Astrophys.* **25**, 425 (1987); S. Faber and J. Gallagher, *ibid* **17**, 135 (1979); *Dark Matter in the Universe*, eds. J. Kormendy and G. Knapp (Reidel, Dordrecht, 1989); E.W. Kolb and M.S. Turner, *The Early Universe* (Addison–Wesley, Redwood City, CA, 1990), Chs. 1–3; J. Binney and S. Tremaine, *Galactic Dynamics* (Princeton Univ. Press, Princeton, 1987), Ch. 10; P.J.E. Peebles, in these proceedings.

2. See e.g., the classic paper, A. Sandage, *Astrophys. J.* **133**, 355 (1961); or E.W. Kolb and M.S. Turner, *The Early Universe* (Addison–Wesley, Redwood City, CA, 1990), Chs. 1–3.

3. For a recent review of the mass estimates for the Galaxy see, M. Fich and S. Tremaine, *Ann. Rev. Astron. Astrophys.*, in press (1991).

3a. D.O. Richstone, *Astrophys. J.*, in press (1991).

4. A. Dressler, D. Lynden-Bell, D. Burstein, R. Davies, S. Faber, R. Terlevich, and G. Wegner, *Astrophys. J.* **313**, L37 (1987).

5. E. Bertschinger, A. Dekel, and A. Yahil, in preparation (1990); M. Rowan-Robinson et al., *Mon. Not. R. astr. Soc.*, in press (1990); M.A. Strauss and M. Davis, in *Large-scale Motions in The Universe: A Vatican Study Week*, eds. V.C. Rubin and G.V. Coyne (Princeton Univ. Press, Princeton, 1988).

6. M. Davis and P.J.E. Peebles, *Astrophys. J.* **267**, 465 (1983); M. Davis, M.J. Geller, and J. Huchra, *ibid* **221**, 1 (1978); P.J.E. Peebles, *Nature* **321**, 27 (1986); P.J.E. Peebles,

The Large-scale Structure of the Universe (Princeton Univ. Press, Princeton, 1980); and references therein.

7. J.H. Oort, *Bull. Astron. Inst. Netherlands* **6**, 349 (1932); J.N. Bahcall, *Astrophys. J.* **276**, 169 (1984); *ibid* **287**, 926 (1984); K. Kuijken and G. Gilmore, *Mon. Not. R. astr. Soc.* **239**, 605 (1989); A. Gould, *ibid* **244**, 25 (1990); F.J. Kerr and D. Lynden-Bell, *ibid* **221**, 1023 (1986).

8. J. Bahcall, *Astrophys. J.* **287**, 926 (1984); J.A.R. Caldwell and J.P. Ostriker, *Astrophys. J.* **251**, 61 (1981); J.N. Bahcall, M. Schmidt, and R.M. Soneira, *Astrophys. J.* **265**, 730 (1983); R. Flores, *Phys. Lett.* **215B**, 73 (1988); D.N. Spergel and D.O. Richstone, in *Proceedings of Les Arcs 88: Dark Matter*, eds. J. Audouze and J. Tran Thanh Van (1989); M.S. Turner, *Phys. Rev.* **D33**, 889 (1986).

9. H. Spinrad and S. Djorgovski, in *Observational Cosmology*, eds. A. Hewitt, G. Burbidge, and Li-Zhi Fang (Reidel, Dordrecht, 1987), p. 129.

10. E. Loh and E.J. Spillar, *Astrophys. J.* **307**, L1 (1986)

11. S. Bahcall and S. Tremaine, *Astrophys. J.* **326**, L1 (1988).

12. L. Cowie, *Physica Scripta*, in press (1991); D.C. Koo and A.S. Szalay, *Astrophys. J.* **282**, 390 (1984); T. Shanks, P.R.F. Stevenson, R. Fong, and H.T. MacGillivray, *Mon. Not. R. astr. Soc.* **206**, 767 (1984); D. Koo, in *The Epoch of Galaxy Formation*, eds. T. Shanks, A.F. Heavens, and J.A. Peacock (Kluwer, Dordrecht, 1989).

13. See e.g., J. Yang, M.S. Turner, G. Steigman, D.N. Schramm, and K.A. Olive, *Astrophys. J.* **281**, 493 (1984); K.A. Olive, D.N. Schramm, G. Steigman, and T.P. Walker, *Phys. Lett. B* **236**, 454 (1990); B. Pagel, G. Steigman, and D.N. Schramm, *Physica Scripta*, in press (1991).

14. G. Steigman, D.N. Schramm, and J. Gunn, *Phys. Lett. B* **66**, 202 (1977); J. Yang, M.S. Turner, G. Steigman, D.N. Schramm, and K.A. Olive, *Astrophys. J.* **281**, 493 (1984).

15. G.S. Abrams et al., *Phys. Rev. Lett.* **63**, 2173 (1989) (Mark II); B. Adeva et al., *Phys. Lett. B* **231**, 509 (1989) (L3); D. Decamp et al., *Phys. Lett. B* **231**, 519 (1989) (ALEPH); M.Z. Akrawy et al., *Phys. Lett. B* **231**, 530 (1989) (OPAL); P. Aarnio et al., *Phys. Lett. B* **231**, 539 (1989) (DELPHI).

16. J. Mather et al. (COBE Collaboration), *Astrophys. J.* **354**, L37 (1990); H.P. Gush, M. Halpern, and E.H. Wishnow, *Phys. Rev. Lett.* **65**, 537 (1990).

17. For discussion of this point see, S. Tremaine and J. Gunn, *Phys. Rev. Lett.* **42**, 407 (1979); Trimble, and Kormendy and Knapp in Ref. 1.

18. M. Milgrom, *Astrophys. J.* **270**, 365 (1983); J. Bekenstein and M. Milgrom, *ibid* **286**, 7 (1984).

19. S. Dimopoulos, R. Esmailzadeh, L.J. Hall, and G.D. Starkman, *Phys. Rev. Lett.* **60**, 7 (1988).

20. J.G. Bartlett and L.J. Hall, *Phys. Rev. Lett.* **66**, 541 (1991).

21. J. Applegate, C.J. Hogan, and R.J. Scherrer, *Phys. Rev. D* **35**, 1151 (1987); C. Alcock, G. Fuller, and G. Mathews, *Astrophys. J.* **320**, 439 (1987); R.A. Malaney and W.A. Fowler, *Astrophys. J.* **333**, 14 (1988).

22. H. Kurkio-Sunio et al., *Phys. Rev. D* **38**, 1091 (1988); *Astrophys. J.* **353**, 406 (1990); K. Sato and N. Terasawa, *Phys. Rev. D* **39**, 2893 (1989); D.N. Schramm, *Physica Scripta*, in press (1991).

23. See e.g., M. Gorenstein et al., *Astrophys. J.* **287**, 538 (1984); C.G. Lacey and J.P. Ostriker, *Astrophys. J.* **299**, 633 (1985); B.J. Carr, *Nucl. Phys.* **B252**, 81 (1985); B.J. Carr et al., *Astrophys. J.* **277**, 445 (1984); in *Inner Space/Outer Space*, eds. E.W. Kolb et al. (Univ. of Chicago Press, Chicago, 1986).

24. For a discussion of the viability of low-mass baryonic objects as the halo dark matter see e.g., D. Ryu, K.A. Olive, and J. Silk, *Astrophys. J.* **353**, 81 (1990); D. Hegyi and K.A. Olive, *ibid* **303**, 56 (1986). Two groups have proposed to use microlensing to search for MACHOs (Massive Astrophysical Compact Halo Objects) through microlensing (Spiro et al. in France, and Alcock et al. in the US). For discussion of this idea see, e.g., K. Griest, Univ. Calif. Berkeley preprint CfPA-TH-90-007 (1990), to appear in *Astrophys. J.*; or B.J. Carr and J.R. Primack, *Nature* **345**, 478 (1990). The first discussion of the use of microlensing to detect halo objects is, B. Paczynski, *Astrophys. J.* **301**, 1 (1986).

25. For a review of the standard cosmology see, E.W. Kolb and M.S. Turner, *The Early Universe* (Addison–Wesley, Redwood City, 1990), Ch. 3; or S. Weinberg, *Gravitation and Cosmology* (Wiley, New York, 1972), Ch. 15.

26. For a textbook discussion of freeze out see e.g., E.W. Kolb and M.S. Turner, *The Early Universe* (Addison–Wesley, Redwood City, 1990), Ch. 5.

27. B.W. Lee and S. Weinberg, *Phys. Rev. Lett.* **39**, 165 (1977).

28. A.D. Dolgov and Ya.B. Zel'dovich, *Rev. Mod. Phys.* **53**, 1 (1981); K.A. Olive and M.S. Turner, *Phys. Rev. D* **25**, 213 (1982).

29. K. Griest and M. Kamionkowski, *Phys. Rev. Lett.* **64**, 615 (1990).

30. See e.g., E.W. Kolb and M.S. Turner, *The Early Universe* (Addison–Wesley, Redwood City, 1990), pp. 56-58.

31. For a review of the minimal supersymmetric extension of the standard model see, H. Haber and G. Kane, *Phys. Rep.* **117**, 75 (1985).

32. K. Griest, *Phys. Rev. Lett.* **61**, 666 (1988); *Phys. Rev. D* **38**, 2357 (1988). The case of a light neutralino that is a pure photino state is treated by, J. Ellis et al., *Nucl. Phys.* **B238**, 453 (1984).

33. K. Griest, M. Kamionkowski, and M.S. Turner, *Phys. Rev. D* **41**, 3565 (1990). The case of a heavy neutralino that is either a pure gaugino or pure Higgsino state is treated by, K. Olive and M. Srednicki, *Phys. Lett. B* **230**, 78 (1989).

34. See e.g., J. Ellis et al., *Phys. Lett. B* **245**, 251 (1990).

35. M. Kamionkowski and M.S. Turner, *Phys. Rev. D* **42**, 3310 (1990); J. D. Barrow, *Nucl. Phys.* **B208**, 501 (1982).

36. The relic abundance of a species with a net particle number is treated by, K. Griest, *Nucl. Phys.* **B283**, 681 (1987).

37. R.D. Peccei and H.R. Quinn, *Phys. Rev. Lett.* **38**, 1440 (1977); F. Wilczek, *Phys. Rev. Lett.* **40**, 279 (1978); S. Weinberg, *Phys. Rev. Lett.* **48**, 223 (1978). For a recent review of the axion see R.D. Peccei, in *CP Violation*, edited by C. Jarlskog (WSPC, Singapore, 1989).

38. For a recent summary of the astrophysical and cosmological bounds to the axion mass see, M.S. Turner, *Phys. Rep.* **197**, 67 (1990); G.G. Raffelt, *ibid* **198**, 1 (1990).

39. M.S. Turner, *Phys. Rev. Lett.* **59**, 2489 (1987).

40. J. Preskill, M. Wise, and F. Wilczek, *Phys. Lett. B* **120**, 127 (1983); L. Abbott and P. Sikivie, *Phys. Lett. B* **120**, 133 (1983); M. Dine and W. Fischler, *Phys. Lett. B* **120**, 137 (1983). The expression for the relic abundance of axions used here is from, M.S. Turner, *Phys. Rev. D* **33**, 889 (1986).

41. R. Davis, *Phys. Lett. B* **180**, 225 (1986); D. Harrari and P. Sikivie, *Phys. Lett. B* **195**, 361 (1987); R. Davis and E.P.S. Shellard, *Nucl. Phys.* **B324**, 167 (1990); A. Dabholkar and J.M. Quashnock, *Nucl. Phys. B* **333**, 815 (1990); C. Hagmann and P. Sikivie, *Nucl. Phys.*, in press (1991); P. Sikivie, *Physica Scripta*, in press (1991).

42. S.-Y. Pi, *Phys. Rev. Lett.* **52**, 1725 (1984); M.S. Turner, *Phys. Rev. D* **33**, 889 (1986); A.D. Linde, *Phys. Lett. B* **201**, 437 (1988); A.D. Linde, Stanford University preprint SU-ITP-883 (1991).

43. M.S. Turner and F. Wilczek, *Phys. Rev. Lett.* **66**, 5 (1991).

44. For a review of magnetic monopoles and their astrophysical/cosmological implications see, J. Preskill, *Ann. Rev. Nucl. Part. Sci.* **34**, 461 (1984); or E.W. Kolb and M.S. Turner, *The Early Universe* (Addison–Wesley, Redwood City, CA, 1990), Ch. 7. There is a very interesting paper that clarifies the production of monopoles when the symmetry breaking pattern is complicated: T.W.B. Kibble and E.J. Weinberg, *Phys. Rev. Lett.*, in press (1991). For a review of other topological defects see, A. Vilenkin, *Phys. Rep.* **121**, 263 (1985).

45. See e.g., G. Rosen, *J. Math. Phys.* **9**, 996 (1968); R. Friedberg, T. D. Lee, and A. Sirlin, *Phys. Rev. D* **13**, 2739 (1976); S. Coleman, *Nucl. Phys. B* **262**, 263 (1985); E. Copeland, K. Lee, and E.W. Kolb, *Nucl. Phys. B* **319**, 501 (1989); B. Holdom, *Phys. Rev. D* **36**, 1000 (1987); E. Witten, *Phys. Rev. D* **30**, 272 (1984); T.D. Lee, *Phys. Rev. D* **35**, 3637 (1987); J.A. Frieman, G. Gelmini, M. Gleiser, and E.W. Kolb, *Phys. Rev. Lett.* **60**, 2101 (1988); E. Copeland, E.W. Kolb, and K. Lee, *Phys. Rev. D* **38**, 3023 (1988); G. Gelmini, M. Gleiser, and E.W. Kolb, *Phys. Rev. D* **39**, 1558 (1989).

46. T. Kephart and T. Weiler, *Phys. Rev. Lett.* **58**, 171 (1987); M.S. Turner, *Phys. Rev. Lett.* **59**, 2489 (1987).

47. M.S. Turner, *Phys. Rev. Lett.* **59**, 2489 (1987).

48. M. Bershady, M.T. Ressell, and M.S. Turner, *Phys. Rev. Lett.* **66**, 1398 (1991); M.T. Ressell's contribution to these proceedings.

49. M.S. Turner, G. Steigman, and L. Krauss, *Phys. Rev. Lett.* **52**, 2090 (1984); P.J.E. Peebles, *Astrophys. J.* **284**, 439 (1984).

50. A. Vilenkin, *Phys. Rev. Lett.* **53**, 1016 (1984); however, see M.S. Turner, *Phys. Rev. Lett.* **54**, 252 (1985).

51. M.S. Turner, G. Steigman, and L. Krauss, *Phys. Rev. Lett.* **52**, 2090 (1984); D.A. Dicus, E.W. Kolb, and V.L. Teplitz, *Phys. Rev. Lett.* **39**, 168 (1977); M. Fukugita and T. Yanagida, *Phys. Lett. B* **144**, 386 (1984); G. Gelmini et al., *Phys. Lett. B* **146**, 311 (1984).

52. See e.g., E.W. Kolb and M.S. Turner, *The Early Universe* (Addison–Wesley, Redwood City, CA, 1990), Ch. 9.

53. K. Rajagopal, M.S. Turner, and F. Wilczek, *Nucl. Phys. B*, in press (1991); and references therein.

54. S.D.M. White, C. Frenk, and M. Davis, *Astrophys. J.* **274**, L1 (1983); *ibid* **287**, 1 (1983); J. Centrella and A. Melott, *Nature* **305**, 196 (1982).

55. For a discussion of the cold dark matter paradigm for structure formation see e.g., G. Efstathiou, in *Physics of the Early Universe*, eds. J.A. Peacock, A.F. Heavens, and A.T. Davies (Adam Higler, NY, 1990); G. Blumenthal, S.M. Faber, J.R. Primack, and M.J. Rees, *Nature* **311**, 517 (1984); E.W. Kolb and M.S. Turner, *The Early Universe* (Addison–Wesley, Redwood City, CA, 1990), Ch. 9; C.D.M. Frenk, *Physica Scripta*, in press (1991).

56. M.S. Turner, in *IUPAP Conference on Primordial Nucleosynthesis and the Early Evolution of the Universe*, ed. K. Sato (Kluwer, Dordrecht, 1991).

57. G.R. Blumenthal, A. Dekel, and J.R. Primack, *Astrophys. J.* **326**, 539 (1988).

58. J. Charlton and M.S. Turner, *Astrophys. J.* **313**, 495 (1987).

59. N. Vittorio and J. Silk, *Astrophys. J.* **285**, L39 (1984); N. Vittorio et al., *Astrophys. J.* **341**, 163 (1989); N. Vittorio et al., *Astrophys. J. (Lett.)*, in press (1990); A.C.S. Readhead et al., *Astrophys. J.* **346**, 566 (1989); J.R. Bond and G. Efstathiou, *Astrophys. J.* **285**, L44 (1984); *Mon. Not. R. astr. Soc.* **226**, 655 (1987); J.R. Bond, in *Frontiers in Physics: From Colliders to Cosmology*, eds. B. Campbell and F. Khanna (WSPC, Singapore, 1990); J.R. Bond, G. Efstathiou, P. Lubin, and P. Meinhold, *Phys. Rev. Lett.*, in press (1991); J.R. Bond's contribution to these proceedings.

60. N. Vittorio and M.S. Turner, *Astrophys. J.* **316**, 475 (1987).

61. See e.g., N. Kaiser and O. Lahav, *Mon. Not. R. astr. Soc.* **237**, 129 (1989); A. Dekel, E. Bertschinger, and S. Faber, *Astrophys. J.*, in press (1990); E. Bertschinger, "Large-scale Motions in the Universe: A Review," to published in the *Proceedings of the 1990 Moriond Workshop on Astrophysics*, in press (1991); *Large-scale Motions in The Universe: A Vatican Study Week*, eds. V.C. Rubin and G.V. Coyne (Princeton Univ. Press, Princeton, 1988).

62. S. Coleman, *Nucl. Phys* **B310**, 643 (1988), and in these proceedings; also see, S. Hawking, *Phys. Lett. B* **134**, 403 (1984); E. Baum, *Phys. Lett. B* **133**, 185 (1983).

63. M.S. Turner and B.J. Carr, *Mod. Phys. Lett. A* **2**, 1 (1987).

64. See e.g., E.W. Kolb and M.S. Turner, *Ann. Rev. Nucl. Part. Sci.* **33**, 645 (1983); *The Early Universe* (Addison–Wesley, Redwood City, CA, 1990), Ch. 6; or Rubakov's contribution to these proceedings.

65. See, e.g., M. S. Turner, in *Dark Matter in the Universe*, eds. J. Kormendy and G. Knapp (Reidel, Dordrecht, 1989); or J. Primack and B. Sadoulet, *Ann. Rev. Nucl. Part. Sci.* **38**, 751 (1989).

66. As reported by the SAGE collaboration at Neutrino '90 (Geneva, 1990).

67. See e.g., J.N. Bahcall and H. Bethe, *Phys. Rev. Lett.* **65**, 2233 (1990), and references therein.

67a. D. Sciama, *Phys. Rev. Lett.* **65**, 2839 (1990).

68. See e.g., *Particle Astrophysics: Forefront Experimental Issues*, ed. E.B. Norman (WSPC, Singapore, 1989); J. Primack and B. Sadoulet, *Ann. Rev. Nucl. Part. Sci.* **38**, 751 (1989); D.O. Caldwell, *Mod. Phys. Lett. A* **5**, 1543 (1990); P.F. Smith and J.D. Lewin, *Phys. Rep.* **187**, 203 (1990).

69. S. DePanfilis et al., *Phys. Rev. Lett.* **59**, 839 (1987); *Phys. Rev. D* **40**, 3153 (1989); C. Hagmann, P. Sikivie, N.S. Sullivan, and D.B. Tanner, *Phys. Rev. D (RC)* **42**, 1297 (1990); *Rev. Sci. Instrum.* **61**, 1076 (1990); S. Inagaki et al., in *Proceedings of the Workshop on Cosmic Axions*, edited by C. Jones and A. Melissinos (WSPC, Singapore, 1990). Many, if not most, of the issues and experiments involving cosmic axions are discussed in *Proceedings of the Workshop on Cosmic Axions*, edited by C. Jones and A. Melissinos (WSPC, Singapore, 1990). Also see Sikivie's contribution to these proceedings.

70. P. Sikivie et al., *Proposal to the DOE and LLNL for an Experimental Search for Dark Matter Axions in the* $0.6 - 16 \mu eV$ *Mass Range* (submitted 20 July 1990); K. van Bibber's contribution to these proceedings.

71. D.O. Caldwell et al., *Phys. Rev. Lett.* **61**, 510 (1988); S.P. Ahlen et al., *Phys. Lett. B* **195**, 603 (1987); D.O. Caldwell et al., *Phys. Rev. Lett.* **65**, 1305 (1990).

Baryonic Dark Matter

George Lake
Departments of Astronomy and Physics
University of Washington
Seattle WA 98195

ABSTRACT

Cosmic nucleosynthesis calculations require far more baryons than appear in current inventories of stars and gas. There are even enough baryons to make the dark matter in galaxies. Previously, I have shown that the dark matter densities in extreme dwarf galaxies are difficult to achieve without dissipation. I enumerate several other reasons to consider baryonic dark matter and construct a simple model of a baryonic halo for the Milky Way. The next generation IR observatories could detect dark matter in other galaxies and discover the freely floating brown dwarfs in our galaxy. Finally, I note the many desirable features of a cosmological model with baryonic dark matter that is "flattened" using a cosmological constant.

I. INTRODUCTION

In the standard picture of the Universe, only 1–10% of the matter is visible. *Cold dark matter* (CDM) is a leading theory, but its successes are based on a density fluctuation spectrum (Frenk *et al.* 1986) rather than the existence of an exotic particle. When Parker (1990) asked numerous astrophysicists: "What is the Dark Matter?", all but two responded: "low mass stars". There are numerous reasons to favor low mass stars:

- Baryons and low mass stars exist. They are "non-exotic" dark matter.
- There are "missing baryons" (§II.a).
- The best motivated exotic CDM candidate, Weakly Interacting Massive Particles are being painted into unreasonable corners of parameter space by results from LEP and laboratory detectors (Schramm 1990).
- If the flatness of the Universe owed to Hot Dark Matter (light "-inos") or a cosmological constant (cf. Fukugita *et al.* 1990 and §III), some or all of the dark matter in galaxies must be baryonic (Lake 1989).
- Collapsing proto-galaxies were thermally unstable at large radii, ~ 80 kpc (Lake 1987, 1990c). How did they collapse leaving nothing behind?
- The high spatial and phase densities of dark matter in the dwarf spheroidals is difficult to reconcile without dissipation (Lake 1990ab and §II.b)
- Dwarf spheroidals link dark matter to low metallicity, while globular clusters show a link between low metallicity and low mass stars (§II.c)
- Using the globular clusters as a "Pop III" tracer provides a good model for the distribution of dark matter (§II.d).
- An extrapolation of the observed spheroidal mass function to $\sim 0.02 M_\odot$ is sufficient to explain the halo dark matter at the solar radius (§II.d)
- The case against baryonic dark matter has been overstated (§II.e)
- In some very favorable scenarios, the IRAS database may afford a detection of the dark matter. It will be difficult for such dark matter to elude detection with the next generation of Infrared Space Observatories, ISO and SIRTF. The MACHO project will afford yet another opportunity to study this form of dark matter. ISO will be launched in 1993 and MACHO collaboration will start taking data several months earlier. In short order, the theory will stand or fall based on observations.

II THE CASE FOR LOW MASS STARS

II.a A Baryonic Accounting

The observed abundances of light elements limits the current density in baryons to the range:

$$0.02 < \Omega_{baryon} < 0.06, \quad \text{for a Hubble constant,} \quad H_o \subset (55, 80) \text{km s}^{-1} \text{Mpc}^{-1} \quad (1)$$

(cf. Olive et al. 1990). I get an upper limit to Ω_{stars} by combining the recent luminosity functions of Efstathiou et al.(1988) with the maximum M/L_{stars} consistent with the kinematical data (Schechter 1980, Kent 1986, 1987, 1988):

$$\Omega_{stars} < 0.003 \quad (2)$$

In clusters of galaxies, half of the baryons are in the X-ray emitting plasma. Even if this is the efficiency of formation for all galaxies, most of the baryons are missing. Finally, I note that there is 10–20 times as much dark matter as luminous matter within galaxies and clusters. While eqn. (1) excludes a universe with $\Omega_{baryon} \equiv 1$ [†]. it does allow for baryonic dark matter to have 10-20 times the mass of luminous stars. Dark halos could be baryonic.

II.b Dwarf Spheroidals are too dark for exotic particles

It's well known that the high phase densities of dark matter in dwarf galaxies (or even the very existence of dwarf galaxies) are inconsistent with Hot Dark Matter unless some of the dark matter is baryonic. However, I've recently shown that the high *spatial* densities of dark matter in the extreme dwarfs of the local group, GR8, Draco and Ursa Minor spells trouble for Cold Dark Matter (Lake 1990b). If the dark matter is dissipationless, then there is a simple relation between the redshift of turnaround z_{turn} and its current mean density. If the dark matter in these extreme dwarfs has the same distribution as the light, then the redshift of formation, $z_{turn} > 30$. This is far higher than in expected in the CDM picture where dwarfs don't form until $z_{turn} < 6$. Extended density distributions can bring about marginal consistency with the CDM density fluctuation spectrum ($z_{turn} < 10$), but in that case, the global mass-to-light ratio exceeds $1,000 \, M_\odot/L_\odot$. Since both of these alternatives are extreme, it seems more reasonable to allow that the dark matter in these dwarfs was baryonic and experienced dissipation.

Taken together with the problem of "missing baryons", it is difficult to escape the conclusion that baryonic dark matter is out there. I argue that it's hiding in the form of low mass stars and brown dwarfs.

II.c Dark Matter and Low Metallicity

Two pieces of evidence point to an association of dark matter with low metallicity and a link between low metallicity and low mass stars. Figure 1 (from Lake 1990b) shows the M/L of the dwarf spheroidals versus luminosity and metallicity. The correlation with metallicity is better and suggests an explanation of the trend seen with luminosity. If low mass stars were copiously produced at low metallicity, both correlations would result.

Observations of globular clusters provide the link between low mass stars and low metallicity. McClure et al. (1986) found that the mass function,

$$\Phi(m) \propto m^{-(1+x)} \quad (3)$$

[†] Even with inhomogeneous nucleosynthesis, it now looks like $\Omega_{baryon} \equiv 1$ is out of the question (Hogan, private communication)

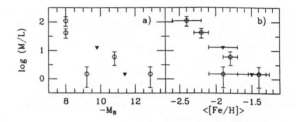

Figure 1: M/L ratios for the dwarf spheroidals versus metallicity and luminosity. Dwarfs with $[Fe/H] \lesssim -2$ have very high M/L values.

of globular clusters varies systematically with metallicity. The lowest metallicity clusters have $x = 2.0 - 2.5$. Pryor *et al.* (1986) found that the correlation persists after correction for mass segregation, but the values of x are lower. This correlation is now disputed. However, several clusters still show a steep rise in the mass function near the hydrogen burning limit (Richer *et al.* 1990).

The details of these particular correlations can easily be disputed. However it is clear that there are numerous dwarf spheroidals, irregulars and regular galaxies with M/L's in excess of 100 (Aaronson and Olszewski 1987, Lake and Schommer 1984, Carrignan and Freeman 1988, Giovanelli and Haynes 1989, Lake Schommer and van Gorkom 1990). It is the small, metal-poor systems that show large amounts of dark matter. This is a strong signal that any baryonic dark matter must be low mass stars not the remnants of massive objects that would have produced metals (cf. Silk 1990).

Draco and Ursa Minor are extreme but still detectable. We should allow for the possibility that their are objects with even lower surface brightness that are correspondingly "darker". Indeed, there are two candidate dark clusters. Sommer-Larsen and Christiansen (1987, hereafter SLC) and Doinidis and Beers (1989) find five horizontal branch stars clumped together in position, magnitude and velocity. The "chance probability of such associations is 1 in 10 million" (SLC). Using the quoted velocity dispersions and mean radii, the central column densities are $\sim 10^5 M_\odot \, pc^{-2}$ with a core radius of $\sim 0.5^\circ$. They are wonderful observational targets.

II.d The Mass Distribution in the Milky Way

Eight years ago, a meeting in Tucson was held on the topic: "Halos of Galaxies". There were really two meetings, one on "halo stars" and the other on "dark halos". Time and again, speakers went out of their way to emphasize that there was no possible connection between these two "halos". It was claimed that there spatial distributions were completely different and it was generally held that the "dark halos" were made of exotic material.

For the purpose of illustration here, I will construct a model of the dark halo which has the following properties: *the dark halo is composed of stars with the mass function and spatial distribution of the metal-poor stellar halo.* This model has some very desirable properties, is consistent with all observations and is amenable to observation by ISO and SIRTF. It is one of many models that must be systematically constructed and analyzed.

For the spatial distribution of the halo, I use the observed distribution of metal-poor globular clusters (those with $[Fe/H] < -0.8$ in Zinn's (1985) sample) as tracers of the mass distribution of the first generation of metal poor objects. Their number density falls to half its central value at $r_{core} \sim 7$ kpc and declines asymptotically as $r^{-3.5}$. Various dynamical processes have reshaped the distribution of clusters (c. f. Aguilar, Hut and Ostriker 1988). The main destructive mechanism is dynamical friction coupled with tidal disruption by the disk. The dynamical wakes in the disk provide most of the drag at radii

$\lesssim$ 20 kpc and are more effective on direct orbits than retrograde. Hence, past destruction can be estimated by comparing the distributions of clusters with direct and retrograde orbits (Armandroff and Lake, in preparation). We find that for clusters with $[Fe/H] < -0.8$, the ratio of direct to retrograde is $\sim$ 2:1 inside the solar radius, R_0 and 1.1:1 outside R_0. The retrograde clusters have $r_{core} \sim 10$ kpc.

In Figure 2, I replace the isothermal halo in the Ostriker-Caldwell (1983) model with the density distribution of the retrograde clusters. The half mass radius of this halo is 73 kpc. The local escape velocity is 545 km s^{-1} and the local density is $\rho_{local} = 7 \times 10^{-3}$ M$_\odot$pc^{-3}. The total mass is 1×10^{12}M$_\odot$, in complete agreement with value recently determined by Zaritsky *et al.* (1989).

Figure 2: A mass model for the galaxy giving the dark matter the same distribution as retrograde metal-poor clusters and using the disk and spheroid Ostriker and Caldwell's (1983) model. The dotted line shows the halo's contribution while the solid line is the sum of all components. The half mass radius of the halo is 73 kpc. The local escape velocity is 545 km s^{-1} and the local density is $\rho_{local} = 7 \times 10^{-3}$ M$_\odot$pc^{-3}. The total mass is 1×10^{12}M$_\odot$ in accord with Zaritsky *et al.* (1989).

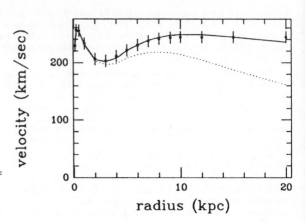

I anticipate objections to the $r^{-3.5}$ model based on extended rotation curves of other galaxies. Figure 3 shows an $r^{-3.5}$ halo model fit to the most extended rotation curve to date, NGC 3198 (van Albada *et al.* 1985).

Finally, I note that the isothermal models fit to the rotation curves of galaxies typically have core radii of $\sim$ 10 kpc and then must be truncated at radii of $\sim$ 50 kpc to integrate to reasonable total masses. Since a hard truncation is unphysical, there is little dynamic range where an r^{-2} profile would really be seen. The $r^{-3.5}$ model not only integrates to a finite mass, but when the normalization is set by the local rotation curve, it integrates to the current estimate for the total mass of the galaxy.

I now examine how the observed halo stars might relate to the unseen low mass stars. Schmidt (1975) finds that the local mass density of halo subdwarfs with $m \gtrsim 0.25$M$_\odot$, is $\rho_{subdw} = 0.8 \times 10^{-4}$ M$_\odot$pc^{-3} and that the luminosity function is best represented with $x \sim 2$ (see eqn. 3). Chiu (1980) finds a higher normalization and x. Regrettably, those two papers still represent the state of the art for the halo luminosity function. (Gilmore [1990] has presented recent work on the low mass stars in the disk.)

I adopt $\rho_{subdw} = 0.8 \times 10^{-4}$ M$_\odot$pc^{-3} and $x = 2.5$. This value of x is intermediate between the values found by Schmidt (1975) and Chiu(1980) and is typical of the values found in the most metal poor globular cluster (McClure *et al.* 1986). With these values, the mass function must continue to $m_{low} \sim 0.02$M$_\odot$ to account for the local density of

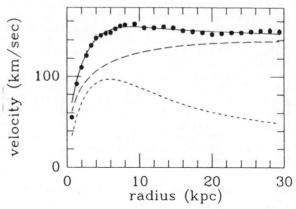

Figure 3: A model for the rotation curve of NGC 3198 using a halo with a density profile $\rho \propto r^{-3.5}$ beyond a core radius. The short dashes show the contribution of the disk, the long dashes are the dark halo and the solid line is the sum.

dark matter. Surprisingly, this continuation is not in conflict with searches for low mass stars. Boeshaar and Tyson (1983) surveyed a 400 pc^3 volume and found 4 M stars. With my adopted mass function, I expect 0.9 stars in their survey. Taken at face value, Boeshaar and Tyson's results argue for a larger value of x which would allow the cutoff mass m_{low} to be larger. The proper motions of their 4 stars would be of great interest.

This simple model could ultimately run afoul with the "prediction" of giants at large distances. However, this can be claimed as a success at the moment. Olszewski (private communication) has looked for giants in an area of 0.5 square degrees toward Ursa Minor and Draco. He looked for stars that fit on the color magnitude diagram of the clusters. He found 3 giants that fit his CM criteria, but were interlopers (their velocities were very different than the DSph's). If I assume that his selection criteria chose halo interlopers with distances ±20 kpc of the dwarf galaxies and normalize to the local counts of giants (Ratnatunga and Freeman 1989), I find that Olszewski's 3 interlopers fits my $r^{-3.5}$ model.

It is shocking that a simple model for a stellar halo derives some support from observations of nearby M stars and distant K giants. It is important to systematically explore the parameter space and constrain such possibilities.

II.e The Mass Distribution of Dark Matter in External Galaxies

The case against baryonic dark matter has been based on comparing optical and infrared surface brightness with the projected mass density of a singular isothermal sphere,

$$\rho = \frac{v^2_{max}}{4\pi G\, r^2} \tag{4}$$

(c. f. Hegyi and Olive 1986). This overestimates the column density since it assumes that the dark matter produces the entire rotation curve and it's singular at $r = 0$. Using the more realistic model sequences of NGC 3198 (Lake and Feinswog 1989), we found that the typical central column densities of bright spirals are $100 - 250$ M$_\odot$ pc^{-2}.

The recalculated constraints on M/L_{halo} are shown in Table 1. The singular isothermal sphere overestimated M/L by ~ 5. The most stringent limits are $(M/L)_V \gtrsim 500$,

$(M/L)_I \gtrsim 40$ and $(M/L)_K \gtrsim 24$. These could be lowered further by allowing for the halo to have a modest flattening. We can compare these values to the two late M dwarfs Wolf 359 and VB10 (Greenstein, Neugebauer and Becklin 1970). The former has $(M/L)_V > 510$, $(M/L)_I > 91$ and $(M/L)_K > 19$ while the latter has $(M/L)_V > 3600$, $(M/L)_I > 530$ and $(M/L)_K > 34$. Main sequence stars have not been eliminated as the constituents of dark halos.

Note that the characteristic central brightness of a disk galaxy corresponds to $I_0 \sim 150$ L$_\odot$pc^{-2}. Since $M/L_{disk} \sim 3$M$_\odot$/L$_\odot$, the column density of the dark matter just begins to dominate at $\gtrsim 2$ disk scalelengths. Extensive modeling of the dwarf spheroidal galaxies, Draco and Ursa Minor (see Table 2 from Lake 1990a) shows that their central surface brightness is only ~ 4 L$_\odot$pc^{-2} while the central mass column densities are very high $\sim 100 - 500$M$_\odot$ pc^{-2} (see Table 2). *So the column density of the dark matter exceeds that in bright galaxies, with little confusion from stellar sources.* We intend to exploit this fact in our searches for dark matter with ISOCAM.

Table 1. Constraints on $(M/L)_{halo}$ in External Galaxies.

Galaxy NGC#	Band	M/L sing. r^{-2} model	Radius disk scale-lengths	M/L realistic models VF[†]	H50[‡]	Ref.
4565	I	76	5-10	58	38	1
	K	38	0.5-2.5	23	15	2
2683	V	420	6	168	168	3
	K	61	6	24	24	3
4244	V	630	1	137	156	3
	K	33	1	7	8	3
5907	V	2000	1	676	496	3
	K	64	0.7	15	11	3

[†]VF uses the distances of Aaronson *et al.* (1982)
[‡]H50 uses a uniform Hubble flow of 50 km s^{-1}Mpc^{-1}
Refs: (1) Hegyi and Gerber (1977)
    (2) Boughn, Saulson and Seldner (1981)
    (3) Skrutskie, Shure and Beckwith (1985)

II.f Detecting the Dark Matter

To calculate some fluxes and surface brightnesses, I use the brown dwarf models of Nelson, Rappaport and Joss (1986). These models assume solar abundances. Nelson (1990) reported new calculations of brown dwarfs of primordial abundance which would result in slightly higher fluxes at a fixed column density. For the fiducial model with a mass function:

$$N(M) \propto M^{-1.5}, \quad m \subset (0.01\text{M}_\odot, 0.08\text{M}_\odot) \tag{5}$$

the column density thresholds for detection with IRAS (or SIRTF or ISO) can be calculated. In this mass range, the luminosities of the brown dwarfs scale as $M^{-2.34}$ and (age)$^{-1.22}$. I adopt ages of 10 Gyr. I find that in each pixel of the 10 micron survey, $(15")^2$ IRAS can detect $\sim 10^6$M$_\odot$ pc^{-2} (at high galactic latitude) . This is a depressing number given that dark halos need only have column densities of ~ 200M$_\odot$ pc^{-2}. However, SIRTF and ISO will have sensitivities of 10^4M$_\odot$ pc^{-2} in each of $(4")^2$ pixels. Since

Table 2. Properties of Draco and Ursa Minor. The King Models assume that the Mass Follows the Light. Isothermal Models fix the stellar distribution function to be isothermal and uses star counts to constrain the density distribution

Parameter	Draco	Ursa Minor
Basic Data		
[1]Distance (kpc)	76 ± 7	79 ± 9
[2]Luminosity,V-band $(10^5 L_\odot)$	1.8	2.2
[3]velocity dispersion (km s^{-1})	9 ± 2	11 ± 3
King (1962) Models		
Tidal radius (arcmin)	$55^{+\infty}_{-30}$	$59^{+\infty}_{-25}$
Half-brightness radius (arcmin)	$6.3^{+1.2}_{-1.1}$	$9.4^{+1.2}_{-1.1}$
[2]Central Surface Brightness $(L_\odot \ pc^{-2})$	4.5	4.1
[4]Mass Column Density $(M_\odot \ pc^{-2})$	210^{+150}_{-100}	280^{+100}_{-180}
[4]M/L_V $(M_\odot/L_\odot)$	45^{+35}_{-20}	50^{+45}_{-25}
Isothermal Plummer Models		
Mass Column Density $(M_\odot \ pc^{-2})$	140^{+200}_{-50}	300^{+170}_{-130}

[1]Sources: Draco, Stetson (1979); Ursa Minor, Nemec *et al.* (1988).

[2]From Hodge's (1964ab) counts with his M3 scaling. Errors are not quoted but estimated to be at least 30%.

[3]From Aaronson and Olszewski (1987)

[4]Extrema are using one sigma errors on σ and R_{HB} and do not include errors in the luminosity and distance nor uncertainties in the velocity ellipsoid.

all stars that we know of form in clusters with column densities $\gtrsim 10^4 M_\odot pc^{-2}$, the detection thresholds may not be so prohibitive. That is, we might expect to detect structure at this level rather than a diffuse glow. The other possibility is to try and detect "free floaters". This is more encouraging with for SIRTF than with ISO. With the mass function in equation 5 and the limiting flux for a one hour exposure with the camera on ISO, there is roughly 1 detectable brown dwarf per square degree. Unfortunately, ISO has a field of view that is $\lesssim 10^{-2}$, so digging the dwarfs out of the data will be tough. To close on a hopeful note....if we adopt the mass function in §II.d, both the diffuse brightness and the free floaters will be detectable with ISO.

III. THE SILENCE OF THE Λ

A decade ago, astrophysical cosmology was defined as "a search for two numbers", Ω and H_o. With inflation and the cold dark matter fluctuation scenario, the search shifted to H_o and the bias parameter. What ever happened to the cosmological constant, Λ? During the conference, there were a number of passing references to a cosmological constant and the recent literature has been signalling its resurection. The reasons for it's original burial are unclear. The claim that Einstein (1970!!) referred to it as "the greatest blunder" of his career is oft quoted (I find that quote four times in Misner, Thorne and Wheeler [1970, hereafter MTW]). However, the source of this remark is always traceable to Gamov's (1970) autobiography, not the most reliable of sources. Einstein (1954) went on record just a year before his death in the revised appendix of *The Meaning of Relativity* with the mild comment that the observed dynamics does not necessitate the introduction of Λ (cf. Pais 1982). Other relativists were far more reluctant to give up the cosmological constant. Eddington (1932) states that he "would as soon think of reverting to Newtonian theory as of dropping the cosmological constant". McVittie, Lemaître and Weyl have all offered similar testimonials. On the next page after one of their invocations of Einstein's ghost, MTW report the work of Zeldovich's (1967) interpretation of Λ as the stress-energy of the vacuum. In Gliner (1969), one finds the first realization of an inflationary phase. Λ's resurgence began before its burial.

Why is the cosmological constant becoming increasing popular? Let's define $\lambda = \Lambda/(3H_o^2)$ so that a flat universe is one where $\Omega + \lambda = 1$ and consider the current state of four problems:

1) **Ages.** Without a cosmological constant, the age of the Universe (τ) is just:

$$\tau = 9.8h^{-1}\text{Gyr} \quad \text{for} \quad \Omega = 0$$

$$\tau = 6.5h^{-1}\text{Gyr} \quad \text{for} \quad \Omega = 1$$

where h is the Hubble constant in units of $100 \,\text{km s}^{-1}\text{Mpc}^{-1} = 1$. Values of h are clustering above 0.7, while ages of globular clusters have are 13 ± 1 Gyr (Vandenberg, Bolte and Stetson 1990). Since globular clusters "know" that they are in galaxies, it seems unreasonable to assume that they formed before the galaxies at least turned around and began collapse. The turnaround epoch of galaxies is $(1 + z) \sim 4$ if $\Omega = 1$ and $(1 + z) \sim 6$ if $\Omega = 0$ (cf. Lake 1987). (Peebles [1989] has summarized some of the arguments in favor of a high redshift for galaxy formation. Thus, this could be enumerating as a separate argument for low Ω.) Taking a lower bound of 12 Gyr for the cluster ages and the above formation redshifts, I find:

$$\tau > 14.4\text{Gyr}, \quad h < 0.68 \quad \text{for} \quad \Omega = 0$$

$$\tau > 13.7\text{Gyr}, \quad h < 0.47 \quad \text{for} \quad \Omega = 1$$

The ages are inconsistent for a Universe flattened by Ω. On the other hand, if I adopt $\Omega = 0.1$ and $\lambda = 0.9$, I find that an $h > 0.7$ still allows an age of 18 Gyr.

2) **Volume.** There are a variety of classical tests that depend on the behavior of the volume with redshift. The simplest of these are number counts. For a very nice recent treatment, I recommend Fukugita *et al.* (1991). The counts of faint galaxies are observed to rise to such large values that an $\Omega = 1$ and $\lambda = 0$ model cannot be made consistent with any amount of evolution. The model with $\Omega = 0.1$ and $\lambda = 0.9$ provides a good fit with "standard evolution". Evolution can be "tuned" to force a fit with the open model with $\Omega = 0.1$ and $\lambda = 0$.

3) **Large Scale Structure.** The inflationary model naturally leads to scale-invariant Gaussian fluctuations (cf. Bardeen, Steinhardt and Turner 1983). The evolution of these fluctuations leads to a break in the power spectrum at a scale proportional to $(\Omega h^2)^{-1}$. This break occurs for both hot and cold dark matter. Until recently, it was (barely) possible to reconcile large scale structure with $\Omega = 1$ at the expense of an uncomfortably small Hubble constant ($h \sim 0.4$-0.5). However, the IRAS data (Saunders *et al.* 1991) and the APM survey (Maddox *et al.* 1990) the reality of large-scale structure as demonstrated earlier by the Great Attractor (Lynden-Bell *et al.* 1988). One possibility is to deny inflation and allow an open Universe with $\Omega = 0.1, \lambda = 0$. This increases the large scale power within the Gaussian scale-invariant paradigm, but removes the justification for this spectrum (making all spectra arbitrary, but making the addition of more power on large scales no less arbitrary than all the rest). This

schemes runs afoul of attempts to reconcile large scale structure with lack of observed structure in the Cosmic Microwave Background (cf. Bardeen, Bond and Efstathiou 1987). A more sensible approach is to adopt $\Omega = 0.1$ or 0.2 and $\lambda + \Omega = 1$. This increases the large scale power to acceptable levels while maintaining consistency with the absence of CBR fluctuations *without* biasing (Bardeen, Bond and Efstathiou 1987, Efstathiou, Sutherland and Maddox 1990).

4) **Inflation.** The previous arguments for a low value of Ω are supported by numerous analyses of the peculiar motions of galaxies (cf. Davis and Peebles 1983) without recourse to biasing. The easiest way to reconcile Ω and inflation is to flatten with λ.

There are, of course, several reasons to exhibit skepticism toward the model with $\Omega = 0.1$ and $\lambda = 0.9$. The first is the "unnatural value" of $\lambda = 3(1 - \Omega)^{1/4}meV$. However, all known particle masses are "unnatural", so I'm not impressed by the imposition of this standard to cosmology without its appearance elsewhere. As a cosmologist, I might take some heart in the fact that Λ is rapidly moving toward its natural value, $3H^2$. There is also the "special epoch" problem. At a redshift of 1, the cosmological constant was still nearly irrelevant. So we live in the "special" time when the cosmological constant has appeared. But, didn't we already know that these are the best of times for cosmology!

It is a pleasure to acknowledge conversations with T. Armandroff, K. Ashman, C. Cesarsky, T. Chester, L. Feinswog, C. Hogan, S. Love, L. Nelson, J. Peebles, and H. Richer. This work has been supported by NASA and a Dudley Award.

REFERENCES

Aaronson, M. and Olszewski, E. 1987, *IAU Symposium 177, Dark Matter in the Universe*, ed. J. Kormendy and J. Knapp, p. 153, (Dordecht: Reidel).

Aaronson, M. *et al.* 1982, *Ap. J. Suppl.*, **50**, 241.

Aguilar, L., Hut, P. and Ostriker, J. P. 1988, *Ap. J.*, **335**, 720.

Bardeen, J., Bond, J. R. and Efstathiou, G. 1987, *Ap. J.*, **321**, 28.

Bardeen, J., Steinhardt, P. J. and Turner, M. S. 1983, *Phys. Rev. Lett.*, **28**, 679.

Baum, W. A. 1986, in *Astrophysics of Brown Dwarfs*, ed. M. C. Kafatos, R. S. Harrington and S. P. Maran (Cambridge: Cambridge Univ. Press).

Boeshaar, P. C. and Tyson, J. A. 1983, in *The Nearby Stars and the Stellar Luminosity Function*, ed. A. G. D. Philip and A. R. Upgren, p. 85.

Boughn, S. P., Saulson, P. R. and Seldner, M. 1981, *Ap. J. (Letters)*, **250**, L15.

Carignan, C. and Freeman, K. 1985, *Ap. J.*, **294**, 494.

Chiu L. T. 1980, *A. J.*, **85**, 812.

Davis, M. and Peebles, P. J. E. 1983, *Ap. J.*, **267**, 465.

Doinidis, S. P. and Beers, T. C. 1989, *Ap. J. Lett*, **340**, L57.

Eddington, A. 1932, *The Expanding Universe*, (Cambridge: Cambridge Univ. Press) p. 24.

Efstathiou, G., Ellis, R. S. and Peterson, B. A. 1988, *M.N.R.A.S.*, **232**, 431.

Efstathiou, G., Sutherland, W. J. and Maddox, S. J. 1990, *Nature*, **348**, 705.

Einstein, A. 1954, *The Meaning of Relativity*, 4th ed., (Princeton: Princeton Univ. Press).

Fukugita, M., Takahara, F, Yamashita, K. and Yoshii, Y. 1990 *Ap. J. (Letters)*, **361**, L1.

Frenk, C. S., White, S. D. M., Efstathiou, G. and Davis, M. 1986, *Nature*, **317**, 595.

Gamov, G. 1970, *My World Line*, Viking Press, New York, p. 44.

Gilmore, G. 1990, *Baryonic Dark Matter*, Gilmore, G. and Lynden-Bell, D 1990, eds. (Cambridge, Cambridge Univ. Press).

Giovanelli, R. and Haynes, M. 1989, *Ap. J. (Letters)*, **346**, L5.

Gliner, E. B. 1969, *Zh. Zksp. Theor. Fiz.*, **49**, 542.

Greenstein, J. L., Neugebauer, G. and Becklin, E. E. 1970, *Ap. J.*, **161**, 519.

Hegyi, D. and Gerber, G. L. 1977, *Ap. J. (Letters)*, **218**, L7.

Hegyi, D. and Olive, K. A. 1986, *Ap. J.*, **303**, 56.

Hodge, P. 1964ab, *A. J.*, 69, 438; **69**, 853.

Kent, S. M. 1986, 1987, 1988, *A.J.*, **91**, 1301; **93**, 816;**96**, 514.

King, I. R. 1962, *Astron. J.*, **67**, 471.

Lake, G. 1987, in *Structure and Dynamics of Elliptical Galaxies*, ed. T. de Zeeuw, (Dordecht: Reidel).

Lake, G. 1989, *Astron. J.*, **98**, 1253.

Lake, G. 1990a, *M.N.R.A.S.*, **244**, 701.

Lake, G. 1990bc, *Ap. J. (Letters)*, **356**, L43; **364**, L1.

Lake, G. and Feinswog, L. 1989, *A. J.*, **98**, 166.

Lake and Schommer 1984, *Ap. J. (Letters)*, **279**, L19.

Lake, G., Schommer, R. A. and van Gorkom, J. 1990, *A. J.*, **99**, 547.

Lynden-Bell, D. *et al.* 1988, *Ap. J.*, **326**, 19.

Maddox, S. J. *et al.* 1990 *M.N.R.A.S.*, **242**, 43p.

McClure, R. D. *et al.* 1986, *Ap. J. Letters*, **307**, L49.

Nelson, L. 1990, in *Baryonic Dark Matter*, Gilmore, G. and Lynden-Bell, D 1990, eds. (Cambridge, Cambridge Univ. Press).

Nemec, J. M., Wehlau, C. and Mendes de Oliveira, A. 1988, *Astron. J.*, **96**, 528.

Olive, K. A. *et al.* 1990, *Phys. Lett. B.*, **236**, 454.

Ostriker, J.P. and Caldwell, J. R. 1983, *Kinematics, Dynamics and Structure of the Milky Way*, p. 249, (Dordecht: Reidel).

Pais, A. 1982, *'Subtle is the Lord...': The Science and Life of Albert Einstein*, page 288, (Oxford: Oxford University Press).

Parker, B. 1990, *Dark Matter and the Fate of the Universe*, (Plenum: New York).

Peebles, P. J. E. 1989, in *The Epoch of Galaxy Formation*, ed. C. Frenk *et al.* (Dordecht: Kluwer).

Pryor, C., Smith, G. H. and McClure, R. D. 1986, *Astron. J.*, **92**, 1358.

Ratnatunga, K. and Freeman, K. 1989, *Ap. J.*, **339**, 126.

Saunders, W. *et al.* 1991, *Nature*, **349**, 32.

Schechter, P. 1980, *A. J.*, **85**, 801.

Schmidt, M. 1975, *Ap.J.*, **202**, 22.

Schramm, D. 1990, Fermilab Astrophysics preprint.

Silk, J. 1990 in *Baryonic Dark Matter*, Gilmore, G. and Lynden-Bell, D 1990, eds. (Cambridge, Cambridge Univ. Press).

Skrutskie, M. F., Shure, M. A. and Beckwith, S. 1985, *Ap. J.*, **299**, 303.

Sommer-Larsen, J. and Christensen, P. R. 1988, *M.N.R.A.S.*, **225**, 499.

Stetson, P. B. 1979, *Astron. J.*, **84**, 1149.

Vandenberg, D. A., Bolte, M. and Stetson, P. B. 1990, *Astron. J.*, **100**, 1445.

van Albada, T. *et al.* 1985, *Astrophys. J.*, **295**, 305.

Zaritsky, D, *et al.* 1989, *Ap. J.*, **345**, 759.

Zeldovich, Ya. B. 1967, *Zh. Eksp. & Theor. Fiz. Pis'ma*, **6**, 883.

Zinn, R. 1985, *Ap. J.*, **293**, 424.

The Search for U(1)′ Dark Matter[1]

David E. Brahm

California Institute of Technology
Pasadena, CA 91125

ABSTRACT

We consider models with a gauged U(1)′, to which is coupled a fermionic standard-model singlet ψ. If ψ's close the universe, a Lee-Weinberg calculation gives the four-fermion coupling G' from the mass m_ψ. We can then calculate the germanium scattering cross-section as a function of m_ψ. The results are nearly model-independent and close to experimental limits for $10\,\text{GeV} \le m_\psi \le 100\,\text{GeV}$. In the region $400\,\text{GeV} \le m_\psi \le 40\,\text{TeV}$ the scattering cross-section depends on $M_{Z'}$, and again approaches experimental limits if $M_{Z'} \le 2\,\text{TeV}$.

I. Introduction

In addition to their standard-model interactions, the known quarks and leptons may interact through an extra U(1)′ sector, with a heavy gauge boson Z′. Such additional gauge groups are predicted by many Grand Unified Theories, for example. We postulate that the universe is closed by a fermionic standard-model singlet ψ which interacts with ordinary matter only through Z′ exchange, with coupling strength

$$G' = \frac{\sqrt{2}}{8} \frac{(g_1')^2}{M_{Z'}^2} \tag{1}$$

A freezeout calculation, in which we set $\Omega_\psi = 1$, fixes the annihilation cross-section $\langle \sigma_A v \rangle$ (Fig. 1a). This is related by crossing symmetry to the elastic cross-section with germanium σ_{el} (Fig. 1b). We end up with a definite prediction for σ_{el} as a function of m_ψ, which can be compared with experimental results from germanium[2,3] (or superconducting granule[4]) detectors.

II. The Basic Calculation

As a concrete example, we will consider the breakdown of SO(10) → SU(5) ⊗ U(1)$_X$, with SU(5) subsequently breaking to the standard model. In this case, the "charge" $X = \sqrt{\frac{5}{8}}(B\text{-}L - \frac{4}{5}Y)$. The $\underline{16}$ of SO(10) contains the known Weyl fermions plus a standard-model singlet N^c. We couple N^c, not to the neutrino (as is often done), but to an SO(10)-singlet N, to form a massive Dirac fermion ψ. These particular choices of U(1)′ and ψ only affect the final results by a multiplicative factor Φ, typically $\frac{1}{2} \le \Phi \le \frac{3}{2}$.

A Lee-Weinberg analysis[5] gives the relic abundance of ψ particles from their annihilation cross-section $\langle \sigma_A v \rangle$. We assume no particle-antiparticle asymmetry for ψ[6], and we take $m_\psi \ll \frac{1}{2}M_{Z'}$. Define

$$Z \equiv \sqrt{\frac{45}{4\pi^3 g^*}} m_\psi M_P \langle \sigma_A v \rangle \tag{2}$$

54

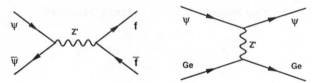

Figure 1: a) Annihilation, b) Elastic Scattering

where g^* is the effective number of relativistic particle degrees of freedom at the freezeout temperature T_F. Above .2 GeV, g^* approximately obeys[7]

$$g^*(T_F) = 90 - \sqrt{(225\,\text{GeV})/T_F} \tag{3}$$

We set

$$\Omega_\psi = \left(\frac{\ln Z}{Z}\right)\left(\frac{12}{g^*}\frac{1}{2.75}\right)T_0^3 m_\psi/\rho_0 = 1 \tag{4}$$

The numerical factors in the second term arise from a depletion in g^*, and a consequent rise in temperature, associated with the QCD phase transition and the electron freezeout. In eq. (2) we use the annihilation cross-section from Fig. 1a in the non-relativistic limit[8,9],

$$\langle\sigma_A v\rangle = \frac{4}{\pi}(G'm_\psi V_\psi)^2\left(\sum_f S_f^2\right) \tag{5}$$

V_ψ is the (vector) charge of ψ under the U(1)$'$ symmetry. The last term is a sum of U(1)$'$ charges over all kinematically allowed final states, three generations of standard-model particles except possibly the top quark. In our SO(10) model, for $m_\psi > m_t$, we have $\sum_f S_f^2 = 4.125$. Combining these equations, we find that $(G'm_\psi V_\psi)$ is approximately a constant,

$$(G'm_\psi V_\psi)^2 = \frac{7.0\times10^{-10}\,\text{GeV}^{-2}}{\left(\sum_f S_f^2\right)}\left(\frac{.25}{\Omega h_0^2}\right)\sqrt{\frac{90}{g^*}}\left(\frac{\ln Z}{24}\right) \tag{6}$$

We take the normalized Hubble parameter $h_0 = \frac{1}{2}$. From lower limits on $M_{Z'}$[10] and the perturbative requirement $g_1' < \sqrt{4\pi}$, we require $m_\psi > 1$ GeV.

From Fig. 1b, we calculate the germanium cross-section[8]

$$\sigma_{\text{el}} = \frac{2}{\pi}(G'm_\psi V_\psi)^2\left(\frac{M_{\text{Ge}}}{m_\psi + M_{\text{Ge}}}\right)^2 V_{\text{Ge}}^2 \tag{7}$$

where $M_{\text{Ge}} = 68$ GeV. When we insert our Lee-Weinberg result, eq. (6), we find

$$\sigma_{\text{el}} = (2.15\times10^{-10}\,\text{barns})\,\Phi\left(\frac{M_{\text{Ge}}}{m_\psi + M_{\text{Ge}}}\right)^2\left(\frac{.25}{\Omega h_0^2}\right)\sqrt{\frac{90}{g^*}}\left(\frac{\ln Z}{24}\right)\left(\frac{A}{72.6}\right)^2 \tag{8}$$

Φ is a numerical factor which depends on the U(1)$'$ symmetry, and on whether m_ψ lies above or below the top mass; it is normalized to unity for our SO(10) model in the region $m_\psi > m_t$:

$$\Phi \equiv 4.24\,\frac{(V_{\text{Ge}}/A)^2}{\left(\sum_f S_f^2\right)} \tag{9}$$

Note our result (8) depends only on m_ψ (and the U(1)$'$ factor Φ), not on $M_{Z'}$, g_1', or even V_ψ. A plot of $\sigma_{\rm el}$ vs. m_ψ appears in Fig. 2. The solid line represents our SO(10) symmetry, while other lines represent other "popular" U(1)$'$ symmetries. A discontinuity appears at our postulated top quark mass of 100 GeV. The shaded region is excluded by current germanium detector data[2], under the assumption that ψ's have the galactic halo density and velocity distribution.

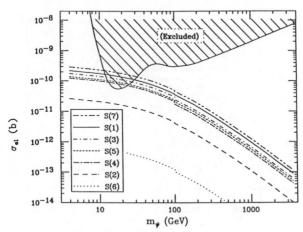

Figure 2: $\sigma_{\rm el}$ vs. m_ψ, Various U(1)$'$s.

III. A Few Complications

Heavier dark matter candidates, with $m_\psi \geq \frac{1}{2} M_{Z'}$, require modification of the above calculations, resulting in a germanium cross-section which now depends on $M_{Z'}$. Eq. (5) is modified by the pole factor[8]:

$$P_{Z'} = \frac{M_{Z'}^4}{(4m_\psi^2 - M_{Z'}^2)^2 + \Gamma_{Z'}^2 M_{Z'}^2} \tag{10}$$

For $m_\psi > M_{Z'}$ a new annihilation channel opens, $\psi\bar\psi \to Z'Z'$. In the non-relativistic limit, this contribution is[11]

$$\langle\sigma_A v\rangle_{Z'Z'} = \frac{R^2}{64\pi} \frac{(g_1' V_\psi)^4}{m_\psi^2} \left[\frac{m_\psi(m_\psi^2 - M_{Z'}^2)^{3/2}}{(m_\psi^2 - \frac{1}{2}M_{Z'}^2)^2} \right] \tag{11}$$

where $R = 1$ in all cases of interest to us[1]. In the limit $m_\psi \gg M_{Z'}$, the last term goes to unity, and $\langle\sigma_A v\rangle \sim (g_1')^4/m_\psi^2$. The Lee-Weinberg calculation ($\langle\sigma_A v\rangle \approx$ Constant) and the perturbative requirement $g_1' < \sqrt{4\pi}$ thus restrict $m_\psi < 40\,{\rm TeV}$. In this limit, the germanium cross-section becomes

$$\sigma_{\rm el} = (1.24 \times 10^{-11}\,{\rm barns}) \left(\frac{m_\psi}{M_{Z'}}\right)^2 \left(\frac{1\,{\rm TeV}}{M_{Z'}}\right)^2 \tag{12}$$

up to factors of nearly unity, including a new group-dependent factor analogous to Φ.

Mass mixing between the Z and the Z' is known to be small[10,12,13], and is unimportant except near $m_\psi = 45\,\mathrm{GeV}$[14]. At the Z resonance, the increase in the annihilation rate must be compensated by a reduction in G'. Since the interaction with germanium is t-channel, no resonance occurs, and the germanium cross-section is dramatically reduced.

In Fig. 3 we plot σ_{el} against m_ψ for various values of $M_{Z'}$, with $\theta_{\mathrm{mix}} = 0.03$.

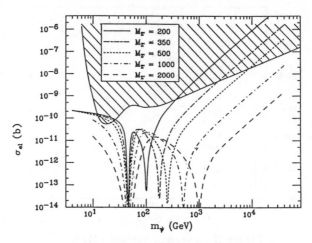

Figure 3: σ_{el} vs. m_ψ, Various $M_{Z'}$, $\theta_{\mathrm{mix}} = 0.03$

IV. Conclusions

As long as $m_\psi \ll M_{Z'}$, we can predict σ_{el} vs. m_ψ for a dark-matter U(1)'-coupled Dirac fermion, without knowing any details about the Z' mass, the coupling g_1', or even the fermion charge V_ψ. Uncertainty in the Hubble parameter and in the form of the low-energy U(1)' introduce small uncertainties in σ_{el}. In this regime, the predicted germanium cross-section is close to experimental limits for $10\,\mathrm{GeV} < m_\psi < 100\,\mathrm{GeV}$.

Another window of experimental detection opens for $m_\psi > \frac{1}{2}M_{Z'}$, anywhere in the range $400\,\mathrm{GeV} < m_\psi < 40\,\mathrm{TeV}$, but now the predicted cross-section depends on $M_{Z'}$.

Under the fairly general assumptions that the dark matter is a U(1)'-coupled fermion, with no particle-antiparticle asymmetry, and with the local density and velocity distribution of the galactic halo, we have shown that detection is within the reach of dark matter experiments over a wide mass range. Depending on whether one comes to this conference from the Astro or the Particle side, this could be considered a prediction for a dark-matter candidate, or a probe of the U(1)' sector.

REFERENCES:

[1] Condensed from D.E. Brahm and L.J. Hall, *Phys. Rev.* **D41**:1067 (1990).

[2] D.O. Caldwell *et. al.*, *Phys. Rev. Lett.* **61**:510 (1988);
D.O. Caldwell *et. al.*, in *Dark Matter: Proc. of the XXIIIrd Rencontre de Moriond*, Les Arcs, Savoie, France, March 8–15, 1988, ed. J. Audouze & J. Tran Thanh Van;
B. Sadoulet, private communication.

[3] S.P. Ahlen *et. al.*, *Phys. Lett.* **195B**:603 (1987).

[4] A. Drukier & L. Stodolsky, *Phys. Rev.* **D30**:2295 (1984).

[5] B.W. Lee & S. Weinberg, *Phys. Rev. Lett.* **39**:165 (1977);
P. Hut, *Phys. Lett.* **69B**:85 (1977);
K. Sato & M. Kobayashi, *Prog. Theor. Phys.* **58**:1775 (1977);
M.I. Vysotskii, A.D. Dolgov & Ya.B. Zel'dovich, *Pis'ma Zh. Eksp. Teor. Fiz.* **26**:200 (1977) [*JETP Lett.***26**:188 (1977)];
D.A. Dicus, E.W. Kolb & V.L. Teplitz, *Phys. Rev. Lett.* **39**:168 (1977);
G.L. Kane & I. Kani, *Nucl. Phys.* **B277**:525 (1986);
E.W. Kolb & K.A. Olive, *Phys. Rev.* **D33**:1202 (1986) & **D34**:2531 (1986).

[6] See G.G. Ross & G.C. Segrè, *Phys. Lett.* **B197**:45 (1987) for an interesting cosmion model using particle-antiparticle asymmetry.

[7] K. Olive, D. Schramm & G. Steigman, *Nucl. Phys.* **B180[FS2]**:497 (1981).

[8] K. Griest & B. Sadoulet, in *Proc. of the Second Particle Astrophysics School on Dark Matter*, Erice, Italy, 1988 (FERMILAB-Conf-89/57-A).

[9] M.W. Goodman & E. Witten, *Phys. Rev.* **D31**:3059 (1985).

[10] G. Costa *et. al.*, *Nucl. Phys.* **B297**:244 (1988).

[11] K. Griest, M. Kamionkowski, & M.S. Turner, *Phys. Rev.* **D41**:3565 (1990), eq. (B7).

[12] J. Ellis *et. al.*, *Nucl. Phys.* **B276**:14 (1986);
E. Cohen *et. al.*, *Phys. Lett.* **165B**:76 (1985).

[13] E. Jenkins, *Phys. Lett.* **192B**:219 (1987);
F. Del Aguila, M. Quirós & F. Zwirner, *Nucl. Phys.* **B284**:530 (1987);
F. Del Aguila, M. Quirós & F. Zwirner, *Nucl. Phys.* **B287**:419 (1987).

[14] K. Enqvist, K. Kainulainen & J. Maalampi, *Nucl. Phys.* **B316**:456 (1989). Also contains a good discussion of solar capture.

58

Indirect Detection Of Heavy Supersymmetric Dark Matter

MARC KAMIONKOWSKI

Department of Physics and Enrico Fermi Institute,
The University of Chicago, Chicago, IL 60637-1433
and

NASA/Fermilab Astrophysics Center, Fermi
National Accelerator Laboratory, Batavia, IL 60510-0500

ABSTRACT

If neutralinos reside in the galactic halo they will be captured in the Sun and annihilate therein producing high-energy neutrinos. Present limits on the flux of such neutrinos from underground detectors such as IMB and Kamiokande II may be used to rule out certain supersymmetric dark-matter candidates, while in many other supersymmetric models the rates are large enough that if neutralinos *do* reside in the galactic halo, observation of a neutrino signal may be possible in the near future.

The idea that a heavy (meaning more massive than the $W^{\pm}$) neutralino[1,2] — a linear combination of the supersymmetric partners of the photon, Z^0, and Higgs bosons—makes up the bulk of the dark matter in the Universe and in the galactic halo has been the focus of much theoretical and experimental research recently.[2,3] Here we address the possibility of indirect detection of heavy neutralinos by observation of high-energy neutrinos from WIMP annihilation in the Sun.[4]

First let us briefly review the supersymmetric model. The neutralino field may be written

$$\tilde{\chi} = Z_{n1}\tilde{B} + Z_{n2}\widetilde{W}^3 + Z_{n3}\tilde{H}_1 + Z_{n4}\tilde{H}_2, \tag{1}$$

where $(Z)_{ij}$ is a real orthogonal matrix that diagonalizes the neutralino mass matrix and depends only on the gaugino mass parameter M, Higgsino mass parameter μ, and the ratio of Higgs vaccuum expectation values $\tan\beta$. In Fig. 1 we plot neutralino mass contours (broken curves) and contours of $Z_{n1}^2 + Z_{n2}^2$ (solid curves), the gaugino fraction, for $\tan\beta = 2$ (plots for other values of $\tan\beta$ are similar). As noted originally by Olive and Srednicki[2] in much of parameter space where the neutralino is heavier than the W, the gaugino fraction is greater than 0.99 and the neutralino is almost pure B-ino, and in much of parameter space where the neutralino is heavier than the W, the gaugino fraction is greater than 0.99 and the neutralino is almost pure B-ino, and in much of parameter space, the gaugino fraction is less than 0.01 and the neutralino is almost pure Higgsino.

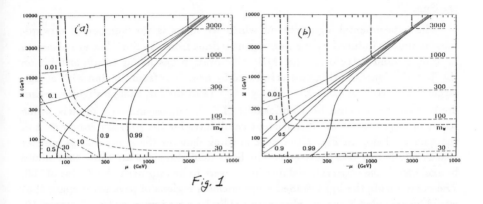

Fig. 1

There are also three neutral Higgs bosons and the lightest, H_2^0, is always less massive than the Z. The masses of the superpartners of the quarks and leptons, which we will collectively refer to as squarks, are all undetermined, but for simplicity we give them all the same mass $M_{\tilde{q}}$ which, assuming the neutralino is the LSP, is greater than $m_{\tilde{\chi}}$.

In most cases of interest accretion of neutralinos from the halo[6] onto the Sun and depletion from the Sun by annihilation come to equilibrium on a time scale much shorter than the solar age in which case the annihilation rate is given by the rate of capture. The capture rate for a specific model is determined by the elastic scattering cross section of the neutralino off of the nuclei in the Sun. Capture of neutralinos that are mixed gaugino/Higgsino states occurs by a coherent interaction with the heavier nuclei in the Sun in which the lightest Higgs boson is exchanged;[6] since the lightest Higgs boson is always lighter than the Z capture of such neutralinos is efficient even for heavier neutralinos. If the squark is not much heavier than a neutralino that is primarily gaugino, then such a gaugino is captured by a spin-dependent interaction in which a squark is exchanged[7] with the hydrogen in the Sun. Neutralinos that are primarily Higgsino (or, if the squark is much heavier than the neutralino, neutralinos that are primarily gaugino) are captured by coherent scattering off of heavy nuclei, but capture of such neutralinos rapidly becomes increasingly inefficient with increasing purity.

To obtain the neutrino spectrum from neutralino annihilation in the Sun one must determine the differential energy flux of neutrinos at the surface of the Sun resulting from the injection of the particles (*e.g.*, pairs of gauge or Higgs bosons or fermion-antifermion pairs) into which the neutralino annihilates in the core of the Sun.[4,8]

The experimental signature on which we focus is the number of upward-moving muons induced by high-energy neutrinos from the Sun that are observed in underground detectors. The IMB collaboration has found an upper limit on the flux of upward-moving muons induced by neutrinos from the Sun with energy larger than 2 GeV of 2.65×10^{-2} m^{-2} yr^{-1},[9] (and similar, though slightly weaker limits have been found by Kamiokande II[10]). Supersymmetric models which result in larger fluxes are inconsistent candidates for the primary component of the galactic halo. In Fig. 2 the dark shading denotes the regions of parameter space excluded by this constraint. The light shaded regions are those that would be excluded if the observational flux limits were to be improved by a factor of 100. The curve inside the light shaded areas encloses regions of parameter space that would be excluded if current observational limits were improved by a factor of 10. To indicate the sensitivity of these results to uncertainties in the calculation, the dashed curve inside the excluded region indicates the region excluded if the true neutrino rate is only 1/5 as large as our calculations indicate. In (a) $\tan \beta = 2$, $m_{H_2^0} = 20$ GeV, the squark mass is taken to be infinite and $\mu > 0$, and (b) is similar except that $\mu < 0$. In (c) $\tan \beta = 2$ and $m_{H_2^0} = 20$ GeV, in (d) $\tan \beta = 2$ and $m_{H_2^0} = 35$ GeV, and in (e) $\tan \beta = 25$ and $m_{H_2^0} = 35$ GeV. In (c), (d), and (e), the squark mass is assumed to be 20 GeV greater than the neutralino mass and only regions of positive μ are shown.

From Fig. 2, we see that limits on energetic neutrino fluxes from the Sun already exclude many supersymmetric models with heavy mixed-state neutralinos lighter than about a TeV when the lightest Higgs is light and $\tan \beta$ is small [Fig. 2(a), (b), and (c)], or when $\tan \beta$ is large [Fig. 2(e)], independent of the squark mass. Current neutrino-flux bounds are ineffective in ruling out neutralinos that are almost pure Higgsino or B-ino; however, if the observational bounds are improved by a factor of ten, far more supersymmetric dark-matter candidates would be observable. The rates from heavy B-inos are sensitive to the squark mass while the rates from heavy B-inos are relatively insensitive to the squark mass as may be seen by comparing Fig. 2(a) and Fig. 2(c).

To conclude, we note that the properties of the heavy neutralino in many

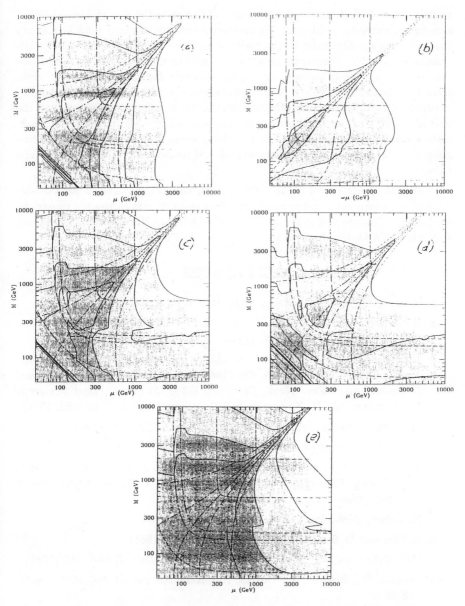

2. Regions where the neutralino is excluded as the primary component of the galactic halo by limits on the flux of upward-moving neutrino-induced muons from the Sun. The dark shaded regions are those excluded by current IMB limits. The light shaded regions are those that would be excluded if current observational limits were improved by a factor of 100. See text.

models are such that their capture and annihilation in the Sun yields an observable flux of energetic neutrinos. We also point out that in many models, a heavy neutralino may easily make up the primary component of the galactic halo while remaining invisible to neutrino detectors, so null results from energetic neutrino searches are not likely to rule out supersymmetric dark matter. Nevertheless, given the present uncertainty as to the nature of the dark matter, the popularity of supersymmetry in particle physics, and the interesting "coincidence" that the relic abundance of the LSP in most supersymmetric models falls near the dark-matter window, it is clear that the search for energetic neutrinos from the Sun holds considerable promise for discovery, should neutralinos reside in the galactic halo.

I wish to thank David Seckel, Kim Griest, and Michael Turner for valuable discussions, comments, and suggestions. This research was supported in part by the DoE (at Chicago and at Fermilab), by NASA (grant NGW-1340 at Fermilab), and by the NASA Graduate Student Researchers Program.

REFERENCES

1. H. E. Haber and G. L. Kane, *Phys. Rep.* **117**, 75 (1985).

2. K. A. Olive and M. Srednicki, *Phys. Lett.* B **230**, 78 (1989); K. Griest, M. Kamionkowski, and M. S. Turner, *Phys. Rev. D* **41**, 3565 (1990).

3. For recent reviews of dark matter and its detection, see V. Trimble, *Ann. Rev. Astron. Astrophys.* **25**, 425 (1989); J. R. Primack, B. Sadoulet, and D. Seckel, *Ann. Rev. Nucl. Part. Sci.* **B38**, 751 (1988); *Dark Matter in the Universe*, eds. J. Kormendy and G. Knapp (Reidel, Dordrecht, 1989).

4. M. Kamionkowski, Fermilab Report No. FERMILAB-Pub-90/181-A.

5. A. Gould, *Astrophys. J.* **321**, 571 (1987).

6. R. Barbieri, M. Frigeni, and G. F. Giudice, *Nucl. Phys.* **B313**, 725 (1989); T. P. Cheng, *Phys. Rev. D* **38**, 2869 (1988); H.-Y. Cheng, *Phys. Lett.* B **219**, 347 (1989).

7. K. Griest, *Phys. Rev.* **D38**, 2357 (1988).

8. S. Ritz and D. Seckel, *Nucl. Phys.* **B304**, 877 (1988).

9. IMB Collaboration: J. M. LoSecco *et al.*, *Phys. Lett.* B **188**, 388 (1987).

10. Totsuka, Y., Institute for Cosmic Ray Research Report No. ICR-Report-192-89-9.

Neutrinos from Dark Matter Annihilation

David Seckel

Bartol Research Institute

University of Delaware, Newark, DE 19716

Abstract

A leading candidate for Dark Matter in our Universe is Weakly Interacting Massive Particles (WIMPs). This paper reviews the search for the high energy neutrino signal that should exist if WIMPs are captured in the Sun or Earth and subsequently annihilate. Comparisons are made of various search strategies. Recent work on WIMPs with mass below that of the W-boson is summarized. Expectations for WIMPs with mass greater than 100 GeV are discussed.

1 Introduction

The talks by M. Turner[1] and J. Peebles[2] addressed the question of whether the Universe is open, closed, or has a critical density of $\Omega = 1$. Many of us favor the possibility that $\Omega = 1$, but realize that constraints from big bang nucleosynthesis[3] would then imply that most of the mass in the Universe is not baryonic. Even aside from the question of Ω, there is ample evidence that $\sim 90\%$ of the mass in galaxies is not readily observable through electromagnetic radiation[4]. It is just possible that the dark matter in our galaxy may be some form of condensed baryonic material, *e.g.* planet sized objects or black hole remnants of supermassive stars; but if so then one would have to either give up $\Omega = 1$, or else invoke a second non-standard form of matter to close the Universe. Although these are viable options, Occam's razor suggests a single form of dark matter that both pervades our own galaxy and supplies the mass to critically close the Universe.

There are, of course, many particle physics candidates for dark matter[5]. A general class of these is weakly interacting massive particles, or WIMPs. This presentation concerns one method for testing the WIMP hypothesis: that WIMPs may be captured in the Sun or Earth and annihilate there, producing some high energy neutrinos, which may be detected in any of a number of large experiments capable of doing neutrino astronomy.

What makes the WIMPs ideal candidates for dark matter is that their abundance[6,7,8,9,10,11] may naturally give $\Omega_\chi \sim 1$, where χ denotes 'WIMP'. When the temperature of the Universe drops below m_χ, the thermal equilibrium abundance of WIMPs becomes Boltzman suppressed. For a short time thereafter $\chi - \chi$ annihilation proceeds quickly enough to maintain approximate thermal equilibrium, but soon the WIMP abundance drops to the point where annihilation is suppressed by the low number density, at which point we say that annihilation freezes out. Thus, the mass density of WIMPs today is determined almost exclusively by the annihilation cross-section at freezeout,

$$\Omega_\chi h^2 = \frac{\langle \sigma v \rangle_a}{1 - 2 \times 10^{-26} \text{ cm}^{-3} \text{ sec}} \tag{1.1}$$

where h is the Hubble parameter in units of 100 km/sec/Mpc, and $\langle \sigma v \rangle_a$ is the thermal averaged annihilation cross-section. The necessary values of $\langle \sigma v \rangle_a$ occur fairly naturally for particles with mass in the range $10-1000$ GeV and with weak strength interactions.

Since WIMPs engage in weak interactions, and weak interactions are observed in a variety of astrophysical and laboratory experiments, one might ask if dark matter in the form of WIMPs might be observable. In general, the answer is yes, but the observations are difficult[12,13]. There are many possibilities: production of WIMPs in accelerator experiments, 'direct detection' of WIMPs in our galactic halo via their elastic scattering off nuclei in a laboratory apparatus[14] the production of an exotic spectrum of cosmic rays[15] or photons[16] from $\chi - \chi$ annihilation in the halo of our Galaxy, and a number of consequences that may ensue if WIMPs are captured in astronomical bodies such as the Earth, Sun, or other stars. If WIMP annihilation is suppressed today (e.g. due to a cosmic asymmtry or perhaps a velocity suppressed annihilation cross-section) then the last category includes observable effects on stellar evolution[17] as well as a possible modification of the solar model which could provide a solution to the solar neutrino problem[18]. If WIMP annihilation proceeds normally, then WIMPs captured in the Sun[19,20,8,22,23,24] or Earth[25,26,27,28,29] may produce neutrino events in large detectors[30,31,32,33,34,35,36] - the subject of this presentation.

The annihilation and production signals in this list are constrained by Ω_χ and Eq. 1.1, although there is a fair amount of model-building flexibility. It is also possible to keep the halo density of WIMPs constant and still choose $\Omega_\chi \sim .1 - .2$. In this case, all signals increase because of the implied rise in cross-sections[37], but not without cost. To do this requires that one gives up on $\Omega = 1$; or that one accepts that WIMPs are clumped in galaxies, whereas most of the mass of the Universe is in a still yet to be discovered form of Even Darker Matter.

Much of this paper focuses on a particularly compelling WIMP candidate - the gaugino which is the lightest supersymmetric particle[38,39] (LSP). Other WIMPs have been discussed as candidates for dark matter; most notably massive Dirac or Majorana neutrinos, but also sneutrinos[40,41]. These possibilities have temporarily gone out of fashion, as a combination of accelerator experiments, laboratory searches for WIMPs, and cosmological arguments have pushed the mass for such objects into the TeV region. The parameter space for LSP's is also being closed down, but the number of possibilities is large enough, even within the minimal supersymmetric extension to the standard model discussed by most authors, that LSP WIMPs are still an exciting possibility. Even if $\Omega_{LSP} \neq 1$, if one believes in supersymmetry then it is hard to avoid a contribution of LSP WIMPs at the level of at least[39,42,43] $\Omega_{LSP} = .01$, and they may still give interesting signals.

2 The Annihilation Signal from the Sun

There are three steps to the calculation of a high energy neutrino signal from WIMP annihilation: capture in the Sun or Earth, production of a source spectrum of neutrinos from WIMP annihilation, and the efficiency of detecting that neutrino flux in a particular detector. I will discuss these issues in detail for the Sun, and then discuss the Earth by comparison.

Capture occurs when a halo WIMP passes through the sun and loses energy in a collision with a nucleus of solar material. After many collisions the WIMP in question will settle down into the core of the sun, where it will become a member of a population of WIMPs in approximate thermal equilibrium with the material at the solar core. LSP WIMPs have the property that they are Majorana fermions[38,39]. Unlike Dirac fermions, there is no possibility of a particle-antiparticle assymetry that will suppress annihilation. As a result, after some time, τ_{eq}, there will be an equilibrium between capture and annihilation[19]. For most WIMP parameters $\tau_{eq} \ll t_{\odot}$, the age of the sun, and so the annihilation rate today should be given by the capture rate. The differential flux of neutrinos of flavor i is then

$$\frac{d\phi_i}{dE} = \frac{\Gamma_C}{8\pi r_\oplus^2} \frac{dY_i}{dE} \qquad (2.1)$$

where Γ_C is the capture rate and $r_\oplus$ is an astronomical unit. The differential yield per annihilation, dY_i/dE, depends upon the details of the WIMP model.

The flux of neutrinos given by Eq. 2.1 may be detected in one of two ways[44,45,46]. 'Contained events' are those where a charged current event *in* the detector produces a charged lepton and a spray of hadrons. Note that this signal is inefficient for τ neutrinos since the 1.784 GeV τ mass suppresses the charge exchange cross-section[20]. 'External muons' occur when a muon produced by a charge exchange event in the rock/water *outside* the detector has sufficient energy to pass into the detector and be identified as a muon. In addition, there is also the possibility of detecting neutral current inter-actions contained in the detector; however, the neutral current cross-section is smaller than the charged current one by about 1/3, and backgrounds are more severe due to the missing lepton momentum. For a flux of τ neutrinos this signal could dominate, if identifiable.

The rates for the two signals are[23]

$$R_C = 72HC \quad \sum_f \Gamma_f \langle Yz \rangle_f \quad \epsilon \text{ kt}^{-1} \text{ yr}^{-1}$$
$$R_E = 7.7HC \quad m_\chi \sum_f \Gamma_f \langle Yz^2 \rangle_f \quad \epsilon \text{ A}_6^{-1} \text{ yr}^{-1} \qquad (2.2)$$

where

$$H = \rho_{.3} \left\langle \frac{1}{v_{300}} \right\rangle \quad \text{and} \quad C = F_C \sum_N X_N \sigma_{\chi N38} F_N(m_\chi) \qquad (2.3)$$

The rates in Eq. 2.2 have already been summed over neutrino flavors. In estimating the detection rates, deep inelastic neutrino-nucleus cross sections rising linearly with E_ν have been isospin averaged as if the detector were equal parts protons and neutrons.

I have broken the rates up into several terms having to do with halo parameters of the dark matter (H), capture efficiency (C), neutrino yields per annihilation, and detector characteristics. The list below gives a description of the individual quantities in Eq. 2.2, with an orientation towards WIMPs with $m_\chi < m_W$. An * beside a quantity indicates some uncertainty in the quantity, perhaps less than 50%; whereas ** indicates a great deal of uncertainty, perhaps an order of magnitude or more.

$\rho_{.3}^{**}$ - $\rho_\chi / .3$ GeV cm^{-3}, where ρ_χ is the local density of WIMPs. If ρ_χ accounts for all the local dark matter, then $\rho_{.3} = 1 \pm .5$ and the uncertainty[47] in $\rho_{.3}$ would rate only one *. However, one must allow that WIMPs may make up only a fraction of the halo.

m_χ^{**} - mass of WIMP in GeV. Masses of $10 - 1000$ GeV are discussed.

v_{300}^* - velocity of WIMP / 300 km sec^{-1}.

F_C - a fudge factor to account for the possibility that capture and annihilation are *not* in equilibrium. For example, for $m_\chi \lesssim 3$ GeV, equilibrium is reached between capture and evaporation[8,48], and the annihilation signals are much supressed. Or, the assumption of equilibrium would not be valid if capture and annihilation rates were sufficiently suppressed. For example, this can happen in some LSP models where squarks are made very heavy and capture is suppressed[23,24].

X_N - fractional abundance, by number, of element N in the sun.

$\sigma_{\chi N38}^{**}$ - elastic cross-section for WIMP-N scattering / 1×10^{-38} cm^2. The fiducial value is a typical WIMP-proton cross-section. There is a lot of model building flexibility here[42]. There is also some uncertainty due to our imprecise knowledge about the structure of nucleons. Many LSP WIMPs couple mostly to the spin of the nucleus. For these cases it is important to note that the spin content of nucleons and nuclei is not completely understood[49,50] and for some extreme cases there may be an order of magnitude uncertainty in $\sigma_{\chi N38}$. In addition to the axial coupling there is also a scalar coupling of LSP's to the mass of the nucleus[42,43]. This coupling is also sensitive to the details of nucleon structure, in this case the fraction of nuclear mass carried by strange quarks[51,29]. For lighter LSPs, the scalar coupling can dominate capture in the Earth[28], but is unimportant in the sun unless one of the Higgs particles is very light. For heavy LSPs, kinematic effects suppress the efficiency of capture by hydrogen, and then for a large part of parameter space capture is dominated by nuclei with mass of 12-20 AMU.

Another issue concerns heavy nuclei as targets. For many purposes it is sufficient to calculate the cross-sections in the limit of zero momentum transfer, *i.e.*, $r_N^2 q^2 << 1$, where r_N is the nuclear size; however for nuclei heavier than helium serving as targets in the sun this approximation breaks down, and the cross-sections calculations must account for nuclear structure[21,24,52]. In the Earth, the WIMP velocities are significantly smaller and the $q^2 = 0$ approximation is reasonable.

$F_N(m_\chi)^*$ - probability that elastic scattering off N will result in capture. The sun's gravitational potential is much deeper than that of the galaxy. As a result, a WIMP interacting in the sun will have a kinetic energy which is dominated by the escape energy from the sun. Even if only a small fraction of that kinetic energy is lost the WIMP becomes bound. Therefore, for relatively light WIMPs the capture efficiency is near 1 for all nuclei of interest. However, if m_χ becomes much greater than the nuclear mass, the energy transferred to the nucleus becomes small compared to the WIMP's kinetic energy, and capture off the lighter elements is suppressed. Capture efficiencies have been calculated in great detail by Gould[27] for both the Earth and sun.

Γ_f^* - branching ratio to particle f. For lighter WIMPs, most work has concerned annihilation into fermions[23]. With the recent interest in heavier WIMPs, annihilation into W^+W^- and Z^0Z^0 must now be considered[24]. Annihilation into Higgses is also possible and may dominate, even for relatively light WIMPs if the Higgs channels are kinematically open[29]. Annihilation to gluons can also be significant through loop diagrams[53]. LSP WIMPs are Majorana fermions which implies that annihilation into fermions is suppressed by a factor of $(m_f/m_\chi)^2$, and direct annihilation into neutrinos is forbidden. Heavy Dirac neutrinos and sneutrinos may annihilate directly into neutrinos.

Essentially all WIMPs have annihilation modes that lead to significant high energy neutrino yields. Although the branching ratios may vary a lot from model to model, this is generally not a great source of uncertainty, since with a few exceptions the yields do not vary greatly from channel to channel. The exception is sneutrino annihilation directly to neutrinos for which the yields are much higher. In fact, a halo made up entirely of μ or e sneutrinos seems to be ruled out[20,54,55].

$\langle Yz \rangle_f$ - average neutrino yield per annihilation into particle f, weighted by neutrino energy; $z \equiv E_\nu/m_\chi$. Weighting by E_ν is relevant for contained events, and accounts for the nearly linear increase in the neutrino absorption cross-section with energy. The neutrinos are usually secondary particles that arise from the decay of primary annihilation products. As long as the primaries are relativistic, the yields dY/dz are nearly independent of m_χ. Yields for c, b, and t quarks, and τ leptons have been estimated[23] using the Lund Monte Carlo[56] to handle hadron masses, branching ratios, and fragmentation effects. These yields have been used by Gelmini, Gondolo, and Roulet[29], and by Kamionkowski[24] to estimate the yields from the annihilation channels involving scalar and vector bosons. Generally, the b and c spectra are softer than those from τ, t or gauge bosons. The Higgs channels tend to be even softer - the neutrinos being demoted to tertiary products instead of secondaries.

$\langle Y z^2 \rangle_f$ - second energy moment of the neutrino yield: appropriate for discussing external events. One power of E_ν accounts for the increase in cross-section, the second accounts for the increased range of the muon. In the formula for R_E a constant $dE/dx = 2.5 \times 10^{-3}$ GeV/(gm/ cm^2) was assumed. For high WIMP masses a more accurate[57] dE/dx should be incorporated by the experimentalist calculating his detector's efficiency.

ϵ - detector efficiency. This must be calculated by the experimentalist (!) using appropriate differential ν fluxes and differential isospin *dependent* deep inelastic cross-sections to correct for the simplistic form of Eq. 2.2. The efficiency factor is especially important for R_T, where a high threshold energy for muon identification and the requirement that signal can only be collected at night may result in efficiencies of only a few % of that estimated from a projected cross-section of the detector.

kt - detector mass / kilotonnes.

A_6 - detector area / 1×10^6 cm^2

Before deciding whether or not the rates in Eq. 2.2 are detectable one must consider the background processes. The most serious background comes from neutrinos produced in cosmic ray air showers[44,45,46,58]. Detailed background calculations should be done on an experiment by experiment basis. Perhaps the safest technique is to take backgrounds from measurements on and off source[55].

A rough guide to backgrounds is[8]

$$B_C \simeq 3/E_T \text{ kt}^{-1} \text{ yr}^{-1} \text{ sr}^{-1} \; (E_T > 1 \text{ GeV}) \qquad (2.4)$$

$$B_E \simeq 6 \; A_6^{-1} \text{ yr}^{-1} \text{ sr}^{-1} \; (E_T = 4 \text{ GeV}). \qquad (2.5)$$

Eq. 2.4 is the background for contained muons; the electron background is less than a third this amount. In Eq. 2.4 the quantity E_T is the threshold for the neutrino energy; while in Eq. 2.5 it refers to the muon energy at the edge of the detector. Eq. 2.5 is not very sensitive[46] to E_T. For the contained event background both spectral and angular information might be used to enhance signal to noise, while for the external signal only angular cuts are very useful. These background rates give only a handful of events per year, even for a large detector such as IMB or Kamiokande. Nevertheless we are approaching the time when high energy solar neutrino searches will become background limited instead of signal limited.

2.1 Contained vrs. external events

It is interesting to compare the prospects for detecting a WIMP annihilation signal using either the contained events or the external muons as a signal. To help decide this issue, consider the ratio

$$\frac{R_E}{R_C} \approx \frac{m_\chi}{20 - 40 \text{ GeV}} \frac{\epsilon_E}{\epsilon_C} \frac{A_6}{M_{kt}}, \qquad (2.6)$$

which is a fair comparison of techniques if the detectors are signal limited. The range $(20 - 40 \text{ GeV})$ arises from plausible variations in branching ratios to different annihilation channels and considering the yields for the different channels. The ratio A_6/M_{kt} is roughly one for a 1 kiloton water Čerenkov detector. Given the need to develop larger detectors for the smaller signals associated with heavier WIMPs, it seems that external muons offer a better opportunity. Before coming to that conclusion, however, one should consider detector efficiency.

Detector efficiency is properly in the province of the experimentalist, but still a few general remarks may be tendered. An important consideration is the energy threshold for detecting an event. This is especially important for detection of a signal from low mass WIMPs, for which the yield will be dominated by the decay of heavy quarks and τs. The spectra for these is quite soft: $\langle z \rangle$ = .138, .125, .295; for c, b and τ, respectively. (Note that the spectrum from b decays includes subsequent decays of c quarks and this softens the b spectrum, but increases the yield) So, for example, a contained event detector with a threshold of 2 GeV would have ϵ_C = .68 for detecting 20 GeV photinos, but raising the threshold to 10 GeV reduces the efficiency to .16. For high mass WIMPs threshold is less of a concern, since for most operational and designed detectors, the thresholds would be small fraction of m_W (DUMAND[36] would be an exception). In addition, these WIMPs often have gauge bosons as the dominant signal and these produce hard neutrinos, $\langle z \rangle$ = .5

Contained event detectors may have very low thresholds, $e.g.$ IMB[30] and Kamiokande[32]; however, larger detectors, $e.g.$ DUMAND, will likely have higher thresholds. Even detectors with low event thresholds may find it useful to make an effective threshold by cutting their data to reduce background and/or improve angular resolution, $e.g.$, the analysis from Frejus[31]. Detectors designed with an eye towards external muons, $e.g.$ MACRO[34] and GRANDE[33], will almost certainly have significant energy thresholds. In addition, this detection scheme cannot collect data during the day due to the background from punch through muons from cosmic ray cascades. For a large area, planar detector (GRANDE) geometry will play a further role in reducing the signal when the sun is near the horizon. Comparing to the example above, a planar detector at 35° latitude would have efficiency to 20 GeV photinos of .13 and .005 for thresholds of 2 GeV and 10 GeV, respectively.

2.2   Comparison of Earth and Sun

Besides looking at the Sun, one can also look for neutrinos from dark matter annihilation in the Earth[25,26,27,28]. It is instructive to compare various aspects of the two signals. First, there is the obvious proximity of the Earth so that the neutrino flux per annihilation is larger by $(r_\oplus / R_\oplus)^2$, just due to geometry. Next, the gravitational potential of the Earth is shallow compared to the sun or the galaxy and this suppresses capture whenever there is not a good match between the WIMP mass and some relatively abundant nucleus. The Earth is made up mostly of heavy spin 0 nuclei, whereas the sun is mostly hydrogen and helium with an abundance of metals of only a few percent. Therefore, WIMPs with only axial couplings to nuclei have their capture in the Earth suppressed; however, WIMPs with scalar or vector couplings have large $\sigma_{\chi N}$ to the heavy elements in the Earth. For moderately heavy WIMPs a good match between WIMP and nuclear masses allows for efficient energy loss and capture; however, for very heavy WIMPs ($m_\chi > 100$ GeV) the shallowness of the Earth's gravitational potential is again dominant and the Earth is not efficient at capture. In fact, because the Earth's potential well is shallower than the solar potential at 1 AU one might worry that at some point capture by the Earth is impossible because the low velocity orbits required for capture cannot be populated by galactic WIMPs. Gould has shown that this worry is spurious - these orbits will be populated by gravitational scattering of WIMPs by Jupiter and Venus [59]. Finally, although the sun is in equilibrium between capture and annihilation for almost all models, the same cannot be said of the Earth. In calculating the Earth signal one may have to reduce the flux by a factor of $\sim (\tau_{eq}/t_\oplus)^2$.

The background issues differ as well. The solar core, where annihilation takes place is truly a point source for neutrino detection, whereas the angular size of the terrestrial annihilation signal could be as much as 5° near the evaporation mass limit [27] of $9 \lesssim m_\chi \lesssim 13$ GeV. Both detectors have the background from cosmic ray cascades in the Earth's atmosphere, but the sun has an additional background from cascades in the solar atmosphere, which could be a factor for very low signal detection [60]. Finally, the because of the background of downward going muons one can only look at the sun during the night, whereas the center of the Earth is always located at the nadir, so the Earth's on time is at least twice that of the sun for the external muon signal.

After all the dust settles, one finds that the signal from the Earth wins only for medium heavy WIMPs $m_\chi \approx (20 - 100$ GeV) with significant scalar or vector couplings [28,29].

3 Results for WIMPs with $m_\chi < M_W$

Comparing event rates (Eq. 2.2) and backgrounds (Eq. 2.4 and 2.5), we see that for $\Omega_\chi = 1$, a detectable signal in a several kiloton detector is not unreasonable. There is; however, enough flexibility in the model building aspects that failure to see a signal does not yet generate tight constraints for LSP WIMPs made of a combination of photino, higgsinos, and zino.

Several experimental groups have produced findings on the flux of high

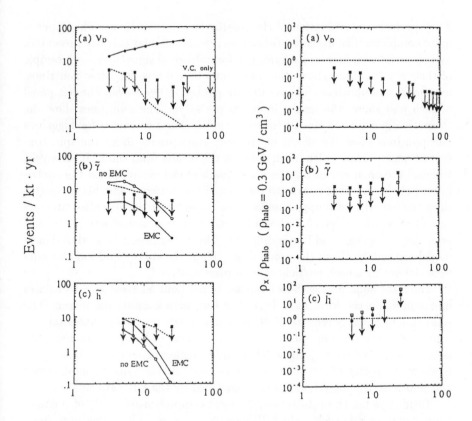

Figure 1: Constraints on dark matter candidates derived by the Kamiokande collaboration for a) Dirac Neutrinos, b) Photinos and c) Higgsinos. See text for details.

energy neutrinos from the sun - notably the IMB, Kamiokande and Frejus collaborations. IMB has published constraints[30] on both contained events and external muons. The Frejus experiment[31] gives a constraint on contained events for which $E_\nu > 2$ GeV.

The data from these experiments is sufficient to severely constrain μ or e sneutrinos and Dirac neutrinos as the dominant form of halo material[20,8,54,55]. It should be remembered that other constraints on Dirac neutrinos and sneutrinos exist from nuclear recoil experiments[61], and the results from LEP[62]. The data does not give significant constraints on higgsinos or photinos[50]. An analysis[29] of more general LSP models is summarized below.

The most recent data presentation and analysis comes from the KII collaboration[32]. Their results include contained events and external muons.

They perform an evaluation of the sensitivity to τ neutrinos. The paper is quite complete. The decisions behind energy and angular cuts are presented, as well as information on backgrounds for individual signals. As an example of their results Fig. 1 shows the constraints they derive for Direc neutrinos, higgsinos, and photinos. The results are plotted in two ways. The left panel of each pair shows the results in events per kt-yr. The solid lines show the theoretically expected signals for each model. The two curves for higgsinos and photinos show the effects of different assumptions about the spin content of nuclei. Those with black dots use the results from the EMC analysis, whereas the open circles use the naive quark model estimates for the contribution of the different quarks to the spin of the nucleon. The black squares show the experimental upper limits for each candidate particle after subtraction of background, which is shown for each case as a dashed line. Note that each dashed background line is different. This is because they analyzed each model separately, 'from the ground up'. Thus, the energy and angular cuts were different for each particle at each mass analyzed.

The right hand panels show how each upper limit on event rate translates into a limit on the local halo density of the particle under discusson. The solid (open) squares refer to EMC or naive quark model estimates for the capture rates in the sun. Clearly, a halo consisting of heavy Dirac neutrinos is not allowed. Of course, this was known from other experiments, but it goes to show the power of the solar neutrino signal. At the same time higgsinos and photinos are not yet seriously constrained by the KII data set.

Turning to the theoretical side, the most comprehensive study of the neutrino signal for relatively light LSP models $(m_\chi < m_W)$ has been performed by Gelmini, Gondolo and Roulet [29]. They consider the general problem of the nature of the LSP in the minimal supersymmetric extension to the standard model. They confined themselves to $m_\chi < m_W$, since they did not consider annihilation of WIMPs into gauge bosons, which are important at higher masses. They did, however, take into account the possibility of annihilation into Higgs bosons, and the effects that Higgs exchange would have on the elastic scattering cross-sections. Their Higgs exchange cross-sections include the larger Higgs couplings to nuclei which arise from the strange quark sea. They calculated signals both from the Earth and the Sun. In addition, they considered the constraints from Ω_χ, accelerator experiments, and direct detection.

A sample of the Gelmini, *et al.* results is reproduced in Fig. 2. The

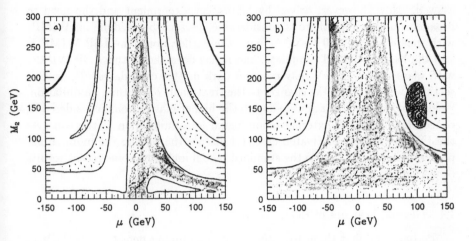

Figure 2: Constraints on light LSP WIMPs. a) Constraints from considering Ω_χ, b) Limits based on the neutrino signal from the Earth, accelerator, and direct detection data. Adapted from Gelmini, *et al.*[29]

graphs are for models in which the relevant Higgs parameters are $\tan\beta = 8$ and $m_H = 50$ GeV. The left hand panel shows contours of $\Omega_\chi h^2$ in the $M_2 - \mu$ plane, where M_2 and μ are the usual supersymmetric mass parameters (see Sec. 4 or, better, the references for definitions of these parameters). The bold line shows the mass contour $m_\chi = m_W$. The lightly shaded regions show $\Omega_\chi > 1$, whereas the speckled regions show $\Omega_\chi < .001$. The right hand panel shows constraints from accelerator experiments (light shading), direct detection (heavy shading), and the neutrino signal from the Earth using the IMB data (speckled). It is interesting that the direct detection experiments and neutrino searches yield similar limits.

4 Prospects for WIMPs with $m_\chi > 100$ GeV

The first papers to discuss the WIMP neutrino annihilation signal concentrated on WIMPs with masses in the mass range of $m_\chi = 3 - 30$ GeV. Thus, much of the discussion about the details focussed on these mass scales as well. Now, however, driven by laboratory experiment and the need to broaden our horizons there is considerable research being done on WIMPs with masses above 100 GeV. As with the earlier work, the LSP WIMP is still the best motivated candidate, and it has been shown that LSP WIMPs with mass up to $\sim$ TeV may be good dark matter candidates[63,64]. In this section I will highlight the changes to the treatments of capture, annihilation, yields and detection that ensue due to the higher masses being considered[23], concentrating on the external muon signal for annihilation in the sun. As the LSP is the best motivated case, I'll discuss the LSP in the minimal supersymmetric extention to the standard model including some recent results by Kamionkowski[24].

4.1 General Features of Heavy WIMPs

Perhaps the first thing to note is that the number density of WIMPs in the galactic halo scales as ρ_χ/m_χ, but the detection cross-section and muon range increase with mass so keeping everything else equal the event rate goes up with mass. Of course everything isn't equal. Typically the capture cross-section $\sigma_{\chi N}$ decreases as model masses increase. Even keeping $\sigma_{\chi N}$ constant the kinematic efficiency for capture decreases as $1/m_\chi$ for WIMPs more than ~ 20 times the nuclear mass. In addition, the momentum transfer to nuclei with $m_N \gtrsim 10$ AMU is no longer small, with the result that form factors must be calculated, and capture by these nuclei is suppressed. The result of these two competing features is that the light metals (C, N, O) play a more important role for capture of heavy WIMPs than they do for WIMPs with $m_\chi < m_W$.

In some extreme cases the capture rate is so suppressed that the abundance of WIMPs in the sun will not have reached equilibrium during the sun's lifetime, and this will further suppress the signals from the values given in Eq. 2.2. In practice this is not yet a major concern for the sun, since if signals get that low they are not going to be detectable for a long time anyway. For the Earth, the equilibrium time scale is shorter and this effect must be included.

For heavy WIMPs the discussion of yields will be modified from that given for the lighter WIMPs. First, one must allow for the inclusion of gauge bosons and Higgses as open annihilation channels. The two body decays, $W \to l\nu$ and $Z \to \nu\bar{\nu}$, will produce a relatively hard spectrum of neutrinos, thus enhancing the detection rate when the gauge boson channels are favored. As mentioned in Sec. 2, Higgs' channels may dominate, but the resulting neutrino spectra are slightly softer.

A more significant difference in yield calculations is the role played by the solar medium as WIMP masses increase. As a high energy neutrino passes through the sun to the surface there is the possibility that it will be absorbed by the solar medium in a charged current scattering event, or lose energy in a neutral current event. For low energy neutrinos these probabilities are small and the absorption effect may be neglected. The absorption cross-sections increase with energy, however, and this effect cannot be neglected for very massive WIMPs. The probability for a neutrino to escape the sun is given by $P(E_\nu) = \exp(-E_\nu/E_{abs,\nu})$, where $E_{abs,\nu} = 198$ GeV and the corresponding energy for antineutrinos is 296 GeV. Thus, only a fraction of the neutrino flux produced by the annihilation of TeV WIMPs will escape the sun. The neutral current cross-sections are smaller and play a less important role. High energy tau neutrinos behave slightly differently in that the τ produced in a charged current event will decay before it stop, producing another tau neutrino. Thus, for ν_τ's the absorption process acts like an energy loss mechanism. In addition, those decays may produce some e and μ neutrinos.

Another effect of the solar medium is the hadronic stopping power of heavy quarks. The lifetime of relativistic bottom and charmed hadrons is stretched out by the relativistic γ factor to the point where they may interact with the solar medium before they decay. An estimate of the stopping power suggests that most b (c) quarks will be slowed to $\sim 470(250)$ GeV before decaying. The effect of stopping power would not affect the more massive top quarks, the non-hadronic τ leptons, or the short lived scalar and vector bosons; and, of course, stopping completely suppresses any neutrino signal from the light quarks, muon, or gluon channels even when the WIMP is light.

The effects of neutrino absorption are totally unimportant for the signal from the Earth, whereas the effect of stopping will kick in at higher energies because the density of the Earth's core is 1/12 that in the solar core.

The last subject is detector efficiency and background rejection. As WIMP masses and neutrino energies increase detection cross-sections also increase. However, due to absorption and stopping neutrino fluxes from the sun are sharply curtailed above a few hundred GeV. Further, the overall lower signals expected as m_χ increases will lead us into the regime where experiments may be background limited. Thus, probing the next decade in some theorists model parameter space may require as much as 100 times the detector size. If use may be made of the energy spectrum, then discrimination against the cosmic ray cascade background may be improved.

4.2 Heavy LSP WIMPs

To begin, it is necessary to give a quick summary of the minimal supersymmetric extension to the standard model[39,42,43,63,64] (MSSM). In the MSSM there are four neutral spin 1/2 particles that have odd R-parity. They are the partners of the photon, Z boson, and the two neutral Higgses. The LSP will then be a linear combination of $\tilde{\gamma}$, $\tilde{Z}$, $\widetilde{H}_1$ and $\widetilde{H}_2$. This basis is particularly convenient for light LSPs, where the VEVs of the Higgs fields dominate the mass matrix. However, for LSP's with mass above 100 GeV the Higgs VEVs are perturbations and it is more convenient to deal with a basis of $\tilde{B}$, $\widetilde{W}_3$, $\widetilde{H}_A$, and $\widetilde{H}_S$; the supersymmetric partners of the U(1) and SU(2) gauges bosons, and the antisymmetric and symmetric linear combinations of the Higgsinos.

In this basis the mass matrix has all its big terms along the diagonal, and it is easy to see that the LSP will be nearly a pure $\tilde{B}$, $\widetilde{H}_A$, or $\widetilde{H}_S$ depending upon whether $M_1 < |\mu|$, or $M_1 > |\mu|$ and the sign of μ (as usual M_1 is the U(1) gaugino mass term, and μ is the mass coupling the two Higgs doublets). Specifically, the LSP can be written as

$$\chi = b_B \tilde{B} + b_3 \widetilde{W}_3 + b_A \widetilde{H}_A + b_S \widetilde{H}_S \qquad (4.1)$$

and one of b_B, b_A, or b_S will be close to 1, and the others will be small. The only exception to this rule is when $M_1 \approx |\mu|$.

The importance of these remarks is that most of the scattering processes that lead to capture are proportional to the product of two of the b's. For example, LSP coupling to the Z boson, which leads to spin dependent scattering off nuclei, is proportional to $b_A b_S$; and coupling to Higgses, which leads to a scalar interaction with nuclei, is proportional to a linear combination of the products $b_B b_A$ and $b_B b_S$, neither of which may be large over most of parameter space. The one important coupling that is not necessarily suppressed is the axial coupling to nuclei through an intermediate squark. If the squarks are close in mass to the LSP then this process may be important;

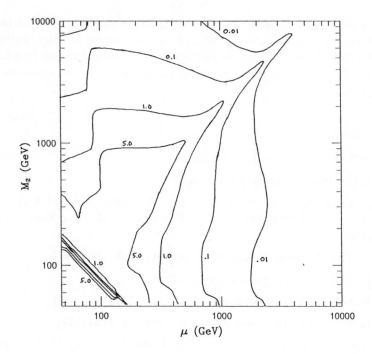

Figure 3: Constraints on high mass LSP WIMPs. Adapted from Kamionkowski[24]

however, the only important nucleus for this process is hydrogen, and efficiency of capture for heavy WIMP scattering from hydrogen is suppressed by kinematics. Thus, of the heavy LSP parameter space only a fairly narrow range leads to efficient capture of WIMPs.

Even so, some interesting results may be derived. Kamionkowski[24] has considered the heavy LSP in great detail, including the effects on capture and yields outlined above. Fig. 3 shows a contour plot adapted from his paper showing the expectations for external muons in the $M_2 - \mu$ plane, with μ taken to be positive. Other parameters are $\tan \beta = 2$, a Higgs mass of 20 GeV, and squark masses taken to be infinite. The solid curves are labeled by the muon flux they produce, in units of the published IMB constraint[30]. Thus a detector sensitive to .01 times the IMB constraint would rule out essentially all LSP WIMPs with mass less than 1 TeV for this choice of $\tan \beta$ and Higgs mass. The contours are sharply peaked along the line $M_1 = \mu$, where capture is most efficient.

Other graphs from the Kamionkowski's paper show similar features but with varying predictions for the muon flux depending upon the choice of Higgs parameters. The value of $\tan \beta = 2$ is reasonably conservative: generally capture increases with larger $\tan \beta$. On the other hand the value of $m_H =$ 20 GeV is very optimistic. Since the capture cross-section for Higgs exchange goes as $1/m_H^4$ the muon fluxes in this figure could easily decrease by a factor of a 100, by increasing the Higgs mass to 60 GeV, without changing any other parameter.

References

1. M.S. Turner, These proceedings.

2. P.J.E. Peebles, These proceedings.

3. J. Yang, M.S. Turner, G. Steigman, D.N. Schramm and K.A. Olive, *Astrophys. J.* **281**, 493 (1984); T.P. Walker, *et al.*, *Ohio State Preprint* (1990)

4. V. Trimble, *Ann. Rev. Astron. Astrophys.* **25**, 425 (1987)

5. L. Hall, in *SLAC Summer School*, Stanford (1988)

6. B. Lee and S. Weinberg, *Phys. Rev. Lett.* **39**, 165 (1977)

7. G. Steigman, *Ann. Rev. Nucl. Part. Sci.* **29**, 313 (1979)

8. K. Griest and D. Seckel, *Nucl. Phys.* **B283**, 681 (1987); (E)*Nucl. Phys.*, **B296** 1034 (1988)

9. M. Srednicki, R. Watkins, and K.A. Olive, *Nucl. Phys.* **B310**, 693 (1988)

10. K. Griest and D. Seckel, *Phys. Rev.* **D**, in press (1991)

11. P. Gondolo and G. Gelmini, *UCLA Preprint* (1991)

12. J. Primack, D. Seckel, and B. Sadoulet, *Ann. Rev. Nucl. Part. Sci.* **38**, 751 (1988)

13. P.F. Smith and J.D. Lewin, *Rutherford Appleton Lab. Preprint* RAL-88-045 (1988)

14. M.W. Goodman and E. Witten, *Phys. Rev.* **D31**, 3059 (1986); I. Wasserman, *Phys. Rev.* **D33**, 2071 (1986); A.K. Drukier, K. Freese, and D. Spergel, *Phys. Rev.* **D33**, 3495 (1986); B. Sadoulet, These proceedings.

15. J. Silk and M. Srednicki, *Phys. Rev. Lett.* **53**, 624 (1984); J. Ellis, *et al.*, *Phys. Lett.* **214B**, 403 (1989); F. Stecker and A. Tylka, *Astrophys. J. Lett.* **336**, L51 (1989); A. Tylka, *Phys. Rev. Lett.* **63**, 840 (1989); M. Kamionkowski and M.S. Turner, *Phys. Rev.* **D43**, 1774 (1991);

16. M. Srednicki, S. Thiessen, and J. Silk, *Phys. Rev. Lett.* **56**, 263 (1986); S. Rudaz, *Phys. Rev. Lett.* **56**, 2128 (1986); L. Bergström, *Phys. Lett.* **225B**, 372 (1989); S. Rudaz, *Phys. Rev.* **D39**, 3549 (1989); G.F. Giudice and K. Griest, *Phys. Rev.* **D40**, 2549 (1989); A. Bouquet, P. Salati, and J. Silk, *Phys. Rev.* **D40**, 3168 (1989)

17. A. Renzini, *Astron. Astrophys.* **171**, 121 (1987); D.N. Spergel and J. Faulkner, *Astrophys. J. Lett.* **331**, L21 (1988); D. Dearborn, G. Raffelt, P. Salati, J. Silk, A. Bouquet, *Center for Particle Astrophysics, Berkeley Preprint* CfPA-TH-89-009 (1989);

18. G. Steigman, C.L. Sarazin, H. Quintana, and J. Faulkner, *Astrophys. J.* **83**, 1050 (1978); D.N. Spergel and W.H. Press, *Astrophys. J.* **294**, 663 (1985); R.L. Gilliland, J. Faulkner, W.H. Press, and D.N. Spergel, *Astrophys. J.* **306**, 703 (1986);

19. J. Silk, K. Olive, and M. Srednicki, *Phys. Rev. Lett.* **55**, 257 (1985)

20. T.K. Gaisser, G. Steigman, and S.Z. Tilav, *Phys. Rev.* **D34**, 2206 (1986)

21. A. Gould, *Astrophys. J.* **321**, 571 (1987)

22. M. Srednicki, K. Olive, J. Silk, *Nucl. Phys.* **B279**, 804 (1987)

23. S. Ritz and D. Seckel, *Nucl. Phys.* **B304**, 877 (1988)

24. M. Kamionkowski, *Fermilab Preprint* FERMILAB-Pub-90/181-A (1991)

25. L.M. Krauss, M. Srednicki, and F. Wilczek, *Phys. Rev.* **D33**, 2079 (1986)

26. K. Freese, *Phys. Lett.* **167B**, 295 (1986)

27. A. Gould, *Astrophys. J.* **328**, 919 (1988)

28. G.F. Giudice and E. Roulet, *Nucl. Phys.* **B316**, 429 (1989)

29. G. Gelmini, P. Gondolo, and E. Roulet, *Nucl. Phys.* **B**, in press (1991)

30. J.M. Losecco *et al.*, *Phys. Lett.* **188B**, 388 (1987); R. Svoboda, *et al.*, *Astrophys. J.* **315**, 420 (1987); D. Casper, Ph.D. Thesis, Univ. of Michigan, (1991)

31. Frejus Collaboration, presented by B. Kuznik, *Orsay Preprint* LAL-87-21 (1987)

32. Y. Totsuko, *Tokyo Preprint* UT-ICEPP-87-02 (1987); N. Sato, *et al.*, *Phys. Rev.* **D**, in press (1991)

33. R. Ellsworth, *et al.*, *Irvine Preprint* UCI Neut No. 87-38 (1987)

34. MACRO Collaboration, *Nucl. Instrum. Methods*, **A264**, 18 (1988)

35. Y. Totsuka, *7th Workshop on Grand Unification, ICOBAN '86* Toyama, Japan (1986)

36. The DUMAND Collaboration

37. S. Rudaz and F.W. Stecker, *Astrophys. J.* **325**, 16 (1988)

38. H. Goldberg, *Phys. Rev. Lett.* **50**, 1419 (1983)

39. J. Ellis, J.S. Hagelin, D.V. Nanopoulos, K.A. Olive, M. Srednicki, *Nucl. Phys.* **B238**, 453 (1984)

40. J.S. Hagelin, G.L. Kane, and S. Raby, *Nucl. Phys.* **B241**, 638 (1984)

41. L.E. Ibanez, *Phys. Lett.* **137B**, 160 (1984)

42. K. Griest, *Phys. Rev.* **D38**, 2357 (1988)

43. R. Barbieri, M. Frigeni, and G.F. Giudice, *Nucl. Phys.* **B313**, 725 (191989)

44. T.K. Gaisser, T. Stanev, S.A. Bludman, and H. Lee, *Phys. Rev. Lett.* **51**, 223 (1983)

45. T. Gaisser and T. Stanev, *Phys. Rev.* **D30**, 985 (1984)

46. T. Gaisser and T. Stanev, *Phys. Rev.* **D31**, 2770 (1985)

47. R. Flores, *CERN Preprint* CERN-TH.4736/87 (1987)

48. A. Gould, *Astrophys. J.* **321**, 560 (1987)

49. J. Ashman, *et al.*, *CERN Preprint* CERN-EP/87-230 (1987)

50. J. Ellis, R.A. Flores, and S. Ritz, *Phys. Lett.* **198B**, 393 (1987)

51. T.P. Cheng, *Phys. Rev.* **D38**, 2869 (1988); H.Y. Cheng, *Phys. Lett.* **219B**, 347 (1989); J.F. Gunion, H.E. Haber, and S. Dawson, *The Higgs Hunter's Guide*, SCIPP-89/13 (1989)

52. J. Engel, *Bartol Preprint* (1991)

53. S. Rudaz (1989), in Ref. 16

54. K. Ng, K. Olive, and M. Srednicki, *Phys. Lett.* **188B**, 138 (1987)

55. K. Olive and M. Srednicki, *Minnesota Preprint* UMN-TH-636/87 (1987)

56. T. Sjostrand, *Univ. Lund Preprint* LU TP 85-10 (1985)

57. W. Lohman, *et al.*, *CERN Yellow Report 83-03*

58. D.H. Perkins, *Ann. Rev. Nucl. Part. Sci.* **34**, 1 (1984)

59. A. Gould, *Inst. for Adv. Study Preprint* (1990)

60. D. Seckel, T. Stanev, and T.K. Gaisser, *Bartol Preprint* (1991)

61. D. Caldwell, *et al.*, *Phys. Rev. Lett.* **61**, 510 (1988)

62. M.Z. Akrawy, *et al.*, (Opal Collaboration), *Phys. Lett.* **240B**, 261 (1990); B. Adeva, *et al.*, (L3 Collaboration), *Phys. Lett.* **233B**, 530 (1989); D. Decamp, *et al.*, (Aleph Collaboration), *Phys. Lett.* **236B**, 86 (1990)

63. K.A. Olive and M. Srednicki, *Phys. Lett.* **230B**, 78 (1989)

64. K. Griest, M. Kamionkowski, and M.S. Turner, *Phys. Rev.* **D41**, 3565 (1990)

II. PHYSICS AT THE PLANCK SCALE, INFLATION, AND BARYON NUMBER VIOLATION

PHYSICS AT THE PLANCK SCALE

Mary K. Gaillard

Department of Physics and Center for Particle Astrophysics,
University of California
and
Physics Division, Lawrence Berkeley Laboratory, 1 Cyclotron Road,
Berkeley, California 94720

Abstract

Effective supergravity theories suggested by superstrings can be explored to determine their potential for successfully describing both observed physics at zero temperature and an inflationary cosmology. An important ingredient in this study is the dynamics of gaugino condensation, which has been the subject of recent activity.

INTRODUCTION

We are accustomed to the well-established Standard Model of particle physics, namely the $SU(3)_c \times SU(2)_L \times U(1)$ gauge theory whose three (scale dependent) gauge coupling constants apparently converge, when appropriately extrapolated, to a common value at a momentum scale, Λ_{GUT}, that is a few orders of magnitude below the Planck scale. Grand unified theories assumed that above that scale the theory would be (at least up to the Planck scale) a conventional, renormalizable quantum field theory (RQFT). The more modern view is that the theory above the GUT scale is a superstring theory [1], possibly in as many as ten dimensions, and that just below it the effective theory is a supergravity theory that coincides with the standard model (or some RQFT extension of it) when the energy scale is low enough that nonrenormalizable couplings, that are suppressed by inverse powers of the Planck mass $m_{Pl} = (8\pi G_N)^{-\frac{1}{2}}$, can be ignored.

The vacuum configuration that defines the effective renormalizable quantum field theory is determined by the vacuum expectation values (vevs) of scalar fields φ of the effective supergravity theory that are generally of order $< \varphi > \sim m_{Pl}$. Obviously, truncation to the effective renormalizable theory cannot be made without first determining these scalar vevs. These in turn determine:

● the value of the cosmological constant;

● the scales of supersymmetry (SUSY) and hence of gauge symmetry breaking–more specifically whether or not the possibility exists of generating a sufficiently large hierarchy between the observed scale of electroweak symmetry breaking and the fundamental scale m_{Pl};

● the shape of the effective scalar potential at temperatures not far below the Planck temperature, which determines whether or not a successful inflationary scenario is possible in the context of the theories considered.

In this talk I will focus on work done mainly in collaboration with Pierre Binétruy that draws on features common to a broad class of effective supergravity theories obtained as classical vacua of the heterotic superstring theory [2]. The idea is to use constraints from the presumed underlying superstring theory to determine the effective "low energy" supergravity theory, much as one uses constraints from the QCD theory of quarks and gluons to determine an effective low energy theory of hadrons.

THE HETEROTIC STRING

According to the presently most popular hope for a fully unified theory, the Standard Model is an effective theory that is the low energy limit of the heterotic string theory [2]. Starting from a string theory in 10 dimensions with an $E_8 \times E_8$ gauge group, one ends up, at energies sufficiently below the Planck scale, with a supersymmetric field theory in 4 dimensions [3], with a generally smaller gauge group $\mathcal{H} \times \mathcal{G}$. $\mathcal{H}$ describes a "hidden sector", that has interactions with observed matter of only gravitational strength, and $\mathcal{G} \supset SU(3)_c \times SU(2)_L \times U(1)$ is the gauge group of observed matter. Part of the gauge symmetry may be broken (or additional gauge symmetries may be generated) by the $10 \rightarrow 4$ dimensional compactification process itself, and part of it may be broken by the Hosotani mechanism [4], in which gauge flux is trapped around space-tubes in the compact manifold. There are now many more examples of effective theories from superstrings than one once thought could emerge. For illustrative purposes, I will stick to the original "conventional" scenario, in which the "observed" E_8 is broken to E_6, long known to be the largest phenomenologically viable GUT, by the compactification process. Then the observed sector is a supersymmetric Yang-Mills theory, with gauge bosons and gauginos in the adjoint representation of $\mathcal{G} \subset E_6$, coupled to matter, i.e., to quarks, squarks, leptons, sleptons, Higgs, Higgsinos, ...

The hidden sector is assumed to be described by a pure SUSY Yang-Mills theory, $\mathcal{H} \subset E_8$, which is asymptotically free, and therefore infrared enslaved. At some energy scale Λ_c, below the compactification scale Λ_{GUT} at which all the gauge couplings are equal, the hidden gauge multiplets become confined and chiral symmetry is broken, as in QCD, by a fermion condensate. In this case the fermions are the gauginos (denoted by λ or $\tilde{g}$) of the hidden sector:

$$< \bar{\lambda}\lambda >_{hid} \sim \Lambda_c^3 \neq 0. \tag{1}$$

The condensate (1) breaks SUSY [5], and by itself would generate a positive cosmological constant. If a single gaugino condensate were the only source of SUSY breaking, and of a cosmological constant, the condensate would be forced dynamically to vanish, due to the condition that the vacuum energy be minimized.

Another source of SUSY breaking is the (quantized) vacuum expectation value of an antisymmetric tensor field H_{LMN}, that is present in 10-dimensional supergravity:

$$H_{LMN} = \nabla_L B_{MN}, \quad L, M, N = 0, \ldots, 9,$$

$$\int dV^{lmn} < H_{lmn} >= 2\pi n \neq 0, \quad l, m, n = 4, \ldots, 9. \tag{2}$$

The vev (2) can arise if H-flux is trapped around a 3-dimensional space-hole in the compact 6-dimensional manifold, in a manner analogous to the Hosotani mechanism for breaking the gauge symmetry. When (1) and (2) are both present, λ and H_{LMN} couple in such a way [6] that the overall contribution to the classical cosmological constant vanishes. There are other potential sources of SUSY breaking, such as a gravitino condensate [7], that might play a similar role.

The particle spectrum of the effective four dimensional field theory includes the gauge supermultiplets $W^a = (\lambda_L^a, F_{\mu\nu}^a - i\tilde{F}_{\mu\nu}^a)$ (that is, gauginos $\tilde{g}^a = \lambda^a$ and gauge bosons g^a) and the chiral supermultiplets $\Phi^i = (\varphi^i, \chi_L^i)$ that contain the matter fields (φ^i = squarks, sleptons, Higgs particles, ..., χ^i = quarks, ...). In the "conventional" scenario these are all remnants of the gauge supermultiplets in ten dimensions:

$$A_M \to A_\mu + \varphi_m, \quad \mu = 0, \ldots, 3, \quad m = 4, \ldots, 9. \tag{3}$$

Thus for each gauge boson A_M in ten dimensions, there are potentially one gauge boson A_μ and six scalars φ_m (and their superpartners) in four dimensions. However not all of these are massless. In the "conventional" picture ($E_8 \to E_6$ in the observed sector) the massless 4-vectors are in the adjoint of E_6, while the massless scalars are in $(27 + \overline{27})$'s that make up the difference: $(\text{adjoint})_{E_8} - (\text{adjoint})_{E_6}$.

In addition there is the supergravity multiplet containing the graviton G and the gravitino $\tilde{G}$, and a number of gauge singlet chiral supermultiplets. One of these, $S = (s, \chi^S)$, includes the "dilaton" Res, where s is the (complex) scalar component of S, whose *vev* determines the inverse squared gauge coupling constant at the GUT scale:

$$g^2(\Lambda_{GUT}) = < (\text{Re}s)^{-1} > . \tag{4}$$

Among the other gauge singlet chiral multiplets that may be present are the so-called moduli $T_i = (t_i, \chi_i^T)$, whose scalar components t_i correspond to quantum fluctuations in the topology of the compact manifold. One of these corresponds to fluctuations in the overall size of the manifold and its *vev* determines the compactification, or GUT, scale, that is, the inverse radius of compactification R:

$$\Lambda_{GUT}^2 = R^{-2} = m_{St}^2 < (\text{Re}t)^{-1} > = m_{Pl}^2 < (\text{Re}s\text{Re}t)^{-1} >, \tag{5}$$

where m_{St}^2 is the string tension. The total number of moduli is equal to the number of matter generations ($\#27's - \#\overline{27}'s$) and is determined by the detailed topology of the compact manifold.

These gauge nonsinglet scalar fields and their axion superpartners $\text{Im}s$, $\text{Im}t_i$, are important because classically their values are flat directions in the effective potential. The quantum effects that lift these degeneracies determine the overall vacuum configuration and therefore the scales of the elementary particle theory. For the same reason these fields are prime candidates for inflatons.

EXAMPLES OF INFLATIONARY SCENARIOS

Because the S field couples to all the Yang-Mills fields and their superpartners, condensation of the hidden gauginos induces a potential for s [6]. For illustrative purposes, Fig. 1 shows the potential in the effective theory obtained by Dine *et al.* [6] for the field variable $\phi = -\ln\text{Re}(s)/\sqrt{2}$, which is the canonically normalized scalar field that rolls according the classical equations of motion. The shape of the potential depends on the ratio $\hat{c} = |\tilde{c}/\tilde{h}|$, where

$$\tilde{c} \propto \int dV^{lmn} < H_{lmn} >, \quad \tilde{h} \propto < \bar{\lambda}\lambda >_{hid}, \tag{6}$$

are parameters of the effective theory that reflect the *vevs* of hidden dynamical variables. The potential has vanishing vacuum energy, and for $\hat{c} > 1.21$ the only global minimum is the SUSY conserving one: $m_{\tilde{G}} = 0$. For $\hat{c} < 1.21$ the vacuum is degenerate in the t direction and local SUSY breaking is possible. It turns out [8,9] that slow rollover inflation with a suitable number of e-foldings can occur only for the potential of Fig. 1(e), with the parameter fine-tuned to the value $\hat{c} = .937 \pm (3 \times 10^{-4})$. For $\hat{c}$ just above this value the field ϕ would have to tunnel to the true vacuum, entailing the problems of the old inflationary scenario, and for values just below it there is no plateau from which a slow rollover can start. In addition one must find a mechanism for stabilizing ϕ at the required initial value; temperature corrections cannot do this [8]. One suggested mechanism [9] is to take into account that condensation does not occur instantaneously at a critical temperature $T_c \sim \Lambda_c$. Ellis *et al.* [9] parameterized the turn-on of condensation by $\tilde{h} = \tilde{h}(T) = \tilde{h}_0(1 - T/T_c)^n$ for temperatures $T < T_c$. This effectively turns Fig. 1 into a moving picture with the potential evolving from 1(a) to 1(e); it then becomes plausible that the field ends up in the second minimum in 1(b) and 1(c) and evolves to the plateau in 1(e) from which it rolls to the true vacuum. Aside from the fine tuning problem, one needs to understand better the dynamics of gaugino condensation to see whether such a "moving picture" is plausible. Gaugino condensation has been the subject of much recent work [10–14] which I will describe below.

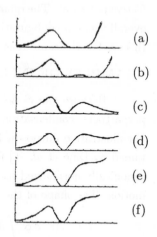

(a)
(b)
(c)
(d)
(e)
(f)

Figure 1. The shape of the potential of Dine *et al.* [6] for the variable $\phi = -\ln(\text{Re}s)/\sqrt{2}$, plotted in the region $-2.3 < \phi < 3.7$, and for (a) $\hat{c} = 1.21$, (b) $1 < \hat{c} < 1.21$, (c) $\hat{c} = 1$, (d) $.937 < \hat{c} < 1$, (e) $\hat{c} = .937$, (f) $\hat{c} < .937$. In all cases the potential vanishes at the SUSY conserving $(m_{\tilde{G}} = 0)$ configuration $\text{Re}s \to \infty$ ($\phi \to -\infty$); this is the only zero of the potential for $\hat{c} > 1.21$.

The fine tuning problem could be evaded in a chaotic inflationary picture [15] as I will point out later. In addition to the scalar field ϕ or $\text{Re}s$ there is its axion (pseudoscalar) superpartner. We found [8] that its potential as determined by the effective theory of Dine *et al.* does not allow for sufficient inflation. A recent reexamination [16] of axion-induced inflation suggests that axions with characteristic scale parameter f of order of the Planck scale can lead to sufficient inflation; however the specific values (including the height of the potential, which here is proportional to the gravitino mass) required in that analysis are not satisfied by the potential for Ims in this effective theory.

The potential for the modulus field t is flat at the tree level of the effective theory of Dine *et al.*, and the degeneracy in Ret is lifted at the one loop level. The earliest attempts [17–19] to evaluate the one-loop corrections resulted in a potential with a negative (and even unbounded [17,19]) cosmological constant. Maeda *et al.* [20] studied the case [18] of a bounded potential by simply adding a constant energy density to fix the global minimum at zero, and found that sufficient inflation could occur. Their key observation was that the t field is driven to a value away from its true vacuum when (as in a chaotic early universe, for example) the other fields are not at their vacuum values. A problem with this analysis is that one cannot simply simply add a constant energy density to the potential; in superstring theory everything is determined by the *vevs* of scalar fields. Another problem is that their ansatz violates general results that show that the SUSY conserving vacuum is always a (possibly degenerate) global minimum of the potential if it is bounded from below.

With Dawson and Hinchliffe we [21] subsequently showed that additional quantum corrections from loop momenta between the condensation and compactification scales could restore the boundedness of the effective one-loop potential, and that local SUSY breaking could occur for reasonable values of the relevant parameters, in which case the potential vanishes at its minimum which remains degenerate in one direction in parameter space. The resulting potential, shown in Fig. 2 in the ($\tilde{c}$,Ret) plane with the other *vevs* fixed at their ground state values, does not yield sufficient inflation when the t field is considered in isolation. On the other hand all observable SUSY breaking effects are found to vanish along the minimum, which is good for phenomenology, since nonvanishing effects at this order would be much too large. The origin of this suppression of observable SUSY breaking has been identified [10]; it is related to a classical symmetry of the effective theory that I will display below.

These last results rely on the assumption that the true vacuum is a minimum with respect to all *vevs*, including the parameter $\tilde{h}$ that reflects the size of gaugino condensation. To determine whether such an assumption is justified requires a better understanding of the dynamics of gaugino condensation, as does a full exploration of possible inflationary scenarios. Gaugino condensation, will be the subject of the remainder of the talk, with the end result that, when constraints from string theory are fully implemented, the picture sketched above may be considerably altered.

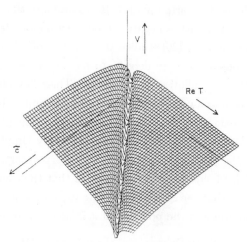

Figure 2. The one loop potential [21] in the $(\mathrm{Re}t, \tilde{c})$ plane, with the remaining parameters fixed at their ground state values, for the effective theory of Dine *et al.* [6]. The continuous degeneracy is reduced to a discrete one when the quantization condition (2,6) for $\tilde{c}$ is taken into account.

CONSTRUCTION OF THE EFFECTIVE LAGRANGIAN

The general supergravity lagrangian was first given in its complete form by Cremmer *et al.* [22]. Its symmetry structure is most readily manifest in the superfield formulation later developed by Binétruy *et al.* [23], in which the Lagrangian can be expressed by just three terms:

$$\mathcal{L} = \mathcal{L}_E + \mathcal{L}_{pot} + \mathcal{L}_{YM}. \tag{7}$$

The first term

$$\mathcal{L}_E = -3 \int d^2\Theta \mathcal{E}\mathcal{R} + h.c. \tag{8}$$

is the generalized Einstein term. It contains the pure supergravity part as well as the (noncanonical, i.e. including derivative couplings) kinetic energy terms for the chiral supermultiplets. The second term:

$$\mathcal{L}_{pot} = \int d^2\Theta e^{K(\Phi,\bar{\Phi})/2} W(\Phi) + h.c., \tag{9}$$

contains the Yukawa couplings and the scalar potential, and the third term

$$\mathcal{L}_{YM} = \frac{1}{4} \int d^2\Theta f(\Phi) W_\alpha^a W_a^\alpha + h.c. \tag{10}$$

is the Yang-Mills lagrangian. Here Θ is a complex two-component fermionic variable in superspace: $x \to x, \Theta, \bar{\Theta}$. Integration over the anticommuting variable Θ is equivalent to differentiation. The expansion of the above expressions in terms of component fields includes derivatives that are covariant with respect to general coordinate, gauge and Kähler transformations. A Kähler transformation is a redefinition of the Kähler potential $K(\Phi, \bar{\Phi}) = K(\Phi, \bar{\Phi})^\dagger$ and of the superpotential $W(\Phi) = \overline{W}(\bar{\Phi})^\dagger$ by a holomorphic function $F(\Phi) = \bar{F}(\bar{\Phi})^\dagger$ of the chiral supermultiplets $\Phi = (\varphi, \chi)$:

$$K \to K' = K + F + \overline{F}, \quad W \to W' = e^{-F} W, \tag{11}$$

Since this transformation changes $e^{K/2}W$ by a phase that can be compensated by a phase transformation of the integration variable Θ, the theory defined above is classically invariant [22,23] under Kähler transformations provided one transforms the superfields $\mathcal{R}$ and W_α^a by the same phase; for example

$$W_\alpha^a \to e^{-\mathrm{Im}F/2} W_\alpha^a. \tag{12}$$

This last transformation, which implies a chiral rotation on the left-handed gaugino field λ_L^a:

$$\lambda_\alpha^a \to e^{-\mathrm{Im}F/2} \lambda_\alpha^a, \tag{13}$$

is anomalous at the quantum level, a point that will be important in the discussion below. (Here a is a gauge index and α is a Dirac index.).

The theory is completely specified by the field content, the gauge group and the three functions K, W and f of the chiral superfields. One can fix the "Kähler gauge" by a specific choice of the function F. In particular, choosing

$F = \ln W$ casts the lagrangian in a form [22] that depends on only two functions of the chiral superfields, f and $\mathcal{G} = K + \ln |W|^2$. The gauge coupling constant is determined by the vev

$$< \mathrm{Re} f(\varphi) >= g^{-2}. \tag{14}$$

I will consider a prototype [24] supergravity model from superstrings, with just one modulus T and one matter generation. The generalization to three moduli and three matter generations is straightforward [14]. The functions K, W and f are given in terms of the superfields $\Phi = \Phi^i$, S and T by

$$f = S, \tag{15a}$$

$$K = -\ln(S + \bar{S}) - 3\ln(T + \bar{T} - |\Phi|^2), \quad |\Phi|^2 = \sum_i \Phi^i \bar{\Phi}^{\bar{i}}, \tag{15b}$$

$$W(\Phi) = c_{ijk}\Phi^i\Phi^j\Phi^k + \tilde{c}. \tag{15c}$$

where the last term in the superpotential W parameterizes the nonperturbative SUSY breaking induced by the vev (2,6) of the antisymmetric tensor field strength. Comparison of (15a) with (14) yields the relation (4), and using this with (5) gives for the vev of the Kähler potential ($< |\varphi^i|^2 > \ll <\mathrm{Re}t>$)

$$16 < e^K >\approx< (\mathrm{Re}s)^{-1}(\mathrm{Re}t)^{-3} >= g^{-4}\Lambda_{GUT}^6/m_{Pl}^6. \tag{16}$$

For $\tilde{c} = 0$, the supergravity theory defined above is classically invariant [25] under the nonlinear transformations

$$T \to T' = \frac{aT - ib}{icT + d}, \quad \Phi^i \to \Phi^{i'} = \frac{\Phi^i}{icT + d}, \quad S \to S' = S,$$

$$ad - bc = 1, \quad a, b, c, d \text{ real}, \tag{17}$$

that form the three-parameter group $SU(1,1)$ or $SL(2,\mathcal{R})$ which is the noncompact version of the rotation group $SU(2)$ or $O(3)$. Eq.(17) effects a Kähler transformation (11) with

$$F = 3\ln(icT + d), \tag{18}$$

under which the full lagrangian is invariant provided the gaugino fields undergo the chiral transformation (12,13). The group of transformations (17) includes a subset with $a = d = 0$, $bc = -1$, under which:

$$t \to b^2/t, \tag{19}$$

where b is a finite, continuous, real parameter. From (5) we see that when the theory is expressed in string mass units m_{St}, (19) corresponds to an inversion of the radius of compactification. For the special case of integer b, this is the well known "duality" transformation that leaves the string spectrum invariant.

When we allow $\tilde{c} \neq 0$, the $SU(1,1)$ symmetry can be formally maintained by allowing this parameter to transform like the superpotential in (11):

$$\tilde{c} \to \tilde{c}' = e^{-F}\tilde{c}. \tag{20}$$

It can be shown [10] that (20) corresponds to the correct transformation property of the antisymmetric tensor field $< H_{LMN} >$ in (2). The $SU(1,1)$ symmetry has recently been shown [10] to be at the origin of the cancellation of observable SUSY breaking effects, found [21] previously by explicit calculation in perturbation theory.

Just as one can construct low energy effective Lagrangians for pseudoscalar mesons that are $q\bar{q}$ bound states using the symmetries of QCD and the chiral and conformal anomaly, one can use [10] $SU(1,1)$ and its anomalies, together with supersymmetry [26], to construct an effective lagrangian for the lightest hidden-sector chiral supermultiplet, denoted $H = (h, \chi^H)$, that is a bound state, with mass m_H, of the hidden gauge supermultiplet. Here the relevant classical symmetries are the chiral transformation (12,13) and the scale transformation

$$\Lambda_{GUT} = m_{Pl}(2g)^{\frac{2}{3}} < e^{K/6} > \to e^{\mathrm{Re}F/3}\Lambda_{GUT}, \tag{21}$$

which follows from (17), (18) and the identification (16) for the cut-off Λ_{GUT} of the effective theory. Scale and chiral invariance are broken at the quantum level by anomalies. The variation of the one-loop quantum corrected SUSY Yang-Mills lagrangian is well-known [26]:

$$\mathcal{L}_{YM}^{\mathrm{Quantum}} \to \mathcal{L}_{YM}^{\mathrm{Quantum}} - \frac{2b_0}{3} \int d^2\Theta F(T)W_\alpha^a W_a^\alpha + h.c., \tag{22}$$

where b_0 is the group theory number that determines the β-function for the Yang-Mills theory. We wish to construct an effective potential for a composite chiral multiple H that represents the lightest bound state of the confined hidden Yang-Mills sector. Thus we identify [27,10]:

$$\mathcal{L}_{YM}^{\text{Quantum}} \Rightarrow \mathcal{L}_{pot}^{eff} = \int d^2\Theta e^{K/2} 2b_0 \lambda e^{-3S/2b_0} H^3 \ln(H/\mu) + h.c.$$

$$= \int d^2\Theta [SU + U \ln(4Ug^2(S)/\Lambda_{GUT}^3(\Phi)\lambda\mu^3) + h.c.], \tag{23}$$

where λ and μ are constants of order unity, U is the interpolating field for the Yang-Mills composite operator:

$$\frac{1}{4} W_\alpha^a W_a^\alpha \Rightarrow U = \lambda H^3 e^{K/2} e^{-3S/2b_0}, \tag{25}$$

and the factor

$$< 2b_0 \ln(H/\mu) >= \frac{1}{g^2}\left[1 + \frac{2b_0 g^2}{3} \ln(4 < U > g^2/\Lambda_{GUT}^3 \lambda\mu^3)\right]$$

$$= \frac{1}{g^2}\left[1 + \frac{2b_0 g^2}{3} \ln(4e^{-1}\Lambda_c^3/\Lambda_{GUT}^3)\right] \tag{26}$$

is the one-loop Yang-Mills field wave function renormalization from the compactification scale to the condensation scale. The composite field (25) and the effective lagrangian defined by (23) transform according to (12) and (22), respectively, provided we impose

$$H \to H' = e^{-F/3}H = \frac{H}{icT + d} \tag{27}$$

and take as Kähler potential (which determines the H-superfield kinetic energy) [10]:

$$K = -\ln(S + \bar{S}) - 3\ln(T + \bar{T} - |\Phi|^2 - |H|^2). \tag{28}$$

We can now solve the effective theory for the condensate vev:

$$< H >= h_0 = \mu e^{-1/3}, \quad \text{or} \quad < \bar{\lambda}\lambda >_{hid} = 4 < U >= \frac{\lambda h_0^3}{g^2}\Lambda_c^3. \tag{29}$$

If we fix H at its ground state value, we obtain an effective theory for Φ^i, S, T and the observable-sector Yang-Mills fields that is defined by $(15a, b)$ and the superpotential

$$W(\Phi) = c_{ijk}\Phi^i\Phi^j\Phi^k + \tilde{c} + \tilde{h}e^{-3b_0 S/2}, \quad \tilde{h} = -\frac{2b_0}{3}\lambda\mu^3 e^{-1}, \tag{30}$$

which is precisely the theory of Dine *et al.* studied previously. This is a reasonable approximation because the (SUSY conserving) mass $m_H \sim \Lambda_c > m_{\tilde{G}}$ of the composite supermultiplet H is larger than the other masses generated at the classical level of the effective theory defined by (23). As mentioned earlier, the effective theory defined by (30) has the following properties at the classical level [6] and at the one-loop [21] level: the cosmological constant vanishes, the gravitino mass $m_{\tilde{G}}$ can be nonvanishing, so that local supersymmetry is broken, in which case the vacuum is degenerate, and there is no manifestation of SUSY breaking in the observable sector. Note that with $\tilde{h}$ now determined by (29,30), minimization of the effective potential with respect to the parameter $\tilde{h}$ is equivalent to minimization with respect to the parameter μ (or λ). The presence of an undetermined parameter reflects [10] an additional degree of freedom of the underlying theory, namely the gauge field strength $F_{\mu\nu}$, which does not appear explicitly as an independent propagating field in the composite supermultiplet formulation.

Including loop correction from the H-sector, one finds that masses are generated for the gauginos of the observable sector that are of order [10]

$$m_{\tilde{g}} \sim \frac{1}{(16\pi^2 m_{Pl}^2)^2} m_{\tilde{G}} m_H^2 \Lambda_c^2 < 10^{-15} m_{Pl} \simeq 2 \; TeV.$$

$$\text{for} \; m_{\tilde{G}} < m_H \simeq \Lambda_c \sim 10^{-2} m_{Pl}. \tag{31}$$

The factor $(4\pi)^{-4}$ appears in (31) because the effect arises first at two-loop order in the effective theory, the factor $m_{\tilde{G}}$ is the necessary signal of SUSY breaking, the factor m_H^2 is the signal of $SU(1,1)$ breaking, and Λ_c^2 is the effective cut-off. This last factor arises essentially for dimensional reasons: the couplings responsible for transmitting the knowledge of symmetry breaking to the observable sector are nonrenormalizable interactions with dimensionful coupling constants proportional to m_{Pl}^{-2}. Once gauginos acquire masses, gauge nonsinglet scalars (in particular the Higgs particle, whose mass governs the scale of electroweak symmetry breaking and hence the gauge hierarchy) will acquire masses $m_\varphi \sim \frac{\alpha}{4\pi} m_{\tilde{g}}$ at the next loop order in the renormalizable gauge interactions.

This theory thus appears capable of generating a plausible gauge hierarchy, which in turn suggests that one should take seriously the study of its possible inflationary scenarios. Specifically, in the context of chaotic inflation [15], we should not set the gauge nonsinglet scalars φ^i in (30) to zero at temperatures near the Planck temperature. We can parameterize their effect by letting $\tilde{c} \to \tilde{c}(\varphi) = \tilde{c} + c_{ijk} < \varphi^i\varphi^j\varphi^k >$, which is temperature dependent. Then one recovers the "moving picture" scenario for Fig. 1 [28]. Moreover, one need not fine tune the zero-temperature parameter $\hat{c}$, since the the desired exponential expansion will occur if the field Res sits for a sufficiently long time in, for example, the second (false) minimum of Fig. 2d. No tunneling is required as the "graceful exit" is provided [29] by the potential changing its shape due to the evolution of the φ^i fields to their ground state values that are equal to zero (or nearly so with respect to the scales in question). This analysis has not been pursued further, because recent developments in understanding the symmetry of effective theories from superstrings appear to change the picture appreciably.

RESTORATION OF MODULAR INVARIANCE

In the formalism presented above, the continuous classical symmetry $SU(1,1)$ or $SL(2,\mathcal{R})$ is broken by anomalies at the quantum level. However the discrete subgroup $SL(2,\mathcal{Z})$ [a,b,c,d integers in (17)] of $SL(2,\mathcal{R})$ is known [30] to be an exact symmetry to all orders in string perturbation theory. This so-called "modular invariance" is restored by adopting, instead of (23), the effective lagrangian [12]

$$\mathcal{L}_{YM}^{\text{Quantum}} \Rightarrow \mathcal{L}_{pot}^{eff} = \int d^2\Theta e^{K/2} 2b_0\lambda e^{-3S/2b_0}H^3\ln(H\eta^2(T)/\mu) + h.c., \quad (32)$$

where

$$\eta(T) = e^{-\pi T/12}\prod_{m=1}^{\infty}(1 - e^{-2m\pi T}) \quad (33)$$

is the Dedekind η-function. This is the unique function of the chiral superfields that has the required analyticity and $SL(2,\mathcal{Z})$ transformation properties [12]. This additional contribution to the Yang-Mills wave function renormalization was found independently [31] from the direct calculation of finite threshold corrections to the leading log approximation that arise from heavy string mode loops. The result (32) has been generalized to the cases of several gaugino condensates [13,14] and of several moduli [14].

The effective scalar potential [11,12] for the theory defined by (32) is unbounded from below. If one appeals to some unknown dynamics, such as wormholes, to restore positivity and erase the (infinitely negative) cosmological constant, one loses all predictive power because the same dynamics can affect other features; specifically one can make no statements about the gauge hierarchy (which involves expanding in quantum fluctuations about the true vacuum) nor about inflation (which involves tracing the evolution of the field configurations toward the true vacuum).

A cure for this disease has recently been found [14]. It suffices to reinterpret the above formalism, which is based on one-loop results for the Yang-Mills field renormalization in SUSY Yang-Mills theories. In the construction (23) or (32) the field renormalization is incorporated into a redefinition of the superpotential. We can instead incorporate it into a redefinition of the Kähler potential, which is in fact the interpretation that is supported by known renormalization effects in supersymmetry and in supergravity [32,10]. That is, instead of (32) and (28), we define the effective composite theory by

$$\mathcal{L}_{YM}^{\text{Quantum}} \Rightarrow \mathcal{L}_{pot}^{eff} = \int d^2\Theta S e^{K/2} \lambda e^{-3S/2b_0} H^3 + h.c., \qquad (34a)$$

$$K = -\ln(S + \bar{S})$$

$$-3\ln(T + \bar{T} - |\Phi|^2 - |H|^2[1 - \tfrac{1}{5}b_0 g^2(S)\ln(H\eta^2(T)/\mu) + \text{h.c.}]). \qquad (34b)$$

The two theories are identical (after appropriate redefinitions of field variables) to first order in the loop expansion parameter b_0; indeed it is a term of order b_0^2 that drives $< V > \to -\infty$ in the theory defined by (32). At its classical level, the theory defined by (34) has once again a vanishing cosmological constant and (for $\tilde{c} \neq 0$) a degenerate vacuum with local SUSY breaking ($m_{\tilde{G}} \neq 0$) possible, and again no SUSY breaking appears in the observable sector at the classical level of this effective theory. [14]

What remains to be done is to reexamine the issues of quantum corrections and inflationary potentials for this modified theory. The analysis is considerably more complicated in this case, and one can expect some qualitative differences from the scenarios described above. For example, one generally finds [33] a different hierarchy of masses, namely

$$m_S > m_H \sim m_{\tilde{G}}, \qquad (35)$$

with SUSY-invariant masses for the S-supermultiplet up to terms of order $m_{\tilde{G}}$. This means, for example, that it makes no sense to fix the H-supermultiplet at its vev and to study the effective potential for s.

It may also be the case that the theory defined by (34) is sufficiently complex that a realistic picture will emerge only when all the moduli fields are included. For example, much of the degeneracy of the simple model defined by (15,23) is lifted by the appearance of the η-function (33). On the other hand, including three moduli (and three matter generations) will increase the degree of degeneracy. The degeneracy in the moduli directions plays an important role in suppressing observable SUSY breaking effects. Finally, in contrast with the simple model (15,23), which remains degenerate in the $\mathrm{Im}t$ direction to all orders in perturbation theory, the axion components of the moduli superfields may develop nontrivial potentials at the quantum level; this might provide additional interesting possibilities for inflation [16].

Acknowledgement. This work was supported in part by the Director, Office of Energy Research, Office of High Energy and Nuclear Physics, Division of High Energy Physics of the U.S. Department of Energy under Contract DE-AC03-76SF00098 and in part by the National Science Foundation under grant PHY-85-15857.

References

1. M. Green and J. Schwarz, *Phys. Lett.* **B149:** 117 (1984);

2. D. Gross, J. Harvey, E. Martinec and R. Rohm, *Phys. Rev. Lett.* **54:** 502 (1985).

3. P. Candelas, G. Horowitz, A. Strominger and E. Witten, *Nucl. Phys.* **B258:** 46 (1985).

4. Y. Hosotani, *Phys. Lett.* **129B:** 75 (1985).

5. H.P. Nilles, *Phys. Lett.* **115B:** 193 (1982).

6. M. Dine, R. Rohm, N. Seiberg and E. Witten, *Phys. Lett.* **156B:** 55 (1985).

7. K. Konishi, N. Magnoli and H. Panagopoulos, *Nucl. Phys.* **B309**: 201 (1988).

8. P. Binétruy and M.K. Gaillard, *Phys. Rev.* **D34**: 3069 (1986).

9. J. Ellis, K. Enqvist, D.V. Nanopoulos, M. Quiros, *Nucl.Phys.* **B277**: 231 (1986).

10. P. Binétruy and M.K. Gaillard, *Phys. Lett.* **232B**: 83 (1989),Annecy-CERN preprint LAPP-TH-273/90, CERN-TH.5727/90 (1990).

11. A. Font, L.E. Ibáñez, D. Lüst and F. Quevedo, CERN-TH.5726/90 (1990).

12. S. Ferrara, N. Magnoli, T. Taylor and G. Veneziano, CERN-TH.5744/90 (1990).

13. T. Taylor, CERN-Northeastern preprint CERN-TH.5839/90, NUB-3001 (1990).

14. P. Binétruy and M.K. Gaillard, LBL-29548, UCB-PTH-90/40, LAPP-TH-274/90 (1990), to be published in *Phys. Lett.*

15. A.D. Linde, *Phys. Lett.* **129B**: 177 (1983).

16. K. Freese, J.A. Frieman, and A.V. Olinto, "Natural Inflation with Pseudo-Nambu-Goldstone Bosons", MIT-FermiLab preprint (1990), to be published in *Phys. Rev. Lett.*

17. J.D. Breit, B.A. Ovrut and G. Segré, *Phys. Lett.* **162B**: 303 (1985).

18. P. Binétruy and M.K. Gaillard, *Phys. Lett.* **168B**: 347 (1986).

19. Y.J. Ahn and J.D. Breit, *Nucl. Phys.* **B273**: 75 (1986); M. Quiros, *Phys. Lett.* **B173**: 265 (1986).

20. K. Maeda, M.D. Pollock, C.E. Vayonakis, *Class. Quant. Grav.* **3**: L89 (1986).

21. P. Binétruy, S.Dawson, M.K. Gaillard and I. Hinchliffe, *Phys. Lett.* **192B**: 377 (1987), and Phys. Rev. **D37**: 2633 (1988).

22. E. Cremmer, S. Ferrara, L. Girardello, and A. Van Proeyen, *Nucl. Phys.* **B212:** 413 (1983).

23. P. Binétruy, G. Girardi, R. Grimm and M. Muller, *Phys. Lett.* **189B:** 83 (1987).

24. E. Witten, *Phys. Lett.* **155B:** 151 (1985).

25. S.P. Li, R. Peschanski and C.A. Savoy, *Phys. Lett.* **178B:** 193 (1986) and *Phys. Lett.* **194B:** 193 (1987).

26. G. Veneziano and S. Yankielowicz, *Phys. Lett.* **113B:** 231 (1982).

27. T.R. Taylor, *Phys. Lett.* **164B:** 43 (1985).

28. Pierre Binétruy, private communication.

29. See A.D. Linde, these proceedings.

30. A. Giveon, N. Malkin and E. Rabinovici, *Phys. Lett.* 220B: 551 (1989); E. Alvarez and M. Osorio, *Phys. Rev.* **D40:** 1150 (1989).

31. L. Dixon, V. Kaplunovsky and J. Louis, SLAC-Texas preprint SLAC-PUB-5138, UTTG-36-89 (1990).

32. P. Binétruy and M.K. Gaillard, *Phys. Lett.* **220B:**, 68 (1989).

33. P. Binétruy and M.K. Gaillard, in progress.

INFLATION AND AXIONS

Andrei Linde[1]

Department of Physics, Stanford University, Stanford CA 94305-4060, USA[2]

ABSTRACT

We illustrate various features of inflationary cosmology by considering the problem of the cosmological constraints on the axion mass. It is argued that in the context of inflationary cosmology the constraint $m_a \gtrsim 10^{-5} eV$ can be avoided even when the axion perturbations produced during inflation are taken into account. It is shown also that in most axion models the effective parameter f_a rapidly changes during inflation. This modifies some earlier statements concerning isothermal perturbations in the axion cosmology. A hybrid inflation scenario is proposed which combines some advantages of chaotic inflation with specific features of new and/or extended inflation. Its implications for the axion cosmology are discussed.

1 Introduction: Inflation and Axions

The standard Big Bang hot universe theory asserts that the universe was born at some moment $t = 0$ about 15 billion years ago, in a state of infinitely large density. With the rapid expansion of the universe, the average energy of particles, given by the temperature, decreases rapidly, and the universe becomes cold. This theory became especially popular after the discovery of the microwave background radiation, which is the main topic of this conference.

However, in the end of 70's it was understood that this theory is hardly compatible with the present theory of elementary particles (primordial monopole problem, Polonyi fields problem, gravitino problem, domain wall problem) and it has many internal difficulties (flatness problem, horizon problem, homogeneity and isotropy problem, etc.).

---

[1]On leave of absence from Lebedev Physical Institute, Moscow 117924, USSR
[2]Bitnet address LINDE@SLACVM

Fortunately, all these problems can be solved simultaneously in the context of a relatively simple scenario of the universe evolution - the inflationary universe scenario [1] - [6]. The invention of this scenario has modified considerably the standard cosmological paradigm [7]. The main idea of this scenario is that the universe at the very early stages of its evolution expanded quasi-exponentially (the stage of inflation) in a state with energy density dominated by the potential energy density $V(\phi)$ of some scalar field ϕ. This rapid expansion made the universe flat, homogeneous and isotropic and decreased exponentially the density of monopoles, gravitinos and domain walls. Later, the potential energy density of the scalar field transformed into thermal energy and still later, the universe was correctly described by the standard hot universe theory predicting the existence of the microwave background radiation.

The main idea of the inflationary universe scenario is very simple and it can be realized in a wide class of realistic theories of elementary particles. On the other hand, despite many efforts, no other solution of all problems mentioned above has been suggested during the last 10 years. Therefore many cosmologists believe that something like inflation actually has taken place at the very early stages of the evolution of the universe. One should note though, that *inflation* is not a magic word which can save all theories of elementary particles from being in a contradiction with cosmology. In many theories it is impossible to obtain inflation. Such theories typically lead to unacceptable cosmological consequences and, perhaps, should be abandoned. Many other theories can support inflation but still lead to disastrous cosmological consequences. Therefore cosmological considerations can be very useful for imposing strong constraints on various parameters of the theories of elementary particles and even on the classes of the cosmologically acceptable theories.

Historically, there were several different versions of the inflationary universe scenario [1]-[6]. One of the most important stages of the development of the inflationary cosmology was related to the *old inflationary universe scenario* of Guth [3]. This scenario was based on three fundamental propositions:

1. The universe initially expands in a state with a very high temperature, which leads to the symmetry restoration in the early universe, $\varphi(T) = 0$, where φ is some scalar field driving inflation (the inflaton field).

2. The effective potential $V(\varphi, T)$ of the scalar field φ has a deep local minimum at $\varphi = 0$ even at a very low temperature T. As a result, the universe remains in a supercooled vacuum state $\varphi = 0$ (false vacuum) for a long time. The energy-momentum tensor of such a state rapidly becomes equal to $T_{\mu\nu} = g_{\mu\nu}V(0)$, and the universe expands exponentially (inflates) until the false vacuum decays.

3. The decay of the false vacuum proceeds by forming bubbles containing the field φ_o corresponding to the minimum of the effective potential $V(\varphi)$. Reheating of the universe occurs due to the bubble-wall collisions.

Unfortunately, as it was pointed out by Guth in [3], this scenario had a major defect. If the rate of the bubble formation is bigger than the speed of the universe expansion, then the phase transition occurs very rapidly and inflation does not take place. On the other hand, if the vacuum decay rate is small, then the universe after the phase transition becomes extremely inhomogeneous, and in a considerable part of the physical volume of the universe the phase transition to the minimum of $V(\varphi)$ never completes. This is called the graceful exit problem.

The main idea of the old inflationary universe scenario was very simple and attractive. However, all attempts to solve or avoid the graceful exit problem and to suggest a successful inflationary universe scenario failed until cosmologists managed to surmount a certain psychological barrier and renounce all three of the aforementioned assumptions, while retaining the main idea of ref. [3] that the universe has undergone inflation during the early stages of its evolution. The invention of the *new inflationary universe scenario* [4] marked the departure from the assumptions (2), (3). Later it was shown that the assumption (1) also does not hold in all realistic models of inflation, for two main reasons. First of all, the time which is necessary for the field φ to roll down to the minimum of $V(\varphi, T)$ typically is too large, so that either inflation occurs before the field rolls to the minimum of $V(\varphi, T)$, or it does not occur at all. On the other hand, even if the field φ occasionally was near the minimum of $V(\varphi, T)$ from the very beginning, inflation typically starts very late, when thermal energy drops from M_p^4 down to $V(0, T)$. In all realistic models of inflation $V(0, T) < 10^{-10} M_p^4$, hence inflation may start in a state with $\varphi = 0$ not earlier than at $t \sim 10^4 M_p^{-1}$. During such a time a typical closed universe would collapse before the conditions necessary for inflation could be realized [7].

The assumption (1) was finally given up with the invention of the *chaotic inflation* scenario [5]. The main idea of this scenario was that instead of making the *assumption* that the scalar field from the very beginning should be in a state corresponding to a minimum of $V(\varphi, T)$, one should study various initial distributions of the scalar field φ and investigate in which case the inflationary regime may occur. The result of this investigation is that there exists a wide class of theories where inflation occurs under quite natural initial conditions during a slow rolling of the scalar field down to the absolute minimum of its effective potential. This class includes all theories of a scalar field φ minimally coupled to gravity, with a polynomial effective potential $V(\varphi)$, or even with an exponential effective potential $V \sim \exp \frac{c\varphi}{M_p}$ with $c \lesssim 5$.

A most surprising property of chaotic inflation is that inflationary domain of the universe containing sufficiently large scalar field φ permanently produce new inflationary domains. Therefore the evolution of the universe in the inflationary scenario has no end and may have no beginning (*eternal chaotic inflation* [10,11]). After inflation the universe becomes divided into different exponentially large domains inside which properties of elementary particles and dimension of space-time may be different[7].

The graceful exit problem in the chaotic inflation scenario just does not appear. It is possible to solve the graceful exit problem in the old inflation scenario as well, in the context of the Jordan-Brans-Dicke theory [9]. This scenario was called *extended inflation* [6]. It has an intermediate position between the old and the chaotic inflation scenario, but it is slightly more complicated since it requires a modification of the standard Einstein gravity theory and consideration of the theories containing at least two different scalar fields. In its original form this scenario was based on the theory of hight-temperature phase transitions, which leads again to the problem of initial conditions. Recently it was shown that the most natural realization of the extended inflation scenario can be achieved along the lines of chaotic inflation, and an eternal extended inflation scenario was proposed [12]. It was shown also that the graceful exit problem can be solved in the theories containing at least two different scalar fields even without any modification of the Einstein gravity theory [12],[13]. The corresponding scenario combines some advantages of chaotic inflation with specific features of new and/or extended inflation. Therefore it was called *hybrid inflation* [13].

In this article we will not give a detailed review of the present status of inflationary cosmology. One can find it in [7,14]. Rather we will try to illustrate some unusual properties of inflationary cosmology by investigation of cosmological consequences of the axion theory[15].

One of the most remarkable results obtained in axion cosmology was the famous cosmological constraint on the axion mass, $m_a \gtrsim 10^{-5}$ eV [16], corresponding to the "radius" of the axion potential $f_a \lesssim 10^{12}$ GeV (see below). Originally this constraint was obtained in the context of the standard Big Bang cosmology. Later it was argued that this constraint can be even strengthened, $m_a \gtrsim 10^{-3}$ eV [17] (see, however, [18]). This, together with the astrophysical constraints $m_a \lesssim 10^{-2} - 10^{-3}$ eV [19], almost completely closed the so-called "axion window" between the astrophysical and cosmological bounds on the axion mass in the standard Big Bang cosmology. Moreover, the vast majority of axion models are incompatible with the standard Big Bang cosmology since they lead to the domain wall problem [20].

The same results remain true in inflationary cosmology if the Peccei-Quinn symmetry breaking (after which axions appear) occurs after the end of inflation. However, if the Peccei-Quinn symmetry breaking occurs during inflation, then the domain wall problem disappears. Moreover, as it was argued in ref. [21] (see also [22]), in such a case there are no cosmological constraints on the axion mass and the "axion window" is widely open. This issue is not of a purely academic interest since the existing programs of experimental search of axions are oriented to the investigation of the "axion window" 10^{-2} eV $\gtrsim m_a \gtrsim 10^{-5}$ eV [23], whereas the search outside of the "window", being technically more difficult, may also give very important results. For example, a discovery of an axion with $m_a \ll 10^{-5}$ eV would serve as a new important argument in favor of validity of inflationary cosmology.[3]

Recently it was argued [25] that it is difficult to avoid the constraint $m_a \gtrsim 10^{-5}$ eV even with the help of inflation. The reason is that in those inflationary models where the rate of expansion of the universe H at the end of inflation is very large, $H \gg 10^9$ GeV, the isothermal density perturbations produced at that time due to quantum fluctuations of the axion field are unacceptably large. The assumption $H \gg 10^9$ GeV is quite realistic. For example, in the simplest models of chaotic inflation the typical value of H at the end of inflation is $\sim 10^{13}$ GeV [7]. The simplest way to overcome this problem is to assume that there were no axions during

---

[3]One should note, though, that there exist some models where the cosmological constraints on m_a can be relaxed without the help of inflation, due to hypothetical processes of a large entropy increase at late stages of the evolution of the universe, see e.g. [24].

inflation, i.e. that the Peccei-Quinn symmetry was broken only after inflation. This returns us to strong cosmological constraints on the axion theories discussed above and to the half-closed "axion window".

One of our main purposes here will be to re-examine this question, as well as the whole issue of isothermal perturbations produced in the axion inflationary cosmology.

For a long time it was believed that the only density perturbations produced during inflation are adiabatic perturbations with a flat (scale-independent, or Harrison-Zeldovich) spectrum [26]. Even though it was clear that in the axion models isothermal perturbations can be produced as well, it was usually believed that they are negligibly small [27]. However, a more detailed investigation of this question has shown that isothermal perturbations produced in the axion theory can be large [28] - [31] and that their spectrum may be different from the flat one [28,30,31]. Recently it was shown also that inflation may lead to formation of big axionic domain walls [32]. All these results, as well as the results of ref. [25], were obtained in the context of simplest axion models in which the degree of the Peccei-Quinn symmetry breaking, which is given by the parameter f_a, does not change at the last stages of inflation and after it. In what follows we will argue that this assumption is valid only in rather restricted class of theories and we will discuss possible implications of the time-dependence of f_a in inflationary cosmology. It will be shown, in particular, that in some versions of the axion theory the existence of light axions with $m_a \lesssim 10^{-5}$ eV does not lead to large isothermal perturbations even if the rate of the universe expansion H at the end of inflation is several orders of magnitude larger than 10^9 GeV.

On the other hand, it is known that in new and extended inflation the value of H may be of the order of 10^9 GeV or even smaller. However, as we already emphasized, all versions of the new inflationary universe scenario (and many models of extended inflation) are plagued by the problem of initial conditions [7,12]. In this paper we will discuss a simplest version of the hybrid inflation scenario[13]. It is based on the chaotic inflation scenario, but the last stages of inflation in this scenario can be similar to the last stages of new or extended inflation. In this scenario inflation occurs under natural initial conditions, and the isothermal density perturbations produced at the end of inflation can be very small even for $m_a \ll 10^{-5}$ eV.

2 Cosmological Constraints on the Axion Mass

To describe the Peccei-Quinn symmetry breaking we shall consider a simple model of the complex scalar field Φ with the effective potential

$$V(\Phi) = -m_\Phi^2 \Phi^* \Phi + \lambda_\Phi (\Phi^* \Phi)^2 + V_o . \tag{1}$$

The field Φ can be represented in the form

$$\Phi(x) = \frac{f_a(x)}{\sqrt{2}} \exp\left(\frac{ia(x)}{f_a(x)}\right) , \tag{2}$$

After the Peccei-Quinn symmetry breaking the radial component of the field Φ acquires a non-vanishing classical part, $f_a = \frac{m_\Phi}{\sqrt{\lambda_\Phi}}$ and the field $a(x)$ in eq. (2) becomes a massless Goldstone scalar field with a vanishing effective potential, $U(a) = 0$.

The main difference between this simple model and the theory of the axion field $a(x)$ is that the axion field at the late stages of the evolution of the universe, when the temperature of the universe after reheating drops down to less than $T_c = O(100)$ MeV, acquires a non-vanishing potential energy density

$$ U(a) = C m_\pi^4 \left(1 - \cos \frac{Na}{f_a} \right) \equiv C m_\pi^4 \left(1 - \cos \theta \right). \tag{3} $$

Here $C = O(1)$ and N is an integer that depends on the detailed structure of the theory. In what follows we will consider the simplest case N = 1. In this case the field $a(x)$ is a pseudo-Goldstone field with the standard kinetic term and the effective mass $m_a \sim 0.6$ eV $\times (f_a^{-1} \times 10^7$ GeV); θ is the corresponding angular variable, $\theta(x) = \frac{a}{f_a}$.

The main line of reasoning which leads to the constraint $m_a \gtrsim 10^{-5}$ eV (and to an equivalent constraint $f_a \lesssim 10^{12}$ GeV) is rather simple. A typical initial value of the axion angle θ in the early universe is $|\theta_1| \sim \pi/2$ (so-called misalignment angle). At a sufficiently small temperature $T \sim T_c$ the field θ start oscillating near its equilibrium value $\theta = 0$ and its energy density decreases as $R^{-3}(t)$ due to the growth of the scale factor of the universe $R(t)$, just as the energy density of nonrelativistic particles. Starting from this time the ratio of the axion energy density ρ_a to the baryon energy density ρ_b does not change and can be estimated by the following expression [16],

$$ \frac{\rho_a}{\rho_b} = O(10^2) \, \theta_1^2 \left(\frac{10^{-6} eV}{m_a} \right)^{1.18}. \tag{4} $$

For small m_a and big $|\theta_1|$ axions may give a dominant contribution to the total energy density of the universe, $\rho_{tot} = \rho_a + \rho_b \sim \rho_a$. In inflationary cosmology this does not lead to a closure of the universe; $\Omega_{tot} = \Omega_a + \Omega_b = 1$ and $\rho_{tot} \sim 10^{-29} g.cm^{-3}$ at $t \sim 10^{10}$ years independently of m_a and $|\theta_1|$. However, from (4) it follows that at the present moment, when $\rho_{tot} \sim 10^{-29} g.cm^{-3}$, the energy density of baryons should be smaller than its observed value $\rho_b \gtrsim 10^{-31} g.cm^{-3}$ unless the initial value of the misalignment angle θ_1 is smaller than some critical value θ_c,

$$ |\theta_c| = B \left(\frac{m_a}{10^{-6} eV} \right)^{0.59}, \tag{5} $$

where the coefficient B = O(1) accounts for several theoretical uncertainties appearing in derivation of eq. (4). For the "natural" value $|\theta_1| \sim \pi/2$ one obtains the constraint $m_a \gtrsim 10^{-5} - 10^{-6}$ eV ($f_a \leq 10^{12} - 10^{13}$ GeV) [16].

However, in inflationary cosmology the notion of "natural" initial value of $|\theta_1|$ does not have a well-defined meaning. During inflation long-wave perturbations of the axion field θ are generated [34]. As a result, after inflation the universe becomes filled with the field θ which looks almost homogeneous on the horizon scale $l \sim 10^{28}$ cm, but on a much bigger scale this field takes all possible values, gradually changing from $-\pi$ to π in different parts of the universe [21]. In principle, we can safely live in an exponentially big domain with $|\theta_1| < |\theta_c|$, so that in the observable part of the universe we will have no problems with axions mentioned above even for very small m_a [22,21]. At first glance such a possibility seems very unnatural. Indeed, it seems improbable that we were born in an atypical domain of the universe with $|\theta_1| < |\theta_c| \ll \pi/2$, and just due to this happy accident we see no contradiction between the theoretical predictions and the observational data.

This problem was further investigated in ref. [21] and a rather unexpected answer was obtained. First of all, it was shown (for $m_a \ll 10^{-5} eV$) that the total number of baryons in the parts of the universe with $\theta_1 > \theta_c$ is proportional to θ_1. This apparently confirmed the naive expectation that it is very improbable to live in a part of the universe with $|\theta_1| < |\theta_c| \ll \pi/2$. However, it was shown also that the properties of the domains with different values of θ_1 crucially depend on the value of θ_1. In particular, the density of matter inside galaxies exhibits an extremely sharp dependence on θ_1: $\rho \sim \theta_1^8$. For example, the total number of baryons in the parts of the universe with $\theta_1 \sim 10\theta_c$ is 10 times bigger that in domains with $\theta_1 \sim \theta_c$. However, the density of matter in galaxies formed in domains with $\theta \sim 10\theta_c$ proves to be *eight* orders of magnitude bigger than that in the domains with $\theta \sim \theta_c$! There is no *a priori* reason to believe that it is more natural for us to live and make observations in the parts of the universe filled with pathologically dense galaxies where stable stellar and planetary configurations are hardly possible, even if these parts of the universe contain ten times more baryons than the parts with $\theta_1 \lesssim \theta_c$. [4]

This argument is based on the anthropic principle, which, roughly speaking, says that a fish finds itself in the water not because the entire universe is filled with water, but because it can only live in water and there is enough water on the earth. To avoid misunderstandings we must emphasize, however, that *we do not use anthropic arguments* in order to disprove the existence of the cosmological constraint $m_a \gtrsim 10^{-5}$ eV. We just show that our universe contains exponentially big parts with $\theta_1 > \theta_c$ as well as parts with $\theta_1 < \theta_c$, and those who wish to prove that it is more "natural" to live in the part of the universe with $\theta_1 \gg \theta_c$ have no other way rather than to use anthropic arguments and prove that it is much more probable for life as we know it to exist in the galaxies with properties which are crucially different from the properties of our own galaxy. Unless such a proof can be given (which seems very unlikely), there is no reason to believe that the constraint $m_a \gtrsim 10^{-5}$ eV ($f_a \lesssim 10^{12}$ GeV) should be actually valid in inflationary cosmology.

---

[4]Note, that from this argument it follows also that with a smaller initial value of $\delta\rho/\rho$ one would obtain galaxies of our type for a bigger value of θ_1, i.e. for a smaller value of Ω_b. This suggests that the small value of Ω_b ($\Omega_b \sim 10^{-2}$) is not an independent small parameter, but it may be related to the small amplitude of density perturbations $\delta\rho/\rho \sim 10^{-5}$ [21].

3   Axion Perturbations with Time-Independent f_a

There is one potential problem with these arguments. The same mechanism which leads to formation of domains with different values of the field θ, leads also to isothermal density perturbations and, consequently, to perturbations of the microwave background radiation (MBR). Indeed, according to [34], during inflation all minimally coupled massless scalar fields, including the field $a(x)$, have quantum fluctuations characterized by amplitude $H/2\pi$, where H is the expansion rate, $H = \left(\frac{8\pi V(\varphi)}{3M_p^2}\right)^{1/2}$. This leads to fluctuations

$$\delta\theta = \frac{H}{2\pi f_a} \tag{6}$$

and, consequently, to anisotropies of the MBR temperature, $\delta T/T \sim \delta\rho/\rho \sim 2\delta\theta/\theta_1$. In our part of the universe $\theta_1 \sim \theta_c$, which is necessary to have the observed value of baryon density, $\Omega_b \sim 10^{-2}$. This gives

$$\frac{\delta T}{T} \sim \frac{\delta\rho}{\rho} \sim \frac{2\delta\theta}{\theta_c} \sim \frac{H}{\pi f_a}\left(\frac{10^{-6}eV}{m_a}\right)^{0.59}. \tag{7}$$

From the experimental constraint $\delta T/T \lesssim 3 \cdot 10^{-5}$ it follows that at the last stages of inflation the parameter H should be sufficiently small,

$$H \lesssim 10^9 GeV\left(\frac{10^{-6}eV}{m_a}\right)^{0.41} \sim 10^9 GeV\left(\frac{f_a}{10^{13}GeV}\right)^{0.41}. \tag{8}$$

For a more detailed and accurate derivation of a similar constraint see the earlier paper by Lyth [33]. This inequality by itself says only that in order to get small density perturbations one should have very small H at the last stages of inflation, which is quite possible. On the other hand, in the simplest models of inflation one typically has $\delta T/T \sim \delta\rho/\rho \sim 10^{-5}$ for $H \sim 10^{13} - 10^{14}$ GeV [7]. In such models eq. (8) implies that m_a should be smaller than 10^{-16} eV [25]. Consequently, the misalignment angle $\theta_1 \sim \theta_c$ in our part of the universe should be extremely small, see eq. (5). This was considered in [25] as a problem, since it may seem unnatural to live in the parts of the universe with $\theta_1 \sim \theta_c \ll 1$. However, this problem coincides with the problem addressed (and, we believe, resolved) in the previous section: The density of matter in galaxies is proportional to θ_1^8, which explains the reason why we cannot live in the parts of the universe with $\theta_1 \gg \theta_c$ [21].

On the other hand, eq. (8) shows that in the axion models where Peccei-Quinn symmetry breaking occurs before the end of inflation, either the Hubble constant H at the end of inflation should be very small [33] or the axion mass m_a should be extremely small [25]. Either of these conclusions would be very important. However, in the derivation of eq. (8) we operated under an implicit assumption that the radial part f_a of the field Φ does not change during the last stages of inflation. As it will be shown in the next section, this assumption (and, consequently, eq. (8)) is valid only in a rather restricted class of axion models, and in many other cases eq. (8) requires strong modification. To illustrate this statement we will consider in the next section several different models of axions which exhibit different behavior of f_a during inflation. In all models considered there we will presume that the Hubble parameter H at the last stages of inflation was big, $H \sim 10^{13} - 10^{14}$ GeV, since otherwise the problem raised in [25] does not appear at all. We will return to the discussion of models of inflation with a small H in the Section 5.

4 Peccei-Quinn Symmetry Behavior During Inflation

4.1 Peccei-Quinn field as an inflaton

We will start with a discussion of a simplest and most paradoxical possibility. Let us assume that $f_a(x)$, being the radial component of the Peccei-Quinn field Φ in the theory (1), is the inflaton field responsible for inflation. Of course, this is a strong assumption, but we need to have an inflaton field anyway, so it is not unreasonable to start our discussion with the most economical model. A similar possibility have been considered earlier by Pi [22]. The main difference which will be very important for us is that in [22] a model of new inflation was proposed, in which the radial component $f_a(x)$ of the field Φ *grows* during inflation. However, all models of new inflation suggested so far suffer from the problem of initial conditions [7]. The model (1) which we will consider describes chaotic inflation during which the field f_a slowly *decreases*, rolling down to its equilibrium value $f_a = \frac{m_\Phi}{\sqrt{\lambda_\Phi}}$ from the region with a very big f_a: $f_a > M_p$. Inflation in this model leads to density perturbations $\delta\rho/\rho \sim 10^{-5}$ for $\lambda_\Phi \sim 10^{-14}$. (In principle, this coupling constant can appear due to radiative corrections if the field Φ interacts with some other fields with the interaction constants $\sim 10^{-6}$, see [35], [22], [12].)

At the last stages of inflation in the model (1), when perturbations responsible for the galaxy formation were produced, the inflaton field f_a was still very big, $f_a \gtrsim 4M_p$ [7]. This means that in order to estimate the contribution of these perturbations to $\delta T/T$ one should substitute into eq. (7) this value instead of the value of f_a corresponding to the minimum of $V(\Phi)$:

$$\frac{\delta T}{T} \sim \frac{2\delta\theta}{\theta_1} \lesssim \frac{H}{4\pi M_p} \left(\frac{10^{-6} eV}{m_a}\right)^{0.59} . \tag{9}$$

The value of the Hubble constant at that time was $H = \sqrt{\frac{2\pi\lambda_\Phi}{3}} f_a^2/M_p \sim 20\sqrt{\lambda_\Phi} M_p \sim 2 \cdot 10^{13} GeV$. This value of H and eq. (9) are completely consistent with the experimental constraints on $\delta T/T$ for $m_a \gtrsim 10^{-10}$ eV, i.e. for a wide range of values of m_a far outside the "axion window".

4.2 Noninteracting inflaton and Peccei-Quinn fields

Let us discuss now a less radical possibility. We will consider a model describing two noninteracting scalar fields, Peccei-Quinn field Φ and the inflaton filed φ, with the effective potential

$$V(\varphi, \Phi) = V(\varphi) - m_\Phi^2 \Phi^* \Phi + \lambda_\Phi (\Phi^* \Phi)^2 + V_o . \tag{10}$$

For definiteness, we will take the effective potential of the inflaton field in the form

$$V(\varphi) = \frac{\lambda_\varphi}{4}\varphi^4. \tag{11}$$

Now one should distinguish between two different possibilities:

1) $m_\Phi > H$

Under this condition field Φ has enough time to roll down to the minimum of its effective potential at $f_a = m_\Phi/\sqrt{\lambda_\Phi}$ and to relax there. In this case all results obtained in [25] are valid, and we get a constraint on the axion mass $m_a \lesssim 10^{-16}$ eV (see Section 3), which corresponds to $f_a \gtrsim 10^4 M_p$. While this is not impossible, it also places a strong constraint on the coupling constant λ_Φ:

$$\lambda_\Phi \lesssim 10^{-8} \frac{m_\Phi^2}{M_p^2}. \tag{12}$$

2) $m_\Phi < H$

In this case the term $m_\Phi^2 \Phi^* \Phi$ can be neglected in the investigation of behavior of the fields Φ and φ during inflation. The corresponding investigation of the theory $\lambda_\Phi(\Phi^*\Phi)^2 + \frac{\lambda_\varphi}{4}\varphi^4$ was performed in [31]. The result is that during inflation the fields f_a and φ approach an asymptotic regime at which $\lambda_\Phi f_a^2 = \lambda_\varphi \varphi^2$, practically independently of the choice of initial conditions. Therefore instead of $f_a = m_\Phi/\sqrt{\lambda_\Phi}$ one should substitute to eq. (7) $f_a = \sqrt{\frac{\lambda_\varphi}{\lambda_\Phi}}\varphi$. This gives, for $\varphi \sim 4M_p$,

$$\frac{\delta T}{T} \sim 2\sqrt{\lambda_\Phi}\left(\frac{10^{-6}eV}{m_a}\right)^{0.59}. \tag{13}$$

Then, from the experimental constraint $\delta T/T \lesssim 3 \cdot 10^{-5}$ it follows that

$$\lambda_\Phi \lesssim 10^{-10}\left(\frac{m_a}{10^{-6}eV}\right)^{1.18}. \tag{14}$$

4.3 The effect of interaction between the inflaton and Peccei-Quinn field

Now let us consider a more realistic model in which the fields φ and Φ interact with each other. The simplest model of that kind is

$$V(\varphi, \Phi) = \frac{\lambda_\varphi}{4}\varphi^4 - m_\Phi^2 \Phi^* \Phi + \lambda_\Phi(\Phi^*\Phi)^2 - \nu\varphi^2\Phi^*\Phi + V_o. \tag{15}$$

From this equation it follows that the effective mass of the field Φ depends on φ:

$$m_\Phi^2(\varphi) = m_\Phi^2 + \nu\varphi^2. \tag{16}$$

One should consider here two different possibilities, corresponding to different signs of ν:

1) $\nu < 0$

In this case at large φ the effective mass squared $m_\Phi^2(\varphi)$ is positive and bigger that H^2. This means that in the beginning of inflation the Peccei-Quinn symmetry is restored, $\Phi = 0$, and no perturbations of this field are produced. However, at $\varphi = \varphi_c$, where $\varphi_c = m_\Phi/\sqrt{\nu}$, the phase transition with the Peccei-Quinn symmetry breaking takes place. After that perturbations of the axion field are generated. Their maximal amplitude corresponds to the time of the phase transition, since at that time the effective value of f_a in eq. (7) is especially small. The possibility of obtaining a spectrum of density perturbations with a cut-off in the long-range limit and with a sharp maximum at $l \sim \exp(\pi\varphi_c^2/M_p^2)$ cm, corresponding to the moment of the phase transition [7], is extremely interesting. However, one may expect that at the moment of the phase transition many axionic domain walls will be produced [20,32]. There are two ways to avoid undesirable consequences of this effect: to produce domain walls either very early or at the very end of inflation. In the first case the walls will be inflated away, in the second case they will not be big enough to cause any problems. This is a very interesting possibility which deserves separate investigation. We will not consider it here since in this case inflation with large H does not seem to help saving axions with $m_a < 10^{-5} - 10^{-6} eV$.

2) $\nu > 0$

We will assume for definiteness that $|\nu| > m_\Phi^2/M_p^2$. This condition implies that the effective value of the parameter m_Φ^2 during inflation is given by $\nu\varphi^2$, which means that the "radius" of the axion potential is equal to

$$f_a(\varphi) = \sqrt{\frac{\nu}{\lambda_\Phi}}\varphi . \tag{17}$$

Therefore instead of eq. (7) we get

$$\frac{\delta T}{T} \sim \sqrt{\frac{2\pi\lambda_\varphi\lambda_\Phi}{3\nu}} \frac{\varphi}{\pi M_p} \left(\frac{10^{-6}eV}{m_a}\right)^{0.59} . \tag{18}$$

The strongest constraint on the value of the constant ν follows from the condition of vacuum stability in the theory (15): $\nu < \sqrt{\lambda_\varphi\lambda_\Phi}$. For definiteness, we will just take here $\nu \sim \sqrt{\lambda_\varphi\lambda_\Phi}$. In this case eq. (18) at $\varphi \sim 4M_p$ and $\lambda_\varphi \sim 10^{-14}$ reads:

$$\frac{\delta T}{T} \sim 6 \cdot 10^{-4}\lambda_\Phi^{1/4} \left(\frac{10^{-6}eV}{m_a}\right)^{0.59} . \tag{19}$$

Then from the condition $\delta T/T \lesssim 3 \cdot 10^{-5}$ it follows that

$$\lambda_\Phi \lesssim 10^{-5} \left(\frac{m_a}{10^{-6}eV}\right)^{2.4} . \tag{20}$$

5 Hybrid Inflation

As we have already mentioned, all problems related to large isothermal perturbations appear only in the models of inflation where the Hubble parameter H is relatively big at the last stages of inflation: $H \gg 10^9$ GeV. One may argue that this is a typical situation in all models of chaotic inflation. Indeed, if one wishes to obtain adiabatic density perturbations $\frac{\delta\rho}{\rho} \sim 10^{-5}$ in the theory $\frac{\lambda_\varphi}{4}\varphi^4$, one should take $\lambda_\varphi \sim 10^{-14}$. At the last stages of inflation in this model, when perturbations responsible for the galaxy formation were produced, $\varphi \sim 4M_p$ [7], which implies that the Hubble parameter at that time was rather big, $H = \sqrt{\frac{2\pi\lambda_\varphi}{3}} \frac{\varphi^2}{M_p} \sim 2 \times 10^{13}$ GeV. A similar result is true for any model of chaotic inflation with $V(\varphi) \sim \varphi^n$. However, one may consider models with a much smaller amplitude of adiabatic perturbations. In these models the value of the Hubble constant can be very small. Such models may be especially interesting if one can obtain axionic isothermal perturbations $\frac{\delta\rho}{\rho} \sim 10^{-5}$ with a non-flat spectrum (see previous Section).

Another possibility is related to models of new or extended inflation. In some of these models the value of H at the last stages of inflation may be very small. For example, in the Shafi-Vilenkin model [35] the typical value of H is $O(10^9)$ GeV [7], which does not lead to any problems with light axions.

There is a potential difficulty associated with this possibility. Indeed, as is argued in [7], [12], the standard realization of new and extended inflation (due to high-temperature phase transitions), is plagued with the problem of initial conditions. The most natural way to have inflation in the models, which were originally constructed for realization of new and extended inflation, is to consider them in a wider context, along the lines of the chaotic inflation scenario. Namely, one should consider the universe filled by a chaotically distributed inflaton scalar field and then investigate the possibility that some parts of the universe will expand exponentially. For example, if in some parts of the universe the initial value of the inflaton scalar field φ in the Shafi-Vilenkin model is very small and homogeneous from the very beginning, then inflation in these parts of the universe occurs during a slow increase of this field, as in the new inflationary universe scenario. On the other hand, in those parts of the universe where the initial value of the inflaton field is very big, inflation occurs during the slow decrease of this field as in the chaotic inflation scenario in the theory $\frac{\lambda_\varphi}{4}\varphi^4$. But here the main problem arises. In the Shafi-Vilenkin model in domains with a small φ inflation may start only at a very late time $t \sim H^{-1}$, which is ten orders of magnitude bigger than the Planck time M_p^{-1}. During such a time a typical closed universe would collapse before the universe will have any chance to inflate. This is a particular manifestation of a general rule: The main part of the volume of the universe appears as a result of inflation of those domains where it starts as close as possible to the Planck density [7]. Therefore, in many models in which inflation may occur at small φ, as in the new (or extended) inflationary universe scenario, inflation in the main part of the universe actually starts with very large φ and with H close to M_p [7,12], and then it ends with $H \sim 10^{13} - 10^{14}$ GeV, as in the theory $\frac{\lambda_\varphi}{4}\varphi^4$. This leads again to very large isothermal perturbations $\frac{\delta\rho}{\rho}$ (if the conditions discussed in the previous section are not satisfied).

Fortunately, a possible resolution of this problem may be proposed. After the discovery of chaotic inflation, it is quite clear that inflation is a rather general phenomenon which may occur in a wide class of scalar field theories including *any* grand unified theory, the standard model of electroweak interactions, etc. The main problem is not to get inflation *per se*, but to find a field φ with a sufficiently flat effective potential to ensure that the density perturbations generated *at the last stage of inflation* are sufficiently small. This is the role played, e.g., by the Shafi-Vilenkin field φ. But in the Shafi-Vilenkin model there are many other scalar fields which also can drive inflation. These scalar fields (which we will call "heavy fields") have much more steep effective potentials. Therefore they roll down first and do not play any role *at the last stages of inflation* in each particular region of the universe. However, at the beginning of inflation the heavy scalar fields are very important.

Indeed, as was shown in [10], if inflation starts at a sufficiently large energy density, then the universe enters a regime of permanent reproduction of inflationary domains containing all possible values of all scalar fields. In particular, inflation driven by heavy scalar fields produces infinitely many domains with a large (and sufficiently homogeneous) scalar field φ and infinitely many domains with a very small φ [37]. In some sense one may still say that the total volume of domains with a large φ is bigger that the volume of domains with small φ, but it is very hard to compare infinities. What is important though is that in the domains with large φ inflation ends with $H \sim 10^{13}$ GeV, which is four orders of magnitude larger than the value of H in the domains with small φ. This implies that in the domains with small initial values of φ we will obtain reasonably small density perturbations which would lead to formation of galaxies of our type, whereas in the domains with large initial values of φ we will get isothermal density perturbations which will be four orders of magnitude larger than in our domain, see eqs. (7) and (9). The existence of life as we know it in domains with such a huge degree of inhomogeneity does not seem possible at all. If this is true, then the combination of chaotic inflation with heavy scalar fields and of the axion theory with light axions may resolve the problem of initial conditions for some versions of new (and extended) inflation: Even though there are more domains in which $\varphi > M_p$ at the last stages of inflation than the domains with $\varphi \ll M_p$, the total number of exponentially expanding domains with $\varphi \ll M_p$ in a self-reproducing inflationary universe is still enormously large (perhaps, infinitely large), and only in those domains we have conditions which are necessary for the existence of life of our type.

To avoid misunderstandings, we must emphasize that what we are doing is not a resurrection of the standard versions of new (or extended) inflation based on the theory of high-temperature phase transitions. We are just trying to implement the main ideas of chaotic inflation in the models which were originally proposed for realization of other versions of the inflationary universe scenario. That is why we will call the scenario described above *hybrid inflation* [13].

To give another example of hybrid inflation, let us consider a model describing two interacting scalar fields, σ and φ, with the effective potential

$$V(\sigma,\varphi) = \frac{\alpha}{4}(\sigma^2 - \frac{M^2}{\alpha})^2 + \frac{m^2}{2}\varphi^2 + \frac{g^2}{2}\varphi^2\sigma^2 \ . \tag{21}$$

The effective mass squared of the field σ is equal to $-M^2 + g^2\varphi^2$. Therefore for $\varphi > \varphi_c = M/g$ the only minimum of the effective potential $V(\sigma,\varphi)$ is at $\sigma = 0$. At the moment when the inflaton field φ becomes smaller than φ_c, the phase transition with the symmetry breaking occurs. We will assume that $m^2 \ll H^2 \ll M^2$ where $H^2 \sim \frac{2\pi M^4}{3\alpha M_p^2}$ at the time of the phase transition. One can easily verify, that, under this condition, the universe at $\varphi > \varphi_c$ undergoes a stage of inflation, which finishes abruptly at the moment of the phase transition. The difference between this scenario and the simplest versions of the chaotic inflation scenario is that here inflation at its last stages is driven not by the energy density of the inflaton field φ but by the vacuum energy density $V(0,0) = \frac{M^4}{4\alpha}$, as in the new inflationary universe scenario.

The amplitude of the adiabatic density perturbations produced in this theory can be estimated by standard methods [7] and is given by

$$\frac{\delta\rho}{\rho} \sim \frac{2}{5} \sqrt{\frac{2\pi}{3}} \frac{gM^5}{\alpha\sqrt{\alpha}M_p^3 m^2} \, . \tag{22}$$

To give a particular example, let us take $g \sim \alpha \sim 10^{-1}$, $m \sim 10^2$ GeV (electroweak scale). In this case $\frac{\delta\rho}{\rho} \sim 10^{-5}$ for $M \sim 10^{12}$ GeV, and the value of the Hubble parameter at the end of inflation is given by $H \sim 5 \times 10^5$ GeV. Note, that reheating in this model is very efficient; the reheating temperature may be as large as 10^{12} GeV, which is quite sufficient for low-temperature baryogenesis.

Of course, the theory (21) is just a toy model. For further development of this model one must ensure that radiative corrections do not change considerably the shape of the effective potential (21) near $\varphi = \varphi_c$, that one can obtain this model in the context of a realistic theory of elementary particles, etc. (A particular semi-realistic version of this model was suggested in [12].) Our main purpose in considering this model here is to demonstrate that there is nothing in the nature of inflationary cosmology that would make it impossible to have inflation with a very small H.

6 Conclusions

The idea that axions can help us to solve the strong CP violation problem is very popular. However, cosmological constraints make life with axions very difficult. There exist two main possibilities:

1) *Peccei-Quinn symmetry breaks down after inflation.* Most versions of the axion theory where this type of behavior may occur are already ruled out by the combination of astrophysical and cosmological constraints on the axion mass. Due to theoretical uncertainties the "axion window" $10^{-2}eV \gtrsim m_a \gtrsim 10^{-5}eV$ is not completely closed yet, but it seems plausible that even to have axion with the mass in the interval $10^{-3}eV \gtrsim m_a \gtrsim 10^{-5}eV$, i.e. well inside the "axion window", one still should require that the Peccei-Quinn symmetry breaks down before the end of inflation [17].

2) *Peccei-Quinn symmetry breaks down before the end of inflation.* The present investigation shows that in this case there are no *model-independent* cosmological constraints on the

axion mass. However, in each particular model cosmological constraints on some parameters do appear. If one prefers to have relatively simple models of inflation where the expansion rate H at the end of inflation is of the order of $10^{13} - 10^{14}$ GeV, then in all models considered above one should have a very flat Peccei-Quinn potential (small constant λ_Φ). This may seem not very natural, but in the simplest inflationary models mentioned above one should have flat potentials and small coupling constants anyway, so the existence of light axions does not make the situation more complicated.

The results of the investigation performed in Section 4 depend crucially on details of the theory, and in particular on the interaction of the Peccei-Quinn field with the inflaton. In this paper we considered only few oversimplified models, and in realistic models much more possibilities may appear. For example, in the DFSZ invisible axion model [36] the axion field is a combination of three different interacting scalar fields. At present two of these fields are very small as compared with the third one, but this may not be the case during inflation. One may try to think also about the possibility that the U(1) symmetry of the Peccei-Quinn theory is broken in the presence of a (complex) inflaton field φ. In this case the axion becomes a pseudo-Goldstone field only after the end of inflation, when/if the inflaton field φ vanishes. This can be achieved e.g. by adding a term $M^2(\Phi^*\varphi^* + \Phi\varphi)$ or $\gamma(\Phi^*\varphi^* + \Phi\varphi)^2$ to the effective potential. Then the axion angle θ will correlate with the corresponding inflaton angle, and fluctuations of the latter will be small as in the theory where the Peccei-Quinn field itself drives inflation, see eq. (9). In such models it might be possible to have small m_a without having small λ_Φ.

The simplest way to solve the problems discussed in the present paper would be to consider models of inflation with a relatively small H. Our investigation shows that this is quite possible in new inflation, extended inflation and in some versions of chaotic inflation. It can be done in a most natural way in the hybrid inflation scenario which combines some advantages of chaotic inflation with specific features of new and/or extended inflation.

Since there are no strong constraints on the coupling constant λ_Φ in the axion theory and the final version of inflationary cosmology is yet to be elaborated, the best resolution of the uncertainties discussed above would be given by an experimental measurement of the axion mass. Technically, it is very hard to search for axions with masses much smaller than 10^{-5} eV. If the axions with $m_a \gtrsim 10^{-5}$ eV will be discovered [23], this will be undoubtedly extremely important. On the other hand, failure to detect axions in this mass interval will not prove that axions do not exist at all. Such a negative result may serve rather as an additional experimental evidence in favor of inflationary cosmology. It may also help us to choose between different versions of inflation and of axion models. This is a good example of an experiment which may give us important results independently of their sign !

References

[1] E.B. Gliner, Sov. Phys. JETP **22** (1965) 378; Dokl. Akad. Nauk SSSR **192** (1970) 771; E.B. Gliner and I.G. Dymnikova, Pis. Astron. Zh. **1** (1975) 7; I.E. Gurevich, Astrophys. Space Sci. **38** (1975) 67.

[2] A.A. Starobinsky, JETP Lett. **30** (1979) 682; Phys. Lett. **91B** (1980) 99.

[3] A.H. Guth, Phys. Rev. **D23** (1981) 347.

[4] A.D. Linde, Phys. Lett. **108B** (1982); **114B** (1982) 431; **116B** (1982) 335, 340; A. Albrecht and P.J. Steinhardt, Phys. Rev. Lett. **48** (1982) 1220.

[5] A.D. Linde, Phys. Lett. **129B** (1983) 177.

[6] D. La and P.J. Steinhardt, Phys. Rev. Lett. **62** (1989) 376.

[7] A.D. Linde, **Particle Physics and Inflationary Cosmology** (Harwood, New York, 1990); A.D. Linde, **Inflation and Quantum Cosmology** (Academic Press, Boston, 1990).

[8] A.D. Linde, Phys. Lett. **B227** (1989) 352.

[9] P. Jordan, Zeit. Phys. **157** (1959) 112; C. Brans and C.H. Dicke, Phys. Rev. **124** (1961) 925.

[10] A.D. Linde, Phys. Lett. **175B** (1986) 395; Physica Scripta **T15** (1987) 169; Physics Today **40** (1987) 61.

[11] A.S. Goncharov, A.D. Linde and V.F. Mukhanov, Int. J. Mod. Phys. **A2** (1987) 561.

[12] A.D. Linde, Phys. Lett. **B249** (1990) 18.

[13] A.D. Linde, Axions in Inflationary Cosmology, Stanford University preprint SU-ITP-883 (1991), to be published in Phys. Lett.

[14] A.D. Linde, Inflation and Quantum Cosmology, Stanford University preprint SU-ITP-878 (1990), to be published in the Proceedings of the Nobel Symposium on the Birth and Early Evolution of our Universe.

[15] R. Peccei and H. Quinn, Phys. Rev. Lett. **38** (1977) 1440; S. Weinberg, Phys. Rev. Lett. **40** (1978) 223; F. Wilczek, Phys. Rev. Lett. **40** (1978) 271.

[16] J. Preskill, M.B. Wise and F. Wilczek, Phys. Lett. **120B** (1983) 127; L.F. Abbott and P. Sikivie, Phys. Lett. **120B** (1983)133; M. Dine and W. Fischler, Phys. Lett. **120B** (1983) 137.

[17] R.L. Davis, Phys. Lett. **B180** (1986) 225; R.L. Davis and E.P.S. Shellard, Nucl. Phys. **B324** (1989) 167.

[18] D. Harrari and P. Sikivie, Phys. Lett. **B195** (1987) 361; A. Dabholkar and P. Sikivie, Nucl. Phys., in press.

[19] G. Raffelt and D. Seckel, Phys. Rev. Lett. **60** (1988) 1793; M. Turner, Phys. Rev. Lett. **60** (1988) 1797; R. Mayle, J. Wilson, J. Ellis, K. Olive, D.N. Schramm and G. Steigman, Phys. Lett. **203B** (1988) 203; G. Raffelt, Berkeley preprint CFPA-TH 89 011 (1989); J. Ellis and P. Salati, CERN preprint TH-5693 (1990).

[20] P. Sikivie, Phys. Rev. Lett. **48** (1982) 1156.

[21] A.D. Linde, Phys. Lett. **201B** (1988) 437.

[22] S.-Y. Pi, Phys. Rev. Lett. **52** (1984) 1725; M. Turner, Phys. Rev. **D33** (1986) 889.

[23] P. Sikivie et al., Proposal to the DOE and LLNL for an experimental Search for Dark Matter Axions in the 0.6 – 16 μeV Mass Range (submitted 20 July 1990).

[24] G. Lazarides, G. Panagiotakopoulos and Q. Shafi, Phys. Lett. **192B** (1987) 323; S. Dimopoulos and L.J. Hall, Phys. Rev. Lett. **60** (1988) 1899; G. Lazarides, R. Schaefer, D. Seckel and Q. Shafi, Nucl. Phys. **B346** (1990) 193.

[25] M. Turner and F. Wilczek, Phys. Rev. Lett. **66** (1991) 5.

[26] V.F. Mukhanov and G.V. Chibisov, JETP Lett. **33** (1981) 523; S.W. Hawking, Phys. Lett. **115B** (1982) 339; A.A. Starobinsky, Phys. Lett. **117B** (1982) 175; A.H. Guth and S.-Y. Pi, Phys. Lett. **49** (1982) 1110; J. Bardeen, P.J. Steinhardt and M.S. Turner, Phys. Rev. **D28** (1983) 679; D.H. Lyth, Phys. Rev. **D31** (1985) 1792; V.F. Mukhanov, JETP Lett. **41** (1985) 493.

[27] M.S. Turner, F. Wilczek and A. Zee, , Phys. Lett. **120B** (1983) 127; M. Axenides, R. Brandenberger and M.S. Turner , Phys. Lett. **128B** (1983) 178 ; P.J. Steinhardt and M.S. Turner, Phys. Lett. **129B** (1983) 51.

[28] A.D. Linde, JETP Lett. **40** (1984) 1333; Phys. Lett. **158B** (1985) 375;

[29] D. Seckel and M.S. Turner, Phys. Rev. **D32** (1985) 3178.

[30] L.A. Kofman, Phys. Lett. **173B** (1986) 400.

[31] L.A. Kofman and A.D. Linde, Nucl. Phys. **B282** (1987) 555.

[32] A.D. Linde and D.H. Lyth, Phys. Lett. **B246** (1990) 353.

[33] D.H. Lyth, Phys. Lett. **B236** (1990) 408.

[34] A. Vilenkin and L. H. Ford, Phys. Rev. **D 25** (1982) 1231; A. D. Linde, Phys. Lett. **116B** (1982) 335; A. A. Starobinsky, Phys.Lett. **117B** (1982) 175.

[35] Q. Shafi and A. Vilenkin, Phys. Rev. Lett. **52** (1984) 691.

[36] M. Dine, W. Fischler and M. Srednicki, Phys. Lett. **104B** (1981) 199; A.P. Zhitnitskii, Sov. J. Nucl. Phys. **31** (1980) 689.

[37] A.D. Linde, Phys. Lett. **B202** (1988) 194.

SOFT INFLATION AND GENERALIZED EINSTEIN THEORIES

ANDREW L. BERKIN and KEI-ICHI MAEDA

Department of Physics, Waseda University

Okubo 3-4-1, Shinjuku-ku, Tokyo 169, Japan

ABSTRACT. We analyse the details of soft inflationary models, which have two scalar fields: one the standard inflaton, whose potential is exponentially coupled to the other field. Such models are derived from both fundamental theories, and in the conformal frame of generalized Einstein theories (GETs). In the latter case, a nonstandard exponential coupling to the inflaton kinetic term also may arise. We discuss the various theories which give soft inflation, and then consider satisfaction of the inflationary constraints in general models. We then specialize to new and chaotic inflation potentials, with both standard and nonstandard kinetic terms. The density perturbations are reduced sufficiently so that new inflation works well, with the coupling constant near the values allowed by grand unified theories. For chaotic inflation with a massive inflaton, we find successful inflation without fine-tuning of the coupling constant or initial data.

1. *Introduction.* Although the inflationary universe solves many cosmological problems, as yet no fully satisfactory model exists. Suppression of density perturbations forces the self-coupling to excessively small values, killing new inflation and imposing a fine-tuning problem on chaotic inflation. In this work [1,2], the constraints are loosened via the action

$$S = \int d^4x \sqrt{-g} \left[\frac{1}{2\kappa^2} R - \frac{1}{2}(\nabla\phi)^2 - \frac{1}{2}e^{-\gamma\kappa\phi}(\nabla\psi)^2 - e^{-\beta\kappa\phi}V(\psi) \right], \tag{1}$$

where $\kappa^2 = 8\pi G$, $V(\psi)$ is the inflaton potential and β and γ are dimensionless coupling constants, with $\beta < \sqrt{2}$ to guarantee power-law inflation. Such potentials arise in superstring or supergravity models, as well as in the conformally transformed frame of generalized Einstein theories (GETs). Such theories include Jordan-Brans-Dicke, induced gravity, Kaluza-Klein models, R^2 and higher order terms in the action, and theories with non-minimal coupling. As the inflaton rolls down the relatively flat $V(\psi)$, ϕ evolves along the exponential potential, resulting in power-law inflation. The exponential potential acts to produce a much smaller effective self-coupling of the inflaton field, thus "softening" the constraints so that new inflation is possible and chaotic inflation is less fine-tuned.

2. *Soft Inflation.* Assuming a spatially flat Robertson-Walker universe, variation of the action (1) yields the field equations

$$\ddot{\phi} + 3H\dot{\phi} + \frac{\gamma\kappa}{2}e^{-\gamma\kappa\phi}\dot{\psi}^2 - \beta\kappa e^{-\beta\kappa\phi}V(\psi) = 0 \tag{2}$$

$$\ddot{\psi} + 3H\dot{\psi} - \gamma\kappa\dot{\phi}\dot{\psi} + e^{(\gamma-\beta)\kappa\phi}V'(\psi) = 0 \tag{3}$$

$$H^2 = \frac{\kappa^2}{3}\left[\frac{1}{2}\dot{\phi}^2 + \frac{1}{2}e^{-\gamma\kappa\phi}\dot{\psi}^2 + e^{-\beta\kappa\phi}V(\psi) \right], \tag{4}$$

where a is the scale factor of the universe, $H = \dot{a}/a$ is the Hubble parameter, an overdot indicates a time derivative, and a prime denotes differentiation with respect to ψ. Inflation occurs as ψ slowly rolls down $V(\psi)$. During this time, we can neglect second derivatives. Because the potential dominates (4), the Hubble parameter acts to damp any initially large motion of the scalar fields, and hence we also can ignore terms containing two first derivatives.

With these considerations, we thus find the attractor solution

$$\phi - \phi_0 = \frac{\beta}{\kappa} \ln \frac{a}{a_0}, \tag{5}$$

$$f(\psi) = f(\psi_0) - \frac{e^{\gamma\kappa\phi_0}}{\beta\gamma} \left[\left(\frac{a}{a_0}\right)^{\beta\gamma} - 1 \right], \tag{6}$$

with a_0, ψ_0 and ϕ_0 initial values, and

$$f(\psi) \equiv \kappa^2 \int d\psi \frac{V}{V'}. \tag{7}$$

Inflation will last until the breakdown of the slow-rolling approximation, when $\ddot\psi \approx 3H\dot\psi$. At this point, rapid oscillations of the ψ field about the minimum of V should reheat the universe, just as in standard one field models. For V constant, power-law inflation with power $2/\beta^2$ results.

The potential for new inflation may be written as $V(\psi) = V_0 - \lambda\psi^4/4$, where V_0 is the GUT scale. For chaotic inflation, the potential used is $V(\psi) = \lambda_n\psi^n/n$, with n an even integer. Reheating commences at t_f when the slow rolling approximation breaks down. In chaotic inflation, we take inflation to end when the $\dot\psi^2$ terms in (2) and (4) become dominant, as shortly after $\ddot\psi \approx 3H\dot\psi$. Then

$$f(\psi) = \kappa^2 V_0/2\lambda\psi^2 \quad \text{and} \quad \kappa\psi_f = [\kappa^4 V_0/\lambda(1 - \beta^2/6)]^{1/2} \quad \text{for new inflation} \tag{8}$$

$$f(\psi) = \kappa^2\psi^2/2n \quad \text{and} \quad \kappa\psi_f = ne^{\gamma\kappa\phi_f/2}/\sqrt{6} \quad \text{for chaotic inflation} \tag{9}$$

The horizon problem is solved if the amount of expansion is greater than about 65 e-foldings. For reheating, a temperature greater than 10^{10} GeV is required for standard baryogenesis, although models with far lower temperatures exist. We assume efficient reheating, with all potential energy becoming radiation, so that

$$T_{RH} \approx \left(\frac{30e^{-\beta\kappa\phi_f}V(\psi_f)}{\pi^2 g_*} \right)^{1/4} > T_{RH,min} \sim 10^{10}\text{GeV}, \tag{10}$$

where $g_*(T) \sim 100$ is the effective number of particle species. Density perturbations are obtained by generalizing standard one field results to find

$$\frac{\delta\rho}{\rho} = \frac{H^2 \left(|\dot\phi| + |\dot\psi|e^{-\gamma\kappa\phi/2} \right)}{\dot\phi^2 + \dot\psi^2 e^{-\gamma\kappa\phi}}, \tag{11}$$

where the right hand side is evaluated at the time the perturbation originally left the horizon. Observations of the microwave background imply $\delta\rho/\rho < 10^{-4}$ for scales currently entering the horizon, which originally left the horizon at time t_h such that $\alpha_{f/h} \equiv \ln(a_f/a_h) \approx 65$. Equation (11) naturally leads to two regimes, namely

$$|\dot\phi| > |\dot\psi|e^{-\gamma\kappa\phi/2} \Rightarrow \quad \frac{\delta\rho}{\rho} \approx H^2/|\dot\phi| \quad \text{(region A)}, \tag{12}$$

$$|\dot\phi| < |\dot\psi|e^{-\gamma\kappa\phi/2} \Rightarrow \quad \frac{\delta\rho}{\rho} \approx H^2 e^{\gamma\kappa\phi/2}/|\dot\psi| \quad \text{(region B)}. \tag{13}$$

3. *New Inflation.* For new inflation, the constraints on the initial value of the inflaton become

$$\psi_0 < \psi_H \equiv \left[\frac{\beta\gamma\kappa^2 V_0}{2\lambda}(e^{\beta\gamma\alpha_{f/h}} - 1)^{-1} \right]^{1/2} e^{-\gamma\kappa\phi_0/2}, \tag{14}$$

$$\psi_0 > \psi_{RH} \equiv \sqrt{\frac{3\gamma\kappa^2 V_0}{2\lambda}} e^{-\gamma\kappa\phi_0/2} \left[\left(\frac{30V_0}{\pi^2 g_*}\right)^{\gamma/\beta} e^{-\gamma\kappa\phi_0} T_{RH,min}^{-4\gamma/\beta} - 1 \right]^{-1/2}, \qquad (15)$$

$$\psi_0 < \psi_D \equiv \left(\frac{\beta\gamma\kappa^2 V_0}{2\lambda}\right)^{1/2} e^{-\gamma\kappa\phi_0/2} \left[\left(\left(\frac{\delta\rho}{\rho}\right)_{cr}^{-1} \frac{\kappa^2}{\beta}\sqrt{\frac{V_0}{3}}\right)^{2\gamma/\beta} e^{-\gamma\kappa\phi_0 + \beta\gamma\alpha_f/h} - 1 \right]^{-1/2} \qquad (16)$$

for the horizon problem, reheating, and density perturbations. Since $V(\psi)$ is very flat for new inflation, $\dot{\phi}$ is the dominant contribution to density perturbations. For initial conditions, ψ_0 may be calculated from the two point correlation function, with the result

$$\psi_0 = H_0 e^{\gamma\kappa\phi_0/2} = \left(\frac{\kappa^2}{3} V_0 e^{(\gamma-\beta)\kappa\phi_0}\right)^{1/2}, \qquad (17)$$

where the additional exponential factor comes from proper normalization upon second quantization.

The case of $\gamma = 0$ is plotted in figure 1 for $\beta = .1$ and $\phi_0 = 10 m_{Pl}$. In all figures, H, RH, and D correspond respectively to horizon, reheating and density constraints. A wide range of λ and V_0 lead to successful inflation, including the standard SU(5) GUT model, marked by a +. These values of λ are sufficiently large so that thermal equilibrium may be established by the onset of inflation, and so the use of a finite temperature effective potential to localize ψ near 0 is justified. The allowable range of β-ϕ_0 parameter space is shown in figure 2 for the SU(5) GUT values.

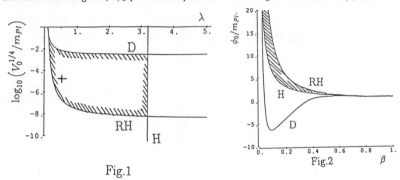

Fig.1

Fig.2

For nonstandard kinetic coupling, $\gamma = \beta/2$, the constraints are given in figure 3, again with $\beta = .1$ and $\phi_0 = 10 m_{Pl}$. Exact numerical results, found using a fourth order Runge-Kutta routine with the full set of equations (2) - (4), are also shown, with dots and x's marking points which do and don't satisfy all criteria. There is quite good agreement, verifying the validity of the approximations. The main difference from the $\gamma = 0$ case is the appearance of the horizon constraint, resulting from the quicker evolution of ψ. λ must be less than about .02, an order of magnitude smaller than typical GUT values, but still many orders better than standard new inflation. Not only is new inflation now possible, but for a wide range of exponential sector parameters, as seen in figure 4, for $V_0^{1/4} = 10^{-5}$ and $\lambda = .01$.

4. *Chaotic Inflation.* For chaotic inflation, the constraints may be written as

$$\psi_0 > \psi_H \equiv \left[\frac{2n}{\beta\gamma\kappa^2}\left(e^{\beta\gamma\alpha_f/h} - 1\right)\right]^{1/2} e^{\gamma\kappa\phi_0/2} \qquad (18)$$

$$\psi_0 \;<\; \psi_{RH} \equiv \left(\frac{2n}{\beta\gamma\kappa^2}\right)^{1/2} \left[\left(\frac{\pi^2 g_* n T_{RH,min}^4}{30\lambda_n}\right)^{-\frac{2\gamma}{n\gamma-2\beta}} \left(\frac{6\kappa^2}{n^2}\right)^{\frac{n\gamma}{n\gamma-2\beta}} e^{-\gamma\kappa\phi_0} - 1\right]^{1/2} e^{\gamma\kappa\phi_0/2} \tag{19}$$

$$\psi_0 \;>\; \psi_D \equiv \left(\frac{2n e^{\gamma\kappa\phi_0}}{\beta\gamma\kappa^2}\right)^{1/2}$$

$$\times \begin{cases} \left[\left(\left(\frac{\delta\rho}{\rho}\right)_{cr}^{-1} \frac{\kappa^2}{\beta}\sqrt{\frac{\lambda_n}{3n}}(\frac{2n}{\beta\gamma\kappa^2})^{n/4} e^{\beta^2\alpha_{f/h}/2}(1 - e^{-\beta\gamma\alpha_{f/h}})^{n/4}\right)^{\frac{4\gamma}{2\beta-n\gamma}} e^{-\gamma\kappa\phi_0} - 1\right]^{1/2} \\[2ex] \left[\left(\left(\frac{\delta\rho}{\rho}\right)_{cr}^{-1} \frac{\kappa^3}{n^{3/2}}\sqrt{\frac{\lambda_n}{3}}(\frac{2n}{\beta\gamma\kappa^2})^{\frac{n+2}{4}} e^{\beta(\frac{\beta}{2}+\gamma)\alpha_{f/h}}(1 - e^{-\beta\gamma\alpha_{f/h}})^{\frac{n+2}{4}}\right)^{\frac{4\gamma}{2\beta-n\gamma}} e^{-\gamma\kappa\phi_0} - 1\right]^{1/2} \end{cases} \tag{20}$$

where the density perturbation conditions are for regions A and B respectively. The reheating and density perturbation constraints are valid for $\beta > n\gamma/2$ and flip inequality signs when the above condition on β is not met. If a quantity in square brackets becomes negative, than that constraint cannot be met for those particular parameter values. The universe will be in regime A at t_h when

$$\frac{1}{\beta\gamma}\ln\left(1 + \frac{n\gamma}{2\beta}\right) < \alpha_{f/h}. \tag{21}$$

As initial conditions, the potential energy is taken to be the Planck scale, $e^{-\beta\kappa\phi_0}V(\psi_0) = m_{Pl}^4$.

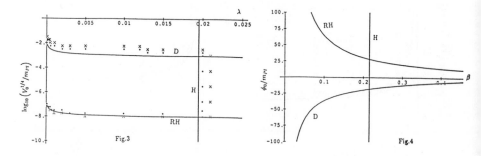

Fig.3 Fig.4

The case of $\gamma = 0$, $n = 4$ is shown in figure 5, for $\phi_0 = 10m_{Pl}$. While quite natural values of λ are allowed, they unfortunately are possible only for a small region of parameter space. Similar results hold for the massive case n=2, which is the other renormalizable value.

Figure 6 displays the case of $n = 4$, for noncanonical kinetic coupling $\gamma = \beta/2$. Just as in the standard one field case, density fluctuations force λ_4 to be excessively small. While one may take comfort that GETs are compatible with this type of chaotic inflation, the continued need for fine-tuning is a disappointment. Because ψ evolves faster, ϕ can change less, and there is little improvement over standard results. The results are ϕ_0 independent due to conformal invariance.

For the massive case $n = 2$, we write $\lambda_2 = m^2$ and plot the constraints on β and m in figure 7, again with $\phi_0 = 10m_{Pl}$. Numerical results are also shown. There is an improvement versus the standard scenario of several orders of magnitude, with m as high as .01 times the Planck mass. Furthermore, as seen in figure 8, for $m = 10^{-4}m_{Pl}$, massive chaotic inflation is possible over a wide range of ϕ_0. The presence of the exponential potential softens density perturbations and allows for a natural inflationary scenario, a significant advance.

5. *Conclusions and Speculations.* An exponential potential multiplying a standard inflaton potential may arise both fundamentally as well as in the conformally transformed frame of generalized Einstein theories. With new inflation potentials, a successful scenario is possible both for standard and nonstandard kinetic coupling. With chaotic inflation, canonical kinetic terms lead to natural values of the coupling constant, but for a slim range of β. Noncanonical kinetic coupling leads to little change in the conformally invariant $n = 4$ case, but massive chaotic inflation allows quite natural values of the mass in a wide range of parameter space.

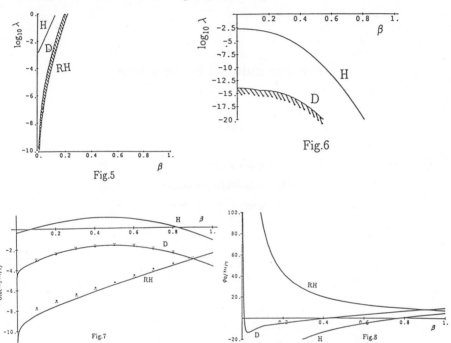

Fig.5

Fig.6

Fig.7

Fig.8

One interesting possibility for future work is to study the full spectrum of density fluctuations. As the expansion is roughly power law, more power at larger scales is expected. In addition, when $\delta\rho/\rho$ changes regions, a natural scale is set in the spectrum at which the behavior of large scale structure may change. Also, if the minimum of V is not zero, then $V_{min}e^{-\beta\kappa\phi}$ acts as an effective slowly changing cosmological "constant". Not only could such a term act as dark matter, but some recent observational results suggest our universe is best fit with a cosmological constant. Note, however, that V_{min} would need to be initially very small but nonzero. Lastly, the use of quantum cosmology or superstrings to obtain an estimate of ϕ_0, the initial value of the exponential field, would be most interesting.

REFERENCES. See the following two papers and citations within:
[1] A.L. Berkin, K. Maeda, and J. Yokoyama, Phys. Rev. Lett. 65, 141 (1990).
[2] A.L. Berkin and K. Maeda, Waseda preprint WU-AP/10/91 (1991).

Gravitino-induced baryogenesis

James Cline

Physics Department
The Ohio State University
Columbus, OH 43210

ABSTRACT: We investigate cosmic baryon generation in the minimal low energy supergravity model with the addition of dimension-four baryon-violating interactions. It is shown that gravitinos, if heavier than squarks, can give rise to a baryon asymmetry as they decay. Assuming that gravitinos are not diluted by inflation in the post-Planckian epoch, we find a realistic baryon-to-photon ratio for various ranges of gaugino and squark masses between 100 GeV and 10 TeV.

I. Introduction

In this work we[*] have taken what was until now an annoying cosmological problem for supergravity, namely the late decays of gravitinos, and turned it into a virtue, by making these gravitinos the origin of the baryon asymmetry of the universe (BAU). I will start with a general review of the problem of baryogenesis and the gravitino problem, and show how both can be solved simultaneously if gravitinos have CP-violating decays into baryons. Our model is constrained by results of particle physics experiments, and also makes several experimental predictions, which I will discuss.

II. Problem: where do the baryons come from?

From the agreement of primordial big-bang nucleosynthesis predictions with observed elemental abundances, we know the ratio of baryons to photons in the universe quite well [2]; it is

$$2.6 \times 10^{-10} \ \lesssim \ \eta \ \lesssim \ 4.3 \times 10^{-10}. \tag{1}$$

How does one explain this small number? If baryon number (n_B) is conserved throughout the history of the universe then (1) results from the initial conditions—a rather unimaginative explanation. Our experience with various extensions of the standard model makes it seem much more likely that baryon number is violated at very high energies. The $\Delta B \neq 0$ reactions would then keep $\langle n_B \rangle = 0$ at sufficiently high temperatures, and the BAU would have to be generated sometime after the Planck era. It was shown long ago [3] that there are *three necessary ingredients for producing the BAU:*

1) Baryon number violating interactions;

2) C and CP violation;

3) Departure from thermal equilibrium.

The first requirement is obvious. Without the second, the $\Delta B \neq 0$ reactions would produce equal numbers of baryons and anti-baryons. The third is needed because in thermal equilibrium, states with opposite baryon number are equally populated (they have equal energy), implying $\langle n_B \rangle = 0$. It is worthwhile to remark that this last condition is usually the most difficult to achieve in models of baryogenesis; one needs either a first order phase transition, or some particle X which is so weakly coupled that its interaction rate falls below the expansion rate of the universe, leaving behind a larger-than-thermal number density of X's.

---

[*] Stuart Raby and I, Ref. [1]. This reference contains the details of calculations described here.

These principles are illustrated by the prototypical framework for baryogenesis, Grand Unified Theories [4]. Here X_μ is a heavy ($\sim 10^{15}$ GeV) gauge boson which has gone out of equilibrium at a temperature greater than its mass. X_μ has B-violating decays into a quark and a lepton. In addition the quarks have CP-violating interactions with Higgs bosons. It is the interference of the tree diagram with the one-loop diagram containing CP-violation, illustrated in Fig. 1, that gives rise to the BAU.

(Fig. 1)

Although one can plausibly expain η (eq.(1)) in this theory, it has some short-comings. All the interesting new physics is taking place at the GUT scale, far out of reach of particle accelerators, and thus far eluding all searches for proton decay. It would be nicer to have a model that could be tested within our lifetime. Furthermore if the BAU is generated at the GUT scale, it is in danger of being erased by $\Delta B \neq 0$ reactions that become effective at lower energies. For example, the standard model itself violates baryon number at temperatures above ~ 10 TeV [5]. The model we propose below is motivated in part by these criticisms.

III. Problem: late-decaying gravitinos in supergravity

Supersymmetry (SUSY) is perhaps the most successful solution of the hierarchy problem [6]. This is the question of why the masses of Higgs and gauge bosons are so small compared to their one-loop corrections, which are quadratically divergent. Supersymmetry introduces a partner for every known particle in such a way that the quadratic part of the divergence is canceled, leaving just a relatively harmless logarithmic divergence. Table I lists the names and spins of the SUSY partners.

Table I

particle	spin	SUSY partner	spin
quark, lepton	1/2	squark, slepton $(\tilde{q}, \tilde{l})$	0
Higgs boson	0	Higgsino $(\tilde{h})$	1/2
gauge boson	1	gaugino (χ)	1/2
graviton	2	gravitino $(\tilde{G})$	3/2

Notice that I have included gravity in the last row. Although gravitons are too weakly coupled to have any effect on cosmology at sub-Planck scales, their superpartners, gravitinos, can have a big effect. Unlike its massless sibling, the gravitino has a large mass that reflects the scale of SUSY-breaking; however its couplings to other particles are just as weak as the graviton's, as shown in Fig. 2.

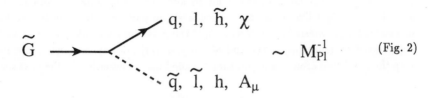

Because they are so weakly coupled, gravitinos go out of thermal equilibrium just below the Planck temperature, and they decay very late in the history of the universe,

$$\tau \sim 200 \left(\frac{m_{\tilde{g}}}{3\,\text{TeV}}\right)^{-3} \text{ sec,} \tag{2}$$

when the temperature is

$$T \sim 0.06 \left(\frac{m_{\tilde{g}}}{3\,\text{TeV}}\right)^{3/2} \text{ MeV.} \tag{3}$$

This creates a number of cosmological problems [7, 8], depending on the mass $m_{\tilde{g}}$:

1 keV $< m_{\tilde{g}} < 10$ MeV. In this case gravitinos would still be present today, giving the universe an energy density far in excess of the critical density. Given the measured value of the expansion rate, we would find that the universe must be much younger than solar system data indicate.

10 MeV $< m_{\tilde{g}} < 10$ TeV. Near the top of this range, the excess energy density of gravitinos will increase the expansion rate of the universe at temperatures ~ 1 MeV, causing weak interactions to freeze out too early, so that the neutron-to-proton ratio is too high and ^{4}He is overproduced. Less massive gravitinos will dilute the baryon-to-photon ratio η due to the entropy produced by their decays, which would spoil the agreement between the nucleosythesis determination of η, eq. (1), and that from direct observations [9]. In addition, the energetic decay products would photodissociate too much ^{2}D [10] and distort the cosmic microwave background.

$m_{\tilde{G}} > 10 \ TeV$. Weinberg [7] gave this as the allowed region for $m_{\tilde{G}}$. However, there is a potential problem even here, if baryogenesis happens before the era of gravitino decay. The entropy produced dilutes η by a factor of 10^6, so that η would had to have been $\sim 10^{-4}$ originally. While this doesn't seem like a lot, it is typically hard to produce such a large asymmetry while remaining within the phenomenological constraints on B- and CP-violation.

One way to avoid these problems is to assume that inflation diluted away the gravitinos [8]. This however will tend to also get rid of other out-of-equilibrium particles one might like to use for baryogenesis; in particular it is difficult to have a successful baryogenesis in SUSY GUT's if the gravitinos are inflated away and not regenerated by reheating at the end of inflation [11]. In any case, we would like to keep the gravitinos around and put them to good use, as described in the next section.

IV. A solution: gravitinos $\longrightarrow$ baryons

I noted in section II that, of the three baryogenesis conditions, departure from thermal equilibrium is often the hardest to achieve, and in the third section we saw that gravitinos create havoc in the early universe because they are so far out of equilibrium when they decay. Could we use gravitinos to produce the baryon asymmetry? In fact it is natural to do so, since SUSY can easily fulfill the other two requirements. C and CP violation are generically present, in the form of phases for the gauge fermion masses, as well as several other complex couplings. And unlike the standard model, it is possible to include renormalizable, baryon-violating operators in the Lagrangian. These are couplings of three squarks or two quarks and a squark, shown in Fig. 3. They arise from a term in the superpotential [12]

$$g_{ijk}\epsilon_{abc}U^c_{ia}D^c_{jb}D^c_{kc}, \tag{4}$$

where U^c (D^c) contains charge $+2/3$ $(-1/3)$, SU(2) singlet quarks and squarks, and i,j,k (a,b,c) are generation (color) indices. Because of the ϵ tensor, g_{ijk} is antisymmetric under $j \leftrightarrow k$; i.e., the charge $-1/3$ fields always involve two different generations. (For concreteness, Fig. 3 shows only the g_{332} coupling between top, bottom, and strange quarks and squarks; A is a complex mass whose magnitude is of the same order as the effective SUSY breaking scale.)

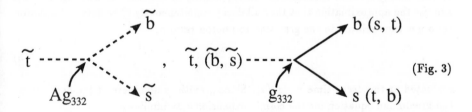

(Fig. 3)

So we see that the ingredients for baryogenesis are present within supergravity. The detailed mechanism is as follows. A gravitino can decay into a squark-antiquark pair ($\tilde{q}\bar{q}$), or into the CP-conjugate state $\tilde{\bar{q}}q$, at tree level. There is a one-loop diagram for the same decay that has a CP-violating, three-squark coupling. These will interfere, with the result that more $\tilde{\bar{q}}q$ than $\tilde{q}\bar{q}$ pairs are produced. So far we have no baryon asymmetry, but we do have what you might call a CP asymmetry. Next the squarks will quickly decay through the baryon-violating interaction of Fig. 3, which converts the CP asymmetry into the desired baryon asymmetry. The whole process is shown in Fig. 4.

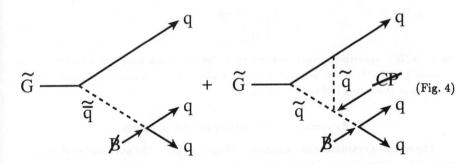

(Fig. 4)

There are also other channels for the decays. The gravitino might decay into a gauge fermion and a gauge boson. The gauge fermion can then decay into quarks and squarks as described above.

The result is that every gravitino decay will on average produce some fractional number ϵ of baryons minus antibaryons, which is determined solely by the diagrams of Fig. 4. Here we will estimate how large ϵ must be to account for the observed BAU, saving for the next section a discussion of experimental constraints on ϵ. Let us assume that gravitinos dominate the energy density of the universe when they decay, and use the approximation that they all decay simultaneously. The baryon-to-photon ratio η is then related to the gravitino-to-photon ratio by

$$\eta = \epsilon \frac{n_{\tilde{G}}}{n_\gamma}, \tag{5}$$

evaluated at the decay time τ, eq. (2). Since gravitinos dominate at the decay time, the Friedmann equation for the cosmic expansion rate implies

$$1/\tau \cong (8\pi G m_{\tilde{G}} n_{\tilde{G}}(\tau))^{1/2} . \tag{6}$$

Furthermore n_γ at the decay time is dominated by photons produced in the decays. Energy conservation gives the decay temperature,

$$\frac{\pi^2}{15} T_D^4 = m_{\tilde{G}} n_{\tilde{G}}(\tau), \tag{7}$$

from which one can determine the photon density

$$n_\gamma(\tau) = \frac{2\zeta(3)}{\pi^2} T_D^3. \tag{8}$$

Putting these results together gives

$$\eta \sim \epsilon \left(\frac{m_{\tilde{G}}}{M_{\mathrm{Pl}}} \right)^{1/2} \sim 10^{-7} \epsilon \tag{9}$$

for a 10 TeV gravitino, which will prove to be the mass range of interest. Thus we need $\epsilon \sim 10^{-3}$. Whether such a value is allowed by experimental limits on CP violation is the subject of the final section.

V. Experimental constraints and predictions

The result of evaluating the diagrams of Fig. 4 is that ϵ, the net number of baryons produced per gravitino decay, has the form

$$\epsilon = g^2 \sum_i \phi_i f_i(\text{mass ratios}), \tag{10}$$

where g is the baryon-violating coupling, the sum is over decay channels, ϕ_i are the CP-violating phases, and f_i are dimensionless functions of the SUSY masses appearing in the loop diagram. Therefore ϵ is potentially constrained by experimental limits on B and CP violation, as well as limits on the masses. I now discuss these in turn.

B violation. Note that there are many B-violating couplings in eq. (4). Not sur-prisingly, the only one which gives rise to observable B violation at low energies is that with the smallest generation indices, g_{112}. This coupling can cause the decay of stable nuclei [12, 13]; for example, a neutron and a proton in ^{16}O could go into K^0, K^+, leaving behind ^{14}N. This puts a fairly stringent limit on g_{112}, but not on the higher generation couplings. We take $g = g_{332}$ to be the strongest B-violating coupling, with $g^2/4\pi = 0.1$ so that it is still perturbative. One can show that the hierarchy $g_{332} \gg g_{112}$ is not upset by radiative corrections, so our assumption is not unnatural.

CP violation. The same CP-violating phases ϕ_i we have used to produce the BAU also give quarks a (chromo)electric dipole moment, which would be manifested in the neutron EDM, d_n. The contributions have been calculated [14, 15] and have the form

$$d_n = \frac{1}{\bar{m}} \sum_i \phi_i g_i (\text{mass ratios}), \tag{11}$$

where $\bar{m}$ is the mass of the gluino or squark. Note the similarity to eq. (10) for ϵ; however since $g_i \neq f_i$, a different linear combination of phases appears in the two expressions. To relate the two, we have assumed that there are no accidental cancellations in the sum, or in other words, that all the phases are approximately equal. This phase can then be bounded by comparing (11) with the measurement[16] $|d_n| = (-0.3 \pm 0.5) \cdot 10^{-25} e$-cm, which gives a 95% c.l. limit of $|d_n| < 1.2 \cdot 10^{-25} e$-cm. We choose the CP-violating phase so that d_n is an order of magnitude smaller, $10^{-26} e$-cm.

The results can be stated as follows. We find that the baryon-to-photon ratio η falls in the allowed range (eq. (1)) for two regimes of the SUSY particle masses. Either 1) $m_{\tilde{q}} > 100$ GeV and $m_{\tilde{\gamma}}, m_{\tilde{z}} > 1\text{-}2$ TeV or 2) $m_{\tilde{q}} > 500$ GeV and $m_{\tilde{\gamma}}, m_{\tilde{z}} < 500$ GeV. In both cases we need a rather heavy ($> 3\text{-}6$ TeV) gluino so that the loop diagrams giving quark EDM's are suppressed, and an even heavier gravitino, $m_{\tilde{g}} > 10\text{-}20$ TeV, so that the gravitino decays before the era of weak interaction freezeout, $T \sim 1$ MeV. Otherwise one may encounter some of the gravitino problems mentioned in section III.

In addition to generating the baryon asymmetry, our model makes some experi-mental predictions. First, we would expect the neutron EDM to be detected in the next generation of searches, since we assumed it was $10^{-26} e$-cm. If it is much smaller than this, we are forced to push the SUSY particle masses up to experimentally less interesting values in order to get a large enough baryon asymmetry. Of course, we

136

should see some of these particles at the SSC; that is another prediction. More specific to our model is the *way* in which we should see them, and that is through the large baryon-violating coupling g_{332}. Fig. 5 illustrates two heavy sea-quarks from a pp collision fusing into a squark, which then decays into heavy quarks. This two-jet event would appear as a resonance at the squark mass. Because of the generation structure of the interaction, the jets would have anomalously high strangeness. We estimate 100 such events should be produced in a year of running at SSC, assuming the beam is tuned to the resonance energy.

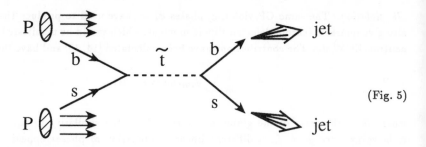

(Fig. 5)

In conclusion, I have argued that gravitino-induced baryogenesis has several advantages: it simultaneously solves the gravitino problem and that of baryogenesis; gravitinos naturally fulfill the baryogenesis requirement of departure from thermal equilibrium; it is a very low temperature scenario for baryogenesis, which makes it more experimentally testable; we predict the neutron electric dipole moment and baryon violating interactions of squarks should be observed; the model is consistent with at least some supersymmetric particles being light ($\sim$ 100 GeV). Unfortunately, we were forced to keep the gravitino itself rather heavy ($>$ 10 TeV) to remain consistent with the standard theory of nucleosynthesis; it would be nicer to have $m_{\tilde{g}}$ near the weak scale (300 GeV) if supersymmetry is to solve the hierarchy problem in a natural way. However it may be possible to get correct elemental abundances even if gravitinos are lighter ($\sim$ 3 TeV) and thus decay after the time when weak interactions keep protons and neutrons in equilibrium. This is a numerical question which is still under investigation [17].

References

1. J. Cline and S. Raby, "Gravitino-induced baryogenesis: a problem made a virtue," Phys. Rev. D, in press (1991).

2. K. Olive, D. Schramm, G. Steigman and T. Walker, Phys. Lett. 236B (1990) 454.

3. A. D. Sakharov, Zh. Eksp. Teor. Fiz. Pis'ma 5 (1967) 32; S. Dimopoulos and L. Susskind, Phys. Rev. D18 (1978) 4500.

4. M. Yoshimura, Phys. Rev. Lett. 41 (1978) 281; A. Yu. Ignatiev, N. V. Krasnikov, V. A. Kuzmin and A. N. Tavkhelidze, Phys. Lett. 76B (1978) 436; J. Ellis, M. K. Gaillard, and D. V. Nanopoulos, ibid. 80B (1978) 360; S. Weinberg, Phys. Rev. Lett. 42 (1979) 850; A. Yu. Ignatiev, V. A. Kuzmin and M. E. Shaposhnikov, Phys. Lett. 87B (1979) 114; S. B. Treiman, F. Wilczek and A. Zee, Phys. Rev. D18 (1978) 4500.

5. V. A. Kuzmin, V. A. Rubakov and M. E. Shaposhnikov, Phys. Lett. 155B (1985) 36; P. Arnold and L. McLerran, Phys. Rev. D36 (1987) 581; ibid. D37 (1988) 1020. E. Mottola and S. Raby, Phys. Rev. D42 (1990) 4202; J. Cline and S. Raby, Phys. Lett. B246 (1990) 163; J. A. Harvey and M. S. Turner, Phys. Rev. D42 (1990) 3344.

6. For a review, see S. Raby, Santa Cruz TASI lectures (1986) 63.

7. S. Weinberg, Phys. Rev. Lett. 48 (1303) 1982.

8. S. Dimopoulos and S. Raby, Nucl. Phys. B219 (1983) 479; J. Ellis, A. D. Linde and D. V. Nanopoulos, Phys. Lett. 118B (1982) 59.

9. K. Olive, D. Schramm, G. Steigman, M. Turner and J. Yang, Ap. J. 246 (1981) 557.

10. J. Ellis, D.V. Nanopoulos and S. Sarkar, Nucl. Phys. B259 (1985) 175; N. Terazawa, M. Kawasaki and K. Sato, Nucl. Phys. B302 (1988) 697; J. A. Frieman, E. W. Kolb and M. S. Turner, Phys. Rev. D41 (1990) 3080.

11. J. Ellis, J. E. Kim and D. V. Nanopoulos, Phys. Lett. 145B (1984) 181.

12. S. Dimopoulos and L. J. Hall, Phys. Lett. 207B (1988) 210; S. Dimopoulos, L. J. Hall, J.-P. Merlo, R. Esmailzadeh and G. D. Starkman, Phys. Rev. D41 (1990) 2099.

13. S. Dimopoulos and L. Hall, Phys. Lett. 196B (1987) 135.

14. J. Polchinski and M. Wise, Phys. Lett. 125B (1983) 393.

15. A. De Rújula, M. Gavela, O. Pène and F. Vegas, Phys. Lett. B245 (1990) 640.

16. K. Smith et al. , Phys. Lett. 234B (1990) 191;

17. J. Cline, S. Raby and R. Scherrer, work in progress.

NATURAL INFLATION

Joshua A. Frieman

NASA/Fermilab Astrophysics Center
Fermi National Accelerator Laboratory
P.O. Box 500, Batavia, IL 60510

ABSTRACT

A pseudo-Nambu-Goldstone boson, with a potential of the form $V(\phi) = \Lambda^4[1 \pm \cos(\phi/f)]$, can naturally give rise to an epoch of inflation in the early universe. Successful inflation can be achieved if $f \sim m_{pl}$ and $\Lambda \sim m_{GUT}$. Such mass scales arise in particle physics models with a gauge group that becomes strongly interacting at the GUT scale, *e.g.*, as is expected to happen in the hidden sector of superstring theories. The density fluctuation spectrum is a non-scale-invariant power law, with extra power on large scales.

In recent years, the inflationary universe has been in a state of theoretical limbo: it is a beautiful idea in search of a compelling model. The idea is simple[1]: if the early universe undergoes an epoch of exponential de Sitter expansion during which the scale factor increases by a factor of at least e^{60}, then a small causally connected region grows to a sufficiently large size to explain the observed homogeneity and isotropy of the universe, to dilute any overdensity of magnetic monopoles, and to flatten the spatial hypersurfaces, $\Omega \equiv 8\pi G\rho/3H^2 \to 1$. As a bonus, quantum fluctuations during inflation can causally generate the large-scale density fluctuations required for galaxy formation.

During the inflationary epoch, the energy density of the universe is dominated by the (nearly constant) potential energy density $V(\phi)$ associated with a slowly rolling scalar field ϕ, the *inflaton* [2]. To satisfy microwave background anisotropy limits [3] on the generation of density fluctuations, the potential of the inflaton must be very flat. Consequently, ϕ must be extremely weakly self-coupled, with effective quartic self-coupling constant $\lambda_\phi < 10^{-12} - 10^{-14}$.

Thus, density fluctuations in inflation are a blessing to astronomers but a curse on particle physicists: although a large number of inflation models have been proposed [4], none of them is aesthetically compelling from a particle physics standpoint. In some cases, the smallness of λ_ϕ is protected against radiative corrections by a symmetry, *e.g.*, supersymmetry. However, the small coupling, while stable (technically natural), is itself unexplained, and is postulated solely in order to generate successful inflation. In recent years, it has become customary to decouple the inflaton completely from particle physics models, to specify an 'inflaton sector' with the requisite properties, with little or no regard for its physical origin. It would be preferable if the small coupling of the inflaton arose dynamically in particle physics models which are *strongly* natural, *i.e.*, which have no small numbers in the Lagrangian.

An example of the kind of thing we want, namely, a scalar field with naturally small self-coupling, is provided by the axion [5]. In axion models, a global $U(1)$ symmetry is spontaneously broken at some large mass scale f, through the vacuum expectation value of a complex scalar field, $\langle \Phi \rangle = f \exp(ia/f)$. (In this case, Φ has the familiar Mexican-hat potential, and the vacuum is a circle of radius f.) At energies below the scale f, the only relevant degree of freedom is the massless axion field a, the angular Nambu-Goldstone mode around the bottom of the Φ potential. However, at a much lower scale, the symmetry is explicitly broken by loop corrections. For example, the QCD axion obtains a mass from non-perturbative gluon configurations (instantons) through the chiral anomaly. When QCD becomes strong at the scale $\Lambda_{QCD} \sim 100$ MeV, instanton effects give rise to a periodic potential of height $\sim \Lambda_{QCD}^4$ for the axion. In 'invisible' axion models [6] with canonical Peccei-Quinn scale $f_{PQ} \sim 10^{12}$ GeV, the resulting axion self-coupling is $\lambda_a \sim (\Lambda_{QCD}/f_{PQ})^4 \sim 10^{-52}$. This simply reflects the hierarchy between the QCD and Peccei-Quinn scales, which arises from the slow logarithmic running of α_{QCD}. Since the global symmetry is restored as $\Lambda \to 0$, the flatness of the axion potential is natural.

Pseudo-Nambu-Goldstone bosons (PNGBs) like the axion are ubiquitous in particle physics models: they arise whenever a global symmetry is spontaneously broken. We therefore choose them as our candidate for the inflaton: we assume a global symmetry is spontaneously broken at a scale f, with soft explicit symmetry breaking at a lower scale Λ; these two scales completely characterize the model and will be specified by the requirements of successful inflation. The resulting PNGB potential is generally of the form

$$V(\phi) = \Lambda^4 [1 + \cos(\phi/f)] . \tag{1}$$

so the potential, of height $2\Lambda^4$, has a unique minimum at $\phi = \pi f$. As we will see below, for $f \sim m_{pl} \sim 10^{19}$ GeV and $\Lambda \sim m_{GUT} \sim 10^{15}$ GeV, the PNGB field ϕ can drive inflation [7]. (Note that in this case, the effective quartic coupling is $\lambda_\phi \sim (\Lambda/f)^4 \sim 10^{-13}$, as required.) These mass scales arise naturally in particle physics models. For example, in the hidden sector of superstring theories, if a non-Abelian subgroup of E_8 remains unbroken, the running gauge coupling can become strong at the GUT scale; indeed, it is hoped that the resulting gaugino condensation may play a role in breaking supersymmetry [8]. In this case, the role of the PNGB inflaton could be played by the "model-independent axion" [9].

For temperatures $T \lesssim f$, the global symmetry is spontaneously broken. Since ϕ thermally decouples at a temperature $T \sim f^2/m_{pl} \sim f$, we assume it is initially laid down at random between 0 and $2\pi f$ in different causally connected regions. Within each Hubble volume, the evolution of the field is described by

$$\ddot{\phi} + 3H\dot{\phi} + \Gamma\dot{\phi} + V'(\phi) = 0 , \tag{2}$$

where Γ is the decay width of the inflaton. In the temperature range $\Lambda \lesssim T \lesssim f$, the potential $V(\phi)$ is dynamically irrelevant, because the forcing term $V'(\phi)$ is negligible compared to the Hubble-damping term. (In addition, for axion models, $\Lambda \to 0$ as $T/\Lambda \to \infty$ due to the high-temperature suppression of instantons.) Thus, in this temperature range, aside from the smoothing of spatial gradients in ϕ, the field does not evolve. Finally, for $T \lesssim \Lambda$, in regions of the universe with ϕ initially near the top of the potential, the field starts to roll slowly down the hill toward the minimum. In those regions, the energy density of the universe is quickly dominated by the vacuum contribution $(V(\phi) \simeq 2\Lambda^4 \gtrsim \rho_{rad} \sim T^4)$, and the universe expands exponentially. Since the initial conditions for ϕ are random, our model is closest in spirit to the chaotic inflationary scenario [10].

To successfully solve the cosmological puzzles of the standard cosmology, an inflationary model must satisfy a variety of constraints.

1) *Slow-Rolling Regime.* The field is said to be slowly rolling (SR) when its motion is overdamped, *i.e.*, $\ddot{\phi} << 3H\dot{\phi}$, and two conditions are met:

$$|V''(\phi)| \lesssim 9H^2 \ , \text{ i.e., } \quad \sqrt{\frac{2\,|\cos(\phi/f)|}{1+\cos(\phi/f)}} \lesssim \frac{\sqrt{48\pi}f}{m_{pl}} \tag{3}$$

and

$$\left|\frac{V'(\phi)m_{pl}}{V(\phi)}\right| \lesssim \sqrt{48\pi} \ , \text{ i.e., } \quad \frac{\sin(\phi/f)}{1+\cos(\phi/f)} \lesssim \frac{\sqrt{48\pi}f}{m_{pl}} \ . \tag{4}$$

From Eqns. (3) and (4) the existence of a broad SR regime requires $f \geq m_{pl}/\sqrt{48\pi}$ (required below for other reasons). The SR regime ends when ϕ reaches a value ϕ_2, at which one of the inequalities (3) or (4) is violated. For example, for $f = m_{pl}$, $\phi_2/f = 2.98$ (near the potential minimum), while for $f = m_{pl}/\sqrt{24\pi}$, $\phi_2/f = 1.9$. As f grows, ϕ_2/f approaches π. (Here and below, we assume inflation begins at a field value $0 < \phi_1/f < \pi$; since the potential is symmetric about its minimum, we can just as easily consider the case $\pi < \phi_1/f < 2\pi$.)

2) *Sufficient inflation.* We demand that the scale factor of the universe inflates by at least 60 e-foldings during the SR regime,

$$N_e(\phi_1, \phi_2, f) \equiv \ln(R_2/R_1) = \int_{t_1}^{t_2} H\,dt = \frac{-8\pi}{m_{pl}^2} \int_{\phi_1}^{\phi_2} \frac{V(\phi)}{V'(\phi)}\,d\phi$$

$$= \frac{16\pi f^2}{m_{pl}^2} \ln\left[\frac{\sin(\phi_2/2f)}{\sin(\phi_1/2f)}\right] \geq 60 \ . \tag{5}$$

Using Eqns. (3) and (4) to determine ϕ_2 as a function of f, the constraint (5) determines the maximum value (ϕ_1^{max}) of ϕ_1 consistent with sufficient inflation. The fraction of the universe with $\phi_1 \in [0, \phi_1^{max}]$ will inflate sufficiently. If we assume that ϕ_1 is randomly distributed between 0 and πf from one horizon volume to another, the probability of being in such a region is $P = \phi_1^{max}/\pi f$. For example, for $f = 3m_{pl}$, m_{pl}, $m_{pl}/2$, and $m_{pl}/\sqrt{24\pi}$, the probability $P = 0.7$, 0.2, 3×10^{-3}, and 3×10^{-41}. The fraction of the universe that inflates sufficiently drops precipitously with decreasing f, but is large for f near m_{pl}. This is shown in Fig. 1, which displays $\log(\phi_1^{max}/f) = 0.5 + \log P$ and ϕ_2/f.

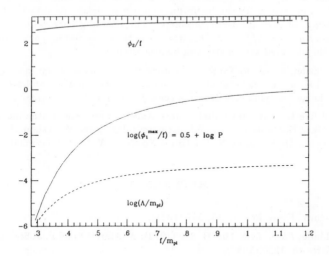

3) *Density Fluctuations.* Inflationary models generate density fluctuations [11] with amplitude at horizon crossing $\delta\rho/\rho \simeq 0.1 H^2/\dot{\phi}$, where the right hand side is evaluated when the fluctuation crossed outside the horizon during inflation. Fluctuations on observable scales are produced 60 - 50 e-foldings before the end of inflation. The largest amplitude perturbations are produced 60 e-foldings before the end of inflation,

$$\frac{\delta\rho}{\rho} \simeq \frac{0.3\Lambda^2 f}{m_{pl}^3} \left(\frac{8\pi}{3}\right)^{3/2} \frac{[1 + \cos(\phi_1^{max}/f)]^{3/2}}{\sin(\phi_1^{max}/f)} . \tag{6}$$

Constraints on the anisotropy of the microwave background [3] require $\delta\rho/\rho \lesssim 5 \times 10^{-5}$, i.e., $\Lambda \lesssim 2 \times 10^{16}$ GeV for $f = m_{pl}$ and $\Lambda \lesssim 3 \times 10^{15}$ GeV for $f = m_{pl}/2$. This bound on Λ as a function of f is also shown in the figure. Thus, to generate the fluctuations responsible for large-scale structure, Λ should be comparable to the GUT scale, and the inflaton mass $m_\phi = \Lambda^2/f \sim 10^{11} - 10^{12}$ GeV.

In this model, the fluctuations deviate from a scale-invariant spectrum: the amplitude at horizon-crossing grows with mass scale M as $\delta\rho/\rho \sim M^{m_{pl}^2/48\pi f^2}$. Thus, the primordial power spectrum (at fixed time) is a power law, $|\delta_k|^2 \sim k^n$, with spectral index $n = 1 - (m_{pl}^2/8\pi f^2)$. The extra power on large scales (compared to the scale-invariant $n = 1$ spectrum) may have important implications for large-scale structure [12].

4) *Reheating.* At the end of the SR regime, the field ϕ oscillates about the minimum of the potential, and gives rise to particle and entropy production. The decay of ϕ into fermions and gauge bosons reheats the universe to a temperature $T_{RH} = (45/4\pi^3 g_*)^{1/4}\sqrt{\Gamma m_{pl}}$, where g_* is the number of relativistic degrees of freedom. On dimensional grounds, the decay rate is $\Gamma \simeq g^2 m_\phi^3/f^2 = g^2\Lambda^6/f^5$, where g is an effective coupling constant. (For example, in the original axion model [5,6], $g \propto \alpha_{EM}$ for

two-photon decay, and $g^2 \propto (m_\psi/m_\phi)^2$ for decays to light fermions ψ.) For $f = m_{pl}$ and $g_* = 10^3$, we find $T_{RH} = 10^8 g$ GeV, too low for conventional GUT baryogenesis, but high enough if baryogenesis takes place at the electroweak scale. Alternatively, the baryon asymmetry can be produced directly during reheating through baryon-violating decays of ϕ or its decay products. The resulting baryon-to-entropy ratio is $n_B/s \simeq \epsilon T_{RH}/m_\phi \sim \epsilon g \Lambda/f \sim 10^{-4} \epsilon g$, where ϵ is the CP-violating parameter; provided $\epsilon g \gtrsim 10^{-6}$, the observed asymmetry can be generated.

In conclusion, a pseudo-Nambu-Goldstone boson, $e.g.$, a heavy (non-QCD) axion, with a potential that arises naturally from particle physics models, can lead to successful inflation if the global symmetry breaking scale $f \simeq m_{pl}$ and $\Lambda \simeq m_{GUT}$.

I would like to thank my collaborators Angela Olinto and Katherine Freese. We also thank Mary K. Gaillard for useful discussions during the Workshop. This work was supported by the DOE and by NASA Grant No. NAGW-1340 at Fermilab.

REFERENCES

[1] A. H. Guth, *Phys. Rev.* D **23**, 347 (1981).

[2] A. D. Linde, *Phys. Lett.* **108 B**, 389 (1982); A. Albrecht and P. J. Steinhardt, *Phys. Rev. Lett.* **48**, 1220 (1982).

[3] See, *e.g.*, D. Salopek, J. Bond, J. Bardeen, *Phys. Rev. D* **40**, 1753 (1989).

[4] For a review, see, *e.g.*, K. Olive, *Phys. Repts.* **190**, 307 (1990).

[5] H. Quinn and R. Peccei, *Phys. Rev. Lett.* **38**, 1440 (1977); S. Weinberg, *Phys. Rev. Lett.* **40**, 223 (1978); F. Wilczek, *Phys. Rev. Lett.* **40**, 279 (1978).

[6] J. E. Kim, *Phys. Rev. Lett.* **43**, 103 (1979); M. Dine, W. Fischler, and M. Srednicki, *Phys. Lett.* **104 B**, 199 (1981); M. Wise, H. Georgi, and S. L. Glashow, *Phys. Rev. Lett.* **47**, 402 (1981).

[7] K. Freese, J. Frieman, and A. Olinto, *Phys. Rev. Lett.* **65**, 3233 (1990).

[8] J. P. Derendinger, L. E. Ibanez, H. P. Nilles, *Phys. Lett.* **155B**, 65 (1985); M. Dine, R. Rohm, N. Seiberg, and E. Witten, *Phys. Lett.* **156 B**, 55 (1985).

[9] E. Witten, *Phys. Lett.* **149B**, 351 (1984); K. Choi and J. E. Kim, *Phys. Lett.* **154B**, 393 (1985). This possibility was briefly considered and rejected by P. Binetruy and M. K. Gaillard, *Phys. Rev.* **D34**, 3069 (1986), but they did not explore the range of allowed parameters in detail.

[10] A. D. Linde, *Phys. Lett.* **129 B**, 177 (1983).

[11] A. H. Guth and S.-Y. Pi, *Phys. Rev. Lett.* **49**, 1110 (1982); S. W. Hawking, *Phys. Lett.* **115B**, 295 (1982); A. A. Starobinskii, *Phys. Lett.* **117B**, 175 (1982); J. Bardeen, P. Steinhardt, and M.S. Turner, *Phys. Rev. D* **28**, 679 (1983);

[12] N. Vittorio, S. Mattarese, and F. Lucchin, *Ap. J.* **328**, 69 (1988).

Is Inflation Observationally "Fine-Tuned"?

BHARAT RATRA

Theoretical Astrophysics and Theoretical Physics,
California Institute of Technology, Pasadena, CA 91125

The small spatial inhomogeneities that are assumed as high redshift initial conditions in the gravitational instability scenario for the formation of cosmological structure might be the remnants of quantum fluctuations during a very early epoch of inflation (a related proposal may be found in Ref. [1] and an early computation is described in Ref. [2]).

One of the first exponential expansion inflation models in which this mechanism was studied is the quartic potential scalar field dominated model, with the scalar field effective potential

$$V(\Phi) = V_0 - \frac{1}{4}\lambda\Phi^4 \tag{1}$$

where V_0 and λ are constant parameters. In this model it is found that the observational upper bound on the late time, large-scale, adiabatic energy density irregularity power spectrum strongly constrains the value of the coupling constant, $\lambda \lesssim 10^{-10}$, [3-6] — i.e., the model needs to be "fine-tuned" to not disagree with large-scale observational data. Over the last half a dozen or so years this result has motivated the search for inflation models that are not observationally "fine-tuned" (some of which might seem to be quite contrived).

We have analyzed the generation and evolution of spatial irregularities in a number of simple scalar field dominated inflation modified hot big bang models, [7-10]. We find that large-scale observational data does not unduly constrain the parameters of these models; in particular, the parameter that plays the role of the coupling constant in these models does not have to be exceedingly small. Given the discussion of the previous paragraph, this result is somewhat surprising, so we briefly summarize why we believe that it is premature to claim that the inflation scenario of the very early universe is observationally "fine-tuned" (a more complete discussion may be found in Ref. [11]).

The reason for the difference in the conclusions of Refs. [3-6] and of Ref. [8] (where the quartic potential inflation model is studied) is the difference in the prescriptions used to implement "reheating", [11]. In the earlier analyses, the approximate expression for the potential during inflation, eqn. (1), is used to determine when inflation ends (this is taken to be the time when the "slow-rolling" approximation breaks down). In our analysis eqn. (1) only holds during inflation and does not determine when inflation ends (which is when the energy density drops to an appropriate value). With the "reheating" prescription of Refs. [3-6] the scalar field takes a time of the order of the Hubble time (at this epoch) to change from the value it has at the end of inflation to the value it has at the minimum of the complete potential — this is the slow "reheating" prescription. With our "reheating" prescription the scalar field reaches the minimum of the potential in a time much shorter than the Hubble time — this is the rapid "reheating" prescription. Using the rapid "reheating" prescription one finds that the quartic potential inflation model cannot be unduly constrained by large-scale observational data, [8,11], in part, because the late time power spectrum depends on both λ and the value of the Hubble parameter at "reheating" (as well as one other free parameter, [8]). With the slow "reheating" prescription the power spectrum only depends logarithmically on the value of the Hubble parameter at "reheating", so the one parameter that it is quite sensitive to, λ, is strongly constrained by the observations.

Only in the case of the quartic potential inflation model does the slow "reheating" prescription result in a late time power spectrum that is not very sensitive to the value of the Hubble parameter at "reheating". If one considers an effective potential, [10],

$$V(\Phi) = M^4 - \frac{1}{4}\lambda M^{4-\alpha}\Phi^\alpha \qquad (2)$$

where M is the energy scale of exponential expansion inflation and α is a parameter, one finds that the $\alpha = 4$ case is nongeneric (in that the dependence of the late time power spectrum on M is only logarithmic). More specifically, the dominant λ and M dependence of the late time baryon dominated epoch power spectrum of large-scale energy density irregularities is, [10],

$$(\lambda)^{2/(\alpha-2)} \left(\frac{M}{m_p}\right)^{2(\alpha-4)/(\alpha-2)}, \qquad (3)$$

where m_p is the Planck mass. (In this model the slow "reheating" prescription may be used only if $\alpha > 2$, [10].) If $\alpha \neq 4$, and not too large, the power spectrum is roughly equally sensitive to λ and M, while the dominant M dependence drops out when $\alpha = 4$. Also if α is sufficiently large the late time power spectrum is almost independent of λ, so observational data cannot require that λ be extremely small (for sufficiently large α the slow "reheating" prescription is somewhat similar to the rapid "reheating" prescription, [10]). If one wishes to have a λ that is not very much smaller than unity (e.g., $\lambda \gtrsim 10^{-2}$) then α needs to be moderately large (e.g., $\alpha \sim 8$) and M needs to be somewhat smaller than is usually considered (e.g., $M \gtrsim 10^{10}$ GeV), [10].

In conclusion we emphasize that it is premature to claim that the inflation scenario of the very early universe is observationally "fine-tuned". We should, however, also emphasize that neither of the "reheating" prescriptions used have been shown to follow from the microphysics of "reheating", [11].

I acknowledge useful discussions with J. Preskill. This work was supported in part by the NSF, grant AST84-51725, by the DOE, contract DE-AC03-81ER40050, and by the California Institute of Technology.

References

1. K. Sato, Mon. Not. R. Astr. Soc. **195**, 467 (1981).

2. V. F. Mukhanov and G. V. Chibisov, JETP Lett. **33**, 532 (1981).

3. S. W. Hawking, Phys. Lett. **115B**, 295 (1982).

4. A. A. Starobinsky, Phys. Lett. **117B**, 175 (1982).

5. A. H. Guth and S.-Y. Pi, Phys. Rev. Lett. **49**, 1110 (1982).

6. J. M. Bardeen, P. J. Steinhardt and M. S. Turner, Phys. Rev. **D28**, 679 (1983).

7. B. Ratra, Princeton preprint PUPT-1102B (1989).

8. B. Ratra, Caltech preprint GRP-217/CALT-68-1594 (1989).

9. B. Ratra, Caltech preprint GRP-229/CALT-68-1666 (1990).

10. B. Ratra, Caltech preprint GRP-243/CALT-68-1702 (1990).

11. B. Ratra, Caltech preprint GRP-242/CALT-68-1701 (1990).

Constraints of Axions from White Dwarf Cooling

Jin   Wang
Department of Astrophysics
University of Illinois
Urbana, IL 61801

Abstract

    In this paper, we perform numerical calculations on the white dwarf
colling process. We found that it provides a constraint on the axion mass. The
upper bound is 0.01 eV.

White dwarfs are the final state of the evolution of stars with initial masses up to a few solar mass (typically less than 8 solar mass). The white dwarf itself is in hydrostatic equilibrium without need of nuclear burning because it is supported by the electric degenerate pressure so that the hydrostatic and thermal properties are largely decoupled. The radius of a white dwarf decreases with increasing mass because in order to support the extra weight, the electrons must be squeezed into higher momentum states. If they are non relativistic, the relation between radius and mass is given [1] [2]:

$$R = 8880 \text{ km } (M_0/M)^{\frac{1}{3}} (\tfrac{2}{\mu e})^{5/3} \qquad (1)$$

If the white dwarf mass becomes so large and radius so small that degenerate electrons become relativistic, there exists no stable configuration. The possible masses of white dwarf are below so called Chandrasekar limit [1] [2]:

$$M_{ch} = 1.457 \, M_0 \, (\tfrac{2}{\mu e})^2 \qquad (2)$$

Observationally it turns out that white dwarf mass distribution is strongly peaked near M = 0.6 solar mass. Since stars of masses up to a few solar mass are believed to become white dwarfs, the excess mass is lost before reaching the white dwarf stage. Most of the mass is ejected when the stars ascends the asymtotic ginat branch, before collapsing to white dwarf. The ejected material forms a planetary nebula, so that the central stars of planetary nebula are identified with nascent white dwarfs. The rate of white dwarf formation as inferred from luminosity function is in good agreement with the observed formation rate of planetary nebulae within the statistical and systematic uncertainties of a factor of 2 [3].

The hottest and brightest white dwarf have a luminosity of $L \sim 10^{-1} L_\odot$, for faintest white dwarf it is $4 * 10^{-5} L_\odot$. They are generally faint. This implies that they can be observed only in the immediate solar neibourhood, typically out to $\sim$ 100 pc for bright white dwarfs. The vertical scale height of the galactic disk of $\sim$ 250 pc [4] is much larger, the observed white dwarfs essentially fill a volume around the Sun. The total density is $\sim 10^{-2} pc^{-3}$. The observed luminosity function is characterized by 3 important features (see Figure 1) its slope which charateriaes the form of the cooling law, its amplitude, which charaterizes the cooling time and white dwarf birth rate, and its sudden break at $log(L/L_\odot) = -4.7$ which charaterizes the beginning of white dwarf formation, i.e., the oldest white dwarfs have not yet reached lower luminosity. From this break we can derive an age for galactic disk of 9.3 ± 2.0 Gyr [5]. These are observational constraints we are going to use for the axions mass upper bound.

Axions were suggested to solve strong CP problem [6]. They are also candidates of dark matter. They interact with matter very weakly so whence they are generated in the hot enviroment such as interior of the star, they normally excape form the star without interacting with normal matter. The net effect of the axions is to carry away the energy from the star, so they are similar to neutrinos as energy dissiapation or carriers (carring away energy from the star) of the star. Since they carry away the energy, they normally accelerate stellar evolution process therefore shorten the life time of the phases of stellar evolution, reduce the number of the stars we can observe (since the number of the stars in particlular phase we observe now is proportional to life time of that phase of the star. It is in this sense we use observation facts of of white dwarf to get the constraint of parameters of the axions. (For axions the only parameter is the axion mass).

A low mass star becomes a white dwarf when its nuclear energy resources have been exhausted: it shines its residual thermal energy. The evolution of white dwarf must be viewed as a cooling process. For hot white dwarfs, the thermal energy is largely stored in the non-degenerate nuclei. At sufficient low temperature, ideal gas law breaks down, eventually nuclei arrange themselves in a crystal lattice. For the situationn we consider, the density of the white dwarfs is small enough that even oldest white dwarfs have not had enough time to reach crystalization phase. Hence ideal gas law is a good approximation.

Now we will derive the bounds of the axion mass, the emission rates of axios were first discussed in [7] and later corrected for ion correlation effects [8]. The emission rates we use here is

$$\epsilon_a = 3.3 \times 10^{23} \, z^2 A^{-1} T_7^4 \alpha_a \qquad (3)$$

where $\alpha_a = 1.6 \times 10^{-23} (m_a/ev)^2$, this is a bremstrulung process, and the axions are DFSZ axions, it is believed that for strongly coupled degenerate plasma such as white dwarf, bremstrulung process dominates the emission rates.

We use the numerical code developed by Kutter, Sparks and Starfield [9] to do the model calculation. Substitute emission rate for axions into energy conservation equation

$$\frac{dL}{dm} = \epsilon_{nuc} + \epsilon_\nu + \epsilon_{grav} + \epsilon_a$$

Note that both ϵ_ν and ϵ_a has minus sign because they loose energy. We evolve the white dwarf from begining of absolute magnitude $M_\nu = 6.5$ to $M_\nu = 14.7$ which covers basically the whole cooling track. The results are shown in Figures. It is in terms of absolute magnitude (luminosity) of white dwarf vs time. We list the standard cooling curve and cooling curve with axions for white dwarf mass 1.0 solar mass.

We first notice that the life time of white dwarf become shorter when the axions emission is included, although the shape of the cooling curves remaints almost unchanged. For standard cooling process, life time of the white dwarf is 9.5×10^9 yrs. For axion mass $m_a \sim$ 0.1 eV, we see that life time $6.\times10^9$ yr. For axion mass 0.01 eV, the life time is 8.9×10^9 yrs . The statistical and systematic uncertainties is factor 2. We see that axion mass less than .1 eV is a good constraint in order not to conflict with observation.

In order to make the contact with observation data we use present white dwarf birth rate in solar neibourhood which is of about 10 yr pc , we assume that it has remained constant during the recent galactic history. (In the past 10^9 years). So $N_p = \dot{N} \times t$. For the present number of white dwarf in a volume of 1000 pc^3 with luminosities between L_1 and L_2 . Also for m = 1.0 solar mass, $m_a \sim 0.1$eV, in the interval between M_1= 10. and M_2= 11. , we get:

$$N_{pred} = 0.034$$

The observational value is 0.08-0.12 [10].
For axion mass m = 0.01 eV, we have:

$$N_{pred} = 0.082$$

So it is safe to say that m_a= 0.01 eV is a conservative upper bound for the axion mass.

We notice that the numerical calculation we have done here is quite close to the analytical estimated that Raffelt got [7].

We also notice that SN 1987A gives more stringent bound on axion mass [11] (upper bound is 10^{-3} eV instead of 10^{-2} eV we get here). But we believe WD cooling result is more reliable since we understand WD physics and related

observation much better than Supernova. WD cooling is hydrostatic process, while SN is hydrodynamic event, there is no satisfied model for supernova yet.
In this paper we perform a numerical calculation on white dwarf cooling process with axions, the constraints for the axion mass we get is 0.01 eV.

References:
1. Shapiro, S. L., and Teukolsky, S. A., 'Black Holes, White Dwarfs and Neutron Stars' (Wiley, New York, 1983).
2. Raffelt, G., 'Stars as Particle Physics Laboratories', LBL preprint 1990.
3. Liebert, J., Dahn, C. C., and Monet, D. G., Ap. J. 332(1988)891.
4. Fleming, T. A., Liebert, J., and Green, R. F., Ap. J. 308(1986)176.
5. Winget, D.E., Hansen, C.J., Liebert, J., Van Horn, H. M., Fontaine, G., Nather, T. E., Kepler, S. O., and Lamb, D. Q., Ap. J. 315(1987)L77.
6. Peccei, R. D., and Quinn, H. R., Phys. Rev. Lett. 38(1977)1440.
7. Raffelt, G., Phys Lett. 166B(1986)402.
8. Nakagawa, M., Kohyama, Y., and Itoh, N., Ap. J. 322(1987)291.
9. Kutter, G. S., and Sparks, W. M., Ap. J. 175(1972)407.
10. Liebert, J., Ann. Rev. Astron. Astrophysics. 18 (1980)363.
   van Horn, H. M., in 'White Dwarfs, IAU Symp. No.42', ed. Luyten, W. J., (Reidel, D., Dordrecht, Holland, 1971).
11. Burrows, A., Turner, M. S., Brinkmann, R. P., Phys. Rev. D39(1989)1020.
   Mayle, R., Wilson, J. R., Ellis, J., Olive, K., Schramm, D. N., and Steigman, G., Phys. Lett. 203B(1988)188, 219B(1989)515.

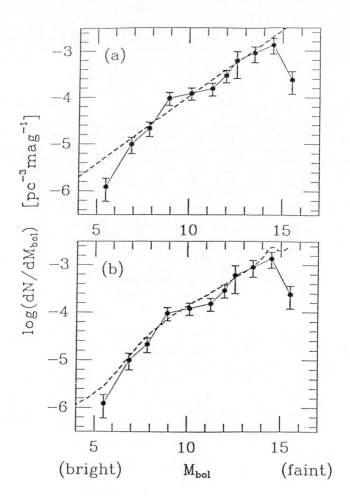

FIGURE 1

Observed luminosity function of white dwarfs as listed in   [1] (a) The dashed line represents Mestel's cooling law with an assumed constant white dwarf birthrate of $B = 10^{-3}\,\mathrm{pc}^{-3}\,\mathrm{Gyr}^{-1}$. (b) The dashed line was obtained from the numerical cooling curve of [.1] for a $0.6\,M_\odot$ white dwarf, assuming the same constant birthrate. Standard neutrino cooling was included in this calculation.

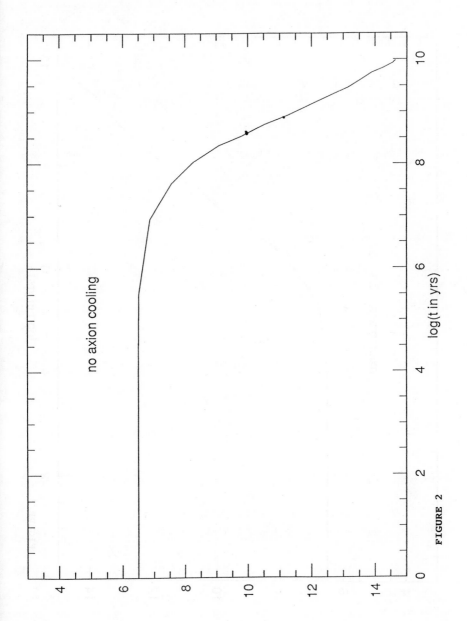

no axion cooling

log(t in yrs)

FIGURE 2

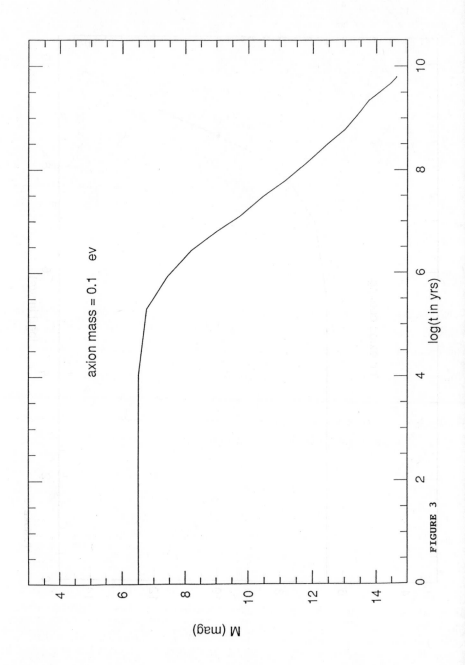

axion mass = 0.1   ev

M (mag)

log(t in yrs)

FIGURE 3

153

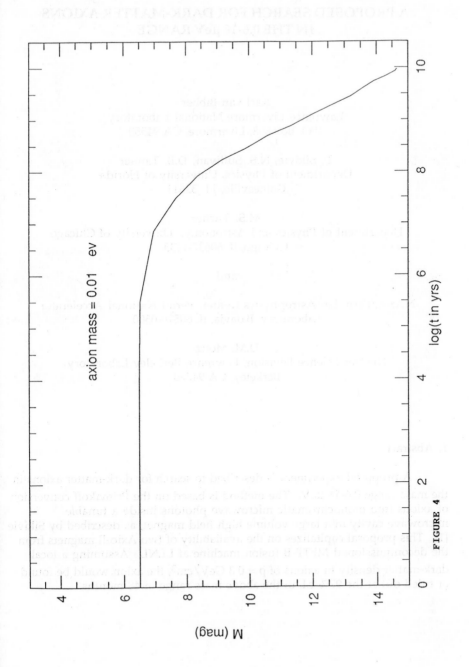

FIGURE 4

A PROPOSED SEARCH FOR DARK-MATTER AXIONS
IN THE 0.6-16 μeV RANGE

Karl van Bibber
Lawrence Livermore National Laboratory
P.O. Box 808, Livermore, CA 94550

P. Sikivie, N.S. Sullivan, D.B. Tanner
Department of Physics, University of Florida
Gainesville, FL 32611

M.S. Turner
Department of Physics and Astronomy, University of Chicago
Chicago, IL 60637-1433

and

NASA/Fermilab Astrophysics Center, Fermi National Accelerator
Laboratory, Batavia, IL 60510-0500

D.M. Moltz
Nuclear Science Division, Lawrence Berkeley Laboratory
Berkeley, CA 94720

1. Abstract

A proposed experiment is described to search for dark-matter axions in the mass range 0.6-16 μeV. The method is based on the Primakoff conversion of axions into monochromatic microwave photons inside a tunable microwave cavity in a large volume high field magnet, as described by Sikivie [1]. This proposal capitalizes on the availability of two Axicell magnets from the decommissioned MFTF-B fusion machine at LLNL. Assuming a local dark-matter density in axions of $\rho = 0.3$ GeV/cm^3, the axion would be found or ruled out at the 97% c.l. in the above mass range in 48 months.

2. Introduction

Generically, the mass of the axion and all its couplings to matter and radiation are proportional to f_A^{-1}. This f_A is the symmetry breaking scale of the Peccei-Quinn symmetry, postulated to provide a mechanism for enforcing CP conservation in the strong interaction. As laboratory experiments and arguments based on stellar evolution constrain the mass of the axion to be less than 10^{-3} eV, it was long thought that such axions, even if abundant from their production in the early universe, would be rendered effectively invisible by the weakness of their couplings. Sikivie [1] showed that this was not necessarily so, and proposed a simple and elegant technique for their conversion and detection as microwave photons in a tunable microwave cavity in a sufficiently strong magnetic field. (A detailed description of this technique is found elsewhere in this volume, as well as results from two pilot experiments [2].) The coherent production of a cold Bose-condensed gas of axions in the early universe has the feature that their contribution to the total mass of the universe goes roughly inversely to the axion mass. From this it is generally argued that closure density would correspond to an axion mass of $m_a < 10$ μeV (subject to an uncertainty estimated to be a factor of 10 either way), which provides a loose lower bound on the axion mass. In inflationary models it is possible that the axion mass could be even smaller than 10^{-6} eV without violating the cosmological bound [3]. Furthermore there is a controversy about whether the overclosure bound on m_a may not be dominated by the mechanism of axion production by radiation from axionic cosmic strings. This is a somewhat technical discussion still under dispute, and only relevant in the case of a non-inflationary scenario or one in which the PQ symmetry breaking would occur after inflation [4]. From an experimental viewpoint one should take the open window to be $10^{-(3-6)}$ eV, and extend the search lower in mass if one can. Two reviews of the status of the axion with brief pedagogical introductions have recently appeared [5,6].

3. Description of the Detector

As mentioned, the proposed experiment is a fairly straightforward scale-up of two pilot efforts (Rochester/BNL/FNAL (RBF) [7], and the University of Florida (UF) [2,8-10]), the discussion of which does not need to be reproduced here. It is clear that the power sensitivity of the pilot experiments, in the mass range which has been measured, falls short of the expected signal by approximately 10^3. The Florida experiment for example had a sensitive volume of 0.008 m^3 and an average field of 7.5 T, and thus $B^2V = 0.45$ T^2m^3. The principal advantage of the proposed experiment will be the very much larger magnetic volume.

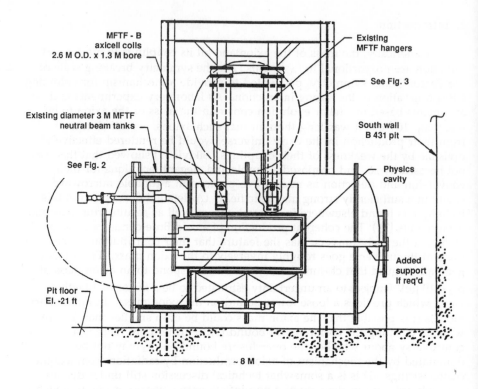

Figure 1. Conceptual design of the axion experiment vessel, showing the magnets, current leads, re-entrant well, microwave cavity, and couplings between the outside world and the cavity. Note the vacuum partition between the magnet and the cavity environments.

3.1. The Axicell magnets

The magnetic volume for this experiment will consist of two Axicell magnets to be removed from the decommissioned MFTF-B at Livermore, and placed flush against one another. The Axicell magnets consist of Nb-Ti conductor, layer-wound on an iron spool-piece, of almost 2000 turns each.

The design operating current is 4238 A, or 8.37×10^6 Amp-turns per magnet. Each Axicell magnet has a clear-bore diameter of 1.34 m and is 1.14 m in depth. The total inductance of the pair is 15 Henries, and the total stored energy is 160 MJ at operating current.

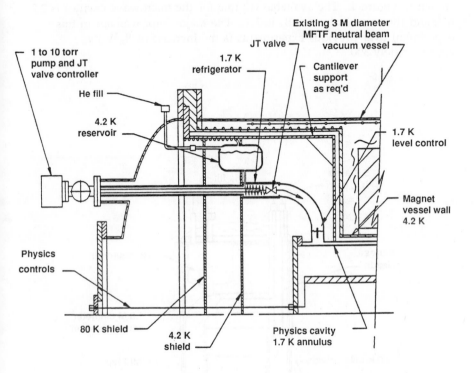

Figure 2. Detail of the refrigerator used for cooling the cavity and cryogenic amplifier.

The magnets will be mounted in a tank (also surplus from the MFTF-B program, a neutral beam injector vessel) with a re-entrant well separating the vacua for the magnets and the copper cavities (Figure 1). This will allow entry and work of short duration on the cavities without breaking vacuum or thermally cycling the magnets. Between the inner radius of the magnet and the cavity packages is a thin annular 1.7K refrigerator and shield (Figure 2). The coil leads will be modified to work in persistent mode with the switch in the lead stack (Figure 3). The cooling will be performed by a helium refrigerator next to the experiment, utilizing a nearby facility to purify, compress and return the boil-off He. An isometric view of the experiment is shown in Figure 4. The available volume for the microwave cavities is 2.7 m^3, and the r.m.s. value for B_z is 7 T. The major improvement in this experiment over previous experiments is the increase of B_z^2V by approximately 300.

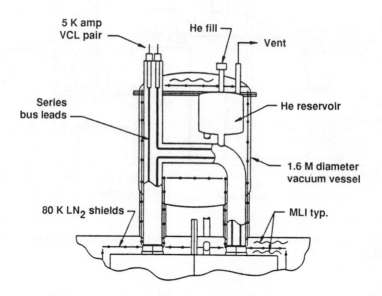

Figure 3. Detail of the current leads and He resevoir tank located above the main tank.

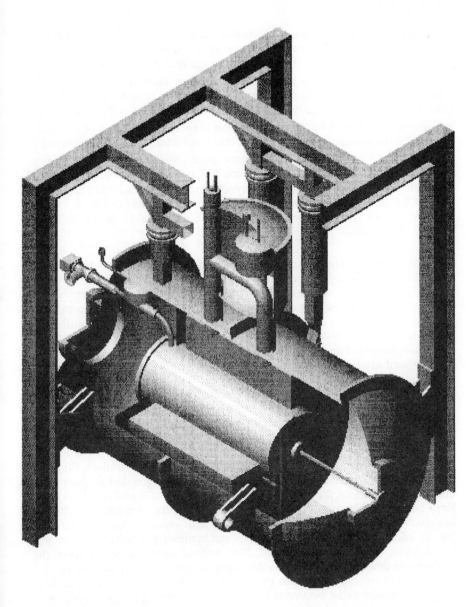

Figure 4. Isometric view of the axion detector.

3.2. The copper cavities

It must be remembered that the diameter of the magnetic volume limits the maximum diameter of a single cavity that may be placed inside, and thus the lowest end of the frequency range that may be searched. The frequency of a right circular cylinder in the TM_{010} mode is given by $v = 115/r$, where n is the frequency in MHz, and r is the radius of the cavity in meters. (The TM_{010} mode is the only really feasible one, and thus will be used exclusively in this work.) As $hv = mc^2$, the mass range that can be covered is limited by the geometry of the cavities that can be fabricated and inserted into the magnet. (A frequency of 1 GHz corresponds to 4.126 μeV, *etc.*) Cavities may be tuned downward 20% or upward 30% or so by radial displacement of a dielectric rod or metal post respectively without too great a loss in C^2Q [9]. As in the RBF and UF experiments, the cavities must be made of OFHC copper for optimum Q. Furthermore adequate care must be taken in the uniformity of their cross-section to avoid mode localization.

In order to cover a continuous range in masses from 0.6 - 16 μeV (148 - 3900 MHz), nine arrays of cavities are planned. All but the first involve multiple cavities whose outputs are combined through standard power splitter/dividers into a single output. Power splitters/combiners combine the outputs of the cavities without introducing cross-coupling between them. (The nature of the power splitter/dividers available requires that the number of cavities be 2^n.) The first three use essentially the entire available magnetic volume, and are built on a right circular cylinder: (1) undivided; (2) divided into 2 cells of semicircular cross-section; and (3) divided into 4 cells of quarter-circular cross-section. The frequencies of the TM_{010} modes for a semicircular cross-sectional cell, and a quarter-circular cross-sectional cell are $f' = 1.56f$, and $f'' = 2.1f$ respectively, where f is that for the cylinder without partition (see Figure 5). The second set of three are built upon 8 cylinders -- 7 cylinders surrounding 1 in the center, each of radius r = 0.30R, where R is the full radius of the available magnetic volume: (4) undivided cylinders; (5) halved; (6) quartered. The third set of three are based on hexagonal-close-packed cells undivided in cross section: (7) r/R = 0.11; (8) r/R = 0.079; and (9) r/R = 0.057. Here however, it becomes necessary to longitudinally segment the cells to minimize problems associated with unfavorable aspect ratios (length/radius) [9]. In addition to making the translational symmetry requirement easier to fulfill (to avoid mode localization), longitudinal segmentation relieves the problem of mode-crowding. It is desirable to minimize the number of crossings of intruder TE modes through the TM_{010} mode. Whenever this happens, mode-mixing occurs and a notch in the tuning range is lost. In order to keep the total lost tuning range to <1%, we choose two-fold segmentation for array (7), and fourfold for (8),(9). With these arrays, and use of dielectric and metal tuning rods, it is possible to cover the proposed mass range continuously (Table I).

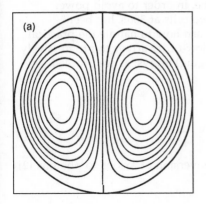
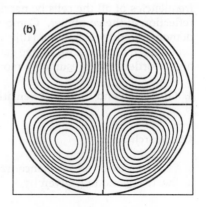

Figure 5. Contours of the TM_{010} mode electric field strength for the cases of (a) semicircular cross-section cavities, and (b) quarter-circular cross-section cavities, made by longitudinal partitioning of a right circular cylindrical cavity.

Power combining of multiple cells involves some care. Prior to each data collection run, one cell (the master) must first be stepped in frequency, and the others (slaves) must be tuned in turn to the frequency of the first. Also it is important that the signal cables of all the cells to the power combiner be of the same length, and the coupling Q_h and the central

Number of cavities	Arrangement	r/R	Packing fraction	Frequency MHz	Tuning Range MHz
1	O	1	1	180	148–233
2	⊖	1	1	281	233–325
4	⊕	1	1	378	325–488
8	O×8	0.30	0.73	595	488–770
16	⊖×8	0.30	0.73	928	770–1070
32	⊕×8	0.30	0.73	1250	1070–1350
128	o×64 × 2	0.11	0.76	1650	1350–1880
512	o×128 × 4	0.079	0.79	2290	1880–2620
1024	o×256 × 4	0.057	0.82	3190	2620–3900

Table I. Parameters of the nine cavity arrays proposed.

frequencies of all the cells be nearly the same, in order to avoid power reflection at the combiner. Combining of two cells at 300K has been demonstrated by Hagmann [10]. We will learn how to handle the practical problems that may arise as one starts from low-order arrays to progressively higher order (and presumably more challenging) arrays, as we proceed.

4. Operation and Proposed Sensitivity

4.1. Search Rate.

The power on resonance from the conversion of axions to photons [2] in the TM_{nl0} mode is given by

$$P_{nl} = \left(\frac{\alpha}{\pi} g\gamma \frac{1}{f_A}\right)^2 V B_z^2 \rho_a C_{nl} \frac{1}{m_a} \text{Min}(Q_L, Q_a) \qquad (1)$$

where C_{nl} is a mode-dependent form factor, Q_L is the loaded quality factor of the cavity and $Q_a \sim 10^6$ is the "quality factor" of the galactic halo axions, *i.e.* the ratio of their energy to their energy spread. This can be rewritten conveniently in terms of benchmark values for all parameters as follows

$$P_{nl} = 3\times10^{-26} \text{Watt} \left(\frac{V}{3 \text{ m}^3}\right)\left(\frac{B_z}{7 \text{ Tesla}}\right)^2 C_{nl} \left(\frac{g\gamma}{0.36}\right)^2 \cdot$$
$$\cdot \left(\frac{\rho_a}{5\times10^{-25} \text{ g/cm}^3}\right)\left(\frac{m_a}{2\pi(1\text{GHz})}\right) \text{Min}(Q_L, Q_a) . \qquad (2)$$

The benchmark value for the axion-photon coupling, 0.36, applies to the simplest axion models, for example that of Dine, Fischler, Srednicki and Zhitnitskii (DFSZ) [11]. In other models, $g\gamma$ can be larger or even smaller.

The search rate (df/dt) involves not only the expected conversion power, but also upon the total noise (the sum of the physical temperature of the cavity and the noise temperature of the microwave receiver), the desired signal-to-noise ratio, s/n, and the cavity Q. The search rate is given by

$$\frac{df}{dt} = \frac{55 \text{ GHz}}{\text{year}} \left(\frac{4}{s/n}\right)^2\left(\frac{V}{3 \text{ m}^3}\right)^2\left(\frac{B_z}{7 \text{ Tesla}}\right)^4\left(\frac{g\gamma}{0.36}\right)^4 \cdot$$
$$\cdot C^2 \left(\frac{\rho_a}{5\times10^{-25}\text{g/cm}^3}\right)\left(\frac{5 \text{ K}}{T_n}\right)^2\left(\frac{f}{1 \text{ GHz}}\right)^2\left(\frac{Q_w}{Q_a}\right) . \qquad (3)$$

The dependence on Q_w is particular to the case of $Q_w < 3Q_a$, which will always pertain in the present experiment. Here Q_w (w = "wall") is the quality factor that would be achieved if absorption in the walls were the only mechanism for removing energy from the cavity. The other contribution is due to the output coupler, Q_h (h = "hole"), and the loaded Q of the cavity, Q_L is given by $Q_L^{-1} = Q_w^{-1} + Q_h^{-1}$. In the limit where $Q_w < 3Q_a$, the search rate is maximized by the choice of $Q_h = Q_w/2$. The intrinsic cavity quality factors have been calculated for oxygen-free high conductivity copper in the anomalous skin-depth limit.

The cavity bandwidth exceeds the anticipated width of the axion line throughout the search range by characteristically an order of magnitude or more. Thus at each central frequency a fast Fourier transform is performed during the integration time of the run, and a power spectrum collected in bins of width approximating the expected width of the axion line. Each spectrum will be searched for 2σ peaks in single bins and combinations of neighboring bins. If a 2σ peak is found, another set of spectra will be taken and averaged with the first. If the peak remains statistically significant this process will be repeated up to a maximum of 5 times, after which the peak will be flagged for later investigation. The total search rate therefore must take into account the rescan frequency (80% of the spectra are expected to have positive peaks exceeding 2σ and will be repeated at least once), and further the degree to which adjacent spectra are overlapped. A more explicit rendering of eqn. (3) is given below, assuming the DFSZ value of g_γ, a local halo density of 5×10^{-25} g/cm^3 (assumed to be entirely axionic), $Q_a = 10^6$, and a volume of 2.8 m^3 available for the cavities

$$\frac{1}{f}\frac{df}{dt} = \frac{4.9\%}{week}\left(\frac{4}{s/n}\right)^2\left(\frac{B_z}{7\ Tesla}\right)^4 P^2 C^2 \left(\frac{5\ K}{T_n}\right)^2 \left(\frac{f}{1\ GHz}\right)\left(\frac{Q_w}{10^5}\right)\left(\frac{1}{N_{rescan}}\right)\left(\frac{\Delta f}{\delta f}\right). \quad (4)$$

In eqn. (4), P is the packing fraction of cavities in the volume, N_{rescan} is the average number of times a given region is rescanned on account of persistent 2σ peaks, Δf is the frequency interval between successive spectra, and $\delta f = f/Q_L$ is the cavity bandwidth. The overlapping of adjacent spectra will be approximately 20%. Table II presents the tuning ranges, expected total noise temperature, quality factors, fractional search rate and total run time for each cavity array. The total time for the experimental program will also include some down time, both for changing of cavity arrays, and also for the tuning of multicavity arrays prior to each run. The scan rate is between 1.3 and 2.1% per week. To achieve a signal-to-noise ratio of 4 for DFSZ axion, the total run time would be under 4 years. If (s/n) were to be raised to 5, the time would be 6 years; if reduced to 3, it would be 2.1 years.

Number of cavities	Arrangement	Tuning Range MHz	T_n K	Q_w	$\frac{1}{f}\frac{df}{dt}$ %/wk	Time weeks
1	◯	148–233	6	530000	1.3	36
2	⊖	233–325	6	370000	1.4	23
4	⊕	325–488	6	300000	1.9	24
8	◯×8	488–770	5	280000	1.8	27
16	⊖×8	770–1070	5	190000	1.9	17
32	⊕×8	1070–1350	5	160000	2.1	11
128	○×64 × 2	1350–1880	6	150000	2.0	16
512	○×128 × 4	1880–2620	7	130000	1.8	18
1024	○×256 × 4	2620–3900	9	100000	1.7	24

Table II. Search rate of the proposed experiment. Rate and time estimates were made with under the condition that a DFSZ axion would be seen with (s/n) = 4, and presumes the intrinsic cavity quality factors and effective total noise temperatures as shown.

4.2. Summary of proposed experiment

The scope of the proposed experimental search is shown in Figure 6. The ordinate is in $g_{a\gamma\gamma}^2$, where $g_{a\gamma\gamma} = \left(\frac{\alpha}{4\pi}g\frac{1}{f_A}\right)$. Shown also are results from the Rochester-Brookhaven-Fermilab, and University of Florida pilot experiments, as well as the DFSZ and KSVZ [12] model predictions between the mass and coupling. We propose to cover the region between 0.6 and 16 μeV, with a s/n = 4. A more complete description of the experiment is found in reference [13].

If the axion is successfully excluded in the region indicated, it would be very interesting to pursue the search upward in mass. As this implies cavities of smaller radii yet, it may be unwieldy to carry this out with a larger number of cavities yet in the same volume. At this point one should consider the use of Axicell insert coils to boost the field to a value of about 14 Tesla, albeit for a diameter of only 30 cm, and a total length of about 100 cm. The overall loss in B^2V of about a factor of 8 would be compensated for in the simplicity of dealing with many fewer cavities.

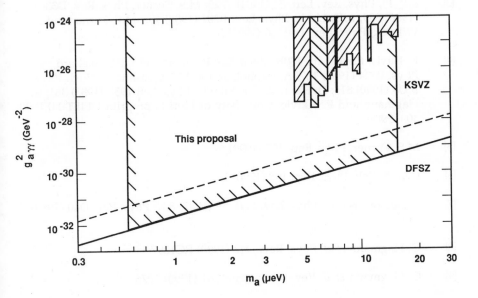

Figure 6. Experimental limits on the electromagnetic coupling $g_{a\gamma\gamma}$ of the axion. Values in the shaded areas at upper right have been excluded by the RBF and UF searches. Values equal to or greater than the DFSZ limit for an axion mass between 0.6 and 16 μeV would be ruled out by this experiment.

We would like to thank John R. Miller, Stewart Shen, Don Slack and Dick Patrick of the Applied Research Engineering Division at LLNL for their design of the magnet and cold-bore facility for this experiment. This work was performed under the auspices of the US Department of Energy at Lawrence Livermore National Laboratory under contract no. W-7405-ENG-48, at the University of Florida under grant no. FG05-86ER-40272, at the University of Chicago under grant no. FG02-90ER-40560, and at Fermilab.

5. References

[1]   P. Sikivie, Phys. Rev. Lett. 51 (1983) 1415

[2]   C. Hagmann, proceedings of this workshop and references therein (1991)

[3]   S.Y. Pi, Phys. Rev. Lett. 52 (1984) 1725; M.S. Turner, Phys. Rev. D33 (1986) 889; A.D. Linde, Phys. Lett. B201 (1988) 437; M.S. Turner and F. Wilczek, Phys. Rev. Lett. 66 (1991) 5

[4]   R. Davis, Phys. Lett. 180B (1986) 225; R. Davis and E.P.S. Shellard, Nucl. Phys. B324 (1989) 167; A. Dabholkar and J.M. Quashnock, Nucl. Phys. B333 (1990) 815; D. Harari and P. Sikivie, Phys. Lett 195B (1987) 361; C. Hagmann and P. Sikivie, University of Florida preprint UFIFT-HEP-90-30 (1990)

[5]   M.S. Turner, Phys. Rep. 197 (1990) 67

[6]   G.G. Raffelt, Phys. Rep. 198 (1990) 1

[7]   S. DePanfilis et al, Phys. Rev. Lett. 59 (1987) 839; Phys. Rev. D40 (1989) 3153

[8]   C. Hagmann et al, Phys. Rev. D42 (1990) RC1297

[9]   C. Hagmann et al, Rev. Sci. Instrum. 61 (1990) 1076

[10]   C. Hagmann, PhD thesis, University of Florida (1990)

[11]   M. Dine, W. Fischler, M. Srednicki, Phys. Lett. 104B (1981) 199; A.P. Zhitnitskii, Sov. J. Nucl. Phys. 31 (1980) 260

[12]   J. Kim, Phys. Rev. Lett. 40 (1977) 223; M.A. Shifman, A.I. Vainshtein, V.I. Zakharov, Nucl. Phys. B166 (1980) 493. In Figure 6, the dashed line is shown for the KSVZ model evaluated with a value of the electromagnetic anomaly of 0, for purely historical reasons. This corresponds to $g_\gamma = 0.97$.

[13]   P. Sikivie et al, "Experimental Search for Dark-Matter Axions in the 0.6-16 μeV Mass Range", Proposal to the US Department of Energy and the Lawrence Livermore National Laboratory (1990)

COSMIC RAYS FROM PRIMORDIAL BLACK HOLES

JANE H. MACGIBBON*

Code 665, NASA/Goddard Space Flight Center,
Greenbelt, MD 20771 USA

Primordial black holes (PBHs) may have formed from initial density perturbations in the early Universe. It is consistent that the emission from PBHs clustered in the Galactic halo could contribute significantly to the extragalactic photon and interstellar e^+, e^- and $\bar{p}$ spectra around $0.1 - 1$ GeV. If PBHs evaporate down to stable relics of order the Planck mass or bigger, in this scenario $\Omega_{relics} \simeq 1$ is also possible[1]. This work was carried out with B.J.Carr of QMWC, University of London, and B.R. Webber of University of Cambridge.

HAWKING EVAPORATION

A black hole with negligible angular momentum and charge emits particles with energy in the range $(Q, Q + dQ)$ at the rate[2]

$$d\dot{N} = \frac{\Gamma_s}{2\pi\hbar} \, dQ \, [\exp \frac{Q}{\hbar\kappa/2\pi c} - (-1)^{2s}]^{-1} \tag{1}$$

per particle degree of freedom, where s is the particle spin, Γ_s is the absorption probability and κ is the surface gravity. The black hole temperature is related to its mass M via $kT = \hbar\kappa/(2\pi c) = \hbar c^3/(8\pi GM)$. $\Gamma_s(M, Q) \longrightarrow 27M^2Q^2/\hbar^2c^6$ as $Q \longrightarrow \infty$ for all s but the $Q \longrightarrow 0$ cutoff depends on s. A $T = 0.3 - 100$ GeV hole will emit particles in the list: $e^\pm$, $\mu^\pm$, $\tau^\pm$, γ, $\nu_{e,\mu,\tau}$, q_u, q_d, q_s, q_c, gluons, q_b, q_t, $W^\pm$ and Z^0. (We omit the small graviton component since we are confining ourselves to only the experimentally verified particles and q_t). These particles will then decay after emission into the astrophysically stable $p\bar{p}$, $e^\pm$, γ and $\nu\bar{\nu}$. To simulate this, we use Bryan Webber's BIGWIG and HERWIG Monte Carlos codes[3] which describe accelerator jet events. (One can show that the decay of the PBH emission is anagolous to the decay of jets created by e^+e^- annihilation in colliders[4,5]). Figure 1 shows the instantaneous emission from a $T = 1$ GeV black hole[5]. The lepton peaks below 1 MeV come from jet neutron β-decay; the $e^\pm$, $\nu\bar{\nu}$ and γ peaks around 100 MeV come from jet $\pi^{\pm,0}$ decay; and the small peaks around 5 GeV are the direct non-decaying emission. (These last peaks were the only ones calculated by previous authors). All protons and antiprotons are jet products. The composition of the emission for $T = 0.3 - 100$ GeV is roughly 1.8% $p\bar{p}$, 19.9% $e^\pm$, 21.3% γ and 57.0% $\nu\bar{\nu}$. Power is emitted in a similar ratio. The emission is dominated by the jet pion decays. From the computer runs, we find that the total flux at these temperatures roughly scales as[5]

$$\dot{N} \simeq 1.4(\pm 0.5) \times 10^{25} \left(\frac{T}{\text{GeV}}\right)^{7/5} \sec^{-1} \tag{2}$$

THE BLACK HOLE LIFETIME

If we equate the energy carried off by the emission to a decrease in the black hole mass, we derive a mass loss rate for the hole of

* NAS/NRC Research Associate

$$\dot{M} \simeq 5.341 \times 10^{25} f(M) M^{-2} \text{ gm sec}^{-1} \tag{3}$$

from Eq. (1). $f(M)$ is normalized to 1 when only massless particles are emitted. In the Standard Glashow-Weinberg-Salam model with 90 $s = 1/2$ modes, 4 $s = 0$ modes and 24 $s = 1$ modes, we have[6] $f(M) \lesssim 15.4$ for $T \lesssim 100$ GeV. Integrating Eq. (3), the black hole lifetime is

$$\tau_{evap} \simeq 6.24 \times 10^{-27} M_i^3 f(M_i)^{-1} \text{ sec} \tag{4}$$

An $M_i \simeq 4 \times 10^{11}$ gm hole evaporates in about 1 yr. If we solve Eq. (4) using the values for the contribution to $f(M)$ per helicity state calculated by Page, Simpkins and Elster[7], the mass of a hole whose lifetime equals the age of the Universe is[6] $M_* \simeq 4.42(\pm 0.01) \times 10^{14} h^{-0.293(\pm 0.003)}$ gm for $\Omega_m = 1.0$. For PBHs forming in the Early Universe, this is then the initial mass of the holes which are just completing their evaporation today.

THE FORMATION OF PBHS

Although PBHs can form in many early Universe scenarios[8] (e.g. from phase transitions, bubble collisions, density perturbations and string collapse), here we will consider the most natural model in which PBHs form from scale-invariant Zeldovich-Harrison density perturbations[9]. The black holes are created with masses of order the horizon size at the formation epoch. This gives an initial mass spectrum for the PBHs of[10]

$$\frac{dn}{dM} = (\alpha - 2) \left(\frac{M}{M_*} \right)^{-\alpha} M_*^{-2} \Omega_{pbh} \rho_{crit} \tag{5}$$

where $\alpha = 2.5$ for formation in the radiation dominated era and Ω_{pbh} is the fraction of the Universe in $M \geq M_*$ holes in units of the critical density ρ_{crit}. The fraction of the Universe going into M holes is[10]

$$\beta(M) \simeq \epsilon(M) \exp \left(\frac{-1}{18\epsilon(M)^2} \right) \tag{6}$$

where $\epsilon(M) = <\delta_{rms}(M)^2>^{1/2}$ is the root-mean-square value of the density fluctuations on scale M. Only if ϵ is independent of M can PBHs form over an extended mass scale. $\Omega_{pbh} \lesssim 1$ corresponds to $\beta \lesssim 10^{-17}$ and $\epsilon \lesssim 0.05$ on the scales associated with the black hole masses. (This ϵ would have to be decoupled from the perturbations on galactic scales which must be of the order of $\epsilon \simeq 0.01$.)

THE EMISSION FROM THE PBH BACKGROUND

To plot the predicted particle backgrounds, we must now integrate the flux (1) over the lifetime of each hole over the initial mass distribution (5). If $dg_{jX}(Q, E)/dE$ describes the decay of particle j with energy Q into particle X with energy E (such that $\int dg_{jX}/dE \, dE = $ number of X particles created in the decay), the present number density of X particles produced by the PBH distribution becomes

$$\frac{dF_X}{dE} = 3.59 \times 10^{-2} (\alpha - 2) \Omega_{pbh} \rho_{crit} M_*^{\alpha-2} \int_{m_{pl}}^{M_*} M_i^{-\alpha} \int_{T(M_i)}^{T(m_{pl})} f(T)^{-1} T^{-4}$$
$$\cdot \sum_j \int_{E'}^{\infty} \Gamma_j(Q, T) \{ \exp \frac{Q}{kT} - (-1)^{2s_j} \}^{-1} \frac{dg_{jX}(Q, E')}{dE'} \, dQ \, dT \, dM_i \tag{7}$$

where E' is the energy immediately after decay which must be redshifted to its present value E. An important feature[5] of the jet fragmentation function $dg_{jX}(Q, E)/dE$ is that it leads to an E^{-1} slope in the $T \simeq 0.02 - 100$ GeV instantaneous spectra around $E \simeq \sqrt{T/\text{GeV}}$ GeV. When we integrate over the PBH distribution, this slope grows faster as E decreases than does the $E^{2-\alpha}$ which we would get from just considering the non-decaying component of the emission[11]. Thus, unlike the work of Page and Hawking[12] and Carr[11] which omitted the jet particle decays, we now predict an E^{-1} slope in the particle backgrounds below $E \simeq T(M_*)$. (These are the backgrounds prior to interactions with other particles and fields in the Universe). Above $E \simeq T(M_*)$, the redshift is negligible and the evaporation in the present epoch gives $dF/dE \propto E^{-3}$ for all species. Figure 2 shows[13] the particle backgrounds (prior to interactions) for $\Omega_m = 1.0$, $h_0 = 0.5$ and $M_* = 4.91 \times 10^{14}$ gm. The low energy slope is not precisely E^{-1} (since the E^{-1} is an approximation to the jet fragmentation function) but depends on the particle species.

Since PBHs should not have extraordinarily high peculiar velocities, it is natural to suppose that they cluster in Galactic halos along with other matter. In this case, the effective Ω_{pbh} for the *charged* emission is enhanced by[14]

$$\zeta' \simeq 9.7 \times 10^5 h_0^{-2} \left(\frac{\Omega_h}{0.1}\right)^{-1} \left(\frac{V_\infty}{220 \text{ kms}^{-1}}\right)^2 \left(\frac{R_h}{10 \text{ kpc}}\right)^{-2} \left(\frac{\tau_{leak}}{\tau_G}\right) \qquad (8)$$

where Ω_h is the halo density in units of ρ_{crit}, V_∞ is the asymptotic circular velocity, R_h is the halo radius, τ_{leak} is the leakage time for charged particles out of the halo and τ_G is the age of the Galaxy. $\tau_{leak} \simeq 10^{-2}\tau_G$ for $0.1 - 1$ GeV antiprotons and possibly 1% to 100% of this value for $e^\pm$. Hence ζ' may be a function of energy E and particle species and charge. When we use this approximation for ζ', we must also perform the M_i integral over M_* to M_G where M_G is the mass of the PBH just expiring as the Galaxy forms. (This approximation is discussed in detail in Refs. 13 and 14.) Figure 3 shows[13] the postgalactic $e^\pm$ and $p\bar{p}$ emission from clustered holes for $\Omega_m = 1.0$, $h_0 = 0.5$, $\tau_G = 12 \times 10^9$ yr, $M_* = 4.91 \times 10^{14}$ gm and $M_G = 2.34 \times 10^{14}$ gm. Again the spectra fall off as E^{-3} above $100 - 300$ MeV.

PARTICLE INTERACTIONS

Since we are interested in matching the γ, $e^\pm$ and $p\bar{p}$ emission to the observed backgrounds around $0.1 - 1$ GeV, we next check that the recent emission at these energies is unaffected by interactions with other particles and fields in the Galaxy and Universe. The following summarizes the results of MacGibbon and Carr[14].

The PBH photons can interact via: (i) ionization and photoelectric absorption; (ii) Compton scattering off electrons; (iii) pair-production off nuclei; and (iv) pair-production off the 3K background photons. The dominant interactions are ionization if the photon energy at the relevant epoch is $E \lesssim 14$ keV; Compton scattering if 14 keV $\lesssim E \lesssim 65$ MeV; pair-production off nuclei if 65 MeV $\lesssim E \lesssim 2 \times 10^3$ GeV; and pair-production off the microwave background if $E \gtrsim 2 \times 10^3$ GeV. The 100 MeV photons are only affected at redshifts greater than a few hundred. At the present epoch, the photons are cut off by pair-production off the microwave background above $E \simeq 10^6$ GeV and ionization below $E \simeq 3$ keV, but those in the Galaxy are otherwise unaffected.

The PBH electrons and positrons can interact and lose energy via: (i) annihilation with the ambient e^+ and e^-; (ii) inverse Compton scattering off the 3K background; (iii) bremsstrahlung off nuclei; (iv) e^+e^- pair-production off nuclei or atomic electrons; (v) synchrotron radiation in ambient magnetic fields; and (vi) ionization and excitation processes. In the most recent pregalactic era, the dominant loss mechanisms are ionization processes for kinetic energies $E \lesssim 2$ MeV and inverse Compton scattering for $E \gtrsim 2$ MeV. All extragalactic $e^\pm$ are cut off by interactions before a redshift of $1 + z \simeq 1.8$. In the postgalactic era, the $E \lesssim 10$ MeV $e^\pm$ lose most of their energy through ionization before they can escape from the Galaxy, while the $E \gtrsim 10$ GeV $e^\pm$ are affected by inverse Compton scattering.

The PBH protons and antiprotons interact and lose energy via: (i) annihilation with the ambient $\bar{p}$ and p; (ii) inverse Compton scattering off the cosmic microwave background; (iii) bremsstrahlung off nuclei; (iv) e^+e^- pair-production off nuclei; (v) other p-p or $\bar{p}$-$\bar{p}$ interactions; (vi) synchrotron emission in ambient magnetic fields; (vii) excitation and ionization processes; and (viii) spallation processes. In the recent pregalactic era, the most important loss mechanisms are the ionization processes for kinetic energy $E \lesssim 150$ MeV; annihilation and other $p - \bar{p}$, $p - p$ and $\bar{p} - \bar{p}$ processes for 150 MeV $\lesssim E \lesssim 4 \times 10^6$ GeV; and inverse Compton scattering for $E \gtrsim 4 \times 10^6$ GeV. Particles with energies between $0.1 - 1$ GeV last interacted at about $1 + z \simeq 200$. The pregalactic emission should not be appreciably modified by synchrotron radiation unless there were extraordinarily strong primordial magnetic fields. Today $E \lesssim 50$ MeV protons and antiprotons will have lost all their energy via ionization before they can escape from the Galaxy, but the high energy protons and antiprotons are only cutoff by pair-production off the microwave background above $E \simeq 7 \times 10^{10}$ GeV.

The solid lines in Figures 4 (i)-(iii) show the maximum redshifts at which particles of energy E at that redshift propagate unaffected. Figure 4(i) shows the interactions for photons - (a) ionization, (b) Compton scattering, (c) pair production off nuclei and (d) pair production off the microwave background. Figure 4(ii) shows the interactions for electrons and positrons - (a) ionization, (b) inverse Compton scattering, (A) Galactic ionization and (B) Galactic inverse Compton scattering. Figure 4(iii) shows the interactions for protons and antiprotons - (a) ionization, (b) p-$\bar{p}$, p-p and $\bar{p}$-$\bar{p}$ processes, (c) inverse Compton scattering, (A) Galactic ionization and (B) Galactic $\pi^{\pm}$ pair production. These Figures apply to all astrophysical scenarios and not just PBH emission. The broken line indicates the energy where the dominant PBH emission is expected to peak at redshift $1 + z$. The shaded lines represent the postgalactic energy cutoffs for charged particles emitted in the Galaxy. Neutrinos only interact substantially during the first second of the Universe.

PHOTONS

Figure 5 shows the observed extragalactic photon spectra after Schonfelder et al[15], together with the (unmodulated) PBH emission. It is interesting that the falloff above 170 MeV resembles the falloff extrapolated from the observations. Even if PBHs are clustered in halos, the γ-rays will escape directly from the galaxies and contribute an isotropic background. Thus the observed γ-ray spectrum puts a limit on Ω_{pbh} independent of clustering: the observed diffuse component at $E \simeq 100$ MeV is[16]

$$\frac{dF_\gamma}{dE} = 1.1(\pm 0.2) \times 10^{-13} \left(\frac{E}{100 \text{ MeV}}\right)^{-2.4 \pm 0.2} \text{ cm}^{-3}\text{GeV}^{-1} \tag{9}$$

implying that[14]

$$\Omega_{pbh} \lesssim 7.6(\pm 2.6) \times 10^{-9} h^{-1.95 \pm 0.15} \tag{10}$$

for $\Omega_m = 1.0$. (The limit is 60% weaker for $\Omega_m = 0.06$.) This corresponds to a limit on the fraction of the Universe going into M_* PBHs of $\beta(M_*) \lesssim 10^{-26}$ and on the amplitude of initial inhomogeneities of $\epsilon \lesssim 0.03$. If the PBHs have the continuous mass spectrum given by Eq. (5), their density is constrained over every mass range. In this case, $M \gtrsim M_*$ PBHs cannot provide the dark matter. However, this need not follow if the PBHs formed over a limited mass range, eg. at a cosmological phase transition after 10^{-23} sec, and the spectrum has a lower cutoff above M_*.

ELECTRONS AND POSITRONS

The origin of the electron and positron cosmic-ray backgrounds measured at Earth has yet to be fully understood. Around 100 MeV the $e^+/(e^+ + e^-)$ ratio appears to approach 0.5. This ratio is not naturally produced in any standard scenario: if the positrons are generated in $p-p$ collisions in the interstellar medium, far more low energy positrons than electrons are produced since the protons can absorb the electrons and become neutrons[17], giving $e^+/(e^+ + e^-) \simeq 0.9$ at $E_{e+} = 100$ MeV. A second electron source is needed in this scenario. On the other hand, PBHs naturally produce $e^+/(e^+ + e^-) = 0.5$ since they emit equal numbers of particles and antiparticles at these energies.

We suggest that the PBH emission may contribute strongly to the interstellar e^- and e^+ spectra around 100 MeV to a few hundred MeV. From observations of non-thermal radio emission and γ-ray bremsstrahlung[18], one infers an local interstellar e^- spectrum with a slope of E^{-2} to E^{-3} between 70 MeV and 1 GeV. At 300 MeV,

$$\frac{dF_{e-}}{dE} \simeq 0.6 - 1.1 \times 10^{-10} \ \text{cm}^{-3} \ \text{GeV}^{-1} \tag{11}$$

If the PBHs are not clustered within galaxies, the amplitudes of the e^+ and e^- spectra are well below the observations. If the PBHs are clustered inside galactic halos, then the PBH emission compared with the interstellar electron background at 300 MeV implies that[14]

$$\Omega_{pbh} \lesssim 1.2 - 4.5 \times 10^{-8} \left(\frac{\Omega_h}{0.1}\right)^{-1} \left(\frac{V_\infty}{220 \ \text{km s}^{-1}}\right)^2 \left(\frac{R_h}{10 \ \text{kpc}}\right)^{-2} \left(\frac{\tau_G}{100 \ \tau_{leak}}\right) \tag{12}$$

The limit is about 65% weaker for $\Omega_m = 0.06$. The last factor in (12) probably lies between $1 - 10$, so the limit is comparable to, and overlaps, the γ-ray limit.

The interstellar e^+ background is more uncertain. An estimate of the combined primary and secondary e^+ spectrum, which fits the observations at Earth reasonably well, gives an interstellar flux of[19]

$$\frac{dF_{e+}}{dE} \simeq 7 \times 10^{-12} \ \text{cm}^{-3} \ \text{GeV}^{-1} \tag{13}$$

at 300 MeV, although this may underestimate the interstellar spectrum by up to an order of magnitude[14]. Eqn (13) implies a limit on Ω_{pbh} of[14]

$$\Omega_{pbh} \lesssim 1.4 - 2.9 \times 10^{-9} h^2 \left(\frac{\Omega_h}{0.1}\right)^{-1} \left(\frac{V_\infty}{220 \ \text{km s}^{-1}}\right)^2 \left(\frac{R_h}{10 \ \text{kpc}}\right)^{-2} \left(\frac{\tau_G}{100 \ \tau_{leak}}\right) \tag{14}$$

for $\Omega_m = 1.0$. Limit (14) is an order of magnitude stronger than (12) if τ_{leak} is independent of particle charge. However, if $e^+/(e^+ + e^-)$ approaches 0.5 at interstellar energies $\lesssim 100$ MeV (where the derived data are more uncertain), then the limits would be equal. Since both limits overlap the γ-ray limit, PBH emission may be contributing significantly to all of the observed (interstellar) e^+, e^- and γ-ray spectra around 100 to a few hundred MeV. If the PBH $e^\pm$ are affected by acceleration in the Galaxy or synchrotron losses in strong magnetic fields in the halo, the limits on Ω_{pbh} in (12) and (14) could be somewhat higher. We can also show that these limits are too stringent for the $M \lesssim M_*$ emission to explain the 0.511 MeV e^+e^- annihilation line from the Galactic center or its diffuse component[14].

ANTIPROTONS

The $n_{\bar{p}}/n_p$ ratio in cosmic rays is $10^{-4} - 10^{-3}$ for kinetic energies $0.1 - 10$ GeV. PBHs could contribute at most $10^{-4} - 10^{-3}$ of the proton background at these energies[20]. The antiprotons, however, are more interesting. The PBAR[21] and LEAP[22] experiments detected no antiprotons at $100 - 640$ MeV and $120 - 860$ MeV, respectively, giving upper limits of $\bar{p}/p < 2.8 \times 10^{-5}$ and $\bar{p}/p < 1.8 \times 10^{-5}$ at the 86% confidence level. A still more recent analysis of the LEAP data by Streitmatter $et \ al.$[23] gives upper limits of $\bar{p}/p < 9 \times 10^{-6}$

172

for $110-600$ MeV and $\bar{p}/p < 6.5 \times 10^{-5}$ for 600 MeV - 1.1 GeV at the 87% confidence level. This experiment also detected two possible antiproton events which, if verified, would correspond to fluxes of $\bar{p}/p \simeq 5 \times 10^{-6}$ at 500 MeV and $\bar{p}/p \simeq 3.5 \times 10^{-5}$ at 700 MeV. Antiproton cosmic rays are usually postulated to be secondary particles, produced by the spallation of the interstellar medium by primary cosmic rays (Gaisser and Levy 1974). The PBAR and LEAP upper limits, though, still exceed the predicted secondary flux by at least an order of magnitude. Can PBHs produce an observable $E \lesssim 1$ GeV antiproton flux?

Using the method of Ref. 14, the LEAP measurement of $\bar{p}/p \lesssim 9 \times 10^{-6}$ at 600 MeV and 1 AU implies a limit on the interstellar antiproton flux of

$$\frac{dF_{\bar{p}}}{dE} \lesssim 6 \times 10^{-16} \quad \text{cm}^{-3} \quad \text{GeV}^{-1} \tag{15}$$

at 1 GeV and 50 AU, with a maximum error of a factor of 2. The antiproton flux at 1 GeV from clustered PBHs is[13]

$$\frac{dF_{\bar{p}}}{dE} \simeq 1.5 - 10 \times 10^{-11} \zeta' \Omega_{pbh} \quad \text{cm}^{-3} \quad \text{GeV}^{-1} \tag{16}$$

for $\Omega_m = 0.06 - 1.0$, $h = 0.3 - 0.7$ and $\tau_G = 10.5 - 19.5 \times 10^9$ yr. (15) then requires that[14]

$$\Omega_{pbh} \lesssim 0.6 - 4.0 \times 10^{-9} h^2 \left(\frac{\Omega_h}{0.1}\right)^{-1} \left(\frac{V_\infty}{220 \text{ km s}^{-1}}\right)^2 \left(\frac{R_h}{10 \text{ kpc}}\right)^{-2} \left(\frac{\tau_G}{100 \, \tau_{leak}}\right) \tag{17}$$

This limit is unchanged if we use the reported (but unverified) LEAP detection of an $\bar{p}$ at 500 MeV. Because of the error in the interstellar $\bar{p}$ calculation and the large uncertainties in the halo parameters and τ_{leak}/τ_G (τ_G/τ_{leak} may be smaller for $\bar{p}$ than e^+ and e^-), (17) is still comparable to the limits from the γ-ray, e^+ and e^- spectra.

CONCLUSIONS

It is possible that all four limits on Ω_{pbh} from the extragalactic γ-rays and interstellar e^+, e^- and $\bar{p}$ spectra overlap, and hence that the PBH emission could contribute significantly to all four spectra. The individual burst emission from each hole, on the other hand, is unlikely to be seen in the standard model[24].

REFERENCES

1. MacGibbon, J. H. 1987, *Nature*, **329**, 308

2. Hawking, S. W. 1975, *Commun.Math.Phys.*, **43**, 199

3. Webber, B. R. 1984, *Nucl.Phys.*, **B238**, 492; Marchesini, G. and Webber, B. R. 1988, *Nucl. Phys.* **B 310**, 461

4. Oliensis, J. and Hill, C. T. 1984, *Phys.Lett.*, **143B**, 92

5. MacGibbon, J. H. and Webber, B. R. 1990, *Phys.Rev.*D41, 3052

6. MacGibbon, J. H. 1990, submitted to *Phys.Rev.D*

7. Page, D.N. *Private Communication*; Elster, T. *J.Phys.*A16, 989(1983); Simkins, R.D. *'Massive Scalar Particle Emission from Schwarzschild Black Holes'* Ph.D.Thesis, Pennsylvania State University (1986); Elster, T. *Phys.Lett.* **92A**, 205(1983)

8. Carr, B. J. 1985, in *Observational and Theoretical Aspects of Relativistic Astrophysics and Cosmology* ed. J. L. Sanz and L. J. Goicoechea (World Scientific), p.1

9. Hawking, S. W. 1971, *Mon.Not.R.ast.Soc.*, **152**, 75

10. Carr, B. J. 1975, *Ap.J.*, **201**, 1

11. Carr, B. J. 1976, *Ap.J.*, **206**, 8

12. Page, D. N. and Hawking, S. W. 1976, *Ap.J.*, **206**,1

13. MacGibbon, J. H. 1991, preprint

14. MacGibbon, J.H. and Carr, B.J. to appear in *Ap.J.*, April 1991

15. Schonfelder, V., *et al.* 1977, *Ap.J.*, **217**, 306

16. Fichtel, C. E. *et al.* 1975, *Ap.J.*, **198**, 163-182

17. Ramaty, R. and Lingenfelter, R. E. 1966, *J.Geophys.Res.*, **71**, 3687

18. Lebrun, F. *et al.* 1982, *Astron.Ap.* **107**, 390; Strong, A. W. and Wolfendale, A. W. 1978, *J.Phys.Nucl.*, *Phys.* **4**, 1793; Webber, W. R. *et al.* 1980, *Ap.J.* **236**, 448

19. Ramaty, R. and Westergaard, N. J. 1976, *Ap.Sp.Sci.*, **45**, 143; Perko, J.S. *Private Communication*

20. Ryan, M. J., Ormes, J. F., Balasubrahmanyan, V. K. 1972, *Phys.Rev.Lett.*, **28**, 985

21. Salamon, M. H. *et al.* 1990, *Ap.J.*, **349**, 78

22. Streitmatter, R. E. *et al.* 1990*b*, *Bull. Amer.Phys.Soc.* **35**, 1066

23. Streitmatter, R. E. *et al.* 1990*a*, *Proc.22th Intl.Conf.Cosmic Rays, Adelaide* in press; Streitmatter, R. E. *Private Communication* 1990

24. Halzen, F *et.al.* 1991, to appear in *Nature*

174

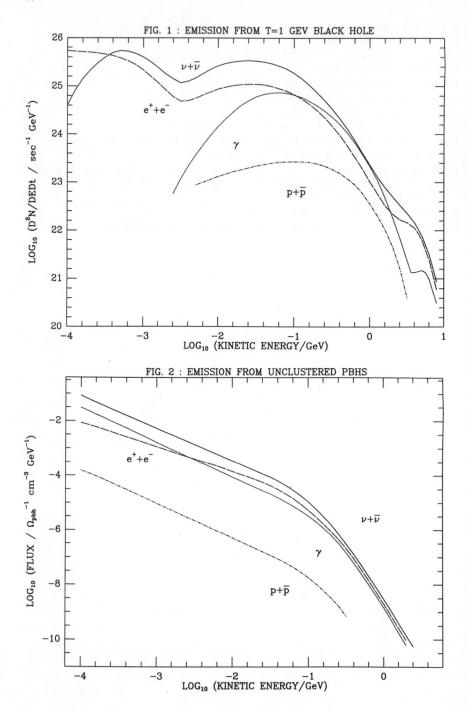

FIG. 1 : EMISSION FROM T=1 GEV BLACK HOLE

$\nu + \bar{\nu}$

$e^+ + e^-$

γ

$p + \bar{p}$

LOG$_{10}$ (D^{2}N/DEDt / sec^{-1} GeV^{-1})

LOG$_{10}$ (KINETIC ENERGY/GeV)

FIG. 2 : EMISSION FROM UNCLUSTERED PBHS

$e^+ + e^-$

$\nu + \bar{\nu}$

γ

$p + \bar{p}$

LOG$_{10}$ (FLUX / Ω_{pbh}^{-1} cm^{-3} GeV^{-1})

LOG$_{10}$ (KINETIC ENERGY/GeV)

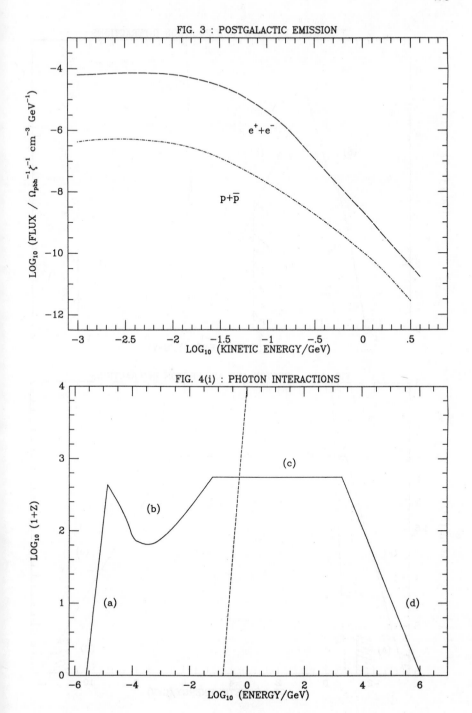

FIG. 3 : POSTGALACTIC EMISSION

$e^+ + e^-$

$p + \bar{p}$

LOG$_{10}$ (FLUX / Ω_{pbh}^{-1} ζ^{-1} cm^{-3} GeV^{-1})

LOG$_{10}$ (KINETIC ENERGY/GeV)

FIG. 4(i) : PHOTON INTERACTIONS

(a) (b) (c) (d)

LOG$_{10}$ (1+Z)

LOG$_{10}$ (ENERGY/GeV)

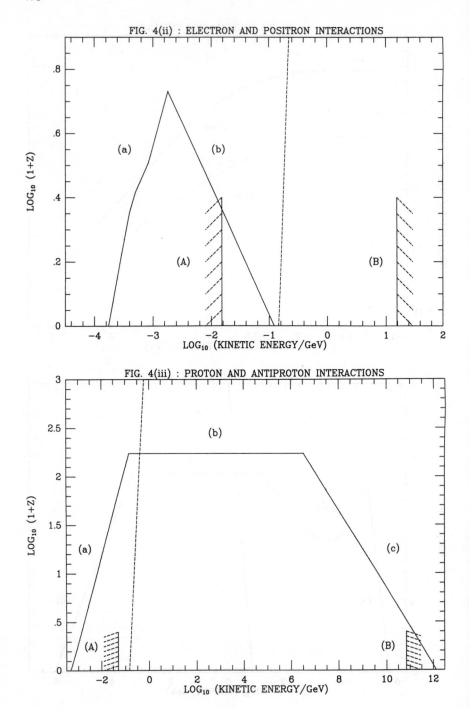

FIG. 4(ii) : ELECTRON AND POSITRON INTERACTIONS

FIG. 4(iii) : PROTON AND ANTIPROTON INTERACTIONS

177

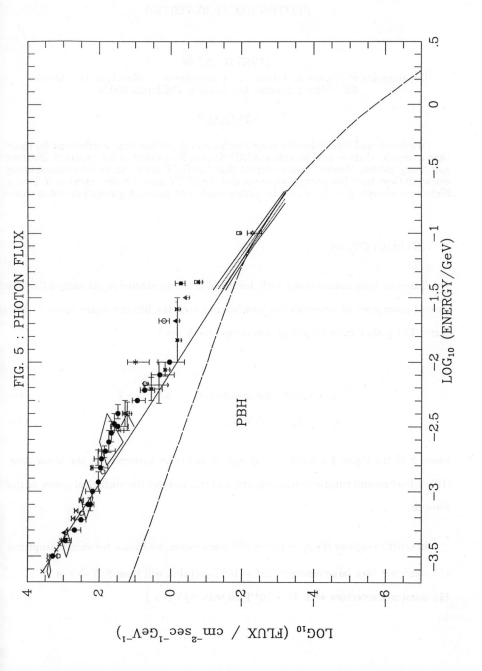

FIG. 5 : PHOTON FLUX

PROTON DECAY REVISITED

DAVID B. CLINE
Departments of Physics & Astronomy, University of California at Los Angeles
405 Hilgard Avenue, Los Angeles, California 90024

ABSTRACT

While the null but dedicated search for proton decay has been carried out for more than a decade, there is new hope that SUSY SU(5) offers a clue to the expected lifetimes and decay modes. Recent results suggest that the GUT scale can be determined using extrapolations from low energy processes and the SUSY particles are expected to have a high mass whereas $p \to K^+ \bar{\nu}$ may be within reach with the next generation of detectors.

1 INTRODUCTION

The search for proton decay with dedicated detectors started in the early 80's. The current generation of detectors has nearly exhausted the lifetime range up to $\sim 10^{32}$ years. The scaling rules for the proton decay search are

$$\tau_p \propto N_p \ [\text{protons in detector}] \qquad S/N \gg 1 \qquad (1a)$$

$$\tau_p \propto \sqrt{N_p} \qquad S/N \lesssim 1 \qquad (1b)$$

where S is the signal for proton decay and N is the background. In the latter case, (1b), a background subtraction is required and this reduces the statistical power of the detector.

In order to explore the $\tau_p \sim 10^{32} - 10^{34}$ years region, we choose between two options

(1) construct very large detectors ($M \sim 10^5$ Tons) that will have $S/N \lesssim 1$ or

(2) construct detectors with $M \sim 10^4$ Tons with $S/N \gg 1$.

In the first category are new water detectors such as Super Kamiokande (Kam II) and in the second are massive *electronic-bubble-chamber-like* detectors such as ICARUS, which uses electron drift imaging techniques.

2 CURRENT LIMITS ON PROTON DECAY

Several complete searches have been carried out for proton decay. Some recent results are given in References 1–10. Two key decay modes are

$$p \rightarrow \pi^0 e^+ \quad \text{(I)} \qquad\qquad (2a)$$

and

$$p \rightarrow K^+ \bar{\nu} \quad \text{(II)} \qquad\qquad (2b)$$

In some ways these two modes *tone* the scale for the entire search for proton decay.

Decay mode (I) is expected to go through normal GUT type processes. Decay mode (II) is expected to go through processes where the GUT-Higgs boson is the key intermediate state. Future searches for proton decay can be judged, in some sense, by how well they are able to search for these two modes.

Table 1 gives the current limit on some decays of type II[6,9] . The lifetime limit for $p \rightarrow \pi^0 e^+$ is in the vicinity of 5×10^{32} years as determined from the IMB and Kam II detectors[8,9] . The fact that τ_p, for $p \rightarrow K^+ \bar{\nu}$, goes like $\sqrt{N_p}$ is stated by Table 1.

3 SUSY AND PROTON DECAY

It is well known that the SU(5) model of Grand Unification disagrees with several precise measurements and the current limits on the proton lifetime. However, for sometime it has also been known that the extrapolation of the three running constants

TABLE 1

Current P Decay Search			
	Signal	Expected Background (BG)	τ/B (No BG Subtr.)
KAM I & II: 3.76 KT-year			
$p \to K^+\bar{\nu}$	9	7.3	5×10^{31}
$\eta \to K^0\bar{\nu}$	0	2.4	8×10^{31}
$p \to \eta\mu$	1	< 0.04	6×10^{31}
$p \to \pi^+\bar{\nu}$	32	32.8	6×10^{30}
			$\hookrightarrow 2 \times 10^{31}$ *(BG Subt.)*
FREJUS: ~ 2 KT-year			
$p \to K^+\bar{\nu}$	1	1.8	1.5×10^{31}
$\eta \to K^0\bar{\nu}$	1	1.8	1.5×10^{31}
$p \to \pi^+\bar{\nu}$	14	11	1×10^{31}

to the GUT scale do not cross in one place (see Fig. 1). One simple remedy for this situation is to invoke a supersymmetric version of SU(5), i.e. SUSY GUTS. Taking this approach has two interesting ramifications

(1) the SUSY particle scale may be beyond the TeV mass range, i.e. squarks, etc., may exist

(2) the decay mode $p \to K^+\bar{\nu}$ is favored through Higgs mediated processes, as shown in Fig. 2[11], however, the lifetime is likely to be between $10^{32} - 10^{34}$ years. (See Table 2 for a supergravity model.)

Thus, SUSY-GUTS presents the next challenge to the proton decay searches.

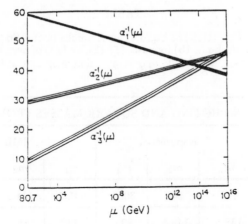

Fig. 1. The running couplings $\alpha_i^{-1}(Q^2)$ (68% c.l. bands)

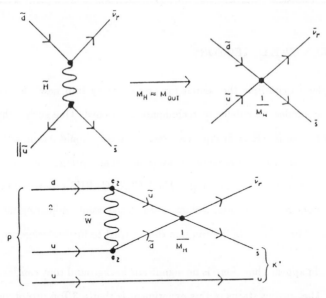

Fig. 2. Proton decay generated by color-triplet Higgsino exchange and W-ino dressing.

TABLE 2

Values of m_1 (GeV) of Eq.(4.4) for Kamioka bound Eq.(2.1a) and IMB bound Eq.(2.1b) for cases (1), (2), and (3) of Sec. IV. ($m_{\tilde{\gamma}}$ = 10 GeV, $m_{\tilde{w}}$ = 40 GeV, α_H = 45°, and M_H = 1 × 10^{16} GeV.) (Ref. 11)

LIMITS ON PHOTINO AND SQUARK MASSES FROM PROTON					
	Kamioka			IMB		
Case	m_1	m_2	m_3	m_1	m_2	m_3
(1)	143	144	3000	142.5	144.5	2050
(2)	138.5	149.5	895	133.5	158	560
(3)	124	220	250	111	–	–

4 ICARUS-LIKE DETECTORS

In order to continue the search for proton decay into new lifetime regions and exotic decay modes, new detector techniques are needed. One such technique is being developed by the ICARUS group. This technique uses liquid Argon and electron-drift imaging of the tracks. If successful, this technique can produce spectacular events that may have no important backgrounds. The UCLA ICARUS group has been simulating the decay mode $p \to K^+\bar{\nu}$. Fig. 3 shows a simulated event, indicating the quality of the images. In addition, dE/dx information can be used to further reference the events.

It would appear that there is no significant background that can fake $p \to K^+\bar{\nu}$ in ICARUS. The current status of this experiment is that a 3 Ton prototype detector has been constructed and is starting to operate at CERN. The next stage is to construct a ~ 1000 Ton detector for Hall C at the Gran Sasso. The ultimate ICARUS detectors could be ~ 10^4 Tons at the Gran Sasso and could extend the lifetime for $p \to K^+\bar{\nu}$ to 10^{34} years.

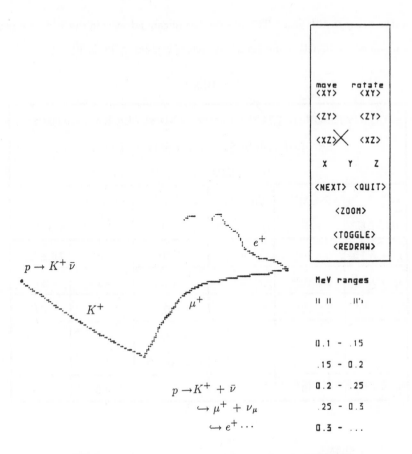

Fig. 3. Simulation of a proton decay event in ICARUS.

5 SUPER KAMIOKANDE-LIKE DETECTORS

At present the best limits for proton decay come from the large water detectors such as IMB and Kam II. A larger version of Kam II is being proposed for Japan[12] . This detector would have a 22,000 Ton fiducial volume. This would be equivalent to approximately 10 times the size of the current Kam II detector. The major question being asked is whether the proton decay search will be extended by a factor of 10 or by $\sqrt{10}$ (Eqs. 1a or 1b). If the latter is correct, the Kam II detector can extend the search for proton decay to 10^{33} years. This is a very significant advance in the field. A possible comparison of the ICARUS and Kam II detectors is given in Table 3[13] .

TABLE 3

COMPARISON OF ICARUS (3000 T) AND SUPER KAM (30.000 T)					
SEARCH FOR SUSY TYPE PROTON DECAY					
– 2 YEAR RUN –					
MODE	ASSUMED τ/B	SIGNAL (ICARUS)	BG (ICARUS)	SIGNAL (S. KAM)	BG S. KAM)
$p \to K^+ \bar{\nu}$	5×10^{32}	4	~ 0	40	56 ± 30?
$\eta \to K^0 \bar{\nu}$	5×10^{32}	4	~ 0	40	20
$p \to \pi^+ \bar{\nu}$	5×10^{32}	4	~ 0	40	280
			(if π^+ range measured)		
$p \to \mu^+ \eta$	5×10^{31}	40	~ 0	400	< 16

6   CONCLUSIONS

It appears that SUSY- GUTS is a viable theory and that three major consequences follow

(1) one Higgs boson may have a mass less than M_Z

(2) the SUSY particles may have masses beyond the TeV range and thus might not be detectable even at the SSC

(3) proton decay, mainly $p \to K^+ \bar{\nu}$ is a crucial test of the theory but the lifetime is expected to be $10^{33} - 10^{34}$ years

These points present an enormous challenge to the proton decay searcher.

REFERENCES

[1] M.L. Cerry, et al., Phys.Rev.Lett. $\underline{47}$ 167 (1986).

[2] M.R. Krishnaswamy, et al., Nuovo Cim. $\underline{9C}$ 167 (1986).

[3] IMB Collaboration, S.Seidel, et al., Phys.Rev.Lett. $\underline{61}$ 2522 (1988).

[4] Kamiokande Collaboration, K.S. Hirata, et al., Phys.Lett. $\underline{B220}$ 308 (1989).

[5] NUSEX Collaboration, G. Battistoni, et al., Nucl.Phys. $\underline{B133}$ 454 (1983).

[6] Frejus Collaboration, Ch. Berger, et al., Nucl.Phys. $\underline{B313}$ 509 (1989).

[7] HPW Collaboration, T.J. Phillips, et al., Phys.Lett, $\underline{B224}$ 348 (1989).

[8] IMB Collaboration, R.M. Bionta, et al., Phys.Rev. $\underline{D38}$ 768 (1988).

[9] Kamiokande Collaboration, K.S. Hirata, et al., Phys.Lett. $\underline{B205}$ 416 (1988).

[10] NUSEX Collaboration, M. Aglietta, et al., Europhysics Lett. $\underline{8}$ 611 (1989).

[11] P. Nath and R. Arnowit, Phys.Rev.D $\underline{38}$ 1479 (1988).

[12] Super Kamiokande Proposal, ICRR report (unpublished).

[13] D. Cline, Internal UCLA Report, unpublished (1990).

III. LARGE SCALE STRUCTURE OF THE UNIVERSE AND STRUCTURE FORMATION

MAPS AND MOTIONS OF GALAXIES IN THE LOCAL UNIVERSE

Alan Dressler, The Observatories of the Carnegie Institution
of Washington, 813 Santa Barbara St., Pasadena, CA 91101

ABSTRACT

The view of large-scale structure in the universe has evolved considerably over the last decade as the prevalence of large superclusters and voids has become more apparent. Highlights of the observational steps that led to this new picture are presented, including efforts to map the distribution of galaxies and, through their peculiar motions, the distribution of dark matter on large scales. Comparison of these two maps for the local volume show rough agreement, though it is still not clear if luminous galaxies trace the distribution of dark matter with any precision.

1. A NEW VIEW OF THE UNIVERSE

Describing the distribution of galaxies and dark matter in the local universe has become a principal goal of observational cosmologists in recent years. This surge in activity is largely hinged on the hope that large scale structure is a reflection of Big Bang physics, a legacy of particles and fields that characterize an energy regime beyond the reach of any earthbound experiment.

The linear relation between recession velocity and distance discovered by Hubble in 1929, based primarily on redshifts from V. Slipher, was a conceptual breakthrough in our understanding of the birth and evolution of the universe, but it also forged a powerful tool for making three-dimensional maps of galaxy distributions. Ironically, it was the consistency of the Hubble relation in various directions of space, established in the 1930's through the 1970's, that helped engrain the notion of a homogeneous, isotropic universe, a paradigm that discouraged the enormous effort that would be required to show otherwise. Reinforced by the discovery of the isotropic distribution of radio sources, the uniform Hubble flow entrained the picture of a quite smooth distribution of galaxies punctuated by rich clusters and groups. Sandage's claim that deviations from uniform Hubble flow were less than 100 km s^{-1}, based on the absence of blueshifts for nearby galaxies, reinforced this notion that the distribution of matter was remarkably smooth.

Thus it was that 30 years after the discovery of the importance of the redshift, this fundamental quantity had been measured for only about 500 galaxies, after 50 years, only about 2000. Until the 1970's, obtaining redshifts was tedious work that gobbled-up large amounts of scarce telescope time; combined with the expectation that redshift maps of a homogeneous, isotropic universe would be rather uninteresting, it is perhaps not surprising that few researchers had been willing to devote much time to the effort. But the proliferation of 2-meter to 4-meter telescopes in the 1970's, added to vast improvements in spectrograph/detector efficiencies, changed the nature of the effort to one where thousands of redshifts could be garnered in a few years with a telescope of modest aperture. The first concerted efforts to map systematically a large volume of space, the Harvard-Smithsonian Center for Astrophysics (CfA) survey (Huchra *et al.* 1983) doubled the number of available redshifts, but it did little to dispel the prevailing notion of a relatively homogeneous universe. The theoretical underpinning came from Peebles (1980), who developed the model that gravitational instability had produced a hierarchical clustering of galaxies from a very homogeneous distribution left by the hot Big Bang. By the 1970's, measurements of the smoothness of the cosmic background radiation (CBR) were beginning to constrain the size of the initial fluctuations, though there was little clue as to their origin. Peebles took them as gaussian and randomly phased, with a "white noise" spectrum - all "initial conditions" put in by hand. As with the N-body simulations of cold and hot dark matter that Peebles' work would sire, the fluctuation amplitude was the principle adjustable parameter with which to reproduce clustering at the present epoch. The "figures of merit" to test the ability of the models to match observations like the CfA survey were the power-law slope and strength of the galaxy-galaxy (two-point) correlation function.

This image of an universe rapidly approaching homogeneity beyond the correlation length of $\sim 5h^{-1}$ Mpc was largely dissolved by the publication of the second CfA survey - the "slice of the universe" - by de Lapparent, Geller, and Huchra, in 1986. Its dense sampling of a large volume of space offered a new portrait of large-scale structure. With its $\sim 50h^{-1}$ Mpc voids and thin, arc-like chains of galaxies, the "slice" raised questions not only about the initial fluctuation spectrum, but even the viability of the bottom-up model of structure formation through gravitational instability. But, as with the fall of most paradigms, there is a strong tendency to credit the decisive blow and forget important, perhaps even compelling, evidence that had previously indicated the prevalence of large-scale inhomogeneity.

Fritz Zwicky and Gerard de Vaucouleurs were on the right track in the 1950's, Zwicky through his prodigious efforts at mapping the distribution over the northern sky and association of all galaxies into superclusters, de Vaucouleurs through his persistent promotion of the idea that our Galaxy sits on the outskirts of a coherent, flattened overdense region - the Local Supercluster. These ideas contradicted the picture of large-scale homogeneity, but they failed to win general acceptance because large numbers of redshifts were not available to defend against the counterclaim that superclustering was a result of projection effects, analogous to stellar constellations. As late as 1980, when Jim Peebles produced a stunning visualization of the Lick counts that showed a motley structure on large-scales, researchers (including Peebles) rejected the notion that these features were real.

The turning of the tide can be located at studies like that Tifft and Gregory (1976), who used new redshifts to argue that the Coma cluster was embedded in a halo of galaxies that extended well beyond what had traditionally been considered to be its boundary. They concluded: "When grouped by redshift and position on the sky, virtually all the galaxies are shown to belong to groups or clusters. Truly isolated galaxies are nonexistent or very rare." Tifft and Gregory also were the first to call attention to large voids, noting "The greater part of the foreground [of the

Coma cluster] is completely devoid of galaxies." A somewhat deeper sampling by Gregory and Thompson (1978) went further, establishing a link from the Coma cluster all the way to the to the next closest rich cluster, Abell 1367 - a distance of $\sim 50h^{-1}$ Mpc. (This structure is part of what has been become known as the "Great Wall".)

The evidence for very large-scale inhomogeneities was markedly strengthened by the discovery of the Great Void in Boötes by Kirshner, Oemler, Schechter, and Shectman (1981). What made this a particularly significant discovery was that it was serendipitous - the survey was aimed at a random part of the sky rather than at one of the filamentary, high-density regions in the sky maps, like the Coma-A1367 or Perseus-Pisces Superclusters. (Since attention had been drawn to these regions by their appearance, the discovery of large-scale structure could be discounted as a selection effect.) The scale of the Great Void is approximately $100h^{-1}$ Mpc, though its shape is controversial and it is now known that it is not completely empty, but rather a region whose galaxy density is somewhere between 4 and 10 times below average.

Geller and Huchra (1989) have added two adjacent slices to the original CfA slice of the universe; this increase in volume has strengthened the case that this survey, too, is now sampling a typical region. There is considerable continuity across all three regions; in particular, they combine to show a long, filamentary structure named the "Great Wall" which has dimensions of roughly $200h^{-1}$ Mpc x $50h^{-1}$ Mpc, and a typical thickness of $10h^{-1}$ Mpc or less! It is not just the total size, but the thinness of such a structure, that challenges models based on the idea that gravitational instability is the dominant mechanism in the formation of large-scale structure. However, it must also be born in mind that these are maps in *redshift space*. Peculiar motions of order 500 km/sec, which appear to be fairly common, can falsely accentuate the thinness of structures like the Great Wall. Viewed perpendicular to the line-of-sight, a thicker structure, its Hubble expansion retarded by the gravity of its greater-than-average density, would appear significantly squeezed.

Margaret Geller consistently makes the point that all surveys have found a structure as large as they *could* find, *i.e.*, a coherent feature that stretches across the entire surveyed volume. To the extent that this is true, it has pressed the need for more extensive surveys, which, by the necessity of limited observational resources, have not been filled surveys like the CfA slice. Pencil beam surveys, which push a basically one-dimensional probe out to large redshifts, are one way to cover huge scales. The results of Broadhurst *et al.* (1990) have been particularly provocative. Not only has this survey appeared to confirm the prevalence of structures on a scale size of $100h^{-1}$ Mpc, the size of the Great Void and the Great Attractor (discussed below), but it has provided evidence for periodicity in the structure. For the first beam in particular, which traced a line from the north to south Galactic caps, both amplitude and phase coherence suggest some sort of an honeycomb-like structure with a modest variation in cell size around $100h^{-1}$ Mpc. Subsequent pencil beams by the group have shown greater variation, and the team has been cautious is claiming too much about the periodicity of the phenomena. However, they believe that the result of a "preferred scale" in large-scale structure is robust.

On the other hand, the early results of a huge new "fan" survey by Kirshner, Oemler, Schechter, and Shectman give little indication of any such cellular structure. When completed, their "unfilled" survey will contain some 10,000 redshifts (typical $z = 0.1$) in two areas. Already it would seem, however, that the distribution of galaxies in these enormous areas, each several times bigger than the CfA slices, is approaching homogeneity, *i.e.*, a structure as big as the entire survey is not readily apparent. Although this might have been forecast from the two-point correlation function and the isotropy of radio counts, the possibility that a fairly homogeneous map may be on the way can understandably be welcomed with a sigh of relief.

2. THE LOCAL VOLUME

The previous discussion carries with it a major assumption that is, as yet, relatively untested: that galaxies are accurate tracers of the dark matter. It is the distribution of this dark component, which contains at least an order-of-magnitude more mass than found in the stars and gas of galaxies, that is predicted N-body simulations based on weakly interacting particles.

The volume defined within 60 h^{-1} Mpc is particularly important in this regard because, even though it contains only a small amount of space compared to the surveys described above, for this volume it is possible to make a map of the <u>mass</u> distribution as well as the galaxy distribution. This is accomplished by measuring the deviations from uniform Hubble flow that are produced by inhomogeneities in the matter distribution, including both luminous and dark. Thus, in this smaller volume we can test the N-body models directly and, by comparing the distribution to fair galaxy samples, determine to what extent and how to use the distribution of galaxies alone to indicate the mass distribution, out to distances that are too great for accurate measurements of peculiar motions.

It is reassuring that this "local volume" includes the salient features seen on the larger scales - several large superclusters and voids. The Local Supercluster, with its remarkably flat distribution of galaxies roughly $25h^{-1}$ across, defines what de Vaucouleurs called the *Supergalactic Plane*. If extended around the sky, the Supergalactic plane intercepts two regions of high galaxy density: the Perseus-Pisces supercluster at a recession ($V \approx 5000$ km s^{-1} and Hydra-Centaurus supercluster ($V \approx 3000$ km s^{-1}) are prominent concentrations on nearly opposite sides of the sky. In addition, there is a sizeable void in front of the Perseus-Pisces supercluster. These structures show up well in optical surveys, but a good deal of attention in recent years has been focussed on the galaxy catalogs produced from the *IRAS* satellite. Advantages afforded by this sample are its uniformity and that the far-IR fluxes of these observations are much less susceptible to dust absorption in our Galaxy; principle disadvantages are that the far-IR flux is not particularly well correlated with galaxy mass or luminosity, and early morphological types (E, S0, Sa), the dominant components of high-density regions, are poorly represented. Nevertheless, the distribution of *IRAS*-selected galaxies shows the same large-scale structure (Strauss and Davis 1988).

It is worth pointing out that redshift surveys over large sky areas containing the Perseus-Pisces (Giovanelli, Haynes, and Chincarini 1986) and Hydra-Centaurus,

Pavo-Indus superclusters (the Great Attractor region, Dressler 1988, 1990) show the same sharp peaks in redshift space as has been seen in the pencil beam surveys by Broadhurst *et al.* All in all, it looks like we live in a fairly representative volume of space.

3. MEASURING PECULIAR MOTIONS

Not only can we measure the space positions of galaxies, we can also measure their *motions*. Although the major component of galaxy motion is due to the expansion of the universe, an overdensity in the matter distribution will impart additional *peculiar velocities*, deviations from pure Hubble flow, in its direction. Thus, galaxies will stream towards overdense regions and away from underdense regions. Averaged over large scales these motions are still small enough that the perturbations can be modeled with linear perturbation theory, which simplifies the recovery of the underlying mass distribution.

Rubin *et al.* (1976) made the first concerted effort to measure departures from uniformity in the Hubble flow by measuring the motions of hundreds of galaxies in various directions, galaxies chosen to be roughly equidistant from the Milky Way. They interpreted the asymmetry they measured in the velocity field as due to a motion of our own Galaxy of V_{pec} ~ 500 km s^{-1}, an unheard of speed according to conventional wisdom. All doubt that the Milky Way was, in fact, in rapid motion (although not in the direction that Rubin *et al.* had found), was soon removed when an unmistakable anisotropy was detected in the temperature of the cosmic microwave background radiation (Smooth, Gorenstein, and Muller 1977, Cheng *et al.* 1979). The detection of this dipole anisotropy, interpreted as the Doppler shift caused by a 600 km s^{-1} peculiar motion of our Galaxy (Smooth and Lubin 1979, Fixen, Cheng, and Wilkensen 1983), raised new questions. Over what range are other galaxies sharing this motion? To what are they streaming?

It is quite simple to measure galaxy motions, of course, at least the component toward or away from us. Spectroscopic observations of galaxies and stars in our own Galaxy are compared to obtain a redshift, assumed to be a measurement of the Doppler shift due to motion of the galaxy. Most of this measured velocity is due to the *expansion* of the universe; the peculiar motion is a small part that can be found only after subtracting the cosmic contribution. This step requires a knowledge of the distance to the galaxy, since, in an Hubble flow unperturbed by a inhomogeneous mass distribution, velocities are a linear function of distance.

Methods for finding distances amount to trying to predict the size or luminosity of a galaxy based on the measurement of some property that can be measured independent of distance. Tully and Fisher (1977) showed that for spiral galaxies the total luminosity is well correlated with galaxy rotation speed. The relation has been tested and calibrated using galaxies in clusters, for which the common distance of all objects means that relative *apparent* brightness indicates true relative luminosity. Tests on some dozens of clusters shows that rotation speed correlates with galaxy luminosity with an rms scatter of about 40%, which translates into a distance error of about 20% per galaxy. Such poor accuracy means that for a galaxy whose distance would correspond to a recessional velocity of 5000 km s^{-1} in a unperturbed

Hubble flow, the subtraction of cosmic contribution will result in an uncertainty in the peculiar motion of order 1000 km s^{-1}. With present techniques, then, we must measure the motions of many galaxies in a particular region of space and combine the results to obtain the needed accuracy of a few hundred kilometers per second.

4. MAPS OF PECULIAR VELOCITIES

In 1982 the Tully-Fisher relation was applied to a sample of spiral galaxies in a fundamental study of the peculiar motions of galaxies in the Local Supercluster. Aaronson *et al.* (1982) measured to distances to several hundred spiral galaxies and subtracted the expansion field component from the observed recession velocity, leaving a field of peculiar motions. Their pioneering study showed a clear pattern of "infall" of galaxies toward the region of highest galaxy density.[1] This confirmed that the higher local density of galaxies was indeed outlining a region of higher *mass* density, and that the effect of gravity over the denser region could be seen. These first measurements of large-scale gravity confirmed that there is at least 10 times as dark matter as can be seen in the visible galaxies.

Davis and Peebles (1983) used this and other measurements when they emphasized the importance of comparing the Milky Way's motion of ~600 km s^{-1} with respect to the CBR to *local* measurements of peculiar motion. More than a dozen studies had measured the peculiar motion toward Virgo at that time; they had found values ranging from 0 - 500 km s^{-1}, to compare with the ~400 km s^{-1} component of Local Group motion with respect to the distant CBR. Davis and Peebles favored the high values, but some of the best determinations of peculiar motion, like the Aaronson *et al.* work, indicated that only 50-75% of the CBR-referenced motion toward Virgo could be attributed to motion that originated within the Local Supercluster (Dressler 1984). Furthermore, this single component is less than half of the total motion indicated by the CBR dipole. These measurements suggested that major sources of gravitational acceleration lay beyond the Local Supercluster, an important clue that mass inhomogeneities were on a very large scale.

A map of peculiar velocities for a larger volume of space became one of the goals of the *Seven Samurai*, the collective name given to a group composed of Burstein, Davies, Faber, Lynden-Bell, Terlevich, Wegner, and myself, who were studying the intrinsic properties of elliptical galaxies. The Seven Samurai measured the velocity dispersions of some 400 elliptical galaxies selected by uniform criteria

---

[1]Actually, the Hubble expansion velocity is larger than the infall, so it is more correct to say that expansion of the Local Supercluster is *retarded* relative to the average expansion rate for the universe. Except for the immediate vicinity of the Virgo and Ursa Major clusters, galaxies are still receding from each other. The Aaronson *et al.* study modeled the infall as a spherical pattern with a fall in peculiar velocity proportion to 1/r, but Burstein (1990) has emphasized that the data are better matched by a shear flow with velocity directly proportional to distance from a "spine" running between the Ursa Major and Virgo clusters.

over the entire sky (Davies *et al.* 1986, Faber *et al.* 1989) Like rotation speed, velocity dispersion a is measured spectroscopically, with the width of absorption lines indicating the characteristic stellar speed. For each galaxy we also made *photometric* observations (Burstein *et al.* 1986), recording the total light intensity from the object and its the run of intensity with radius. It was discovered that in a three-space of velocity dispersion, total brightness, and average (surface) brightness, the distribution of elliptical galaxies is nearly planar (Dressler *et al.* 1987a, Djorgovski and Davis 1987). This allowed the construction of new distance indicator which, like the Tully-Fisher relation, used a distance independent parameter, velocity dispersion, to predict a linear combination of parameters that are distance-dependent. The accuracy of this method for predicting distances was found to be about the same as the method for spirals, about 20%.

When the method was applied to the sample of elliptical galaxies, the result was surprising. A map of the peculiar velocities exhibited a large-scale flow pattern over a region nearly 5 times the size of the Local Supercluster (Dressler *et al.* 1987b). The Samurai confirmed that galaxies within the Local Supercluster streaming along with the Local Group, and that there was no obvious end to the flow, as would be expected if the source of the gravitational pull had been reached. This result was described simply as "bulk flow", though later we became convinced that the data would allow extraction of terms higher than the dipole. In particular, it seemed quite significant that galaxies in the direction of the flow out ahead of the Milky Way were moving more rapidly, typically at 1000 km s^{-1}. This included the rich association of Centaurus clusters, previously considered a likely source of the flow, now found to be racing ahead to some more distant gravitational center. Conversely, galaxies "at our back" were observed to be moving in the same direction but more slowly than the Milky Way and its companions.

This gradient was easily interpreted in terms of a quadrupole contribution to the velocity field, the expected effect of the pull of a distant mass concentration. This quadrupole term had previously been discovered in the velocity field of the Local Supercluster (Lilje, Yahil, and Jones 1986). This evidence encouraged us to model the flow pattern expected for a gravitational source just outside the well-sampled volume. We found that a distributed, spherical mass in the direction Galactic longitude = 307°, Galactic latitude = 9°, whose center lay at a distance of 4350 km s^{-1} from our Galaxy, would induce a peculiar velocity field that matched the observations in surprising detail (Lynden-Bell *et al.* 1987). Normalizing the model to the observed peculiar velocities implied a mass of $\sim 3\text{-}5 \times 10^{16}$ solar masses - equivalent to a huge supercluster. The introduction of this "Great Attractor (GA) had greatly reduced the chi-square with the addition of a handful of free parameters.

An all-sky galaxy map (Fig. 8 of Lynden-Bell *et al.*) of the galaxy distribution in this part of the sky, produced by Ofer Lahav from three galaxy catalogs, showed that the GA region, though partly obscured by dust from the plane of our own galaxy, is the "most luminous" part of the sky. Apparently a substantial overdensity of galaxies accompanies the mass overdensity revealed by the measurements of peculiar motions. At least in a rudimentary sense, galaxies were found to trace the hidden mass.

The GA model offered a serious challenge for the cold-dark-matter (CDM) paradigm. N-body simulations by Bertschinger and Juskiewicz (1988) showed that such a massive and extensive concentration was extremely improbable, especially in the prevalent model where galaxies were assumed to cluster more strongly than the dark matter, (*biasing* factor $b = 2.5$). They pointed out, however, that the GA *model* was not well constrained by the data then available, and a number of less extreme mass distributions would also fit the data. Further tests of the GA model could clearly provide a strong constraint on CDM.

New measurements of peculiar motions in the GA region have become available. Large-amplitude peculiar motions were found in the GA region by Aaronson *et al.* (1989) for cluster spirals, and by Lucey and Carter (1988) for cluster ellipticals. Although these independent observations were generally supportive, they also uncovered anomalous motions that suggest that, not surprisingly, the GA model is an oversimplification.

Sandra Faber and I spent three more observing seasons at the Carnegie Institutions's Las Campanas Observatory and obtained data for 134 elliptical galaxies in the GA region and 117 spiral galaxies, both in and out of clusters (Dressler and Faber 1990a,b; Burstein, Faber, and Dressler 1990) These new data not only represented a manifold increase in the number of galaxies in the region with peculiar velocities, but also probed further into and beyond the putative center of the GA. This is crucial because the unambiguous signature of a region in gravitational contraction is a change in sign of peculiar velocities as the center of mass is crossed. The new data show just such an effect. Compared to the straight line that indicates a uniform Hubble expansion, galaxies on the near side of 4000 km s^{-1} moving too fast for their distance, *i.e.*, they have a net positive peculiar motion. Beyond this the peculiar motions drop rapidly, crossing zero at a distance of approximately 5000 km s^{-1}, and they turn negative beyond. Although the most distant points are subject to greater noise and increasing systematic effects, it would appear that galaxies on the far side of the Great Attractor are seen "falling back". A completely independent data set for spirals by Mathewson and collaborators (private communication) shows the same pattern. The full S-wave pattern, so named because the data describe a full oscillation around the Hubble line, is the expected signature for a region in which the Hubble expansion has been slowed. Hubble diagrams done in exactly the same way for galaxies in other directions in the sky show no such evidence for a large gravitational distortion.

5. DO GALAXIES TRACE THE MASS?

Though the GA model is undoubtedly an oversimplification, the data do seem to bear out the notion that this one large overdensity dominates the velocity field over a region of space as large as $100h^{-1}$ Mpc across. This is still a small volume of the universe, much less than 1%, but it does represent a significant reach to the largest structures yet discovered. Though of similar scale to the Great Wall and other prominent superclusters, the Great Attractor is unique in that it is the first large-scale feature mapped in the *mass* distribution, the more fundamental quantity of large-scale structure.

It is important to emphasize that, although it was clear from 1986 that the Great

Attractor might amount to nothing more than the dark matter associated with several superclusters of galaxies in this region of the sky, without an extensive and complete galaxy map this could not be concluded without prejudice. Thus, the worth of the GA model was to describe the mass distribution *independently* of the galaxy distribution, as a step toward making an unbiased assessment of whether galaxies actually trace the distribution of dark matter. Recently, the British press carried a report of the "death of the Great Attractor", citing a paper by Rowan-Robinson *et al.* (1990) in which they conclude that luminous galaxies in the region, if embedded in a dark matter distribution at the closure density, are sufficient to explain the peculiar motions of galaxies. This work, though an important step in comparing the maps of galaxies and mass, misrepresented the concept of the Great Attractor as due to dark, mysterious matter unassociated with galaxies, and thus failed to emphasize the important point that, based on their survey, light appears to trace mass over the volume of space studied.

Because present techniques limit the accuracy of distance measurements to ~20% per galaxy, it is difficult to determine peculiar velocities in the more distant Great Wall in order to see if a similar underlying *mass* structure is present. No doubt there are many other "great attractors" that are just beyond the reach of present techniques. To compare the distribution of light and mass, then, we must focus our attention on the smaller volume within 6000 km s^{-1}.

Comparisons have been made to galaxy samples done with catalogs using visible and far-infrared light, the latter from observations made with the *IRAS* satellite. Redshifts for several thousand galaxies detected by *IRAS* have been used to construct galaxy maps, with the assumption that this subset is representative of *all* galaxies within the volume (Strauss and Davis 1988, Rowan-Robinson 1989) The results vary somewhat between two different *IRAS* samples, and also for the optically-selected sample, but a basic feature of all surveys is that there are two main galaxy concentrations in the volume, one associated with the Great Attractor and another identified as the Perseus-Pisces supercluster. Estimates on the relative strength of these concentrations range from near parity to a 2:1 ratio in favor of the Great Attractor.

The issue of whether light traces mass in the volume may rest on whether the Perseus-Pisces supercluster is as rich in galaxies as the Great Attractor region. There is a fairly striking dissimilarity between the map of the mass distribution calculated from the velocity field using the POTENT method by Bertschinger and Dekel (1988) compared to one of the IRAS samples. The galaxy map implies that we are caught midway between two comparable overdensities, yet the peculiar velocity field shows no such bimodality - a single mass feature appears to dominate. However, the "one-in-six" sparse IRAS sample done by the Rowan-Robinson group shows the effect less strongly (Perseus-Pisces is less prominent), thereby muddling the issue. Furthermore, data for peculiar motions in the Perseus-Pisces region remain sparse, particularly in the densest concentrations in the direction of the Perseus constellation.

It is crucial to realize that studies using galaxy surveys alone to find the sources of the 600 km s^{-1} Local Group motion with respect to the CBR suffer two serious drawbacks. First, they *assume* that light traces mass; second, by fitting only the dipole term, they are unable to locate the distance of these sources. Shot noise and small-number statistics for distant galaxies in the *IRAS* samples result in very poor knowledge of the scale over which the dipole originates. On the other hand, using the full map of peculiar velocities has the great advantage that gradients in the velocity field locate the sources exactly through a local measurement. It is through the comparison of the full velocity field and the local galaxy counts that one can meaningfully address the questions of whether light traces mass and the value of Ω for clustered matter.

Recently, Willick (1990) has presented preliminary evidence that galaxies on the Pisces side of the supercluster exhibit only a minor infall pattern - less than half as strong as in the GA - and that the entire Pisces region is moving in the general direction of the Local Supercluster/Great Attractor at some 400 km s^{-1}. If this result holds up, and is bolstered by measurements now being processed for galaxies in the Perseus region, it will exacerbate the difference in the distributions of light and mass. Furthermore, since the pull of the Great Attractor, even when combined with the Local Supercluster, should be too weak to pull Perseus-Pisces as well, such a result may imply that a significant part of the motion may arise from a yet more distant, even greater attractor (Scaramella *et al.* 1989, Raychadbury 1989).

6. FUTURE RESEARCH DIRECTIONS

Within the next five years it will be possible to assemble a truly representative galaxy survey and to extend measurements of peculiar motions to thoroughly test the degree to which galaxies trace the mass distribution, and the degree to which the mass distribution is dominated by large concentrations like the Great Attractor. If it turns out that galaxies are good tracers of the underlying mass distribution, we can proceed with confidence to study large-scale structure solely from the more-easily obtained distribution of galaxies. However, the early evidence may offer some reason to believe that galaxies trace the mass poorly, and in a way that may not be easily predictable. Such an important claim requires much better proof, but if true, reading and interpreting large-scale structure of the universe solely from galaxy maps will be significantly complicated or even blocked. On the other hand, discovering that galaxies don't follow precisely the distribution of dark matter would be a powerful clue to the question of how galaxies formed and the seeding of structure.

Observations of many types and from many sources continue to indicate that vast organizational patterns, at least 100 Mpc across, are a common feature of the universe. On the other hand, observations the CBR radiation continue to paint a picture of a remarkably smooth young universe. For example, recent measurements of temperature fluctuations on scales of 0.1 - 1.0 degrees set upper limits for $\Delta T/T$ < 3 x 10^{-5} (Partridge 1989). Yet model-independent, robust calculations (Bertschinger, Gorski, and Dekel 1990) show that the fluctuation that grew into the Great Attractor would have generated a temperature fluctuation $\Delta T/T \geq 1$ x 10^{-5}: photons lose this much energy as they climb out of the gravitational potential well of the proto-Great Attractor. Failure to have yet detected the fluctuations that grew into galaxies, clusters, and superclusters still leaves acceptable models, but the "squeeze" is on (Readhead *et al.* 1989, Meinhold and Lubin 1990). The extent to which these two types of observations are in conflict represents the greatest challenge to those who model the growth of structure in the early universe.

REFERENCES

Aaronson, M., Huchra, J, Mould, J, Schechter, P. L., and Tully, R. B. 1982, *Ap.J.*, **258**, 64.

Aaronson, A. *et al.* 1989, *Ap.J.* **338**, 654.

Bertschinger, E., and A. Dekel 1989, *Ap.J.Letters*, **336**, L5.

Bertschinger, E. Gorski, K. M., and Dekel, A. 1990, *Nature* **345**, 507.

Bertschinger, E., and Juskiewicz, R. 1988, *Ap.J.Letters* **334**, L59.

Broadhurst, T. J., Ellis, R. S., Koo, D. C., and Szalay, A. S. 1990, *Letters to Nature*, **343**, 726.

Burstein, D. 1990, *Rept. Prog. Phys.* **53**, No. 4.

Burstein, D., Davies, R. L., Dressler, A., Faber, S. M., Stone, R. P. S., Lynden-Bell, D., Terlevich, R. J., and Wegner, G. 1987, *Ap.J.*, **64**, 601.

Burstein, D., Faber, S. M., and Dressler, A. 1990, *Ap.J.*, **354**, 18.

Cheng, E. S., Saulson, P. R., Wilkenson, D. T., and Corey, B. E. 1979, *Ap.J.Letters* **232**, L139.

Davies, R. L., Burstein, D., Dressler, A., Faber, S. M., Lynden-Bell, D., Terlevich, R. J., and Wegner, G. 1987, *Ap.J.Suppl.*, **64**, 581.

Davis, M., and Peebles, P. J. E. 1983, in *Annual Reviews of Astronomy and Astrophysics*, **21**, 109.

de Lapparent, V., M. J. Geller, and J. P. Huchra 1986, *Ap.J.Letters*, **302**, L1.

Djorgovski, S., and Davis, M. 1987, *Ap.J.*, **313**, 59.

Dressler, A. 1984, *Ap.J.*, **281**, 512.

Dressler, A. 1988, *Ap.J.*, **329**, 519.

Dressler, A. 1990, *Ap.J.Suppl.*, in press.

Dressler, A., Lynden-Bell, D., Burstein, D., Faber, S. M., Terlevich, R. J., and Wegner, G. 1987a, *Ap.J.*, **313**, 42.

Dressler, A., Faber, S. M., Burstein, D., Davies, R. L, Lynden-Bell, D., Terlevich, R. J., and Wegner, G. 1987b, *Ap.J.Letters*, **313**, L37.

Dressler, A., and Faber, S. M. 1990a, *Ap.J.* **354**, 13.

Dressler, A., and Faber, S. M. 1990b, *Ap.J.Letters*, **354**, L54.

Faber, S. M., Wegner, G., Burstein, D., Davies, R. L, Dressler, A., Lynden-Bell, D., and Terlevich, R. J. 1989, *Ap.J.Suppl.*, **69**, 763.

Fixsen, D. J., Cheng, E. S., and Wilkenson, D. T. 1983, *Phys.Rev.Letters* **50**, 620.

Geller, M. J., and Huchra, J. P. 1989, *Science*, **246**, 897.

Giovanelli, R., M. P. Haynes, and G. L. Chincarini 1986, *Ap.J.*, **300**, 77.

Gregory, S. A., and Thompson, L. A. 1978, *Ap.J.* **222**, 784.

Kirshner, R. P., Oemler, A., Schechter, P. L., and Shectman, S. A. 1981, *Ap.J.Letters*, **248**, L57.

Huchra, J. P., Davis, M., Latham, D., and Tonry, J. 1983, *Ap.J.Suppl.* **52**, 89.

Lilje, P., Yahil, A., and Jones, B. J. T. 1986, *Ap.J.* **307**, 91.

Lucey, J. R., and Carter, D. 1988, *M.N.R.A.S* **235**, 1177.

200

Loitering Universe

Hume Feldman
Canadian Institute for Theoretical Astrophysics
60 St. George St, Toronto, On M5S 1A1, Canada

Observations in cosmology over the past decade have given strong evidence for a certain cosmological model, i.e. a Friedmann-Robertson-Walker (FRW) Universe with a hot big bang. Beyond this, it has become conventional for cosmologists to assume that the present content of the Universe is primarily in some non-relativistic gas of particles, perhaps just baryons, and that the present large-scale distribution of such matter is primarily determined by gravitational instability. Recent observations suggest more inhomogeneity on large scales than is predicted by the 'standard cold dark matter model'[1], while these observations do not conflict with the above-mentioned cosmological prejudices, they have lead a number of researchers to question some of our basic hypotheses.

In this lecture I would like to propound a kind of cosmology which is call a "loitering Universe"[2]. It is an expanding Friedmann cosmology which undergoes a recent phase of very slow growth or 'loitering'. In order to retain the successes of the Hot Big Bang, before loitering the Universe is assumed to have gone through the usual radiation-dominated ($a \propto t^{\frac{1}{2}}$) and matter-dominated ($a \propto t^{\frac{2}{3}}$) phases. The loitering phase comes after this standard Einstein-deSitter phase. As we shall see, after loitering it is generally necessary to go into a phase of rapid expansion in which the scale factor grows more rapidly than proportional to time. A schematic representation of such an expansion law is illustrated in Fig. 1.

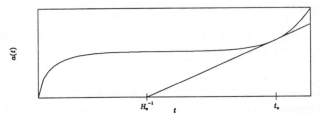

Fig. 1) The scale factor $a(t)$ in the Loitering Universe scenario. Early evolution is matter dominated, followed by a loitering phase and eventually an inflationary phase. The universe is very old i.e. $t_o \gg H_o^{-1}$.

The advantages of such an expansion law are as follows. First we see that such a Universe can be old, much older than H_o^{-1}, where H_o is the Hubble constant. This is illustrated graphically in Fig. 1. This ameliorates the age problem which arises from the possible conflict between the estimated ages of the oldest globular clusters, 1.6×10^{10} yrs[3]. The age problem is a serious one for pundits of flat cosmologies in which $t_o = \frac{2}{3}H_o^{-1}$ and would be a big problem for all standard cosmologies if H_o is closer to 100 km/sec/Mpc as recent observations suggest[4].

The other advantage of a loitering Universe is that one may grow large amplitude inhomogeneities while producing very little anisotropy in the microwave background radiation. In the loitering scenario, a low density of gravitationally clustering matter (CM) is not much of a hindrance to growing lots of structure. This is in sharp contrast to standard cosmologies where a low density of CM severely curtails the amount of growth. In the loitering model it is the CM which dominates during the matter-dominated phase. A low present density in CM means a lower redshift of matter-radiation equality and a larger comoving horizon size at equality, λ_{eq}. The value of λ_{eq} depends only on the present density of CM and the Hubble constant and not on the details of the expansion law. Having a large value of λ_{eq} insures that the Mézáros effect[5] will cause the density perturbations to fall off very rapidly above this scale.

The reason the loitering Universe works so well in terms of growing perturbation is that the inhomogeneities may grow extremely rapidly during the loitering phase so that the amplitude of inhomogeneities at recombination could be extremely small. This means that all of the anisotropies generated at last scattering will be much smaller than in a standard cosmology.

If we have an expansion law with an intermediate semi-static phase, it is not totally clear how we should decide when today is. One important indicator is the ratio of the amount of matter which clusters (CM) to the critical density. For simplicity we assume that the matter which undergoes gravitational clustering has negligible pressure. Let us parameterize the amount of 'cold' matter by

$$\Omega_{cm} = \frac{\rho_{cm}}{\rho_{crit}} \; ; \qquad \rho_{crit} \equiv \frac{3}{8\pi G}\left(\frac{\dot{a}}{a}\right)^2 \tag{1}$$

where ρ_{cm} is the density of CM. While the present value of Ω_{cm} is not known, we use as our working hypothesis that $\Omega_{cm} \geq 0.1$.

Now let us consider the growth of inhomogeneities during a loitering phase. As the expansion of the Universe slows down, the growth of the density contrast: $\delta = \frac{\delta \rho_{cm}}{\rho_{cm}}$ speeds up. This can be readily seen from the equation describing the growth of density contrast in an expanding Universe:

$$\ddot{\delta} + 2\frac{\dot{a}}{a}\dot{\delta} = 4\pi G\rho_{cm}\delta \ . \tag{2}$$

If $4\pi G\rho$ is much greater than both $\frac{\dot{a}}{a}$ and $\frac{\ddot{a}}{a}$ then (2) can be solved in the WKB approximation, giving

$$\delta = \frac{1}{a}\exp\left(\int \sqrt{4\pi G\rho_{cm}}\,dt\right) = \frac{1}{a}\exp\left(\int \sqrt{\frac{3}{2}\Omega_{cm}}\,d\ln a\right) \ . \tag{3}$$

To get rapid growth of perturbations it is necessary that $\Omega_{cm} \gg 1$. This is essentially our definition of loitering. Thus a loitering universe has a quasi-exponential growth rate for density perturbations, approaching the exponential Jeans instability which arises in a non-expanding space. The quasi-exponential growth represents an enormous speed up over the growth rate of perturbations in a matter dominated FRW Universe, in which $\delta \propto a$. The large growth rate implies that perturbations can grow to a much larger value than in a standard cosmology.

We have said that we need $\Omega_{cm} \gg 1$ for loitering. From the Friedmann equation,

$$\left(\frac{\dot{a}}{a}\right)^2 = \frac{8\pi G}{3}\left(\rho_T - \frac{k}{a^2}\right) \ , \tag{4}$$

where ρ_T is the total energy density in matter and k is the curvature constant of the Friedmann cosmology. Equations (1) and (4) tell us that this can only happen if $k > 0$ or if $\rho_{cm} \gg \rho_T$. If we are to obey the weak energy condition then $\rho_{cm} \leq \rho_T$ and a positive k is the *only way* to achieve rapid growth of perturbations. Thus under most circumstances a loitering universe is a closed universe.

Since we are dealing with a closed Universe there must be some form of matter which allows the universe to loiter rather than recollapse. The Friedmann equation (4) tells us that this matter must have an energy density which decreases more slowly than a^{-2}. This form of matter must dominate after loitering or the universe would just collapse.

In order to determine the quality of a particular model for the expansion law, $a(t)$, let us define two parameters:

$$\Delta \equiv \frac{\delta}{(1 + z_{rec})\delta_{rec}} \ ; \qquad \eta \equiv \frac{3}{2}\frac{\dot{a}}{a}t \ , \tag{5}$$

where rec refers to just after recombination. The parameters Δ and η are defined to be unity for a matter-dominated Einstein-deSitter Universe. Since the CM is cold the growth of δ during the Einstein-deSitter epoch is proportional to the scale factor independently of the scale. Δ and η are defined so that small values are 'bad'. The value of η gives the age of the universe relative to H_o^{-1} and a large value is clearly good for the age problem. The value of Δ gives the amount of growth since recombination and thus measures the amount of primordial microwave anisotropy expected for a given amount of inhomogeneity today.

To discuss the loitering mechanism and its cosmological consequences in a conceptually clear way, we develop a simple model that has all the interesting features of the loitering scenario without any of the more complicated dynamics involved in a more sophisticated theory. Consider a closed Universe where, in addition to the conventional matter (ie. matter with both p and $\rho \geq 0$), there exists matter having an unusual equation of state $\bar{p} = \gamma\bar{\rho}$ where $-1 \leq \gamma < -\frac{1}{3}$. The energy density of such matter scales as $\bar{\rho} \propto \frac{1}{a^n}$ where $0 \leq n < 2$. The Einstein equations are

$$\left(\frac{\dot{a}}{a}\right)^2 = \frac{8\pi G}{3}\left(\rho_T - \frac{k}{a^2}\right) = \frac{8\pi G}{3}\left(\frac{\rho_o}{a^3} + \frac{A}{a^n} - \frac{k}{a^2}\right) \ ,$$
$$\frac{\ddot{a}}{a} = \frac{4\pi G}{3}(\rho_T + 3p) = \frac{4\pi G}{3}\left(\frac{\rho_o}{a^3} + \frac{(n-2)A}{a^n}\right) \ . \tag{6}$$

If all the terms in the right hand side of (5) become of comparable magnitudes at late times then a period of stationary expansion may result during which $\dot{a} \simeq \ddot{a} \simeq 0$ (loitering), or $\dot{a} \simeq$ constant $\ll 1$, $\ddot{a} \simeq 0$ (coasting).

Let us consider the question of loitering in a closed FRW cosmological model in slightly greater detail. (5) can be rewritten as

$$\frac{\dot{a}^2}{2} + V(a) = C \ , \tag{7}$$

202

where

$$V(a) = -\frac{4\pi G}{3}\left(\frac{\rho_o}{a} + \frac{A}{a^{n-2}}\right) ; \qquad C = -\frac{4\pi G}{3}k = \text{constant} . \qquad (8)$$

In this form (7) is equivalent to an equation of motion for a point particle moving with conserved energy in a potential $V(a)$. The form of the potential is plotted in Fig. 2a. We see that for the desired range $0 \leq n < 2$ the potential is convex and as n gets larger, the potential becomes flatter.

To illustrate only a few generic possibilities, consider the case when initial conditions are so imposed that the particle is moving towards the potential $V(a)$ from left to right, with a small initial kinetic energy. In this case expansion will be followed by contraction – curve A in Fig. 2b. If the particle is moving towards $V(a)$ from right to left, again with a small kinetic energy then it will rebound, this time towards larger values of a, corresponding to a Universe in which the source term dominates over both the spatial curvature as well as the matter density. This describes a Universe which first contracts and then re-expands – curve B in Fig. 2b. If the kinetic energy is large compared to the potential energy and the particle is moving from left to right, then the kinetic energy of the particle will be sufficient to take it over the potential barrier, which corresponds to the scale factor of the Universe having an inflection point – curve C in Fig. 2b. The case of singular interest from the present standpoint is the intermediate one, when the particle starts off with just enough kinetic energy to take it up to the top of the potential V(a). Cosmologically this represents an asymptotic approach to a static Einstein Universe – curve D in Fig. 2b. Value of the kinetic energy arbitrarily close to this critical value gives rise to a loitering phase of the scale factor (Fig. 1).

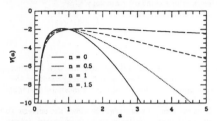

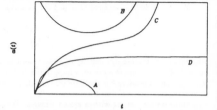

Fig. 2a) The scalar field potential $V(a)$ [Eq. (8)].

Fig. 2b) Expansion laws for different initial conditions.

The dwelling time, i.e. the length of time the Universe remains in the loitering phase can be expressed as

$$\Delta t_d = \rho(a_*)^{-1/2}\ln|\varepsilon| ; \qquad \varepsilon = \frac{\rho_{cm}(a_*)}{\rho_{cm}(a_*) - \rho_{cm*}} . \qquad (9)$$

a_* is the value of the scale factor during loitering and ρ_{cm*} is the value of the matter term (for a given fixed value of the kinetic energy) which will yield the asymptotically static Einstein Universe. For the cosmological model under consideration to look like the observed Universe at late times, we require that $\varepsilon > 0$, so that the Universe re-expands after loitering at a fixed redshift $z_* = \frac{a_o}{a_*} - 1$ for a period Δt_d (a_o is the value of the scale factor today.)

To find the redshift of the loitering phase we solve the Einstein equations (6) by simultaneously imposing the condition $\ddot{a} = \dot{a} = 0$. As a result we obtain

$$\Omega_{cmo} = \frac{2-n}{(1+z_*)^{3-n} - (3-n)z_* - 1} ; \qquad 0 \leq n < 2 , \qquad (10)$$

where Ω_{cmo} is the density parameter of CM now. A plot of z_* vs. n for different values of Ω_{cmo} is given in Fig. 3. If $n > 2$ the universe will eventually always collapse. In our model if $\Omega_{cmo} = 0.1$ today then the redshift of loitering must have occurred at $2 < z < 7.2$, the exact value depends on n.

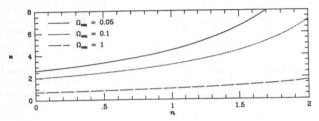

Fig. 3) A plot of redshift at loitering as a function of the power n [Eq.(10)] in the loitering Universe scenario.

The parameters Δ and η defined in (-2) for the case described in this section are presented in Fig. 4. The Universe is quite old and does develop a large growth of inhomogeneities.

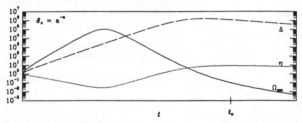

Fig. 4) The evolution of the parameters $\eta(t)$, $\Delta(t)$ and $\Omega_{cm}(t)$ for a source term with $\epsilon_s \propto a^{-n}$ [Eqs. (1), (5)].

One may also obtain an intermediate stationary stage of expansion of the Universe using a dynamical scalar field rather than a static cosmological potential. We consider the scalar field to be homogeneous and effectively massless. The total density and pressure of the CM and the scalar field are

$$\rho_T = \frac{\rho_{cm}}{a^3} + \left[\frac{1}{2}\dot{\varphi}^2 + V(\varphi) \right] ; \qquad p = \frac{1}{2}\dot{\varphi}^2 - V(\varphi) ; \tag{11}$$

where $V(\varphi)$ is the scalar field potential. The equation of motion for the scalar field is

$$\ddot{\varphi} + 3\frac{\dot{a}}{a}\dot{\varphi} + V'(\varphi) = 0 , \qquad V(\varphi) = V_o e^{-\mu\varphi} . \tag{12}$$

The constant V_o is not a fundamental parameter of the theory as it may be changed to any arbitrary positive value by redefining $\varphi \to \varphi + \text{constant}$. The quantity V_o is dimensionful and in the units we have defined above we will take $V_o = 1$. The sign of μ is also arbitrary as it can be changed by redefining the sign of φ. We will always consider $\mu > 0$. The parameter $|\mu|$ is a fundamental parameter.

If the energy density of the scalar field dominates the energy density of the Universe then the Einstein equations (4) and the equation of motion of the scalar field (11) can be solve exactly to obtain the following solutions:

$$a = A\tau^q ; \qquad q = \frac{16\pi G}{\mu^2} ; \qquad \varphi = \frac{2}{\mu}\ln\tau - \frac{1}{\mu}\ln\frac{2(48\pi G - \mu^2)}{\mu^4} ; \qquad \tau = t - t_1 ; \tag{13}$$

where t_1 and A are arbitrary. The requirement that $q > 1$ leads to an upper limit on μ: $|\mu| < \sqrt{16\pi G}$.

There is only one free physical parameter in this model and that is μ. However there are also the initial conditions which must be specified and this amounts to choosing φ and $\dot{\varphi}$ at some initial time. Let us consider the behavior of the scalar field at early times. Before the scalar field potential becomes an important contributer to $\ddot{a}$ it will also not be an important contributer to $\ddot{\varphi}$. Thus at early times the the scalar field obeys the equation of motion

$$\ddot{\varphi} \simeq -3\frac{\dot{a}}{a}\dot{\varphi} \qquad \Longrightarrow \qquad \dot{\varphi} \propto a^{-3}. \tag{14}$$

The kinetic energy density of the scalar field, $\frac{1}{2}\dot{\varphi}^2$ decays rapidly as a^{-6} while the potential energy $e^{-\mu\varphi}$ goes to a constant given enough time. It is thus likely to be a good approximation to ignore the initial kinetic energy of the scalar field and take $\varphi(t = t_i) = 0$. We are thus left with only one parameter to adjust for the initial condition, i.e. φ at $t = t_i$. With these initial conditions the scalar field acts like an effective cosmological constant initially. We will use

$$\Lambda_i \equiv e^{-\mu\varphi}|_{t=t_i} . \tag{15}$$

to parameterize the initial conditions.

If $k = +1$ and $\mu > \sqrt{16\pi G}$ then the Universe recollapses independently of the initial conditions. However, if $k = +1$ and $\mu < \sqrt{16\pi G}$ then there is critical value for Λ_i which we denote $\Lambda_{crit}(\mu)$. If $\Lambda_i < \Lambda_{crit}(\mu)$ the Universe recollapses while if $\Lambda_i \geq \Lambda_{crit}(\mu)$ it does not. If $\Lambda_i \gg \Lambda_{crit}(\mu)$ the solution is essentially the same as the $k = 0$ case while if $\Lambda_i \ll \Lambda_{crit}(\mu)$ the scalar field never plays a significant role in the dynamics. The behavior for the critical value of Λ_i is rather different than for a cosmological constant as it does not approach a strictly loitering solution at all. Rather the solution approaches

$$a = \sqrt{\frac{8\pi G}{3}} \frac{\tau}{\sqrt{\frac{16\pi G}{\mu^2} - 1}} ; \qquad \varphi = \frac{2}{\mu}\ln\tau + \frac{1}{\mu}\ln\frac{\mu^2}{4} ; \qquad \tau = t - t_2 ; \qquad \Omega_{cm} \ll 1 . \tag{16}$$

204

While cosmological expansion has not stopped in the critical case, it is only growing proportional to time with a prefactor which can be very small if μ is much less than its maximal value, $\sqrt{16\pi G}$. Thus we can have a slow-growth phase in this model with which to amplify inhomogeneities. Strict loitering at a fixed redshift has been effectively replaced by a stationary 'coasting' phase, during which $\dot{a} \simeq$ constant $\ll 1$. Thus, whereas in the case of a loitering Universe most of the growth in gravitational instability occurs at a fixed redshift – the redshift of loitering, in the case of a coasting Universe, there is a window in redshift space during which the formation of gravitationally bound systems is considerably speeded up. Indeed existing observational constraints on the cosmological parameter q_o do not rule out the possibility that we may presently be living during a coasting epoch.

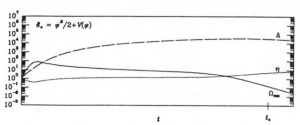

Fig. 5) The evolution of the parameters $\eta(t)$, $\Delta(t)$ and $\Omega_{cm}(t)$ for a scalar field source term (coasting scenario) [Eqs. (1), (5)].

We have seen that an intermediate stationary stage of expansion of the Universe can be very useful. In such a cosmology the age of the universe can be significantly larger than H_o^{-1} thus eliminating any problem with globular cluster ages. Also density inhomogeneities grow in such a way as to produce smaller microwave anisotropies in comparison with the present level of inhomogeneity. A low density of matter which clusters is not a hindrance to the growth of perturbations. Furthermore, any feature in the density power spectrum associated with the transition from radiation-domination to matter-domination is on a larger scale than in an Einstein-deSitter universe. This may be the explanation of the very large-scale inhomogeneities we observe. The ingredients needed to make this model work may not be so appealing. One needs a closed universe of size not that different from the present horizon size. One also needs a form of matter with $p < -\frac{1}{3}\rho$ to prevent a recollapse and one needs to tune the density of this matter to give the proper type of expansion law. Some models for matter with these properties were described in this lecture. While many of these ingredients may seem rather *ad hoc*, I feel the cosmology it produces is worth considering.

References

[1] Lynden-Bell, D., Faber, S.M., Burstein, D., Davies, R., Dressler, A., Terlevich, R.J., and Wegner, R., (1988), Ap.J.326 19; Saunders, W., Frenck, C., Rowan-Robinson, M., Efstathiou, G., Lawrence, A., Kaiser, N., Ellis, R., Crawford, J., Xiao-Yang, X., and Parry, I., (1991) Nature, 349 32.
[2] Sahni, V., Feldman, H.A., and Stebbins, A. CITA preprint 1991, Submitted to Ap.J.
[3] Janes, K., and Demarque, P., (1983) Ap.J. 264 206
[4] Jacoby, G.H., Ciadrdullo, R., and Ford, H.C., (1990) Ap.J., 356 332.
[5] Peebles, P.J.E., (1980) "Physical Cosmology", Princeton University Press, Princeton, NJ.

NATURAL LATE-TIME PHASE
TRANSITIONS AND COSMOLOGY

RICHARD WATKINS

Enrico Fermi Institute
The University of Chicago
Chicago, IL 60637

ABSTRACT: We study the cosmological implications of late time phase transitions in the context of natural particle physics models. In particular, two models are presented to generate the large scale structure in the universe purely from field dynamics and without the use of topological objects.

The idea that the density fluctuations responsible for large scale structure in the universe were generated after the decoupling of matter and radiation is attractive for a number of reasons. First, microwave background anisotropies caused by fluctuations on the surface of last scattering (the Sachs-Wolfe effect) are avoided. These anisotropies usually put strong constraints on models of structure formation. Second, since the scale of the horizon is large at late times, galaxy and cluster scale fluctuations can be made causally. Finally, fluctuations can be made with large initial amplitudes so that non-linear objects can form quickly, making it possible to make galaxies at large redshift. This idea is, however, difficult to implement from a particle physics point of view. In order to make a particle physics model that will act on cosmological scales it is necessary to introduce extremely small mass scales into the theory.

To date, two attempts have been made to implement this idea. Hill, Schramm, and Fry presented a scenario where thick domains walls are formed in a phase transition after decoupling.[1] This scenario is attractive because it is based on a particle physics model in which the small mass scale needed arises naturally and is protected to all orders against renormalization.[2] However, subsequent studies of this scenario have raised questions as to whether domain walls can provide adequate power for structure formation while still avoiding unacceptable microwave anisotropies from horizon sized walls.[3]

Press, Ryden, and Spergel[4] took a very different approach. In their scenario, density fluctuations are generated solely by the dynamics of a complex scalar field with a spontaneously broken global symmetry, without recourse to walls. The potential of the scalar field is constructed so that in its minimum the field will have a very small mass ($m_\phi \sim 10^{-28}$ eV) and a large expectation value ($\phi_0 \sim 10^{17}$ GeV). In addition to explaining the origin of the large scale structure, this model also supplies a critical density of very light particles to be the dark matter. This also explains why dark matter particles do not fall into galaxies or clusters of galaxies: their Compton wavelength is so large that they cannot be confined to the cores of these objects. While potentially successful from a cosmological point of view, this model requires a tremendous amount of fine-tuning. Not only is it required to put a very small dimensionless number ($\sim 10^{-110}$) into the theory by hand, but fine-tuning must be done at each order in perturbation theory to maintain its smallness.

In this talk, I will report on some of the results from work done with Chris Hill and Josh Frieman in which we extend and amplify some of the ideas presented in these models.[5] We study a number of particle physics models which are technically natural and exhibit late-time phase transitions. The cosmological implications of these models are discussed in detail. Because of time constraints, I will concentrate on two of these models.

It has long been known that pseudo-Goldstone bosons can have small masses which appear naturally and are protected against radiative corrections. The best known example of this is the axion, whose mass is given by $m_a = \Lambda_{QCD}^2/f_a$, where Λ_{QCD} is the QCD confinement scale and f_a is the Peccei-Quinn symmetry breaking scale. Hill and Ross have generalized the axion model, and found that in certain theories one could have other axion-like objects called schizons with mass $m_s = m_{fermion}^2/f_s$, where $m_{fermion}$ is the mass of a quark or lepton and f_s is a symmetry breaking scale comparable to the Peccei-Quinn scale.[2] If one takes $f_s \sim 10^{16}$ GeV and $m_{fermion}$ to be the hypothetical mass of a neutrino $m_\nu \sim 10^{-2}$ eV, then one finds $m_s \sim 10^{-29}$ eV, giving the schizon an astronomically large Compton wavelength $\lambda_C \sim$ Mpc.

The potential for the schizon can be written as

$$V(\phi) = C(T)m_\nu^4 \cos(\phi/f_s),$$

where $C(T) \sim 1$ and comes from considering finite temperature corrections. For the moment we will take f_s and m_ν to be free parameters. For small oscillations about the minima, $V(\phi) \simeq \frac{1}{2}(\frac{m_\nu^4}{f_s^2})\phi^2$ so that $m_s \simeq m_\nu^2/f_s$ as advertised. It can be shown that in some models we have the following unusual behavior: For temperatures $T \sim T_c \sim m_\nu$, $C(T) \sim log(T/T_c)$, so that the potential will change sign and flip, causing a phase transition. At $T \ll T_c$, $C(T)$ becomes constant ~ -1.

This property of the potential allows us to construct a scenario somewhat along the lines of Press, Ryden, and Spergel.[4] At a temperature $T_1 > T_c$ such that $m_s(T_1) = 3H(T_1)$, the Hubble damping will become small enough that the field will begin to oscillate about the high-temperature minimum of the potential. We have dubbed this the "pre-slosh". This oscillation will be damped by the expansion, so that when T_c is reached, the field will be localized close to its minimum, which when the potential flips becomes the maximum. Immediately before T_c, the energy density in the universe becomes dominated by the vacuum energy of the field, leading to a short period of rapid expansion which is terminated when once again the field rolls down to the minimum of the potential. The oscillating field creates the required critical density of schizons.

The fluctuations that we need will come from the initial misalignment of the ϕ field. For temperatures above T_1, the field behaves essentially like a massless Goldstone mode, and as such will be uncorrelated on scales greater than the horizon. When the field begins to oscillate at $T = T_1$, this misalignment will become order unity fluctuations in the field on the horizon scale at that temperature, which is roughly the Compton wavelength of the field. As the oscillation is Hubble damped, the fluctuations will shrink with the amplitude of the oscillation. At T_c, when the field again begins to roll, the small fluctuations in the field will be amplified by the negative curvature of the potential, again leading to order unity fluctuations in the field. The scale of the fluctuations will be slightly larger than the Compton wavelength due to the expansion of the universe between T_1 and T_c. This fluctuation in the field produces order unity fluctuations in the energy density of the schizons, which now dominate the energy density of the universe.

One may wonder why we needed the potential to flip at all, since during the pre-slosh we generate order unity fluctuations in the energy density of the schizons. The key point is that the fluctuations produced here are horizon sized, and the smoothness of the microwave background strongly constrains horizon-size fluctuations in the total energy density. In our model, the ϕ field contributes only a small part of the energy density at T_1, so fluctuations in the total energy density are small. During the period when the field is at the maximum of the potential, its energy is not redshifting away, while that of the matter is, so that the ϕ field energy can come to dominate. Meanwhile, the horizon has grown, so that when we produce our final density fluctuations they are well inside the horizon.

The two free parameters in the theory, f and m_ν, are constrained by the following: 1. The final density of schizons must be equal to the critical density. 2. Both the pre-slosh and the final roll down must occur between decoupling and when the earliest objects form ($z \sim 4$). 3. We must make structure on scales of interest, $L \approx 20h^{-1}$ Mpc. 4. The scale on which we have order unity density fluctuations must be far enough inside the horizon to satisfy microwave background constraints. The allowed region of parameter space is shown in Fig. 1. We have not yet included the microwave background constraint in this figure, which will lower the upper bound on f slightly. This is still work in progress.

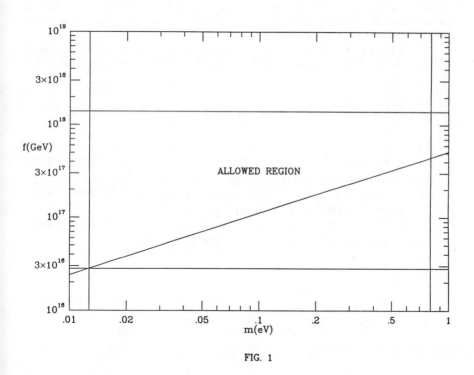

FIG. 1

Late-time phase transitions can also be used to make seeds for structure formation. Villumsen, Scherrer, and Bertschinger have considered a class of models in which the universe is dominated by hot dark matter and density perturbations are generated by randomly distributed seed masses.[6] They find that in order for this model to give the correct spectrum of fluctuations, the seeds must have a mass $m_s > 10^7 M_\odot$ and a number density $n_s < 700\,\mathrm{Mpc}^{-3}$. Such large mass objects are difficult to form before the universe is about a year old, since before this time there is too little mass inside the horizon.

For this scenario, we consider a general pseudo-Goldstone boson θ with a potential $V(\theta) = C(T)m^4 \cos(\theta/f)$, where we leave m and f as free parameters. In this case we take $C(T)$ to behave as in axion models, $i.e.$, $C(T)$ turns on like a power law at $T \sim m$. We normalize so that $C(0) = 1$. This type of potential can arise in a number of simple generalizations of the axion. In analogy with the axion, when the Compton wavelength of the field comes inside the horizon, it will begin to oscillate, creating a cold, coherent density of bosons. However, because the field is initially misaligned on scales of the horizon, there will be fluctuations in the density on this scale, making lumps of bosons.[7] We find that for reasonable choices of $m \sim$ keV and $f \sim 10^{15}$ GeV, it is easy to make seeds with the desired mass and number density.

I would like to acknowledge my collaborators J. Frieman and C. Hill. This work was supported by the DOE (at Chicago and Fermilab) and NASA (NAGW-1340 at Fermilab).

REFERENCES

1.) C.T. Hill, J.N. Fry, and D.N. Schramm, *Comments Nucl. Part. Sci.* **19** (25) 1989.

2.) C.T. Hill and G. Ross, *Phys. Lett.* **B205** (125) 1988; *Nucl. Phys.* **B311** (253) 1988.

3.) W.H. Press, B.S. Ryden, and D.N. Spergel,*Astrophys. J.* **347** (590) 1989.

4.) W.H. Press, B.S. Ryden, and D.N. Spergel,*Phys. Rev. Lett.* **64** (1084) 1990.

5.) J. Frieman, C. Hill, and R. Watkins, in preparation.

6.) J.V. Villumsen, R. Scherrer, and E. Bertschinger, *Astrophys. J.* **367** (37) 1991.

7.) C.J. Hogan and M.J. Rees, *Phys. Lett.* **B205** (228) 1988.

IV. COBE AND THE ANISOTROPY IN THE MICROWAVE BACKGROUND

THE DIFFUSE INFRARED BACKGROUND:
COBE AND OTHER OBSERVATIONS

M. G. Hauser, T. Kelsall, S. H. Moseley, Jr.,
R. F. Silverberg
Laboratory for Astronomy and Solar Physics
Goddard Space Flight Center
Greenbelt, MD 20771

T. Murdock (General Research Corporation)
G. Toller, W. Spiesman, J. Weiland
(General Sciences Corporation)

ABSTRACT

The Diffuse Infrared Background Experiment (DIRBE) on the Cosmic Background Explorer (COBE) satellite is designed to conduct a sensitive search for an isotropic cosmic infrared background radiation over the spectral range from 1 to 300 micrometers. The cumulative emissions of pregalactic, protogalactic, and evolving galactic systems are expected to be recorded in this background. The DIRBE instrument, a 10 spectral band absolute photometer with an 0.7 deg field of view, maps the full sky with high redundancy at solar elongation angles ranging from 64 to 124 degrees to facilitate separation of interplanetary, Galactic, and extragalactic sources of emission. Initial sky maps show the expected character of the foreground emissions, with relative minima at wavelengths of 3.4 micrometers and longward of 100 micrometers. Extensive modelling of the foregrounds, just beginning, will be required to isolate the extragalactic component. In this paper, we summarize the status of diffuse infrared background observations from the DIRBE, and compare preliminary results with those of recent rocket and satellite instruments.

INTRODUCTION

The search for cosmic infrared background radiation (CIBR) is a relatively new field of observational cosmology. Measurement of this distinct radiative background, expected to arise from the cumulative emissions of pregalactic, protogalactic, and galactic systems, would provide new insight into the cosmic 'dark ages' following the decoupling of matter from the cosmic microwave background radiation. Observationally, there have been no corroborated detections of the CIBR, though possible evidence for an isotropic infrared background in data from recent rocket experiments has been reported (Matsumoto et al. 1988a; Lange et al. 1990). The Diffuse Infrared Background Experiment (DIRBE) on the COBE spacecraft is the first satellite instrument designed specifically to carry out a systematic search for the CIBR. In this paper we shall report the status of the DIRBE investigation, which is still in its early phase, and will compare preliminary DIRBE measurements of the infrared sky brightness with those of the recent investigations from instruments on sounding rockets and the IRAS satellite. Current conservative upper limits on the CIBR will be compared with representative theoretical predictions to illustrate the status and challenges of measuring this important cosmic fossil.

COSMOLOGICAL MOTIVATION FOR CIBR STUDIES

Cosmological motivations for searching for an extragalactic infrared background have been discussed in the literature for several decades (early papers include Partridge and Peebles 1967; Low and Tucker 1968; Peebles 1969; Harwit 1970; Kaufman 1976). Both the cosmic redshift and reprocessing of short wavelength radiation to longer wavelengths by dust act to shift the short wavelength emissions of cosmic sources toward or into the infrared. Hence, the wide spectral range from 1 to 1000 micrometers is expected to contain much of the energy released since the formation of luminous objects, and could potentially contain a total radiant energy density comparable to that of the cosmic microwave background radiation (CMBR).

The CIBR has received relatively little attention in the theoretical literature compared to that devoted to the CMBR (Negroponte 1986), which has a central significance to Big Bang cosmology and quite distinctive and definite predictions as to its character. However, advances in infrared instrumentation, and especially the introduction of cryogenically cooled infrared instruments on space missions, has stimulated increasing attention to prediction of the character of the CIBR (Fabbri and Melchiorri 1979; Bond, Carr, and Hogan 1986; McDowell 1986; Fabbri et al. 1987; Fabbri 1988; Bond, Carr, and Hogan 1990). Though it is outside the scope of this paper to review these predictions (some of which are discussed in other papers at this Workshop), it is increasingly clear that measurement of the spectral intensity and anisotropy of the CIBR would provide important new insights into intriguing issues such as the amount of matter undergoing luminous episodes in the pregalactic Universe, the nature and evolution of such luminosity sources, the nature and distribution of cosmic dust, and the density and luminosity evolution of infrared-bright galaxies.

MEASURING THE ABSOLUTE BRIGHTNESS
OF THE INFRARED SKY

Observing the CIBR is a formidable task. Bright foregrounds from the atmosphere of the Earth, from interplanetary dust scattering of sunlight and emission of absorbed sunlight, and from stellar and interstellar emissions of our own Galaxy dominate the diffuse sky brightness in the infrared. Even when measurements are made from space with cryogenically cooled instruments, the local astrophysical foregrounds strongly constrain our ability to measure and discriminate an extragalactic infrared background. Furthermore, since the absolute brightness of the CIBR is of paramount interest for cosmology, such measurements must be done relative to a well-established absolute flux reference, with instruments which strongly exclude or permit discrimination of all stray sources of radiation or offset signals which could mimic a cosmic signal.

Table 1 lists recent experiments capable of making absolute sky brightness measurements in the infrared (for a compilation including some earlier measurements, see Negroponte 1986). Instruments or detector channels designed specifically to measure that part of the spectrum dominated by the CMBR have been excluded. Murdock and Price (1985) flew an absolute radiometer with strong stray-light rejection on a sounding rocket in 1980

Table 1.   ABSOLUTE INFRARED BACKGROUND MEASUREMENTS

REFERENCE	WAVELENGTH (μm)	BEAM SIZE (DEG)	SKY COVERAGE	SOLAR ELONGATION* (DEG)	DATES
Murdock & Price 1985	2 - 30	0.1x0.25	40°-80°rad cone b=-23, l=155	($\lambda-\lambda_\odot$= 22- 180)	8/18/80; 7/31/81
Matsumoto et al. 1988a	0.7 - 5	4	32° rad cone b=55, l=150	($\lambda-\lambda_\odot$= 113- 180)	1/13/84
Lange et al. 1990	100 - 300	7.6	15° rad cone b=35, l=203	140- 170	2/22/87
IRAS (ZOHF, vers. 3)	12 - 100	0.5x0.5	All	80-100; 60-120	1/26/83- 11/22/83
COBE/DIRBE	1 - 300 1 - 3.4 pol.	0.7x0.7	All	64-124	11/18/89- 9/21/90
COBE/DIRBE	1 - 4.5 1 - 3.4 pol.	0.7x0.7	All	64-124	9/21/90- present
COBE/FIRAS (high freq.)	100 - 500	7	All	94	11/18/89- 9/21/90

*In this paper, elongation is the angle between the instrument line-of-sight and the observer-Sun line. The geocentric ecliptic longitude is denoted by λ.

and 1981. Their primary objective was measuring scattering and emission from interplanetary dust, and no attempt was made to extract an extragalactic component. Matsumoto et al. (1988a) flew a near-infrared experiment on a rocket in 1984. They have reported possible evidence for an isotropic residual near 2 micrometers, perhaps in a line feature, for which they cannot account in their models of emission from the interplanetary medium and the Galaxy. This group has flown a modified instrument early in 1990 to investigate further this result (Matsumoto, private communication). Lange et al. (1990) report a detailed analysis of the data in the high-frequency channels of the rocket experiment flown by Matsumoto et al. (1988b) in 1987. After correlating with galactic and interplanetary signals, upper limits and a possibly significant isotropic residual are reported. The IRAS sky survey instrument, though not specifically designed for absolute background measurements, had an approximate zero point calibration, and was, within the limits of long-term stability, capable of good relative total sky brightness measurements. Uncertainties in the IRAS absolute calibration have impeded efforts to extract an estimate of the CIBR (Rowan- Robinson 1986). Table 1 includes two entries for the DIRBE instrument on COBE: one for the period of operation with liquid helium in the dewar, when measurements were made over the 1 - 300 μm spectral range, and one for the period following helium depletion, when the 1 - 4.5 μm sensors continued to operate. The DIRBE experiment is discussed in some detail below. Finally, Table 1 lists the high frequency channel (100 to 500 μm) of the COBE FIRAS experiment. Though results are not yet available from these data, the FIRAS, with its all-sky coverage, excellent stray light rejection, absolute calibration, and high sensitivity, promises to be an important instrument for CIBR studies. Quantitative comparison of the measurements from the experiments listed in Table 1 and a summary of current CIBR limits are discussed below.

THE DIFFUSE INFRARED BACKGROUND EXPERIMENT

The Diffuse Infrared Background Experiment (DIRBE) on the COBE mission is the first space experiment designed primarily to measure the CIBR. The aim of the

DIRBE is to conduct a definitive search for an isotropic CIBR, within the constraints imposed by the local astrophysical foregrounds. The experimental approach is to obtain absolute brightness maps of the full sky in 10 photometric bands (J[1.2μm], K[2.3μm], L[3.4μm], and M[4.9μm]; the four IRAS bands at 12, 25, 60, and 100 μm; and 120-200 and 200-300 μm bands). In order to facilitate discrimination of the bright foreground contribution from interplanetary dust, linear polarization is also measured in the J, K, and L bands, and all celestial directions are observed hundreds of times at all accessible solar elongation angles (depending upon ecliptic latitude) in the range 64 to 124 degrees. The instrument is designed to achieve a sensitivity for each field of view of λI_λ = 10^{-13} W cm^{-2} sr^{-1} (1 σ, 1 year). This level is well below estimated CIBR contributions (e.g., Bond, Carr, Hogan 1986; see Fig. 3a). For further general information about the COBE mission, see the descriptions by Mather (1982) and Gulkis et al. (1990). Early scientific results from the mission have been summarized by Mather et al. (1990).

The DIRBE instrument is an absolute radiometer, utilizing an off-axis folded Gregorian telescope with a 19-cm diameter primary mirror. The optical configuration (Magner 1987) is carefully designed for strong rejection of stray light from the Sun, Earth limb, Moon or other off-axis celestial radiation, or parts of the COBE payload (Evans, 1983). Stray light rejection features include both a secondary field stop and a Lyot stop, super-polished primary and secondary mirrors, a reflective forebaffle, extensive internal baffling, and a complete light-tight enclosure of the instrument within the COBE dewar. Additional protection is provided by the Sun and Earth shade surrounding the COBE dewar, which prevents direct illumination of the dewar aperture by these strong local sources. The DIRBE instrument, which is maintained at a temperature below 2 K within the dewar, measures absolute brightness by chopping between the sky signal and a zero-flux internal reference at 32 Hz using a tuning fork chopper. Instrumental offsets are measured by closing a cold shutter located at the prime focus. All spectral bands view the same instantaneous field-of-view, 0.7 deg x 0.7 deg, oriented at 30 degrees from the axis about which the COBE spacecraft spins at 0.8 rpm. The rotation modulates the solar elongation angle of the DIRBE line-of-sight by 60 degrees during each rotation, and allows the DIRBE to sample fully 50% of the celestial sphere each day. Four highly reproducible internal radiative reference sources can be used to stimulate all detectors when the shutter is closed to monitor the stability and linearity of the instrument response. The redundant sky sampling and frequent response checks provide precise photometric closure over the sky for the duration of the mission. Calibration of the photometric scale is obtained from observations of isolated bright celestial sources. Careful measurements of the beam shape in pre-flight system testing and during the mission using scans across bright point sources allow conversion of point-source calibrations to surface brightness calibrations.

Mapping of the sky with the DIRBE instrument began on November 21, 1989, though checkout and tuning of spacecraft and instrument operations occupied much of the time until December 10. Routine surveying of the sky continued until depletion of the liquid helium on September 21, 1990. The interior of the dewar has subsequently warmed slowly to a present temperature of about 40 K. Though the detectors at wavelengths longer than 5 μm provide no useful data following helium depletion, the instrument mechanisms continue to operate normally and the J, K, L, and M band detectors (all of which are InSb photovoltaic detectors) continue to provide usable data at reduced sensitivity. The present plan is to continue to operate the DIRBE through a second year if the quality of data warrants it. This will allow completion of sky mapping over one full orbit around the Sun to aid modeling of the interplanetary dust foreground, and will enable a search for temporal variation in this foreground during the second year.

The data obtained during the helium-temperature phase of the mission, based upon quick-look examination, are of excellent photometric quality, showing good sensitivity, stability, linearity, and stray-light immunity. Few artifacts are apparent other than those induced by energetic particles in the South Atlantic Anomaly and variations in instrument temperature. Both of these effects will be removed in final data

processing. Strong rejection of off-axis radiation
sources is confirmed by the absence of response to the
Moon (which saturates the response in all detectors when
in the field of view) until it comes within about 3
degrees of the field of view. The sensitivity per field
of view (shown in Figure 3b), based upon noise measured
with the shutter closed and response determined from
measurements of known celestial sources, is better than
the goal stated above except at 150 and 250 μm. The
sensitivities are worse than the design goal by factors
of 10 and 4 respectively at these wavelengths.
Nevertheless, averaging over many high galactic latitude
fields-of-view will allow very sensitive measurements at
these wavelengths. The nuclear radiation environment in
orbit caused very little response change (<1%) in all
detectors except the Ge:Ga photoconductors used at 60 and
100 μm. Thermal and radiative annealing procedures
applied to the Ge:Ga detectors following passages through
the South Atlantic Anomaly will allow response correction
to about 1% at these wavelengths. It is expected that
fully reduced DIRBE sky maps will have photometric
consistency over the sky better than 2% at each
wavelength, nearest-neighbor band-to-band (color)
brightness accuracy of 3% or better, and absolute
intensity scale accuracy better than 20%.

Qualitatively, the initial DIRBE sky maps show the
expected character of the infrared sky. For example, at
1.2 μm stellar emission from the galactic plane and from
isolated high latitude stars is prominent. Zodiacal
scattered light from interplanetary dust is also
prominent. At fixed ecliptic latitude the zodiacal light
decreases strongly with increasing solar elongation
angle, and at fixed elongation angle it decreases with
increasing ecliptic latitude. These two components
continue to dominate out to 3.4 μm, though both become
fainter as wavelength increases. A composite of the 1.2,
2.3, and 3.4 μm images was presented by Mather et al.
(1990). Because extinction at these wavelengths is far
less than in visible light, the disk and bulge stellar
populations of the Milky Way are dramatically apparent in
this image. At 12 and 25 μm, emission from the
interplanetary dust dominates the sky brightness, again
strongly dependent upon ecliptic latitude and elongation
angle. At wavelengths of 60 μm and longer, emission from

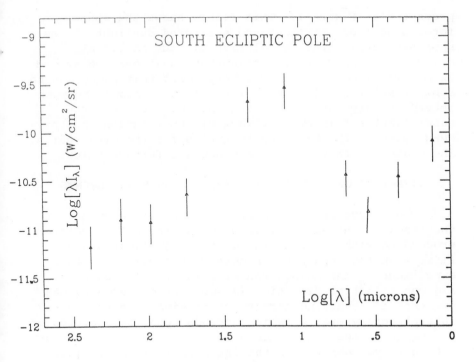

Figure 1. DIRBE measurements of the sky brightness at the South Ecliptic Pole.

---

the interstellar medium dominates the galactic brightness, and the interplanetary dust emission becomes progressively less apparent. The patchy infrared cirrus noted in IRAS data (Low et al. 1984) is evident at all wavelengths longer than 25 μm. The DIRBE data will clearly be a valuable new resource for studies of the interplanetary medium and Galaxy as well as the search for the CIBR.

In searching for the extragalactic infrared background, the most favorable conditions are directions and wavelengths of least foreground brightness. In general, because of the strong interplanetary dust foreground and the relatively modest gradient of that foreground over the sky, the infrared sky is faintest at high ecliptic latitude. A preliminary DIRBE spectrum of the sky brightness toward the south ecliptic pole is

presented in Figure 1. This figure shows the strong
foreground from starlight and scattered sunlight at the
shortest wavelengths, a relative minimum at 3.4 μm,
emission dominated by interplanetary dust peaking around
12 μm, and generally falling brightness from there out to
submillimeter wavelengths. The 3.4 μm "window", between
interplanetary dust scattering and emission, and the
submillimeter "window", between cirrus emission and the
CMBR, present the best opportunities for the CIBR search.
The status of this search is discussed further below.

COMPARISON OF INFRARED SKY BRIGHTNESS MEASUREMENTS

There have been a number of recent measurements of
the absolute infrared sky brightness (Table 1). Before
proceeding with a discussion of their implications, we
first wish to compare DIRBE data with those from other
experiments. The comparison of brightnesses measured in
a given celestial direction is somewhat confused by the
presence of a varying foreground arising from the
interplanetary medium, which depends upon the position of
the Earth at the time of observation, as well as upon any
intrinsic variability in the interplanetary dust cloud.
We have attempted to minimize differences attributable to
foreground changes by selecting DIRBE data obtained on
the same day of year (day number), and hence solar
elongation angle, as that of the comparison experiment.
We emphasize that these comparisons are made between as-
observed sky brightnesses, not components of emission
inferred from the measurements.

The near-infrared data of Matsumoto et al. (1988a)
toward the north galactic pole are compared with DIRBE
data in Figure 2a. The absolute brightnesses are
consistent within the indicated error bars at 2.2 and 4.9
μm. However, at 1.2 μm the DIRBE measurement is a factor
of 1.5 above the rocket data, and at 3.4 μm the DIRBE
result is a factor of 1.6 less. We can with some
confidence conclude that the shape of the energy
distribution over this spectral range is quite different
in the two experiments, with the DIRBE finding a much
steeper slope over the 1.2 to 3.4 μm range. This
conclusion is strengthened by the fact that the indicated
error bars for the DIRBE data are entirely dominated by
the systematic absolute calibration uncertainty. For

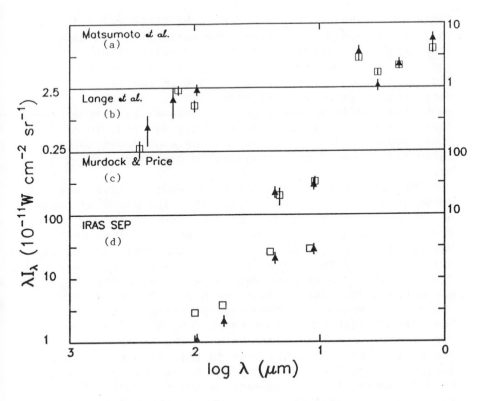

Figure 2. Comparison of absolute sky brightness
measurements made with the DIRBE with those of previous
experiments. In all cases, the DIRBE data are shown as
filled triangles, with error bars including estimated
systematic calibration uncertainties. Table 1 gives
observational parameters of the experiments compared with
the DIRBE results. Note that the scale on the ordinate
is logarithmic in each panel, with labels alternately on
the left and right. (a) Matsumoto et al. (1988a) rocket
instrument observations near the north galactic pole.
(b) Lange et al. (1990) presentation of the high
frequency rocket data of Matsumoto et al. (1988b) near b
= 20 deg, l = 198 deg. (c) Murdock and Price (1985)
rocket observations at the north ecliptic pole.   (d)
IRAS data from Vers. 3 of the Zodiacal Observation
History File observed toward the south ecliptic pole on
19 Feb 83.

these four wavelengths, however, the DIRBE celestial
calibration was derived from observations of the same
star (Sirius), yielding band-to-band calibration
uncertainties much smaller than the error bars shown in
Fig. 2a. Since a steeper 1.2 to 3.4 μm energy
distribution is more consistent with the expected
contributions from starlight and zodiacal light than the
data of Matsumoto et al. 1988a (e.g., see Fig. 10 of
their paper), these DIRBE data raise some doubt about
their inference of an extragalactic component near 2.2
μm.

The far-infrared data of Lange et al. (1990) are
compared with DIRBE data in Fig. 2b. The paper of Lange
et al. presents an analysis of data from the high
frequency channels of the measurements described by
Matsumoto et al. (1988b), including the variation of sky
brightness over the small circle scanned during their
flight. Since their data were obtained at a solar
elongation angle larger than reached by DIRBE, we have
chosen to compare with the measured point at their
smallest elongation (140 degrees), which is near b = 20
deg, l = 198 deg. We have averaged the DIRBE data over
the area of their 7.6-deg beam in this direction. The
largest elongation reached by the DIRBE observations in
this direction is 124 degrees. Since the zodiacal
emission is not bright at long wavelengths, and varies
little with elongation at large angles, no attempt has
been made to correct for the slight difference in
observing conditions. The DIRBE preliminary calibration
is particularly uncertain at 150 and 250 μm; within these
large uncertainties, the two experiments are reasonably
consistent at these wavelengths. At 100 μm the DIRBE
result is a factor of 1.8 brighter, a difference larger
than the estimated 20% uncertainty in DIRBE absolute
calibration.

The north ecliptic polar brightnesses at 10 and 20
μm measured by Murdock and Price (1985) are compared with
DIRBE data in Fig. 2c. These measurements agree within
10%, well within the estimated uncertainties.

IRAS measurements toward the south ecliptic pole are
compared with DIRBE data in Fig. 2d. The two experiments
are seen to agree reasonably well at 12 and 25 μm, but
the DIRBE results are substantially fainter at 60 and 100
μm, reaching a factor of 2.6 at 100 μm. We have made a

more detailed comparison with IRAS data at several points on the sky. By choosing points of different brightness, and assuming the DIRBE photometry to be correct, we can determine both zero point and responsivity 'errors' for the IRAS data. We find evidence for zero point errors, evidently not constant in time, of a few MJy/sr at all IRAS wavelengths. These are most significant (as a fraction of total brightness) at the longest wavelengths. We also find that the adopted IRAS dc responsivity at 60 and 100 μm was evidently too low, making reported sky brightnesses too high. The responsivity problem can be at least qualitatively understood in terms of the detector characteristics and the data reduction procedures used for IRAS image-format data (see the IRAS Explanatory Supplement (1988), pp. IV-9 and IV-10 for a discussion of the dependence of detector responsivity on frequency and incident power). The combined effect of the zero point and responsivity discrepancies is particularly large at 100 μm toward faint sky regions, as illustrated in Fig. 2d. Since IRAS detector responsivity depended on source angular scale (IRAS was a scan-modulated instrument) and source brightness, no simple conversion between the total brightness data from the two experiments is universally applicable. It should be noted that this discussion applies only to surface brightness measurements; point source flux calibration is not subject to similar problems.

CURRENT LIMITS ON THE CIBR

To meet the cosmological objective of measuring the CIBR, the foreground light from interplanetary and galactic sources must be discriminated from the total observed infrared sky brightness. This task requires extensive careful correlation studies and modelling, which in the case of the DIRBE investigation is just beginning. A conservative upper limit on extragalactic light is the total observed brightness in a relatively dark direction. The sky brightness at the south ecliptic pole presented in Fig. 1 is a fair representation of the best current limits from the DIRBE. As noted above, the faintest foregrounds occur at 3.4 μm, in the minimum between interplanetary dust scattering of sunlight and re-emission of absorbed sunlight by the same dust, and

```
================================================================
```
Table 2.   COSMIC INFRARED BACKGROUND LIMITS
```
================================================================
```

REFERENCE	WAVELENGTH (μm)	λI_λ (10^{-11} W cm^{-2} sr^{-1})
```
================================================================
```
Matsumoto et al. 1988a[1]	1.27	<0.6
	2.16	1.0±.37
	3.8	1.1±.34
	5.0	<0.5
Lange et al. 1990[1]	100	<0.32
	135	0.55±0.16
	275	<0.29
DIRBE (SEP)[2]	1.2	8.3±3.3
	2.3	3.5±1.4
	3.4	1.5±0.6
	4.9	3.7±1.5
	12	29 ±12
	22	21 ±8
	55	2.3±1
	96	1.2±0.5
	151	1.3±0.7
	241	0.7±0.4

NOTES:
1. values or upper limits for an isotropic component of the total observed sky brightness.
2. value of the total observed sky brightness in a dark direction, with no separation of components.
```
================================================================
```

longward of 100 μm, where interstellar dust emission begins to decrease. Through careful modelling, we hope to be able to discriminate isotropic residuals at a level as small as 1 percent of the foregrounds. These near-infrared and submillimeter 'windows' will allow the most sensitive search for, or limits upon, the elusive cosmic infrared background.

The DIRBE data at the south ecliptic pole are listed in Table 2 along with CIBR data from the Matsumoto et al. (1988a) and Lange et al. (1990) papers. These investigators have attempted to discriminate the various

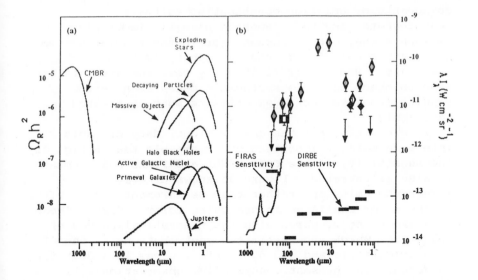

Figure 3. (a) Representative theoretical estimates for backgrounds arising from pre- and proto-galactic objects in a dust-free universe (Carr 1988; Bond, Carr, and Hogan 1986). Ω_R is the energy density in radiation per logarithmic frequency interval, in units of the critical density for closure, $\rho_C = 2 \times 10^{-29}$ h^2 g cm^{-3}, where h is the Hubble constant in units of 100 km/sec per Mpc. (b) Current observational limits on the isotropic extragalactic infrared background radiation. The scale on the ordinate corresponds to that in panel (a). The data are from Table 2: the south ecliptic pole sky brightness measured by the DIRBE (open diamonds); estimated near-infrared isotropic residuals from Matsumoto et al. (1988a) (filled diamonds and upper limits); estimated far-infrared isotropic residuals from Lange et al. (1990) (open square and upper limits). The estimated minimum detectable brightness (1 σ) per instrument field of view by the COBE experiments in a nominal 1 year mission is also shown: a solid horizontal bar for each DIRBE band; the solid curve for the FIRAS high frequency channel (per spectral resolution element).

components of emission contributing to their
measurements, and have reported possible isotropic
residuals they cannot account for in their solar system
or galactic emission models, or upper limits on such
residuals. Because of limited sky coverage and little
time to check possible systematic measurement errors,
results from the rocket experiments must be treated
cautiously.

In order to put the observational data of Table 2
into a cosmological context, they are plotted in Fig. 3b.
The sensitivity of the COBE DIRBE and FIRAS measurements
is also illustrated. Figure 3b shows clearly that
foreground emission, rather than measurement sensitivity,
is the major challenge in searching for the CIBR. These
data are to be compared with the theoretical curves of
Figure 3a, which show estimated contributions to the CIBR
from pregalactic and protogalactic sources in a dust-free
universe. The present conservative observational limits
are beginning to constrain some of the theoretical
models, though in a dusty universe energy from these
sources can be redistributed farther into the infrared
than shown in Fig. 3a. If the foreground components of
emission can confidently be identified, the current COBE
measurements will seriously constrain (or identify) the
CIBR across the infrared spectrum. However, the spectral
decade from about 6 to 60 μm will have relatively weak
limits until measurements are made from outside the
interplanetary dust cloud.

CONCLUSION

The cosmic infrared background radiation promises to
enhance our understanding of cosmic evolution since
decoupling. The high quality and extensive new
measurements of the absolute infrared sky brightness
obtained with the DIRBE and FIRAS experiments on the COBE
mission promise to allow a definitive search for this
elusive background, limited primarily by the difficulty
of distinguishing it from bright astrophysical
foregrounds. Careful photometric reduction and modelling
of the COBE data will yield important new cosmological
and astrophysical insights.

ACKNOWLEDGMENTS

The authors gratefully acknowledge the contributions to this report by their colleagues on the COBE Science Working Group and the many other participants in the COBE Project, especially H. Freudenreich, J. A. Skard, and A. Panitz. We thank A. Lange for permission to quote unpublished results, and useful discussions of his results. The COBE mission is funded by NASA's Astrophysics Division. Mission management, design, development, operation, and data reduction are the responsibility of the Goddard Space Flight Center.

REFERENCES

Bond, J. R., Carr, B. J., and Hogan, C. J. 1986, Ap. J., 306, 428.
Bond, J. R., Carr, B. J., and Hogan, C. J. 1990, "Cosmic Backgrounds from Primeval Dust", Ap. J. (to be publ.).
Carr, B. J. 1988, Comets to Cosmology, A. Lawrence (ed.), Proc. of the Third International IRAS Conference, London, July 6-10, 1987, pp. 265-278, Springer Verlag.
Evans, D. C. 1983, SPIE Proc., 384, Generation, Measurement, and Control of Stray Radiation III, R. P. Breault (ed.), pp. 82-89.
Fabbri, R. 1988, Ap. J., 334, 6.
Fabbri, R. and Melchiorri, F. 1979, Astr. and Ap., 78, 376.
Fabbri, R., Andreani, P., Melchiorri, F., and Nisini, B. 1987, Ap.J. 315, 12.
Gulkis, S., Lubin, P. M., Meyer, S. S., and Silverberg, R. F. 1990, Sci. Am., 262, 132.
Harwit, M. 1970, Rivista del Nuovo Cimento Vol. II, 253.
IRAS Explanatory Supplement 1988, Infrared Astronomical Satellite (IRAS) Catalogs and Atlases, Vol. 1, C. A. Beichman, G. Neugebauer, H. J. Habing, P. E. Clegg, and T. J. Chester (eds.), NASA RP-1190.
Kaufman, M. 1976, Ap. Sp. Sci., 40, 369.
Lange, A. E., Richards, P. L., Hayakawa, S., Matsumoto, T., Matsuo, H., Murakami, H., and Sato, S. 1990, "The 100 to 300 micron Diffuse Background: Spectrum of the Infrared Cirrus and New Limits on the Extragalactic Background", (private comm.).
Low, F. J. and Tucker, W. H. 1968, Phys. Rev. Lett., 21, 1538.
Low, F. J., et al. 1984, Ap. J. (Lett.), 278, L19.
Magner, T. J. 1987, Opt. Eng., 26, 264.
Mather, J. C. 1982, Opt. Eng., 21, 769.
Mather, J. C., et al. 1990, Observatories in Earth Orbit and Beyond, Proc. of IAU Colloquium 123, Y. Kondo (ed.), Kluwer Acad. Publ., Dordrecht.
Matsumoto, T., Akiba, M., and Murakami, H. 1988a, Ap. J., 332, 575.
Matsumoto, T., Hayakawa, S., Matsuo, H., Murakami, H., Sato, S., Lange, A. E., and Richards, P. L. 1988b, Ap. J., 329, 567.

McDowell, J. C. 1986, MNRAS, 223, 763.
Murdock, T. L., and Price, S. D. 1985, Astr. J., 90, 375.
Negroponte, J. 1986, MNRAS, 222, 19.
Partridge, R. B. and Peebles, P. J. E. 1967, Ap. J., 148, 377.
Peebles, P. J. E. and Partridge, R. B. 1967, Ap. J., 148, 713.
Peebles, P. J. E. 1969, Phil. Trans. Royal Soc. London, A, 264,
  279.
Rowan-Robinson, M. 1986, MNRAS, 219, 737.

AUDIENCE QUESTIONS

B. Partridge: Have fluctuations in sky brightness yet
been analyzed in the DIRBE data?

Answer: This has not yet been done, though we shall
attempt to do so. Identifying fluctuations of
cosmological significance will be observationally
difficult due to the patchiness of the foreground
arising from the Galaxy.

EARLY RESULTS FROM THE FAR INFRARED
ABSOLUTE SPECTROPHOTOMETER (FIRAS)

J. C. Mather, E. S. Cheng, and R. A. Shafer
Laboratory for Astronomy and Solar Physics
Goddard Space Flight Center, Greenbelt, MD 20771

R. E. Eplee and R. B. Isaacman
General Sciences Corporation, Laurel, MD

D. J. Fixsen
University Space Research Association, Greenbelt, MD

S. M. Read
STX Corporation, Lanham, MD

S. S. Meyer and R. Weiss
Massachusetts Institute of Technology, Cambridge, MA

E. L. Wright
UCLA Department of Astronomy, Los Angeles, CA

ABSTRACT

The Far Infrared Absolute Spectrophotometer *(FIRAS)* on the Cosmic Background Explorer *(COBE)* mapped 98% of the sky, 60% of it twice, before the liquid helium coolant was exhausted. The *FIRAS* covers the frequency region from 1 to 100 cm^{-1} (10 mm to 0.1 mm wavelength) with a 7° angular resolution. The spectral resolution is 0.2 cm^{-1} for frequencies less than 20 cm^{-1} and 0.8 cm^{-1} for higher frequencies. Preliminary results include: a limit on the deviations from a Planck curve of 1% of the peak brightness from 1 to 20 cm^{-1}, a temperature of 2.735 ± 0.06 K, a limit on the Comptonization parameter y of 10^{-3}, on the chemical potential parameter μ of 10^{-2}, a strong limit on the existence of a hot smooth intergalactic medium, and a confirmation that the dipole anisotropy spectrum is that of a Doppler shifted blackbody. Although there are many unresolved issues about the data processing and analysis of systematic errors, some of which are described here, the instrument is expected to achieve its goal of an accuracy and rms sensitivity for νI_ν better than 10^{-9} W/m^{2}sr, for each 7° beamwidth and each 5% spectral resolution element, for frequencies from 1 to 20 cm^{-1}.

INTRODUCTION

The cosmic microwave background radiation (CMBR) from the Big Bang, and the cosmic infrared background radiation (CIB) from the first objects to form after the Big Bang, are the subjects of NASA's first satellite designed primarily for observational cosmology, the *COBE*. The mission has been described by Mather (1982) and Gulkis *et al.* (1990), and early results have been summarized by Mather *et al.* (1990a,b).

We summarize here the key events in the expanding universe which influence these backgrounds. Processes in the early expansion of the universe such as inflation and subsequent nucleosynthesis, which take place within the first few minutes after the initial explosion, set the scale for the largest anisotropies and inhomogeneities. The number of quanta in the CMBR is essentially fixed when the time scales for photon creation and destruction exceed the expansion time scale, which occurs at a redshift $z \sim 10^6$ when photons outnumber matter particles by a factor of $\sim 10^8$. After this epoch the spectrum of the CMBR can be modified to a Bose–Einstein distribution with a chemical potential μ by such processes as black hole formation, cosmic string annihilation, matter/antimatter annihilation, or the decay of exotic particles. After a redshift of $z \sim 3 \times 10^4$, the processes which re-establish equilibrium become too slow to achieve a Bose–Einstein distribution, and any other energy release into the radiation field mediated by hot electrons produces a mixture of blackbodies at different temperatures, characterized by the parameter y (Zel'dovich and Sunyaev 1969). After the decoupling of matter and radiation at $z \sim 10^3$, the photons are free to move unimpeded, and the major features of the spectrum and angular distribution of the CMBR as now observed should have been established.

The processes leading to the formation of galaxies and large scale structures, as well as the epoch when these processes occurred, are major unknowns in current cosmology. One can bracket the epoch by noting that the earliest quasars have a $z<5$ and that galaxies at their present size would overlap at z between 20 and 100. The formation of galaxies may well be accompanied by the creation of a hot intergalactic medium, which could produce anisotropies and spectral distortions in the CMBR as well as a signature in the infrared. Furthermore, it is expected that precursors to the large scale structures should have left anisotropies and possibly spectral distortions in the CMBR and X-ray background.

Galactic and Solar System foreground radiation limit the ability of solar system based instruments to measure the cosmic background radiation. The primary such background for the *FIRAS* instrument is the interstellar dust, which absorbs starlight and reradiates at a temperature of the order of 22 K. A full sky survey has been made to find the directions where such interference is a minimum and to enable accurate extrapolations to a viewpoint outside our Galaxy.

FIRAS INSTRUMENT DESCRIPTION

The purpose of the *FIRAS* is to compare the spectrum of the CMBR with that of a precise blackbody, enabling the measurement of very small deviations from a Planckian spectrum. The *FIRAS* instrument covers two frequency ranges, a low frequency channel from 1 to 20 cm^{-1} and a high frequency channel from 20 to 100 cm^{-1}. It has a 7° diameter beamwidth, established by a non-imaging parabolic concentrator, which has a flared aperture to reduce diffractive sidelobe responses. The instrument is calibrated by a full aperture, temperature-controlled external blackbody, which can be moved into the beam by command.

The *FIRAS* is the first instrument to measure the background radiation and compare it with such an accurate external full beam calibrator in flight. The spectral resolution is obtained with a polarizing Michelson interferometer, with separated input and output beams to permit fully symmetrical differential operation. One input beam views the sky or the full aperture calibrator, while the second input beam views an internal temperature controlled reference blackbody, with its own parabolic concentrator. Both input concentrators and both calibrators are temperature controlled and can be set by command to any temperature between 2 and 25 K. The standard operating condition is for the two concentrators and the internal reference body to be commanded to match the sky temperature, thereby yielding a nearly nulled interferogram and reducing almost all instrumental gain errors to negligible values.

The instrument's apodized spectral resolution is limited by the maximum stroke length to 0.2 cm^{-1} in the low frequency channels, and by the microprocessor buffer size and telemetry bandwidth to 0.8 cm^{-1} in the high frequency channels. Beam divergence within the instrument limits the spectral resolution to $\sim 1\%$ and shifts the effective frequency scale by 0.4%. The rms sensitivity for frequencies from 2 to 20 cm^{-1} is $\nu I_\nu < 10^{-9}$ W/m^{2}sr per 7° field and per 5% spectral resolution element after one year of operation. There are four large area (0.5 cm^2) composite bolometer detectors, two on each output of the spectrometer. The separation of low and high frequencies is accomplished by a capacitive grid dichroic filter.

The external calibrator determines the accuracy of the instrument for broad band sources like the cosmic background radiation. It is a re-entrant cone shaped like a trumpet mute, made of Eccosorb CR-110 iron-loaded epoxy. The angles at the point and groove are 25°, so that a ray reaching the detector has undergone 7 specular reflections from the calibrator. This can be seen by approximating the circular groove by a straight groove and drawing the multiple images of each surface in the other. The calculated reflectance for this design, including diffraction and surface imperfections, is less than 10^{-4} from 2 to 20 cm^{-1} . Measurements of the reflectance of an identical calibrator in an identical antenna using coherent radiation at 1 cm^{-1} and 3 cm^{-1} frequencies confirm this calculation. The instrument is calibrated by measuring spectra with the calibrator in the sky horn while operating all other controllable sources within the instrument at a sequence of different temperatures. The overall responsivity and the emissivity of each source can be determined relative to the external calibrator by solving a set of coupled linear equations.

The first results of the *FIRAS* (Mather *et al.* 1990a, Cheng *et al.* 1990) may be summarized as follows. The intensity of the background sky radiation is consistent with a blackbody at 2.735 ± 0.06 K. Deviations from this blackbody at the spectral resolution of the instrument are less than 1% of the peak brightness. The quoted uncertainty in temperature is due to an uncertainty in the thermometer calibration; we expect to reduce this uncertainty by additional tests. The measured spectrum is shown in Figure 1, and is converted to temperature units in Figure 2, where it is compared to previous measurements. The deviations can be fitted to the Sunyaev-Zel'dovich form for Comptonization (Zel'dovich and Sunyaev 1969), giving a limit on the y parameter of $|y| < 10^{-3}$ (3 σ). They can also be fitted to a Bose-Einstein distribution with a chemical potential, giving a limit of $|\mu| < 10^{-2}$

(3 σ). There is a strong limit on the existence of a smooth hot intergalactic medium: it can contribute less than 3% of the X-ray background radiation even at a reheating time as recent as $z = 2$. There is no evidence of a distortion of the spectrum, such as that reported by Matsumoto *et al.* (1988), and the measured temperature is consistent with previous reports and the recent rocket result of Gush *et al.* (1990).

The variation of the spectrum with position in the sky as measured by the *FIRAS* is dominated by the dipole pattern of the Doppler shift of the background temperature, plus a variation in the interstellar dust emission. A preliminary analysis of three weeks of data taken near the Galactic poles shows exactly such a variation. A determination of the spectrum of the dipole was made by calculating the average spectra in two large circular regions of the sky each of angular diameter 60°, one centered at $(\alpha, \delta) = (11.1^h, -6.3°)$ and the other at $(23.1^h, 6.3°)$, which lie along opposite ends of the dipole axis as determined by the *DMR*. The difference between these spectra can be fit extremely well by the difference of two blackbodies, and is consistent with a peak dipole amplitude of 3.3 ± 0.3 mK and the assumed dipole direction.

SYSTEMATIC ERROR ANALYSIS AND ULTIMATE ACCURACY

The *FIRAS* data set includes about 500,000 interferograms per detector, each susceptible to a wide variety of possible kinds of errors. We have characterized the possible errors and begun to develop software to deal with all of them on a routine basis, since automation is required for such a large data set. These errors, and the approaches to deal with them, are outlined below. We believe that every one can be understood and dealt with in software to the degree necessary to meet our accuracy requirements. They include:

- telemetry errors
- cosmic rays striking the detectors
- cosmic rays striking the control sensors for the mirror mechanism and stopping the scanning
- errors in thermometer calibration
- possible gradients in temperature in the calibrators
- a shift in the frequency scale
- detector nonlinearities
- stray signals from the Earth when it rises above the top of the shield and shines on the instruments
- stray light from the Moon
- Sun glints from defects in the sunshade edge
- thermal crosstalk from the *DIRBE*
- emission from the *DMR* local oscillators
- vibrations of the moving mirror
- variations of the optical response across the antenna beam and
- multipass phenomena within the interferometer.

Telemetry errors are detected in numerous ways. The data stream from the spacecraft contains words whose values are known, such as frame sync words, clocks, and frame

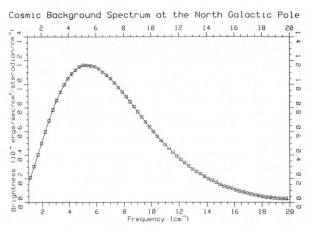

Figure 1. Preliminary spectrum of the cosmic microwave background from the *FIRAS* instrument, compared to a blackbody. Boxes are measured points and show size of assumed 1% error band.

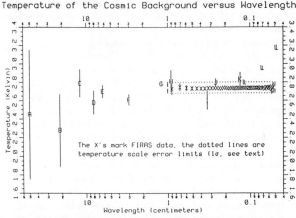

Figure 2. Composite plot of recent measurements of the temperature of the sky. A = Sironi *et al.* (1987), B = Levin *et al.* (1987), C = Sironi and Bonelli (1986), D = De Amici *et al.* (1988), E = Mandolesi *et al.* (1986), F = Kogut *et al.* (1988), G = Johnson and Wilkinson (1987), H = Smoot *et al.* (1985), I = Smoot *et al.* (1987), J = Crane *et al.* (1989), K = Meyer *et al.* (1989), Palazzi *et al.* (1990), L = Matsumoto *et al.* (1988).

counters, which must agree with expectations. Unfortunately there are no overall data checksums. The interferograms do have check sums, so most bad telemetry can be detected in this way. Interferograms are not used in sky maps or calibrations unless they are consistent with others taken under the same instrument conditions and looking in the same direction, but considerable effort is still necessary to understand the criteria for consistency.

Cosmic rays strike the detectors several times per second in normal operation, delivering impulses of heat and producing "glitches" in the interferograms. They are flagged by the on-board microprocessor but not removed; that is reserved for ground software. The ground software recognizes the glitches by comparing multiple observations of each pixel and removing the impulses that exceed a criterion for normal detector noise. This software works well for the nearly null interferograms, but glitches that occur near the peak of the interferogram are harder to discern. This is important in the calibration data, and for observations of the Galactic Plane where the interferograms are not all consistent because the Galaxy is thinner than the beamwidth. Tests of new methods are in progress.

Cosmic rays also strike the control sensors for the mirror mechanism and stop the scanning several times per day, even outside the South Atlantic Anomaly (the SAA, a region of high trapped proton flux) where the scanning is commanded off. Each such event results in a loss of data and a thermal impulse to the cryostat, since the mechanism stops by pushing against a mechanical stop, and the servo drives a high current into the motor. The loss of data is not serious but the temperature impulses cause a major loss of calibration stability and accuracy. New software is being developed to incorporate the measured effects of temperature changes of the scanning mirrors, and to delete data taken when the detector temperature is either out of range or changing too rapidly.

Errors in thermometer calibration are difficult to determine from flight data alone. The photometric calibration data are being examined for consistency, and small adjustments to the temperature scales are being tested to see if the overall calibration residuals can be reduced. Present indications are that adjustments of a few milliKelvin are sufficient at 2.7 K, but adjustments up to 1 K may be required at 20 K. A backup calibrator is being instrumented for further tests on the ground, and backup thermometers have been recalibrated to investigate the possibility of long term drifts in the calibration constants.

The flight data indicate significant gradients in the temperatures of both the internal and external calibrators, amounting to 17 and 6 mK respectively at 2.7 K. These values are the same as before launch, and are the reason why the announced cosmic background temperature uncertainty is 60 mK. The thermal design for the external calibrator predicts that it cannot have such a large gradient, so this will also be investigated in the test of the backup unit. Similar tests for the internal calibrator are also planned.

A shift in the frequency scale is possible for two reasons. First, the reference scale used to govern the sampling of the interferogram is made of glass, and shrinks during the cooldown from room temperature. Second, beam divergence within the interferometer multiplies the effective frequency by the cosine of the angle from the beam center. These effects add up to about 0.5%, an amount that is consistent with the photometric calibration.

Detector nonlinearities may be important for large signal sources, such as the Moon and some of our calibration data. In addition, the detector responsivity is strongly dependent on the operating point, as determined by the DC bias voltage and temperature. Our software models these variations, and preliminary fits to the calibration data show that preflight values for the detector constants are approximately correct.

Stray signals from the Earth when it rises above the top of the shield and shines on the instruments are very important to the *FIRAS*. The heat from the Earth at worst case during the June solstice is sufficient to raise the input horn antenna from 2.7 K to 7 K, and the cooldown time is of the order of 20 minutes. These temperature changes are a serious challenge to the calibration software, and it is not known whether good measurements of the background can be retrieved from them. Fortunately, most of the sky seen during that season was also observed early in the mission without interference from the Earth.

Stray light from the Moon is presently being investigated. It is clearly detectable in the high frequency channel interferogram peak height as far away as 20° from the Moon, approximately as expected from preflight infrared laser measurements of the antenna beam profile. At the present the software makes no attempt to model or remove the effects of the Moon except to cut out data taken too close to the Moon and not include them in the sky maps. About 3% of the observations are taken within 20° of the Moon.

Sun glints are reflections from the Sun into the top of the cryostat, presumably from some small part of the insulation blanket on the sunshield that has come loose, or from an unexpected position of the horn covers for the *DMR* after they successfully deployed. They are detected by their thermal effect on the vapor cooled shields of the cryostat, and occur once per spacecraft rotation, but not always with the same amplitude. Searches of both *DIRBE* and *FIRAS* data have revealed no effect of such glints, probably because the Sun is not a particularly bright far infrared source and because the instrument response is very small at large angles from the line of sight.

Thermal crosstalk from the *DIRBE* is not presently known to be a problem but needs further investigation. The *DIRBE* thermal effects on the *FIRAS* are much smaller than the temperature changes caused by mode changes of the mirror mechanism, cosmic ray events in the mirror mechanism electronics, and temperature controller setpoint changes for the horns and calibrators. In principle the calibration software should handle all such effects without any residual errors.

Emission from the *DMR* local oscillators would be seen as constant spectrum lines in all the sky spectra at the known oscillator frequencies or their harmonics, which would not be present in the calibration data. No such lines have been recognized to date. In addition, the emission from the oscillators might be expected to be modulated at 100 Hz, the frequency of the Dicke switches used by the *DMR*. In that case, the noise level of the *FIRAS* detectors should show a peak at 100 Hz, and there is no evidence for such a peak.

Vibrations of the moving mirror produce errors because they cause the interferograms to be sampled at incorrect locations, creating frequency modulation sidebands of spectral features. Most spacecraft vibrations are not synchronous with the scans, but some vibrations are excited at each velocity change of the moving mirror and occur on every stroke. These errors are still negligible when the interferograms are of small amplitude, as for normal sky measurements, but they are important for calibrations. We see small ripples in the instrument phase function which may be due to these vibrations.

Variations of the optical response across the antenna beam have been determined using the Moon as a source. We find that not only does the efficiency vary with the location of the Moon in the beam, but it varies differently at different frequencies, and the effective center (zero path difference point) of the interferogram also changes. These effects do not cause errors for diffuse sources but limit our ability to measure the Moon and use it for confirming our calibration.

Multipass phenomena within the interferometer were known before launch from tests of breadboard interferometers. They arise from radiation that is reflected from the bolometer or its housing and filter, which then returns to the interferometer, is modulated a second time, reflects from the optics at the input ports, returns through the interferometer and is modulated a third time, and finally reaches the detectors again. Such radiation should not be a problem for the null test, where the sky is compared directly to an external calibrator at the same temperature. However, it may limit the accuracy of the instrument at higher frequencies where Galactic emissions are of interest. Studies are beginning to determine the magnitude of these effects, and preliminary results limit the multipass energy to less than 1% of the total.

The ultimate accuracy of the instrument is limited by the reflectance of the calibrator, by leakage around the edge of the calibrator, by instrument stability between calibrations, by the detector noise, by the time devoted to calibration, and by residual statistical and systematic errors in the determination of the calibration parameters. The reflectance of the calibrator is important because the calibrator is illuminated by radiation from the instrument, including contributions from the antenna (adjusted to match the sky temperature) and from the interferometer (usually around 1.6 K). As described above, we believe the reflectance is less than 10^{-4}. There is also a possibility of leakage around the edge of the calibrator. Tests were done in orbit to investigate the change of the interferogram as a function of the position of the calibrator, but no changes were seen until the calibrator was far out of its normal position. The stability of the instrument is always difficult to analyze, but over the course of the 10 months of operation the instrument transmission function changed by less than 2%, with no evident trend. During the last month of operation, calibration and sky data were interleaved with a 50% duty cycle and a 7 day period to reduce sensitivity to potential instabilities. A total of one month of near-null calibration data were obtained, out of a total of 10 months of operation, so the calibration is potentially accurate to 10^{-10} W/m^{2}sr, or even better if averaged over frequencies.

SUMMARY AND CONCLUSIONS

The detailed analysis of the *FIRAS* data to model and remove systematic errors and local foreground sources is just beginning, but the temperature has been measured as 2.735 ± 0.06 K, and limits on the distortion parameters have been obtained, with $|y| < 10^{-3}$ and $|\mu| < 10^{-2}$ (3 σ). The dipole anisotropy has the expected shape for a Doppler shifted blackbody as measured by *FIRAS*, and the dipole amplitude and direction are consistent with previous measurements. Systematic errors limit the data accuracy, but analysis in progress is expected to reduce them below the goal of 0.1% of the peak CMBR brightness.

ACKNOWLEGEMENTS

It is a pleasure to acknowledge the vital contributions of all those at GSFC who devoted their efforts to making this challenging mission not only possible but enjoyable as well. The National Aeronautics and Space Administration/Goddard Space Flight Center (NASA/GSFC) is responsible for the design, development, and operation of the Cosmic Background Explorer (*COBE*). GSFC is also responsible for the development of the analysis software and for the production of the mission data sets. The *COBE* program is supported by the Astrophysics Division of NASA's Office of Space Science and Applications.

REFERENCES

Cheng, E. C., Mather, J., Shafer, R., Meyer, S., Weiss, R., Wright, E., Eplee, R., Isaacman, R., and Smoot, G. 1990, *Bull. APS*, **35**, 937.

Crane, P., Hegyi, D. J., Kutner, M. L., and Mandolesi, N. 1989, *Ap. J.*, **346**, 136-142.

DeAmici, G., Smoot, G., Aymon, J., Bersanelli, M., Kogut, A., Levin, S., and Witebsky, C., 1988, *Ap. J.*, **329**, 556.

Gulkis, S., Lubin, P.M., Meyer, S.S., and Silverberg, R. F.,1990, *Sci. Am.*, **262**(1), 132-139.

Gush, H.P., Halpern, M., and Wishnow, E.,1990, *Phys. Rev. Lett.*, **65**, 537.

Johnson, D. G., and Wilkinson, D. T. 1987, *Ap. J. (Letters)*, **313**, L1-L4 .

Kogut, A., Bersanelli, G., De Amici, G., Friedman, S., Griffith, M., Grossan, B., Levin, S., Smoot, G., and Witebsky, C. 1988, *Ap. J.*, **325**, 1.

Levin, S., Witebsky, C., Bensadoun, M., Bersanelli, M., De Amici, G. , Kogut, A., and Smoot, G. 1987, *Ap. J.*, **334**, 14.

Mandolesi, N., Calzolari, P., Cortiglioni., S., and Morigi, G., 1986, *Ap. J.*, **310**, 561.

Mather, J. C.,1982, *Opt. Eng.*, **21**(4), 769-774.

Mather, J.C., Cheng, E.S., Eplee, R.E., Jr., Isaacman, R. B., Meyer, S. S., Shafer, R. A., Weiss, R., Wright, E.L., Bennett, C. L, Boggess, N. W., Dwek, E., Gulkis, S., Hauser, M. G., Janssen, M., Kelsall, T., Lubin, P. M., Moseley, S. H. Jr., Murdock, T. L., Silverberg, R. F., Smoot, G. F., and Wilkinson, D. T. :1990a, *Ap. J.*, **354**, L37-L41.

Mather, J.C., *et al.* 1990b, *Proc. IAU Colloquium 123 "Observatories in Earth Orbit and Beyond"*, ed. Y. Kondo, Kluwer Acad. Publ., Dordrecht.

Matsumoto, T., Hayakawa, S., Matsuo, H., Murakami, H., Sato, S., Lange, A. E., and Richards, P. L.,1988, *Ap. J.*, **329**, 567-571.

Meyer, D.M., Roth, K.C., and Hawkins, I., 1989, *Ap. J. (Letters)*, **343**, L1.

Palazzi, E., Mandolesi, N., Crane, P., Kutner, M.L., Blades, J.C., and Hegyi, D.J., 1990, *Ap. J.*, **357**, 111-C3.

Sironi, G.,*et al.* 1987, *Proc. 13th Texas Symposium on Relativistic Astrophysics*, ed. M. P. Ulmer, (Singapore: World Scientific), 245.

Sironi, G., and Bonelli, G., 1986, *Ap. J.*, **311**, 418.

Smoot, G., De Amici, G., Levin,. S., and Witebsky, C. 1985, *Soc. Italiana di Fis. Conf. Proc.*, **1**, 27.

Smoot, G., Bensadoun, M., Bersanelli, M., De Amici, G., Kogut, A., Levin, S., and Witebsky, C. 1987, *Ap. J. (Letters)*, **317**, L45.

Zel'dovich, Ya. B., and Sunyaev, R. A. 1969, *Ap. Space Sci.*, **4**, 301.

AUDIENCE QUESTIONS

Ofer Lahav: An upper limit of 10^{-3} on the Comptonization parameter y is not a small number, in the sense that spectral distortions due to hot IGM could still be hidden in the "perfect" blackbody spectrum. What are the prospects to get an even lower limit on y from COBE and the other experiments?

Answer: The *FIRAS* sensitivity and accuracy should be good enough to measure a y as small as 10^{-4}, since the early result was derived from only 9 minutes of data. The known ultimate limitation of the FIRAS is the external calibrator reflectance of about 10^{-4}.

The Cosmic Background Radiation

Anisotropy Program at UCSB

Peter Meinhold, Todd Gaier, Joshua Gundersen, Timothy Koch, Jeffrey Schuster, Michael Seiffert, Philip Lubin

Department of Physics

University of California

Santa Barbara CA 93106

and

Donald Morris

Morris Research Inc.

Berkeley CA 94704

ABSTRACT: The Cosmic Background Radiation (CBR) is considered to be one of the best probes of structure in the universe on large scales and at early times. Most mechanisms for the production of structure have some effect on spatial temperature variations in the CBR. Models for galaxy and cluster formation, in particular the popular cold dark matter (CDM) models, suggest that the largest power in the anisotropy of the CBR occurs at angular scales of 10 arcminutes to a few degrees, with particular models making specific predictions for the shape and amplitude of the CBR temperature autocorrelation function.(Bond *et al.*1987)

During the past two and a half years we have undertaken several experiments to measure the structure of the CBR on scales from 10 arcminutes to 10 degrees, from both balloon borne and ground based (South Pole and mountain top) sites. We will briefly summarize these measurements and discuss possible future directions. Our current sensitivity is at the 10^{-5} level at 30 arc minutes and below 10^{-5} at 90 arc minutes. The goal is to achieve part per million sensitivity in the next few years.

Introduction

In the past two and a half years we have made CBR anisotropy measurements with two different optical systems and four separate cryogenic detectors. One system is a 1 meter diameter off-axis Gregorian telescope which has been configured with a 90 GHz SIS mixer, a four channel (3, 1.6, 1.1, 0.8 mm) Helium-3 bolometric detector, and a broad band direct amplification HEMT receiver from 25 to 35 GHz. The second system consists of two pairs of corrugated scalar feeds with fixed angular separation, operating at 15 and 23 GHz, also with direct amplification HEMT detector systems. In what follows, we discuss measurements from the 90 GHz system, the two frequency direct amplification system, and the broad band direct amplification system. The bolometer system and results are discussed in Alsop *et al.*1991 (this volume).

Gregorian telescope system

Since 1985 we have designed, built and flown a millimeter wave telescope system for making measurements of anisotropy in the CBR at angular scales from 0.5 to a few degrees. To date, this system has made three balloon flights and two expeditions to the South Pole for ground-based observations, making anisotropy measurements of the CBR as well as galactic emission, at wavelengths from 1 mm to 1 cm, and with Gaussian beamwidths of 0.5 and 1.6 degrees full width at half maximum (FWHM).

The instrument discussed here has operated in several different configurations. The original goal was to fly the telescope at balloon altitudes (from 30 to 42 km) to avoid most of the contributions from the atmosphere, particularly water vapor and oxygen. Figure 1 shows a calculated plot of atmospheric emission at sea level, South Pole (or mountaintop) elevation, and balloon altitude. In particular, at 90 GHz, the atmospheric emission at 30 km is roughly 4 orders of magnitude smaller than at sea level. This required development of a stabilized pointing platform for operation in the balloon environment, with rms tracking stability better than 1/10 of the beamsize for both elevation and azimuth controls. Well into the development of the pointing platform, we decided the same system could be easily adapted to do measurements from the ground at the South Pole, and constructed a servoed rotation stage for azimuth tracking from the ground.

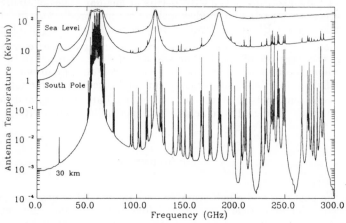

Figure 1: Atmospheric Emission Calculated From JPL Line Catalog and US Standard Atmosphere, for 30 km, 3.6 km (South Pole Barometric Altitude) and Sea Level.

The original detector for the experiment was a Niobium junction SIS (Superconductor-Insulator-Superconductor) mixer based receiver. In addition, we have recently built a wide band direct amplification receiver, using a High Electron Mobility Transistor (HEMT) amplifier operating from 25 to 35 GHz which has also made measurements using this telescope. Following is a brief description of the system and results from some of the detectors. As part of the Center for Particle Astrophysics we are also flying multi-frequency bolometric detectors in this same system in collaboration with Paul Richards and Andrew Lange at UC Berkeley (see Alsop et al.in this volume for further details).

Telescope

Our system is based on an off axis Gregorian telescope with a movable ellipsoidal secondary. The primary mirror is a 1 meter focal length, 1 meter diameter aluminum dish, machined from a single piece of material on a numerically controlled mill. The secondary mirror is placed so that one focus is in the feed horn of the detector system, while the other is in the focal plane of the primary mirror. Rotation of the secondary about the axis of the feed horn moves the second focus in the focal plane of the primary, which moves the beam on the sky. We generally throw the beam by roughly 2 x FWHM to measure temperature differences between patches of sky. For a full description of the system see Meinhold, 1990, Lubin, Meinhold and Chingcuanco, 1990.

The main reason for choosing this particular configuration is the extremely low sidelobe response. The central lobe is well approximated by a Gaussian with 0.5° FWHM. In addition we have made measurements of the sidelobe response to about -85 dB, limited by the antenna test range. We also include a large reflective groundshield during flight, and an even more extensive ground and sun shield for our ground based work, since the sun was continually up during our measurements at the South Pole.

Pointing platform

A schematic drawing of the pointing system is shown in figure 2. The platform is suspended from a 3 million ft^3 Helium filled balloon by a parachute and a 20 meter steel cable "ladder". The system is a basic az-el mount, the azimuth motion accomplished by moving the entire frame, and elevation by moving the inner frame, to which the telescope is mounted. Our requirement is to remove motions associated with the balloon, point to target positions in right ascension and declination, and track these positions over time, compensating for latitude and longitude changes, and obtaining arcminute rms stability.

Attitude sensing is accomplished with a 3 axis gyro and navigation processor, with absolute pointing verification from a real time CCD star tracking camera aligned with the telescope beam. For backup sensors, we have a 3 axis magnetometer for the azimuth and a 16 bit resolver for the elevation. The gondola azimuth is controlled by an active triple race bearing/torque motor combination and a flywheel, incorporated into a PID (proportional, integral, differential) control algorithm. Elevation tracking is done with a ball screw linear actuator driven by a stepper motor. All sensors and servo systems are controlled by an onboard

242

computer, which also does data acquisition. Science and housekeeping data are telemetered down in real time as well as recorded on board.

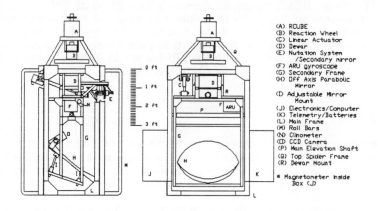

(A)	RCUBE
(B)	Reaction Wheel
(C)	Linear Actuator
(D)	Dewar
(E)	Nutation System /Secondary mirror
(F)	ARU gyroscope
(G)	Secondary Frame
(H)	Off Axis Parabolic Mirror
(I)	Adjustable Mirror Mount
(J)	Electronics/Computer
(K)	Telemetry/Batteries
(L)	Main Frame
(M)	Roll Bars
(N)	Clinometer
(O)	CCD Camera
(P)	Main Elevation Shaft
(Q)	Top Spider Frame
(R)	Dewar Mount

* Magnetometer inside Box (J)

Figure 2: Scale drawing of gondola showing all major telescope and servo elements

Expedition summaries

The first flight occurred in August, 1988, from the National Scientific Balloon Facility in Palestine Texas. The detector system was the 90 GHz SIS (Superconductor Insulator-Superconductor) mixer. The pointing system worked well during the flight, and useful calibration scans of Jupiter and a measurement of emission from the galaxy were obtained, though telemetry problems prevented long CBR integrations.

The same system which was flown in August was taken to the South Pole in November, 1988 and obtained over two hundred hours of data on the CBR. Scanning efficiency and weather reduced this to about 50 hours of effective integration time at a nominal noise level including sky of about 6 mK·sec$^{1/2}$ (Rayleigh-Jeans). In addition interesting results concerning atmospheric fluctuation levels and sky opacity were obtained for this site.

There have also been two flights of the system as part of our Center for Particle Astrophysics collaboration. This experiment (called "MAX" for Millimeter Anisotropy eXperiment) first flew in November, 1989 and for a second time in July, 1990. The system used a different secondary mirror and a multi-frequency bolometric detector system (for spectral discrimination of spurious and foreground sources, primarily galactic and atmospheric). These flights are discussed in detail in another paper in this volume (Alsop et al.).

Our group has just completed another South Pole expedition from November 1990 to January 1991, again using the SIS superheterodyne receiver, as well as a new direct amplification HEMT system at 25-35 GHz with extremely low noise (about 1/10 of the noise of the 1988 SIS system), and a slightly larger (1.6° FWHM) beam. Data analysis of this expedition is underway, and some preliminary results are discussed below.

CBR anisotropy study at 0.5° FWHM

Data taken with the SIS detector during December, 1988, have been used to place an upper limit on fluctuations in the CBR of $\delta T/T < 3.5 \cdot 10^{-5}$, for fluctuations with a Gaussian autocorrelation function of $\sigma=20$ to 30 arcminutes (Meinhold and Lubin 1991). This result places stringent limits on cold dark matter (CDM) theories of structure formation. For theoretical interpretations of this data see Bond *et al.*1991, Vittorio *et al.*1991.

Our measurement strategy consists of throwing the beam sinusoidally on the sky with an amplitude of 0.7° and at a frequency of 8 Hz, giving an effective throw similar to a square wave chop of 0.5°. Using a lockin amplifier, we then integrate the radiometric difference between the two chop positions for one second at a time. The telescope thus makes a single difference measurement of temperature for two points separated by about 1 degree on the sky, with a characteristic size of 0.5° FWHM. We then integrate for approximately 1 minute and slew the entire telescope by one degree and integrate again. We measure 9 differences, spaced so that one beam from each point coincides with a beam from an adjacent point. During the entire measurement, the telescope tracks the earth's rotation, and the final result is a linear scan of points about some fiducial center in celestial coordinates.

The data obtained this way are edited to remove bad weather, then fit to remove long term offsets due to drifts in the detector, optics, electronics and atmosphere. We found it necessary to remove a gradient in right ascension from the data, which appears to be due to atmospheric structure. This changes our sensitivity to CBR anisotropies and is taken into account in calculating the upper limits. Figure 3 is the result of binning the edited and fit data in right ascension. The error bars are

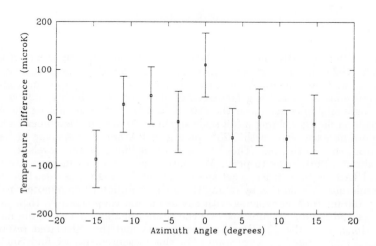

Figure 3: Final South Pole 1988–89 SIS Data Set
Gradient Removed. Errors are 60 microK per point.

"measured", that is they reflect the total scatter of the points in a bin. Our upper limits are calculated by Monte Carlo simulations of Gaussian correlated sky maps, which are then "measured" with the experimental response. By varying the signal to noise, and comparing statistical measures of the simulated data sets to the actual one, we can estimate 95% confidence level upper limits. These limits are at 55% power and are shown in figure 4. In addition to simple Gaussian correlation function models, more realistic CDM models have been calculated, leading to constraints on CDM parameters (Vittorio *et al.*1991; Bond *et al.*1991).

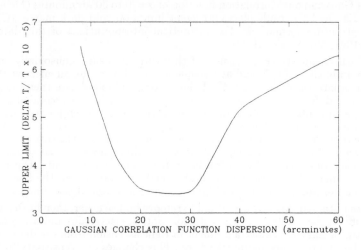

Figure 4: Upper limits on Gaussian fluctuation in the CBR from SIS data taken during 1988−1989 polar summer.

Galaxy data

In addition to CBR anisotropy data, we have made measurements of galactic emission. Galactic emission is a foreground source for CBR measurements, to be minimized and subtracted. At 90 GHz, we have made two separate measurements of galactic emission near the galactic center. The first was a highly undersampled 3 point single difference scan near the galactic center, performed during the August, 1988 balloon flight, yielding a full scale signal of 20 mK. In December 1990, a set of single difference scans with 0.25° spacing in RA and 0.5° spacing in Dec were made from the South Pole. These scans show significant signal levels, the largest being about 12 mK peak to peak. We match these two measurements to the 100 micron IRAS maps, in order to get an estimate of the brightness ratio from 90 GHz to 100 microns. We do this by taking the high resolution (2 arcminute bins) IRAS map at 100 microns, convolving with our beam and chop strategy, then sampling the measured region. Figure 5 shows the results of this for one of the measured declinations, with the 90 GHz measured points and the calculated response on the IRAS map, scaled by a constant. For this measurement, we find 200 MJy/sr at 100 microns corresponds to 1 mK Rayleigh-Jeans at 90 GHz. This is useful in attempting to locate regions of low galactic contamination, and in subtraction of whatever contamination exists. The scaling described is very naive, and says nothing about the difficult problem of extrapolating numbers measured in the galactic plane and near the galactic center to regions far from the plane, where we do our CBR measurements. Specifically, contributions from galactic components such as HII regions are highly localized, and can seriously complicate both the estimates of dust emissivity scaling and the understanding of spectra found off the plane. In fact, there is at least one potentially confusing HII region close to the position where we measure the galactic center emission. To help discriminate galactic contamination from CBR fluctuations, we also need to perform multi-frequency measurements such as our MAX flights.

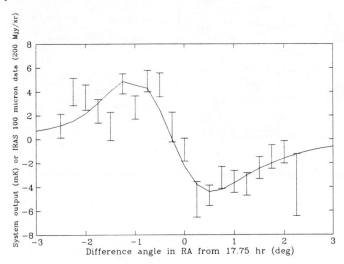

Figure 5: SIS data from South Pole, Dec, 1990. Scan of Galactic center (RA 17.75 hr, Dec −29 deg). Error bars are +/− 1 sigma, and smooth curve is smoothed, differenced IRAS 100 micron map in units of 200 Mjy/sr.

7° HEMT Radiometer

In 1987-1988, we built a HEMT (High Electron Mobility Transistor) radiometer to study the anisotropy of the CBR at larger angular scales. Angular scales between 5 and 15 offer a good test of inflationary models and large-scale structure via the Sachs-Wolfe effect. This is also the scale where a possible anisotropy detection was reported a few years ago (Davies et al.1987).

The instrument operates at 15 and 23 GHz. By making a measurement at two frequencies we hoped to remove possible signals from galactic emission which is expected at these frequencies at a level of 0.1 mK. The optics consist of pair of scalar feed horns with a beam width of 7 deg FWHM and a separation of 10 deg. By keeping the optics simple we achieve high sidelobe rejection. A cryogenic circulating switch modulates the signal between a pair of horns at each frequency. The signal is then amplified by a cryogenic HEMT amplifier. Further amplification is achieved by ambient RF amplifiers. The signal can then be passed through a bandpass filter and converted to a voltage in a detector diode. The signal is then demodulated electronically and stored on a computer. The target field of the instrument can be varied by placing the entire apparatus on a rotation table or by placing a flat movable reflector in front of the antennae.

The noise in both systems is about $3.5 \text{mK} \cdot \text{sec}^{1/2}$ including sky. Noise during runs could be higher depending upon location and seeing conditions. Some excess noise could be removed by wagging the beams periodically on the sky and subtracting signals from adjacent sky positions. This system is shown in figure 6.

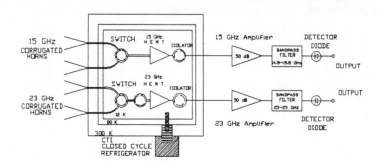

Figure 6: Dual Frequency HEMT Radiometer Schematic

Data acquisition strategy

The instrument was used on three separate occasions to obtain data. In all three cases high, dry sites were chosen to reduce the effects of atmospheric emissions. The instrument was operated at the South Pole during the austral summers of 1988-89 and 1990-91. A northern hemispheric measurement was made from the White Mountain Research Station fall of 1989. Data from the two earlier results will be discussed here. The data from the most recent South Pole expedition have not yet been analyzed.

Our first South Pole strategy was similar to that described by Davies *et al.*. Our instrument, with an output proportional to temperature differences on the sky, is slewed between two azimuth positions, and the field of view changes with the rotation of the earth. Signals from the two positions are subtracted resulting in the familiar three-beam response, sensitive to the second derivative in sky fluctuations. After this second subtraction a small offset of about 1 mK remained. A slight linear trend was also evident across a 9 hour RA stretch of data. Removal of this line has a negligible effect on our sensitivity at our most sensitive angular scale, about 8 deg for Gaussian fluctuations. This is the only fitting that was necessary on this data set. The sky at the South Pole is very stable for long periods of time.

This data set is shown in figure 7, and represents about 24 hours of useful data, which was all we were able to use after editing data from poor weather periods and large systematics. Still we are able to set a good upper limit with this data, and compare it to the result of Davies et. al.. Figure 8 shows such a comparison of relative likelihoods of both our data set and that of Davies et. al. While both upper limits are about the same for the same experimental geometry, we do not detect any significant fluctuations. Their result has a 40 percent probability of being consistent with this data set.

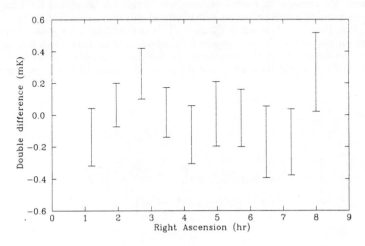

Figure 7: 23 GHz HEMT system. South Pole 1988–89 data set. Binned in 0.5 FWHM bins, linear fit removed.

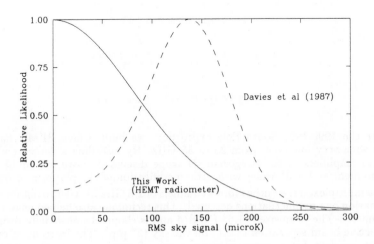

Figure 8: Comparison of relative likelihoods for Davies et al, 1987 data and HEMT radiometer 1988–89 South Pole data.

248

The expedition to White Mountain was staged to allow us to look directly at the anisotropy of Davies et. al. The experimental configuration was not as clean as for our previous expedition. Due to the geometry of the mid-latitude observation, a flat movable mirror was used to provide a slow second chop. The unfortunate effect of such a chop through large elevation differences is a large second chop offset proportional to the DC atmospheric temperature. This offset was about 30 mK in the 15 GHz channel and 70 mK in the 23 GHz channel. Because the DC level of atmospheric emissions varies on time scales of several hours, a linear fit is insufficient for removal of excess atmospheric noise. A bandpass filtering scheme was devised which again preserves sensitivity at our most sensitive scale. A section of data from Dec=40 is shown in figure 9. This is precisely the region where Davies *et al.* saw excess signal. While we see no evidence for this signal in our respective data set, we do not have the sensitivity to fully rule out their detection. The 23 GHz channel could not be used to obtain useful data due to the large atmospheric effect.

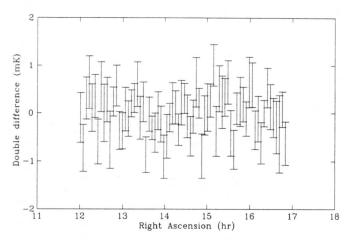

Figure 9: 15 GHz HEMT radiometer White Mountain, 1989 data. Dec 40 degrees. Data binned as in Davies et al (1987).,

25–35 GHz HEMT experiment

For the 1990-1991 South Pole expedition, we built a new HEMT based detector with very low noise from 25 to 35 GHz. By installing a detector with one of these amplifiers in the Gregorian telescope discussed above, we could make a measurement at $1 - 3°$ scales with little modification of the platform or optics.

We multiplexed the output into 4 channels of 2.5 GHz in order to obtain spectral information on any fluctuations detected. This technique also helped us to evaluate systematics. The beamsize is 1.6° FWHM and is chopped sinusoidally resulting in an effective beam separation of 2.1° on the sky (3° p-p). The instrument operated without incident for 3 weeks. During times of good weather our best channel exhibited noise of 1.4mK·sec$^{1/2}$. The combined channels show sensitivity around 700 to 800 μK·sec$^{1/2}$, including sky noise. A sample of ten hours of this data is displayed in raw form (with no fitting) in figure 10. A total of about 200 hours of useful data were obtained after editing. This time is divided into six independent measurements. Each measurement is similar to the SIS measurement strategy described above, consisting of 9 overlapping single difference measurements. During preliminary analysis, scans of this data have exhibited single point errors of less than 10 μK. Further evidence of the instrument sensitivity is shown in figure 11a and b, which show one hour of data taken around the Large Magellanic Cloud. The complete data analysis is still in progress, these results are preliminary.

Conclusion

We have described measurements of CBR anisotropy made over the past few years. Published upper limits on $\delta T/T$ from these experiments are at the $3.5x10^{-5}$ level. New data which is still being analyzed should push the sensitivity below 10^{-5}. Experiments currently underway or being contemplated both in collaboration and within the UCSB group are expected to make measurements with precision of a few $x10^{-6}$ within a few years, either detecting anisotropy or putting even stricter limits on the parameters of current cosmological models. If no anisotropy is found at the 10^{-6} level very few of the current theories appear viable.

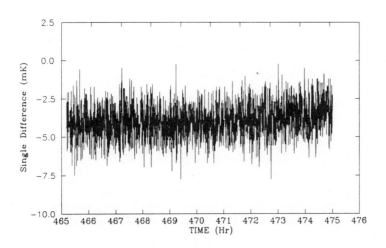

Figure 10: PHEMT Channel 2 raw data, South Pole 1990−91.

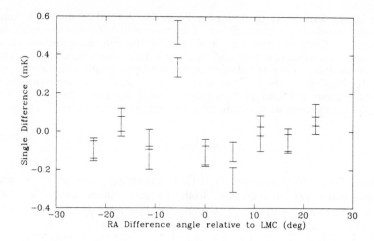

Figure 11a: PHEMT channels 1 and 2. Scan of Large Magellanic Cloud (centered at RA 5.6 hr, Dec −68 deg).

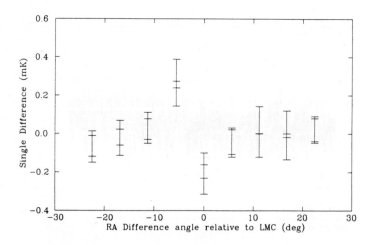

Figure 11b: PHEMT channels 3 and 4. Scan of Large Magellanic Cloud (centered at RA 5.6 hr, Dec −68 deg).

Acknowledgements

This work was supported by the National Aeronautics and Space Administration, under grant NAGW-1062 and GSRP grant NGT-50192, the National Science Foundation Polar grant DPP878-15985, California Space Institute grant CS 35-85, the University of California, the University of California White Mountain Research Station, and the U.S. Army. We also gratefully acknowledge the support of the NSF Center for Particle Astrophysics. We particularly want to thank Nancy Boggess for her vision of the viability of this research and for her continued support. This work would not have been possible without the support and encouragement of Richard Muller, Fred Gillett, Buford Price, and John Lynch. We wish to thank Anthony Kerr and S. K. Pan of NRAO for supplying the exceptional SIS mixer. The Nb/Al-Al$_2$O$_3$/Nb junctions used were supplied by Hypres corporation. Special thanks to Robert Wilson, Anthony Stark, Joe Stack, and Paul Moyer at Bell Labs for machining the primary and secondary mirrors. In particular we would like to thank Bill Coughran, and all of the South Pole support staff for the highly successful 1988-89 and 1990-91 polar summers. We also gratefully acknowledge the HEMT preamplifiers supplied by Mike Balister and Marion Pospieszalski of the NRAO.

References

Bond, J. R., Efstathiou, G., Lubin, P. M., and Meinhold, P. R., 1991, *Phys. Rev. Lett.*, Submitted.

Bond, J. R., and Efstathiou, G., 1987, *M.N.R.A.S.*, **226**, 655.

Davies, R. D. , Lasenby, A. N., Watson, R. A., Daintree, E. J., Hopkins, J., Beckman, J., Sanches-Almeida, J., and Rebolo, R., 1987, *Nature*, **26**, 462.

Lubin, P. M., Meinhold, P. R., Chingcuanco, A. O., 1990, *The Cosmic Microwave Background: 25 Years Later*, ed. N. Mandolesi and N. Vittorio, Kluwer Academic Publishers, The Netherlands. p. 115

Vittorio, N. , Meinhold, P. R., Muciaccia, P. F., Lubin, P. M., and Silk, J., 1991, *Ap. J. Lett.*, submitted.

MAX, A MILLIMETER-WAVE ANISOTROPY EXPERIMENT TO SEARCH FOR ANISOTROPY IN THE COSMIC BACKGROUND RADIATION ON INTERMEDIATE ANGULAR SCALES

D. C. Alsop, A. C. Clapp, D. A. Cottingham, M. L. Fischer,
A. E. Lange, P. L. Richards, G. Smoot
Department of Physics and Space Sciences Laboratory,
University of California, Berkeley, CA 94720

J. O. Gundersen, T. C. Koch, P. R. Meinhold, P. M. Lubin
Department of Physics,
University of California, Santa Barbara, CA 93106

E. S. Cheng
Code 685, NASA Goddard Space Flight Center,
Greenbelt, MD 20771

E. Kreysa
Max Planck Institut fur Radioastronomie,
Auf dem Hugel 69, D-5300 Bonn 1, West Germany

ABSTRACT

We report preliminary results from two balloon flights of an experiment designed to search for anisotropy in the cosmic background radiation (CBR) on angular scales from 0.3 to 3 degrees. The instrument was a multiband, ^{3}He-cooled bolometric photometer installed on a pointed, low background telescope. During the first flight, in November 1989, we measured the spectrum of the brightness of the galactic plane at a galactic longitude of $l_{II}=24°$ and searched for CBR anisotropy at nine points centered near the north celestial pole. The latter data set an upper limit on CBR anisotropy of $\Delta T/T <$ 1. x 10^{-4} (95% CL) for a Gaussian correlation function with a correlation length of 0.3°. The second flight, in July 1990, featured more sensitive detectors. We searched for CBR anisotropy at seven points near the star, Gamma Ursa Minoris. Despite a large, long timescale drift in the signal offsets, detector limited performance was achieved for anisotropies on scales < 5°. The statistical noise per point was almost five times smaller than in the first flight. Analysis of this data is still in progress. We also discuss future plans for improved receivers and a new telescope.

INTRODUCTION

A variety of models for structure formation in the universe, including the standard cold dark matter model, predict the largest anisotropies at angular scales from 0.3 to 3 degrees (e.g. Bond, 1989). These scales correspond to the distances probed by observations of the large scale distribution of galaxies and clusters and to the size of the causal horizon at decoupling. Current upper limits restrict anisotropies on these scales to several parts in 10^5. (Meinhold and Lubin, 1990, Timbie and Wilkinson, 1990.) Though improvements in detector technology have increased the raw sensitivity available to improve upon these observations, strategies for subtracting galactic emission and avoiding increasingly problematic, systematic noise sources are required to realize the potential of these detectors. Even in the millimeter wave region, where the ratio of CBR to galactic anisotropy is highest, galactic emission will foil naive attempts to probe anisotropy to below the 10^{-5} level. Noise created by imperfect subtraction of the uniform but very large background of atmospheric emission, telescope mirror emission and CBR will also rise in proportion to the sensitivity of the receiver. We have developed a multi-frequency bolometric receiver which combines the high sensitivity of bolometric detectors with the increased galactic emission and systematic error rejection capability of simultaneous, multi-frequency observing. The receiver was installed on a low-sidelobe telescope, which we designed to minimize background emission and beam modulation systematics, and flown at high altitude on a balloon to reduce atmospheric emission and noise.

INSTRUMENT

The instrument is a pointed, off-axis telescope with a one meter primary mirror, a chopping secondary mirror and a multi-band bolometric receiver (Meinhold, 1989, Fischer et al. in preparation). The receiver accepts radiation from the sky in a 0.5 degree FWHM Gaussian beam. To minimize the fraction of rays accepted by the receiver which miss either of the mirrors, the mirrors were designed large compared to the beam. Such errant rays would increase the background since they strike baffles which are much more emissive than the mirrors.

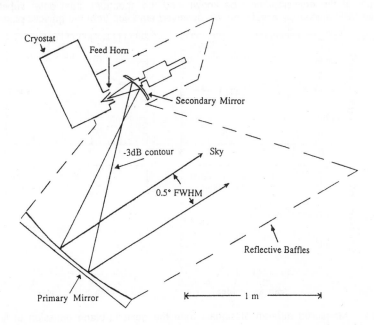

Figure 1. Side view of the telescope. Optical rays indicate nominal -3 dB contours. Dashed lines indicate reflecting baffles which block direct earthshine paths to optical surfaces.

The beam is sinusoidally chopped at 6.1 Hz between two points on the sky separated by 1.2 degrees. This chop is achieved by rotation of the secondary mirror around the receiver feedhorn axis. The chopping mechanism has been optimized for high stability in both amplitude and zero position. The gondola azimuth can be changed continuously or in steps on a several second timescale. After entering the receiver feedhorn and passing through blocking filters, the beam is divided into separate frequency bands by the selective reflection and transmission of mesh interference filters and then concentrated onto the four, separate bolometric detectors. During the first flight, the bands were centered at 3, 6, 9, and 12 cm^{-1}. The 3 cm^{-1} channel was removed for the second flight. The detectors were ^{3}He cooled to below 0.3 mK.

The system can be calibrated during the flight by moving a partially reflecting membrane into the prime focus from one side. The membrane reflects a reproducible fraction of an ambient temperature blackbody source. The reflectivity of the calibrator membrane was measured in the laboratory and ranged from 10^{-3} to 10^{-2} at frequencies from 3 to 12 cm^{-1}. The signals from planets scanned in flight provide a check on the calibration.

GALACTIC EMISSION

In order to subtract galactic emission from a CBR anisotropy observation, the spectrum of the emission must be known and the spectrum must differ substantially from the CBR anisotropy spectrum. We observed emission from the galactic plane

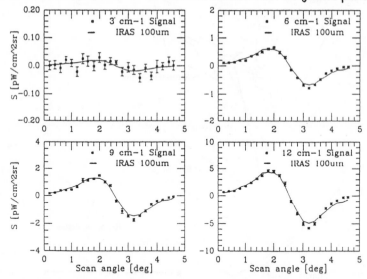

Figure 2. Measured differential signals from the galactic plane emission at l_{II} = 24°. The solid line shows the signal calculated by convolving the 100 μm IRAS signal with the effective shape of the chopped antenna beam. The amplitude of the calculated signal has been fit to the measured data separately for each passband.

during the first flight to obtain a high signal-to-noise measurement of the spectrum of galactic emission. The scan passed across the plane at a galactic longitude of 24° at an inclination of 26° from the normal. The beam was chopped along the scan direction, producing a differential measurement of the sky brightness as a function of scan angle. A model of the chopped beam was also scanned across the IRAS 100μm map for comparison. Fig. 2 shows the correlation between the observed signal and the model IRAS 100μm data. The steeply rising spectrum inferred from the correlations indicates that the signal was dominated by dust emission.

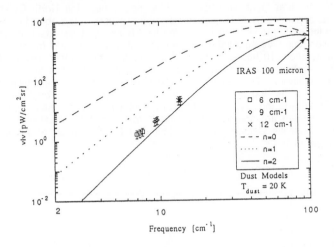

Figure 3. Measured galactic plane differential brightness as a function of frequency. The curves indicate three thermal dust emission models with emissivity indices n=0, 1, and 2, and dust temperature T_{dust} = 20 K.

Fig. 3 shows the peak amplitude of the IRAS correlated component of the measured differential sky brightness as a function of frequency for the 6, 9 and 12 cm^{-1} bands. For comparison, we also show three plausible models for the thermal dust emission, each normalized to the peak amplitude of the 100 cm^{-1} IRAS brightness model described above. The models for dust emission assume a power law frequency dependence for the emissivity with index, n = 0, 1 and 2, and a dust temperature, T_{dust} = 20°K. Because the filter bands have a finite bandwidth, the effective frequency of each passband moves toward higher frequency as the dust spectrum steepens. Three points, corresponding to the three dust spectral indices, are displayed for each of the bands. The 3 cm^{-1} result is not shown because of its low signal to noise ratio.

These data are consistent with an emissivity index, n, between 1 and 2. Other authors, Lubin et. al., 1989 and Page et. al., 1990, report emissivity indices in the same range. Preliminary results from the DIRBE experiment on COBE indicate that the 100μm IRAS calibration for diffuse sources is in error by up to a factor of 2.6 (Hauser et al, 1990). Recalibration may reduce the 100μm emission and therefore decrease the spectral index derived by all authors.

If the spectrum at high galactic latitudes is as steep as the in the plane of the galaxy, dust emission should be easily removed from multi-band, millimeter-wave CBR anisotropy observations. The spectrum at high latitudes may differ, however. Dust at high galactic latitudes is well correlated with HI column density while dust in the plane near the galactic center is better correlated with CO emission (Sodroski et. al., 1987); the different conditions of the dust in molecular clouds could lead to different emissivities and temperatures. As long as the dust emissivity index is positive, however, the spectrum can be readily differentiated from a CBR anisotropy spectrum.

FIRST FLIGHT CBR ANISOTROPY SEARCH

During the first flight we observed nine points near the North Celestial Pole; each point was separated by 1° in scan angle. The telescope was wobbled back and forth in a step-and-integrate mode. The integration time between steps was 1 minute and the total observation time was 40 minutes. The observation was a small fraction of the total balloon flight duration, 10 hours, because the chopper malfunctioned several hours into the flight. Fig. 4 shows the azimuth binned 6 cm^{-1} data as a function of scan angle after an offset was removed. The other bands did not have sufficient sensitivity to provide interesting upper limits and are not shown. The scatter in the 6 cm^{-1} data has a χ^2 probability of 0.26 and thus is consistent with detector noise. The one sigma errorbars for each binned point are approximately 140mK (ΔT of a 2.7K blackbody). A standard likelihood ratio analysis yielded an upper limit on CBR anisotropy of 1.0×10^{-4} for a correlation angle of 0.3°.

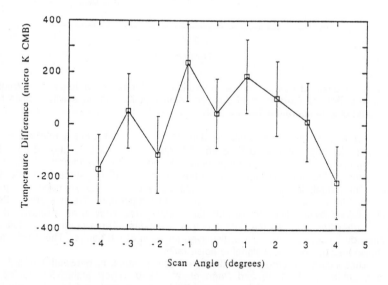

Figure 4. Measured temperature differences in the 6 cm^{-1} band vs. scan angle. These data were measured during the NCP scan of the first flight.

SECOND FLIGHT CBR ANISOTROPY SEARCH

For the second flight, the 3 cm⁻¹ channel was removed in order to increase the sensitivity of the higher frequency channels. New detectors with three times better sensitivity were also added in the 6 and 9 cm⁻¹ channels. The modified receiver achieved three times better sensitivity at 6 cm⁻¹ and 7 times at 9 cm⁻¹. Fig. 5 shows the distribution of the short term noise observed in flight at 6 cm⁻¹ and indicates a 0.65 mK/rt s single chopped sensitivity.

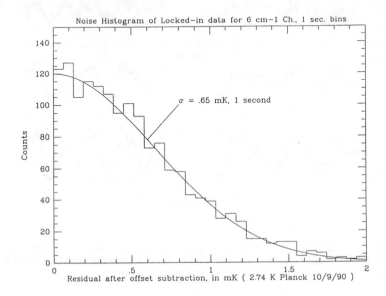

Figure 5. Distribution of short term noise in the 6 cm-1 channel during the second flight.

Though the second flight also lasted 10 hours, a malfunction again limited the CBR observation to a small fraction of the available time. The CBR search began toward the end of the flight because a software error in the guidance system computer prevented the gondola from stabilizing earlier. The search was divided into two 37 minute observations by a 13 minute calibration. The scan pattern consisted of discrete azimuth steps between seven integration points performed at 20 second intervals. The integration points were spaced 1° apart. Seven complete scans were performed in each half of the observation yielding a total of 28 observations per integration point. The scan pattern and the 196 integrations for the 6, 9 and 12 cm⁻¹ bands áre plotted in Fig. 6.

Noise well in excess of detector noise is apparent in all three channels. A crude correlation between the channels allowed the determination of a spectrum for the noise which was steeper than Raleigh-Jeans. The noise was not correlated with azimuth so it could not have been caused by CBR anisotropy or any other source fixed in the sky.

258

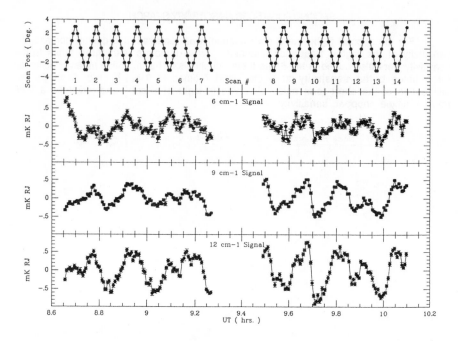

Figure 6. Second flight CBR anisotropy scan pattern and azimuth binned signals for the 6, 9, and 12 cm⁻¹ channels. An offset and a gradient have been removed from each half of the data. The calibration which divides the observation is not shown.

A rapid rotation in azimuth performed later in the flight as a diagnostic produced signal modulations of similar amplitude and spectrum to the noise in Fig. 6 but the modulation rate was much more rapid. This test suggests that the rapid azimuth rotation modulated the signal. Since only the earth, the atmosphere and the balloon do not rotate with the gondola, they are the likely sources. The source could not have been stable on the time scale of a scan because azimuth correlation would have been apparent. The spectrum of the emission was inconsistent with oxygen and water but was marginally consistent with ozone. The possibility that emission from the balloon or the earth was accepted in the sidelobes of the instrument cannot be eliminated. The low frequency noise observed during the first flight was dramatically lower. (Fig. 7)

Because the scan modulation of the observation position was relatively rapid, most of the low frequency noise was outside the frequency range in which CBR anisotropy information would be located. Fig. 8 shows the effect of a highpass filter on the noise in the 12 cm⁻¹ channel; near detector-noise-limited performance can be achieved by filtering the data with a highpass that removes the signal from only the largest angular scale anisotropies. When the data is averaged into seven final data points, one sigma errors per bin are all below 40 mK. The data should have the sensitivity to set a limit below 3×10^{-5} for a 0.3° correlation angle. Data analysis is still in progress.

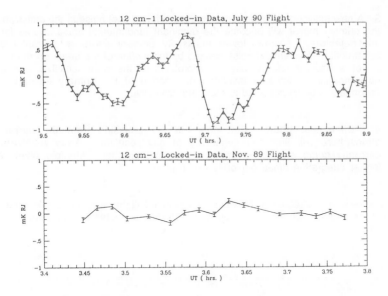

Figure.7 A comparison between the low frequency noises observed in the 12 cm⁻¹ channel during the two MAX flights. A constant offset has been removed from both data sets.

FUTURE PLANS

The two flights of the MAX experiment were part of an ongoing program to develop balloon-borne bolometric receivers for observation of CBR anisotropies at the 10^{-6} level. To achieve this goal will require both laboratory technology development and practical observing experience. The previous MAX flights have provided important feedback on observing strategies and difficulties which is currently guiding future plans. These plans include:

1. Another flight of the ³He detectors in May 1991. The flight will be at higher altitude and the filters will be modified to reduce atmospheric emission. The objectives are to reduce and understand the low frequency noise observed in the second MAX flight and to increase the observation duration. An eight hour observation would have almost triple the sensitivity of the second flight's. An observation of the Sunyaev-Zeldovich effect for the Coma cluster is also planned.

2. A new receiver with detectors cooled to 100 mK by an adiabatic demagnetization refrigerator is nearing completion. These detectors will have 5x greater sensitivity. A flight with this receiver is planned for the coming year. This large improvement in sensitivity will likely produce both exciting data and information on systematics which may limit the sensitivity.

3. A new telescope is under development. Photon shot noise in the emission from the primary mirror will become the dominant fundamental noise source for 100 mK detectors. A new, lower emissivity primary mirror is currently under construction. Systematic noise caused by the motion of the beam on the primary mirror is also expected to become a problem for the more sensitive receiver. A new modulation scheme involving the motion of the primary instead of the secondary is being designed to eliminate this noise.

4. Ultimately, photon shot noise in the emission from the mirror and atmosphere will limit the sensitivity of any receiver. Array receivers are currently being studied to improve sensitivity further. Multiple pixels allow longer integration times per observation point.

Table 1 shows the schedule of previous and planned flights and the upper limits they have set or should set in the absence of interfering systematic problems or real anisotropy.

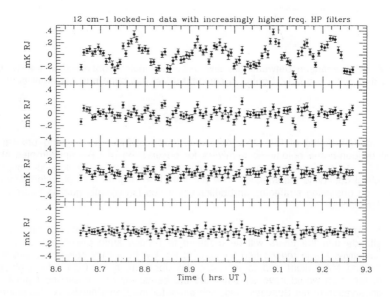

Figure 8. The effect of highpass filtering in time on the excess noise in the 12 cm^{-1} channel. The topmost data have had only signals with frequencies lower than any detectable CBR anisotropy would produce removed. The lower data have had increasingly higher frequencies removed. Only anisotropies with periods < 5° remain in the bottom most data.

Flight Date	Sensitivity Relative to First Flight	Integration Time (Hours)	$\Delta T/T$ upper limit (95%)
Nov. 89	1	0.7	1×10^{-4}
July 90	0.3	1.3	$(< 3 \times 10^{-5})$
Spring 91	(0.3)	(6)	(1×10^{-5})
Fall 91	(0.07)	(6)	(2×10^{-6})

Table 1. Sensitivity of past and future MAX flights. Quantities in parentheses are anticipated but not yet achieved.

We gratefully acknowledge the assistance of the staff of the National Scientific Balloon Facility. M.L.F., D.C.A. and P.R.M. were partially supported by NASA Graduate Student Research Program Fellowships. Work at the University of California, Berkeley was funded by NASA grant NSG-7205 through the Space Sciences Laboratory, and by the Center for Particle Astrophysics, a National Science Foundation Science and Technology Center operated by the University of California under Cooperative Agreement No. AST_8809616. Work at the University of California at Santa Barbara was funded by NASA grant NAGW-1062, and the NSF's Center for Particle Astrophysics.

REFERENCES

Bond, J. R., in Frontiers in Physics - From Colliders to Cosmology , Astbury, A., Campbell, B.A., Israel, W.I., Kamal, A.N. and Khanna, F.C. eds. , World Scientific, 1989

Hauser, M. G., Kelsall, T., Moseley, S.H., Silverberg, R.F., Murdock, T., Toller, G., Speisman, W., Weiland, J., preprint, to be published in the proceedings of the "After the First Three Minutes" Workshop. Oct. 15-17,1990, Univ. of Maryland

Lubin, P. M., Meinhold, P. R., Chingcuanco, A. O., in The Cosmic Background Radiation, 25 Years Later, Mandolesi, N., Vittorio, N. eds., Kluwer Academic Publ., 1990

Meinhold, P. R., Lubin, P.M. 1990 Ap J submitted

Page, L. A. Cheng, E.S., Meyer, S.S., Ap. J. 355, L1-L4, (1990)

Timbie, P.T., Wilkinson, D.T., Ap.J. 353, 140, (1990)

Sodroski, T. J., Dwek, E., Hauser, M.G., Kerr, F.J., Ap. J. 322:101-112, (1987)

BACKGROUND RADIATION INHOMOGENEITIES FROM GAS PROCESSES

J. Richard Bond and Steven T. Myers
CIAR Cosmology Program
Canadian Institute for Theoretical Astrophysics
University of Toronto, ON M5S 1A1, Canada

ABSTRACT *On scales less than a few degrees the primary CMB anisotropies (from linear processes occurring at photon decoupling) and the secondary CMB anisotropies (from nonlinear processes occurring later) arise mainly from gas-driven mechanisms such as Thomson scattering, inhomogeneous Compton cooling, and radiant emission by dust. We describe some of the current experiments which probe these anisotropies and the constraints on models that can be derived from them. We pay special attention to hierarchical Gaussian models of structure formation such as the $\Omega = 1$, $H_0 = 50$, adiabatic CDM model. For these theories, observations currently restrict the biasing factor defining the amplitude of the perturbations to be $b_\rho \gtrsim 0.7$ (for $\Omega_B \gtrsim 0.03$) assuming standard recombination. If there is early reionization, the best limits currently come from all-sky surveys probing the Sachs-Wolfe effect, giving $b_\rho \gtrsim 0.3$. The currently preferred value of the biasing factor (apart from the 'extra large scale power conundrum' of deep redshift surveys) is $b_\rho \sim 1 - 1.5$, so modest experimental sensitivity improvements will test these theories. We also present our methods for generating large catalogues of galaxy groups and clusters using a hierarchical Gaussian peaks formulation. These are used to generate simulated maps of secondary CMB anisotropies due to the Sunyaev-Zeldovich effect and the associated soft X-ray emission viewable with ROSAT. The non-Gaussian nature of the resulting image statistics and the ability to superpose various backgrounds and foregrounds are essential for optimally designing observational tests of these models. Both aspects are well-handled by the map synthesis technique. The prospects for ambient SZ observation depend strongly on the primordial amplitude of the perturbations but radio source confusion will make it difficult at Rayleigh-Jeans wavelengths. We also sketch how the hierarchical peaks method can be applied to the construction of primeval galaxy catalogues and display illustrative far infrared maps of dust-laden star-bursting galaxies at $z \sim 5$ that may be detectable with ground-based sub-mm telescopes.*

1. INTRODUCTION

Background radiations in all wavebands ranging from the X-ray to the radio provide invaluable windows on the universe at moderate to high redshift. The deepest probe is provided by the primordial cosmic microwave background (CMB). To avoid excessive distortion of the $T = 2.74$K blackbody spectrum, most of the energy in it must have arisen prior to a redshift $z_P \sim 10^{6.4}$, when the energy injected could have been thermalized by free-free absorption and double Compton scattering. Energy released between z_P and $z_{BE} \sim 10^{4.6}$ gives rise to a Bose-Einstein distribution characterized by a nonzero photon chemical potential. From COBE's FIRAS and the University of British Columbia's COBRA experiments, we now know that at most 1% of the CMB energy could have been injected in this 'Bose-Einstein' era. Since the medium is still extremely optically thick to scattering at that time, anisotropies would be wiped out up to the time of photon decoupling. If energy is injected later than z_{BE}, the gas will Compton cool, yielding the characteristic distortion shape. FIRAS and COBRA limit this amount of energy to be 0.4% of that in the CMB. As well, photon energy injected at shorter wavelengths than the bulk of the CMB will be redshifted into a secondary background. If it is starlight from the first generation of stars, the redshifted waveband will be the near infrared, unless there is a sufficient dust cover between the sources and us that it gets absorbed and reradiated into a redshifted waveband in the far infrared to submillimetre. In addition, optical, UV and even higher energy backgrounds are expected from starlight and accretion-light generated at lower redshifts and from the gravitational binding energy released in the formation of cosmic entities. The sequence of energy releases will be characterized not only by monopole quantities such as the globally-averaged spectrum, but also by telltale angular patterns that might be used to probe the sort of objects that give rise to the backgrounds and their clustering properties. These backgrounds may be the only window we have to the Universe at moderate redshift.

1.1 Overview of Primary and Secondary Anisotropies

In this paper, we adopt the somewhat arbitrary split between primary and secondary CMB anisotropies given by Bond (1988) to describe these angular patterns. *Primary* anisotropies are those that were generated when the Universe was (predominantly) in the linear regime. They are invaluable as a direct probe of the amplitude and statistics of primordial perturbations without the complications engendered by the nonlinear and dissipative processes that characterize *secondary* anisotropies, which are all the rest. In particular, secondary ones arise once objects collapse and liberate energy. In many models of structure formation, the distinction between primary and secondary is quite blurred. For example, short distance scale structure can go nonlinear shortly after (or even before) photon decoupling and lead directly to anisotropies. This might be expected in cosmic string theories and in scenarios with a large amount of power on small scales such as isocurvature baryon models.

When the photon transport equations are linearized, the terms which act as sources driving the development of primary anisotropies are: (1) the Sachs–Wolfe source, most important on large angles, (2) the photon 'bunching' source, important on intermediate angles and, for isocurvature perturbations, on large angles as well, and (3) the electron velocity source due to the Doppler effect in the Thomson scattering terms, an effect which is significantly diminished by the fuzziness of the last scattering surface (through destructive interference of opposing contributions to $\Delta T/T$ from the troughs and crests that straddle the photon decoupling surface). This damping is important for waves whose size is smaller than the width of the last scattering surface. There is a corresponding 'coherence' angle θ_c, which ranges from $\sim 2' - 10'$ (for normal recombination) to a few degrees (for models with early reionization). This scale θ_c gives the characteristic smoothness of the hills and valleys of the primary $\Delta T/T$ pattern. However, each of the three sources contribute to a rich anisotropy pattern that appears on all higher angular scales (Fig.2.1). The relative magnitude at any given scale depends upon the shape and statistical distribution of the primordial spectrum, upon the global geometry and upon the baryon and dark matter abundances.

One may hope that the statistics of the initial perturbations can be inferred from those of the primary anisotropies. For example, initially Gaussian density perturbations, the form most commonly assumed, breed Gaussian-distributed $\Delta T/T$ patterns. In many respects the most interesting anisotropies are those that arise only from the inhomogeneous rippling of our past light cone associated with metric perturbations. Above the angle subtended by the horizon at photon decoupling, which is a few degrees if recombination is normal (so decoupling occurs at $z \sim 1000$), we expect these Sachs-Wolfe anisotropies to dominate and give us a clean window through which to view the primordial geometry. Whether there is anything to see at large angles, however, depends very much upon the shape of the primordial power spectrum. If we can pin the amplitude of primordial perturbations to the structure we see, and if the fluctuations are initially scale-invariant as inflation theories most often predict, then these large-angle anisotropies should be visible with COBE's DMR. On the other hand, it is also possible that these very large scales, which are unrelated to any of the collapsed structure we see in redshift surveys of any sort, have no significant fluctuations at all, in which case we must be content with what we can learn from the scales over which gas dynamical processes can operate.

Although primary small-scale anisotropies are severely damped if there is early reionization (the photon-decoupling surface is up to Gpc thick rather than 10's of Mpc), quadratic nonlinearities in the Thomson scattering do not suffer dramatic destructive interference, since different wave-modes are coupled (Vishniac 1987). This effect leads to significant anisotropies in some theories such as the isocurvature baryon models (Efstathiou and Bond 1987). However, if the gas moves relative to the primordial density perturbations (non-local biasing), then the quadratic nonlinearity assumption has to be abandoned in favour of a more powerful treatment of Thomson scattering from the nonlinear clumps of ionized flowing gas.

Once matter begins to collapse, the gas can heat up through shocks thermalizing the released gravitational energy or through other heat sources such as thermonuclear or accretion energy. At high redshifts ($z \gtrsim 10$) the main cooling mechanism is Compton cooling. If the Universe is optically thick to Compton scattering at that time then we do not expect to see anisotropies. It is only after the Thomson optical depth falls below unity that a nonuniform pattern emerges to tell us of the state of the heated gas. In explosion and other models with significant energy release beyond the obligatory gravitational binding energy, this can occur early enough that the Sunyaev-Zeldovich (SZ) anisotropies that accompany Compton cooling distortions offer a window to the Universe at $z \sim 10$. In hierarchical models in which only binding energy release heats the gas to X-ray emitting temperatures, SZ anisotropies may tell us only about the $z \lesssim 1$ universe when pancakes, filaments, and clusters form in abundance (§4).

Within the collapsing gas, stars will also liberate potentially observable energy, and blanketing dust may shift the energy into a distortion of the CMB. This dust could be generated in the first stars that form at $z \sim 300 - 10$, or might not appear in significant abundance until galaxies form at $z \sim 20 - 3$ (the respective redshifts being very model-dependent). The anisotropy pattern would be quite different in the two cases. We also address this in §4.

A secondary background is characterized by a monopole, which gives the angle-averaged energy spectrum, by a dipole, largely induced by the motion of the earth relative to the rest frame of the radiation, which therefore should nearly coincide with the CMB dipole, and by higher multipoles which provide a rich storehouse of information on the origin of the background in question. For low multipoles, there will be a Sachs-Wolfe effect similar to that for primary anisotropies, but, given the current COBE constraints on the amplitude of the background, it is unlikely to be competitive with the primary multipoles. Where anisotropies generated by gas or dust processes are most interesting is for multipoles above about 50; *i.e.*, on angular scales below a few degrees.

1.2 Gaussian Fields Are Maximally Random

The initial conditions most often invoked for cosmic structure formation are linear perturbations distributed as a homogeneous and isotropic Gaussian random field. All that is required to fully specify the statistics is a single function, the (linear) density fluctuation spectrum, $d\sigma_\rho^2/d\ln k \equiv \frac{k^3}{2\pi^2}\langle|(\delta\rho/\rho)(k)|^2\rangle$, giving the contribution of (statistically independent) modes of comoving wavenumber $\vec{k}$ to the *rms* linear density fluctuations σ_ρ. Although Gaussian perturbations generated by quantum noise during inflation has served as the standard model of the past decade, non-Gaussian initial perturbations may arise in inflation models with more than one scalar field, or in theories with phase transitions of various sorts.

For Gaussian models, the shape of $d\sigma_\rho^2/d\ln k$ prior to the onset of nonlinearity is determined by the linear evolution of the post-inflation spectrum. The overall normalization amplitude of the linear spectrum is parameterized by a biasing factor b_ρ, which is one if mass traces light and greater than one if galaxies are more clustered than the mass distribution. We technically use J_3 normalization as described in Bardeen *et al.* 1986, §VIf. We also sometimes refer to an alternative normalization amplitude, σ_8, which gives the *rms* mass fluctuations within spheres of radius $8\,h^{-1}$Mpc. The relation between σ_8 and b_ρ varies slightly from model to model; for standard CDM, we have $\sigma_8 \approx 0.93 b_\rho^{-1}$.

Since one of the aims of this paper is to emphasize the difference in observing strategies that one would adopt to test non-Gaussian rather than Gaussian anisotropies, we first describe the fundamental characteristic of Gaussian fluctuations. Consider a general statistical distribution functional $\mathcal{P}[F(\vec{x})]\mathcal{D}F(\vec{x})$ giving the probability of a field configuration $F(\vec{x})$ of a random field F. Define the entropy of this probability to be

$$\text{Entropy}[\mathcal{P}] = -\int \mathcal{P}[F(\vec{x})]\,\ln\left(\mathcal{P}[F(\vec{x})]\right)\,\mathcal{D}F(\vec{x}).$$

Among all of the distributions with a specified spectrum $d\sigma_F^2/d\ln k$, the Gaussian one is the one which maximizes the entropy. This lemma is the basis of a picturesque way of thinking of non-Gaussian fluctuations. Fig. 2.1(b) displays typical 2D angular power spectra for primary anisotropies (for the CDM model of §3) and secondary anisotropies (for the SZ and dust models of §4). The statistics of the primary ones are Gaussian, and so the maps are maximally random distributions of the power available. The best observing strategy is then to concentrate the observing time on just a dozen or so patches of the sky because you are bound to hit something (see Readhead *et al.* 1989, Appendix A, for a quantification of this point). The secondary ones are non-Gaussian and the power in the maps (*e.g.*, Fig. 4.4, 4.7) is more concentrated around the 'hot' or 'cold' spots than in the Gaussian case. In that case, a better observing strategy is to sample many patches at lower sensitivity to look for the regions of high power concentration.

§2 reviews the experimental status, §3 reviews the current constraints on the scale invariant models, in particular the CDM model, from primary anisotropies, and §4 presents new maps of the X-ray background and SZ anisotropies simulated using a hierarchical peaks approach to identify groups and clusters of galaxies. We also show how our method compares with the results of N-body simulations and recent X-ray cluster observations for the CDM model. Application to far-infrared emission from bursting primeval galaxies is also sketched. The non-Gaussian maps of secondary anisotropies for the SZ and dust models shown in Figures 4.4 and 4.7 should be compared with the Gaussian maps of primary CDM and primary isocurvature baryon anisotropies displayed in Bond and Efstathiou (1987) and Bond (1990). We also show the extent to which the OVRO small angle experiments are likely to suffer contamination from secondary anisotropies and find that faint radio sources completely dominate the seondary signal.

In this paper, we let $h \equiv H_0/100 \, \mathrm{km \, s^{-1} \, Mpc^{-1}}$ and Ω_{nr} denote the energy density relative to closure in (stable) non-relativistic particles (*e.g.*, CDM and baryons). Inflation implies that Ω is nearly unity, while the Hubble parameter must be in the 40 to 50 range to avoid a time crisis for globular clusters and nuclear cosmochronology, and Ω_B must be $\lesssim 0.06(2h)^{-2}$ to maintain the successes of primordial nucleosynthesis. We use the minimal CDM model ($\Omega = \Omega_{nr} = 1$, h=0.5, adiabatic scale invariant perturbations) to provide our standard illustration of structure in the CMB. The main free parameter is b_ρ. Currently it is thought to lie in the range 1-2.6, with the 1-1.5 range now preferred over the higher ~ 2 value. However, for many of the calculations in this paper, the defining features depend only upon a relatively limited band of k-space. For example, for the group and cluster catalogues of §4, the important waveband is from $\sim 2 \, h^{-1} \mathrm{Mpc}$ to $\sim 8 \, h^{-1} \mathrm{Mpc}$. We can think of b_ρ as a way of characterizing the primordial amplitude of the perturbations in this waveband. The role of the wavelengths longward of this band is to set the large scale clustering of the objects. To be sure, these correlations may be observable in the background radiations we discuss, but it is the nature and abundance of the objects themselves which in the first instance characterize the radiation pattern. We therefore believe that, although we used the CDM model in §4, our results are robust for a more general class of Gaussian hierarchical clustering theories.

2. ANISOTROPY EXPERIMENTS

It is customary to categorize observations as very small ($< 1'$), small ($1'-30'$), intermediate ($30'-2°$), or large scale ($> 2°$). Large angle measurements do not probe gas dynamical processes very well, but are superb for probing primordial metric fluctuations, thereby providing a direct window on the early universe. They can be either space-based, such as RELICT 1 (Klypin *et al.* 1987) and COBE's DMR (Smoot *et al.* 1991), balloon-borne, such as those of Boughn *et al.* (1990) and Meyer, Cheng and Page (1991), or ground-based, such as the Tenerife experiment (Davies *et al.* 1987) and the interferometric observations of Timbie and Wilkinson (1988, 1990).

Currently, the highest sensitivity intermediate angle experiment is that of Meinhold and Lubin (1991), who took advantage of the superior observing conditions at the South Pole. The data was taken at a wavelength of $\lambda = 3.3$mm in a region at $\alpha = 21^h.5$, $\delta = -73°$ where galactic dust emission as seen by IRAS at $\lambda = 100\mu$ was found to be low, with an expected contribution of 18–36 μK *rms*. Nine fields were linked together in a strip from $\alpha = 20^h.52$-$22^h.48$. Each measurement consisted of a single difference of half cycles of a $1°.4$ sinusoidal chop with a 30' beam (FWHM). A mean level and gradient was removed from each observing cycle through the fields. A total of 43 hours of useful integration time was obtained. For uncorrelated Gaussian fluctuations, the 95% (Bayesian) credible limit is $\delta T/T < 3.4 \times 10^{-5}$. There are a number of other intermediate angle experiments now underway, using balloons, the South Pole, and other extremely cold places such as Saskatchewan, Canada.

Currently, the most sensitive small angle experiment is that of Readhead *et al.*(1989), who used the 40-m radiotelescope at the Owens Valley Radio Observatory (OVRO) at a frequency of 20 GHz. A "double-differencing" switching scheme was used with Gaussian beams (1.8 FWHM) separated by 7.15, for a set of 8 fields near the North Celestial Pole (NCP). The 8 fields were observed for a total of 398 hours; one field was discarded from the subsequent analysis due to the presence of a faint but significant nonthermal extragalactic radio source in the beam. A 95% credible limit of $\delta T/T < 1.7 \times 10^{-5}$ between single uncorrelated beams was derived for the 7 remaining fields.

The Readhead *et al.* experiment pushed for maximum sensitivity to Gaussian fluctuations by observing a relatively small number of fields with low individual noise levels (31 μK 1σ). However, if the anisotropies are non-Gaussian, then the impressive limits we find in §3 for the OVRO 8-field and South Pole 9-field experiments do not apply. To constrain such models, a large statistical sample of fields is required. However, the penalty is that each field will have a higher noise level for the same amount of observing time. An example is the RING experiment, a survey of 96 fields undertaken with the OVRO 20 GHz system (Myers 1990, Myers, Readhead, and Lawrence 1991). The 96 fields are arranged in a ring around the NCP at a declination of $88°10'42''$. Double switching is performed as in the 8-field experiment, and the geometry is such that the reference fields in the switching cycle are the main fields for adjacent observations — thus the RING is interlocked, which provides extra checks on possible systematic errors in the radiometry. With an average 'error bar' of 113μK (1σ), the expected 95% upper limit for 96 fields is $\delta T/T < 1.8 \times 10^{-5}$, nearly as good as was seen in the 8-field experiment; however, an increased variance in the data points is found above the level expected for the noise alone and an upper limit of $\delta T/T < 5.4 \times 10^{-5}$ between uncorrelated single beams is found. Unfortunately, the enhancement of the signal from non-Gaussian fluctuations afforded by the enlarged area coverage also increases the contamination from discrete extragalactic radio sources. We discuss the effect of this 'secondary background' on the RING and 8-field experiments in §4.3. A new Owens Valley experiment with a 7.5 FWHM beam and 24.5 separation is also underway, using a dedicated 5-metre diameter radiotelescope at 32 GHz, which is expected to probe fluctuations below the 10^{-5} level.

On even smaller angular scales, at or below $1'$, the VLA has been used to probe fluctuations at 5 GHz (Martin and Partridge 1988 and Fomalont *et al.* 1988) and at 15 GHz (Hogan and Partridge 1989). Because of the lower observing frequency, contamination from discrete extragalactic radio sources within the imaged field has plagued the 5 GHz efforts. After careful subtraction of the contribution from these objects, 95% limits of $\delta T/T < 6 \times 10^{-5}$ on the scale of $1'$ and $\delta T/T < 1.2 \times 10^{-4}$ on $18''$ have been reached. The 15 GHz limits are somewhat worse due to poorer system performance of the VLA at this frequency. New observations at 8.5 GHz using new low noise receivers have produced limits competitive with the filled-aperture experiments: preliminary 95% limits quoted by Partridge (1991) give $\delta T/T < (1-2) \times 10^{-5}$ on the scale of $1'$ and $\delta T/T < 6 \times 10^{-5}$ on $10''$.

There has also been recent interest in higher frequency experiments at very small scales using sub-millimetre telescopes. Kreysa and Chini (1988) used the IRAM Pic de Valeta dish with a resolution of $11''$ FWHM in a double-differencing three-beam experiment with a throw angle of $30''$ to search, at $\lambda = 1300\mu$, for emission from quasars that had been detected by IRAS at 100μ. Most of their fields gave no quasar signal, and could be treated as blank sky. The 95% credible limit for 21 of their 25 fields is $\delta T/T < 3.2 \times 10^{-4}$. They argued that 4 fields could be rejected since it was assumed the quasars were the sources. The JCMT experiment of Church *et al.* (1990) was a three beam experiment at 800μ, with an 18″ resolution and a throw angle of 40″, which gave a 95% credible limit of 1.4×10^{-3}. In spite of the poorer sensitivity, this JCMT result gives competitive limits on dust-generated anisotropy because the wavelength is shorter than IRAM (§4.4). SCUBA (Cunningham and Gear 1990) is an array instrument to be used on the 15m JCMT in a few years which will give diffraction-limited performance for each pixel at 855μ ($14″$ FWHM, with 37 pixels), and $438\ \mu$ ($7″$ FWHM, with 91 pixels). Although it is similar to the Church $et\ al.$ experiment in resolution and configuration, it promises a vastly improved sensitivity per unit integration time.

The ability of these experiments to probe various theories of anisotropy can be illustrated by comparing the multipole band that the experiments are sensitive to with the angular power spectra of the theories, C_ℓ, as we do in Figure 2.1. If we expand the microwave background pattern in multipole moments, $\Delta T/T(\hat{q}) = \sum_{\ell m} a_{\ell m} Y_{\ell m}(\hat{q})$, where $\hat{q}$ is the angular direction of the incoming photons, then $C_\ell = \langle |a_{\ell m}|^2 \rangle$. For primary anisotropies and Gaussian initial perturbations, only C_ℓ is needed to fully specify a realization of the anisotropy pattern. For secondary anisotropies, which are generically non-Gaussian, an infinite number of higher order (reduced) correlation functions of $a_{\ell m}$ is needed to realize the patterns. The multipole band can be characterized by a filter function W_ℓ such that the rms fluctuations in a given experimental field is $(\Delta T/T)^2_{rms} = \sum_\ell (2\ell + 1) W_\ell C_\ell/(4\pi)$.

The angular spectra shown in Fig.2.1b are for models described in §3 and §4. We refer the reader to these sections for details. We first make some general comments. The South Pole experiment has a multipole range ideal for optimizing the signal from power spectra like cdm-sr but a filter with a beam about 3 or 4 times larger is better for probing cdm-nr. OVRO filters are good for SZ, but smaller scale ones are not unless one is looking at specific clusters, and primeval galaxy backgrounds from dust are best probed with arcsecond resolution: the 18″ is almost too large (§4.4).

The quantities actually plotted in Fig.2.1b are $\ell^2 C_\ell/(2\pi\sigma_T^2)$, the power per log of wavenumber, normalized to have unit area under the curves. The normalization factor σ_T gives the total rms fluctuations for the model in question. The 'cdm-sr' primary anisotropies (for the $\Omega = \Omega_{nr} = 1$, $\Omega_B = 0.1$, h=0.5 CDM model with standard recombination) have $b_\rho \sigma_T = 3.5 \times 10^{-5}$; with no recombination (or very early reionization), $b_\rho \sigma_T = 2.5 \times 10^{-5}$. For the ambient SZ anisotropies, the power spectrum was computed by Fourier transforming a $4° \times 4°$ map similar to Figure 4.4a except unfiltered. For the waves within the box, $\sigma_T = 6 \times 10^{-6}$ (at radio frequencies), significantly smaller than the primary ones.

The angular power from dust-emission in the pregalactic and protogalactic medium is concentrated at sub-arcsecond scales, corresponding to scales just above the size of the sources. The discreteness of the objects leads to a Poissonian 'white noise' power spectrum (du_P) just above these scales. To illustrate this we have chosen a specific model that we use extensively in §4.4 which has primeval galaxies radiating at $z \sim 5$ with comoving number density n_{G*} that are clustered as well. At 800μm, the normalization in $\Delta T/T$ is $\sigma_T = 8.5 \times 10^{-3}\ (0.02(\,h^{-1}Mpc)^{-3}/n_{G*})^{1/2}$, which scales with the abundance in the classic $1/\sqrt{N}$ Poissonian fashion. Thus, if the source density is about that of bright galaxies now, $\sim 0.02(\,h^{-1}Mpc)^{-3}$, then the fluctuation power is huge, and this remains so even when one filters with the JCMT beam. On the other hand, if the objects are dwarf galaxies with a density 100 times greater, seeing this discreteness noise becomes difficult. The dust power extending to smaller ℓ is the contribution from galaxy clustering. For the case shown here, with nonlinear clustering following a $\xi \sim r^{-1.8}$ correlation function attached to a larger scale linear clustering and biasing contribution (the two parts of du_c), the normalization is $\sigma_T = 8.8 \times 10^{-4}$ at 800μm. This would dominate the anisotropy signal in COBE's DIRBE experiment for example.

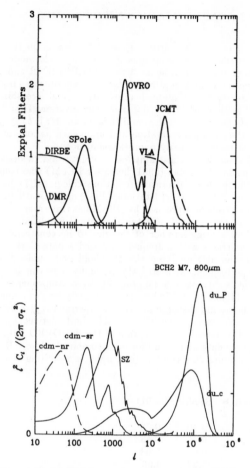

FIGURE 2.1 (a) Experimental filters as a function of the multipole ℓ. When they are multiplied by the angular power spectra shown in (b) and integrated over ln ℓ, they give the variance in the anisotropies expected in each field of the relevant experiment. We can roughly consider ℓ to be in inverse radians, so the ℓ-pole probes angles around 3438/ℓ arcminutes. The COBE FIRAS and DMR filters and the RELICT filter are the same, 7°. The DIRBE filter is a single beam one with a 42' beam. OVRO is Owens Valley with a 1.8' beam and 7.15' throw. The RING experiment has the same filter. JCMT is the Church et al. (1990) experiment, good for probing dust emission anisotropies. VLA roughly corresponds to the VLA in D-mode at 8.5 GHz. The South Pole filter is that for the Meinhold and Lubin (1991) experiment. (b) Angular power spectra $\ell^2 C_\ell/(2\pi)$ are plotted for primary CDM anisotropies with standard recombination (cdm-sr) and with early reionization (cdm-nr), for the SZ effect from groups and clusters (as derived from a 4° × 4° map similar to Fig.4.4a), and for a dust model of primeval galaxies emitting in the sub-mm at z ~ 5 (showing separately the Poissonian piece, du_P, and the continuous clustering piece, du_c). The normalizations σ_T are given in the text.

3. CONSTRAINTS FROM PRIMARY ANISOTROPIES

In this section, we review the current state of the limits on theories of structure formation assuming Gaussian initial conditions such as the CDM model. At the moment the constraints from large angle experiments such as COBE's DMR and the earlier Soviet RELICT 1 satellite experiment and from recent balloon experiments (Boughn et al. 1990, Meyer et al. 1991) do not restrict inflation theories with normal recombination as much as intermediate and small angle experiments do. For scale invariant $\Omega = \Omega_{nr} = 1$ theories, the large-angle 95% confidence limit on the amplitude parameter is $b_\rho \gtrsim 0.2 - 0.4$, with the range depending upon the specific observation and upon such parameters as Ω_B, but not very much upon whether there was early reionization or not.

Intermediate and small angle constraints on scale invariant theories were derived by Bond, Efstathiou, Lubin and Meinhold (1991, hereafter BELM) who compared the South Pole and OVRO data with the theoretical predictions calculated in linear perturbation theory using the methods of Bond and Efstathiou (1987) and Efstathiou and Bond (1986, 1987). The Bayesian analysis procedure used by BELM has three great advantages: Firstly, for Gaussian anisotropies the integral over the theoretical amplitudes can be performed analytically, which allows a large number of theories to be constrained with very little computer time. This contrasts with a frequentist-based Monte Carlo approach used by Vittorio et al. (1991) who also analyzed the Meinhold and Lubin data (and obtained results in quantitative agreement with BELM). Secondly, one can incorporate properly modelled systematic effects such as residual gradients from atmospheric effects, which had to be removed from the Meinhold and Lubin signal. Thirdly, one can combine data sets to place more stringent constraints than the individual experiments give by themselves.

In the following table, we list the 95% credible limits derived by BELM on b_ρ for $\Omega = \Omega_{nr} = 1$, h=0.5, CDM models as a function of Ω_B, for OVRO data alone, the Meinhold and Lubin South Pole data alone, and for both combined. Here, SR denotes standard recombination and NR denotes no recombination, which is a limiting case of early reionization.

	Ω_B	SPole+OVRO	SPole	OVRO
SR	0.01	0.52	0.40	0.28
SR	0.03	0.66	0.45	0.43
SR	0.1	0.95	0.63	0.63
NR	0.1	0.27	0.27	0.06
SR	0.2	1.3	0.93	0.75

We therefore conclude that the experiments are very near to the predicted levels of the SR models, but are still far off the NR models, for which the large angle limits are currently better. This situation will change once the experiments with 1° to 2° beams begin to deliver their data. When we analyzed the 96-field RING data using the Bayesian approach, we found b_ρ limits significantly lower than those obtained for the South Pole and OVRO 7-field observations, due to the larger error bars and residual signal. As we discuss in §4.3, the central issue of source contamination must be well-modelled for the RING to improve the limit.

The angular CMB power spectrum for a massive neutrino-dominated universe is quite similar in shape to that of the CDM model but the overall amplitude must be higher to ensure that enough structure forms. Indeed, the relation between the redshift z_{nl} at which linear theory predicts that the rms fluctuations in the density first reach unity is related to b_ρ by $1 + z_{nl} \approx 1.1/b_\rho$. Thus, we must have $z_{nl} \lesssim 0.7$, which makes it very unlikely that these hot dark matter models will work without some fix such as an isocurvature component from strings, small mass seeds, or whatever.

There are indications that the CDM model does not have enough large scale power in the density fluctuations to explain the observed clustering. Bardeen, Bond and Efstathiou (1987) found that modifying the contents of the universe so that linear evolution of an initially scale invariant power spectrum would lead to more large scale power than the CDM model has has a price: it also invariably leads to higher amplitude small and intermediate angle anisotropies, and therefore even tighter constraints on b_ρ than those in the Table. Breaking initial scale invariance to get more large scale power also gives larger anisotropies and tighter b_ρ limits. The paradox of strong large scale clustering in a universe with low CMB anisotropy is growing. Fixes such as early reionization will only delay the day of reckoning a bit since experiments are being undertaken which optimally probe NR models. Other fixes such as non-local biasing of light compared with the mass distribution may solve the clustering problem, but cannot give CMB isotropy.

4. SECONDARY ANISOTROPIES FROM NONLINEAR STRUCTURE

Current constraints on secondary radiation backgrounds, even those from COBE, still allow reasonably large energy densities Ω_{RT} (in units of the critical energy density). Expected thermonuclear, accretion and other sources must give $\Omega_{RT} \gtrsim 10^{-8}h^{-2}$ and would quite plausibly give Ω_{RT} around $10^{-6}h^{-2}$. (See Bond, Carr and Hogan (1986, 1991), hereafter BCH1 and BCH2.) For comparison, the total CMB energy is $\Omega_{cmb} \approx 25 \times 10^{-6}h^{-2}$. The most popular models for spectral distortions of the CMB are emission from primeval dust, which FIRAS constrains to be $\Omega_{RT}h^2 \lesssim 0.6 \times 10^{-6}$ over the band $500 - 10^4\,\mu$, and Compton cooling of hot gas, which COBE and COBRA constrain to be $\Omega_{RT}h^2 \lesssim 10^{-7}$ (scaling as $\bar{y}/10^{-3}$). COBE could achieve limits on $\Omega_{RT}\,h^2$ as low as $\sim 3 \times 10^{-8}$ for general distortions, and as low as 3×10^{-9} for Compton distortions. DIRBE could detect a background below 10^{-7} from the far to near infrared, if the large foregrounds can be properly subtracted. Thus we should expect that an observable background will be found. However, to unravel its nature the correlations in the background will be fundamental, and these will definitely be non-Gaussian.

The minimal extension of the linear theory of primary anisotropies of §3 to get a non-Gaussian background is obtained by going to second order perturbation theory. These quadratic nonlinearities can play a very important role in constraining models with early reionization such as the classic isocurvature baryon models of the 1970s (Peebles 1987, Efstathiou and Bond 1987). Although there is no well-motivated particle physics model for the generation of these, open universes with initially Gaussian fluctuations in the baryon density with a power law spectrum $d\sigma_{n_B}^2/d\ln k \propto k^{3+n_s}$ are often assumed, with n_s treated as a parameter describing the local spectral shape. Large scale clustering data motivate values in the range $n_s = 0$ (Poisson seed model) to $n_s = -1$. The combination of the Meinhold–Lubin and OVRO data can be used to rule out much of Ω_B-n_s space for $n_s \lesssim 0$ (BELM); e.g. $n_s = -3$, the scale invariant slope, is very strongly ruled out. However, there is little reason to suppose that isocurvature perturbations would be Gaussian and power law. Even if they are, the second order perturbation theory calculations may not describe the anisotropies from the heated matter, although for the constraints derived from the OVRO experiment to be invalidated fairly large scale gas transport at these early epochs is required. A general treatment of Thomson scattering by nonlinear bulk-flow currents would be required to address this, a subject not yet well explored.

The obvious way to solve nonlinear gas problems is with large scale hydrodynamical simulations of the pregalactic and intergalactic medium. Even with the greatly expanded computational power that is being applied to 3D hydrodynamic/N-body cosmology codes, we are still some distance from generating realistic radiation patterns because of resolution limitations. We expect that the future will bring truly large scale simulations many Gpc in depth which will be solved at all of the interesting sites in the volume by spatially adaptive hydro techniques. Until that day it is worthwhile to explore the parameter space using a variety of approximate methods.

4.1 Constructing Catalogues for Secondary Anisotropies

The most often used method for estimating average backgrounds utilizes the Press-Schechter (PS) mass function for Gaussian hierarchical theories. In spite of the fact that this has recently been put on a firmer theoretical foundation and shown to accord reasonably well with the mass function of groups found in N-body studies (*e.g.*, Bond, Cole, Efstathiou and Kaiser 1991 [BCEK]), the PS mass function does not give information on fluctuations. What is required is an assignment scheme for a point process. One attempt within the basic PS framework is the 'brick' model used by Cole and Kaiser (1989) for calculating SZ from clusters. Schaeffer and Silk (1988) used the Schaeffer ansatz for the form of the entire nonlinear hierarchy of n-point functions of galaxies to estimate SZ anisotropies for clusters.

Bond (1988) advocated shot noise models consisting of a catalogue of objects (the shots) with assumed profiles for the gas and dust surrounding the shots. For a Gaussian hierarchical model such as the CDM scenario, the shots were taken to be density peaks of the initial Gaussian random field filtered on a variety of scales. In shot noise models, anisotropies are a result of two effects: the Poisson noise associated with the finite number of sources and the continuous clustering of the shots. For SZ anisotropies, groups and clusters associated with various filtering scales were used. Variants of the same method were used to estimate SZ anisotropies in explosion models as well, with the shots being defined by the centres of non-overlapping voids. Bond (1988) and BCH2 used shot noise models to estimate angular power spectra and *rms* anisotropies from dust emission.

These calculations can give the average distortion and the *rms* anisotropies as a function of beam size semi-analytically, but to get the complete statistical distribution in practice requires Monte Carlo simulations. The limited hydrodynamical (Ryu *et al.* 1990, Cen *et al.* 1990) and sticky-particle (*e.g.*, Thomas and Carlberg 1990, Klypin and Kates 1991) studies that have been performed to date have given maps from which the full statistical information could be drawn, except that the single 3D boxes used in the simulations were of relatively small volume. So far one cannot satisfy at the same time the competing needs to treat large scale clustering, to have high resolution in the collapsed objects, and to deal with the rare events that give the largest excursions in the anisotropy signals.

We now describe the current status of a program that we have undertaken to make realistic secondary background maps, applying a significant improvement in the Gaussian peaks approach to create simulated catalogues of objects for hierarchical Gaussian models of structure formation such as the CDM model.

The major obstacle in the Bond (1988) approach was the cloud-in-cloud problem of overcounting objects of smaller mass that would have merged into higher mass ones. We avoid this in the following way: We find the (linear) density peaks on a hierarchy of filter scales $R_{f1}, ..., R_{fn}$ that are above some threshold f_v associated with complete collapse of at least the central neighbourhood of the peak. From each peak we then go out in sequential radial shells and determine the average overdensity within that radius. When this value has fallen below f_v, we let that define the radius r_{pk} of the peak. The threshold we chose, $f_v = 1.686$, is that value for which the shell of matter at that radius will have collapsed to a point in a spherical model. The mass at larger radii will still be infalling, while that at smaller radii will have passed through the centre, and so, presumably, will have virialized with the rest of the interior mass. Although there could be a sudden increase in overdensity at some shell at a radius $r > r_{pk}$, we expect this to be picked up as part of a peak on a larger filtering scale. The locally sphericalized profile can also be used to determine the binding energy per mass within the shell and thus the final virial velocity and temperature associated with the collapsed entity. The cloud-in-cloud problem is solved by having peaks of larger extent annihilate peaks of smaller extent.

Considerable progress can also be made analytically. One can evaluate the average number of peaks per unit filtering radius which upcross through the critical f_v contour (Bond 1989). The mass for the peaks can be determined by estimating r_{pk} using, instead of the local profile, the statistically-averaged profile of the field subject to the constraint that one has a peak with height f_v. This same profile can be used to estimate the binding energy of the peaks. The upcrossing peaks with no exclusion will clearly provide an upper bound for the mass function. Since the annihilation depends on non-local properties of the field, it is not feasible to do full calculations analytically. However, the assumption that only Poisson correlations of peaks be included yields a semi-analytic result which agrees well with the Monte Carlo method we use in this paper. The comparison of the analytic and simulated mass functions is shown in Figure 4.2. Although the analytics does quite well for average properties, correlations can be treated semi-analytically only in the weak clustering limit, while the full patterns require full simulations.

The physics defining the profiles about those peaks that survive in the catalogue is currently put in by hand, motivated by observed profiles and by the gross parameters such as binding energy found in the spherical model. Evolution in the peak population with redshift translates into evolution of the shots, their clustering and their profiles. Additional dynamics is included in the simulations by moving the peaks according to the Zeldovich approximation. To avoid the problems using this when there is too much nonlinear dynamics (caustics wash out), we use the displacement field smoothed over the Gaussian scale $R_f = H_0^{-1}\sigma_v/\sqrt{3}$, where σ_v is the current 3D rms velocity dispersion determined from linear theory; this choice is motivated by the Zeldovich filtering scale found by Bond and Couchman (1987).

A reasonable next step is to make better use of the initial conditions around each peak by doing local hydrodynamic collapses. This is not feasible for the number of peaks we have to deal with, but what is reasonable, and for the future, is to simulate generic local collapses and derive the peaks' properties from these, along the lines of Evrard's (1990) 3D hydro simulations of clusters.

To illustrate our approach, we describe our construction of group and cluster catalogues for the CDM model with biasing factors $b_\rho = 1$, 1.4 and 1.7. We choose some maximum angle over which we wish to construct maps, here 5°. We then lay down an inverted pyramid of boxes which contains within it a 5° cone. To cover the region out to redshift 1.5, beyond which there are very few groups, eleven layers of boxes of size $200\,h^{-1}$Mpc are needed. For each box, we construct a linear density F_k in Fourier space. For each of 15 Gaussian filtering scales R_{fj}, we FFT the filtered $F_k \exp[-(kR_{fj})^2/2]$, forming $F(\vec{r}; R_{fj})$. If there is significant power beyond the fundamental mode in the box, we add by brute force a contribution to F from very long wavelengths. For the CDM model in boxes of this size, this was not necessary. We find the peaks of each $F(\vec{r}; R_{fj})$ realization. Upon completion of the peak-finding for all 15 filters, we then determine the radii r_{pk} subtended by the peaks in Lagrangian (initial condition) space. The radius r_{pk} is defined to be that r at which the (unfiltered) volume-averaged value $\bar{F}(< r)$ first drops below f_v. The mass associated with

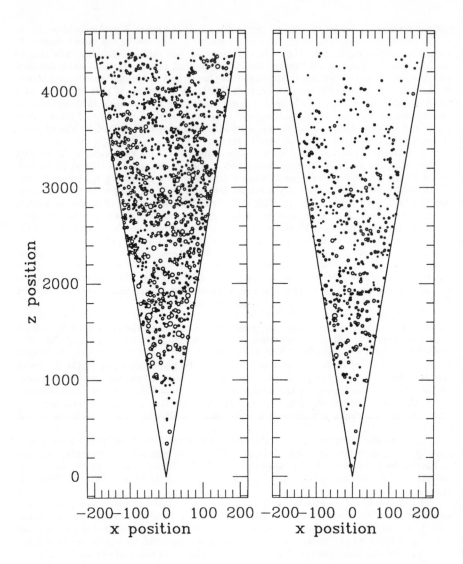

FIGURE 4.1 *Wedges (5° × 1°) taken from the $b_\rho = 1$ (left panel) and $b_\rho = 1.4$ (right panel) Monte-Carlo peaks simulations. The radius of the circle is proportional to the Lagrangian top-hat radius r_{pk} of the cluster.*

the peak is $M = (4\pi/3)\rho_{cr}\Omega_{nr}\,r_{pk}^3$, where ρ_{cr} is the comoving critical density. Our filters cover a mass range from $10^{13.5}$ up to 10^{16} $M_\odot$.

Assuming a sphericalized profile, we can also determine the binding energy per mass within r_{pk} (assuming as well that $\Omega = \Omega_{nr} = 1$, in which case E/M is independent of redshift, unless there is mixing (shell crossing through r_{pk}) or radiation loss or gain):

$$[E/M](< r_{pk}) = -\frac{1}{2}(Har_{pk})^2 S(r), \quad S(r) = r_{pk}^{-5}\int_0^{r_{pk}} \bar{F}(< r)\,dr^5; \quad (4.1a)$$

$$\bar{n}_B = 4.2\times 10^{-4}\Omega_B h^2 [f_v(1+z)]^3\,\text{cm}^{-3}, \quad T_V = \frac{m_N(-2[E/M](< r_{pk}))}{3Y_T}. \quad (4.1b)$$

The average density of baryons within r_{pk} is given by eq.(4.1b), as is the temperature of the uncooled baryons T_V, assuming the dark matter within r_{pk} is relaxed to a hydrostatic equilibrium with a uniform virial velocity which the uncooled gas shares. Here, m_N is the nucleon mass and Y_T is the number of particles per baryon in the gas (1.69 if it is fully ionized).

We then determine the displacement vectors $s_i(\vec{r})$ by FFT of $s_i(\vec{k}) = -ik^{-1}\hat{k}F_k$ $\exp[-(kH_0^{-1}\sigma_v/\sqrt{3})^2/2]$ for each peak to use in the Zeldovich approximation. We then move on to the next box.

After all boxes are completed, we trim the catalogue by having the peaks found using large filter radii annihilate those found using smaller filter radii that overlap with them. Among overlapping peaks of the same filter radius, the bigger ones annihilate the littler ones. We then move the peaks using 'Zeldovich dynamics' with the $s_i(\vec{r})$ of the peaks calculated for each box.

Slices through the final $b_\rho = 1$ and 1.4 simulations are presented in Figure 4.1. The clusters found within a $5° \times 1°$ wedge taken from the center of the constructed cones are shown as open circles, where the diameter of the circle is proportional to the Lagrangian radius $r_{pk} \propto M^{1/3}$ of the cluster. The scale is in Mpc (comoving).

4.2 Comparing Hierarchical Peaks with N-body Groups and X-ray Clusters

We are in the process of calibrating our results by detailed comparison with group catalogues, currently constructed from CDM N-body simulations of $(400\,\text{Mpc})^3$ boxes with 128^3 particles, computed using the P^{3}M method, by Couchman, and the PM method, by Villumsen. Figure 4.2 compares our results for the same initial conditions that Couchman used with his N-body groups, found with the most popular cluster-finding algorithm, percolation. To percolate groups out of a particle distribution, one 'top-hat' smooths the particle number density over radius $p\bar{n}^{-1/3}$, where $\bar{n}$ is the average particle density and p is the percolation parameter, then identifies groups with the connected regions of space lying above a fractional smoothed overdensity $n/\bar{n}$ cut of 2. One does not differentiate between those islands of particles so generated that are long and stringy and the more compact ones we are interested in, so improvements in the group-finding algorithm will be necessary. Nonetheless, we have found that for the Couchman and Villumsen simulations, the positions and masses of their groups and our peaks both in initial Lagrangian space and final Eulerian space (after the dynamics have been applied) agree rather well visually, for percolation parameters p between 0.2 and 0.3, the conventional range taken by N-body simulators. The average density in objects above mass M (in units of the critical density), $\Omega(> M)$, shown in Figure 4.2 shows one aspect of this agreement. Note that the PS method does quite well also, as anticipated by the N-body comparisons for power law spectra given by BCEK, but now showing the success for the very important rare events on the extreme Gaussian tail of the mass function.

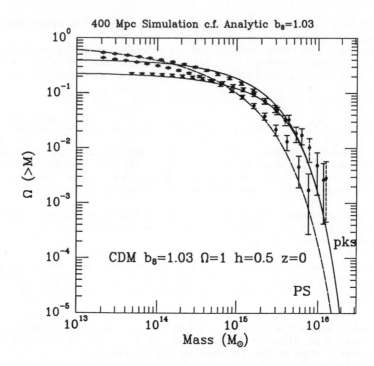

FIGURE 4.2 *The mass function* $\Omega(>M)$ *for a* $b_\rho = 0.96$ $(b_8 \equiv \sigma_8^{-1} = 1.03)$ *CDM model constructed with our Monte Carlo hierarchical peak simulation method (circles, with Poisson error bars dashed) is compared with those for Couchman's groups defined by percolation parameters* $p = 0.2$ *(lower circles, solid error bars) and* $p=0.3$ *(upper circles, solid error bars) found in his* $(400\,\mathrm{Mpc})^3$ *box* P^3M *simulation. The semi-analytic hierarchical peak results assuming only Poisson correlations is the lower solid curve, while the upper solid one is the result with no exclusion for overlapping peaks. The analytic Press-Schechter formula using the BCEK mass assignment is the long-dashed curve. The continued rise to small masses of the percolation group results may be due to extended unvirialised objects at low mass that our hierarchical peaks method is not designed to find.*

Hydrodynamical models of collapsing peaks are needed to find the shape of the cluster profiles and how they evolve. Here we assume the gas relaxes into the profiles inferred from cluster observations (*e.g.*, Mushotsky 1988). That is, we keep T_V uniform throughout the cluster and adopt a 'β-model' for the (uncooled) cluster gas density, with $\beta = 2/3$: $n_B = n_{Bc}/(1 + r^2/r_{core}^2)$ which we also truncate at r_{pk}. We have adopted two models for the core radius, a fixed $r_{core} = 300$ kpc, which is used ihe all of the Figures here, and a (self-similar) fractional radius $r_{core} = f_X r_{pk}$, with $f_X = 0.1$. The $f_X = 0.1$ luminosity agrees with that for the fixed core size case at 10^{15} $M_\odot$, and is quite similar for all masses above 10^{14} $M_\odot$. Even within the simple spherical model assumption, the true variation of r_{core} with mass and redshift is as yet quite unstudied.

A crucial test of a given model such as CDM is whether the theoretical cluster catalogues agree with observed cluster and group catalogues. Although it has been known since the beginnings of the biased CDM model that high biasing factors yield anemic clusters, and that this could serve as a powerful test of the model, it has been difficult to quantify because of uncertainties in whether peak and PS models are accurate, the small number of clusters found in N-body studies, and the somewhat sketchy observational data on clusters. In particular, there has been a strong suspicion that the optically determined velocity dispersions of clusters were often overestimated, apparently confirmed by Edge *et al.* (1990) using X-ray catalogues.

X-Ray surveys so far provide the best observational comparison samples for our simulations. The compilation of all-sky survey data by Edge *et al.* (1990) provides a flux-limited nearby cluster sample with X-Ray temperature information which we compare with our Monte-Carlo hierarchical peaks simulations and the analytic Press-Schechter and hierarchical peaks formulas: panels (a), (b) and (c) of Figure 4.3 show the flux counts, luminosity and temperature distributions, respectively. All are obtained for the 2–10 keV flux limit of 1.7×10^{-11} erg cm^{-2} s^{-1} used by Edge *et al.* Our numerical results used 27 contiguous $(400 \, \text{Mpc})^3$ boxes arranged in a cubical configuration, which gives as well a redshift limit of $z \leq 0.108$ which we also applied to the Edge *et al.* data (thus keeping 44 out of their 46 clusters that were within such a volume limit). As in Figure 4.2, the error bars, when present, are assigned assuming Poisson statistics for the cumulative and binned differential counts. Note the continuation of the remarkable agreement between our 'exact' numerical results and 'approximate' analytical results found in Figure 4.2 to these cases, for which the temperature as computed using the binding energy enters as well as the mass. The PS curves always lie below the peaks curves, although they do have similar shapes. The data tend to lie between the $b_\rho = 1$ and $b_\rho = 1.4$ peak curves for the flux counts, with the correct slope. The X-Ray temperatures from the Edge *et al.* sample are consistent with the $b_\rho = 1$–1.4 hierarchical peaks models, although it should be noted that the substantial temperature measurement errors have not been folded into the error bars shown for the Edge data points. We find the level of agreement between the theory and data encouraging, given that no parameters were adjusted to force the fit.

The luminosity functions of Figure 4.3(b) shows the biggest discrepancy between CDM model predictions and the X-Ray observations. Again, the PS result tends to underestimate the counts relative to the hierarchical peaks model, but both fall off more steeply than the Edge *et al.* luminosity function above 10^{44} erg s^{-1}, with an effective cutoff at around 10^{45} erg s^{-1} for even the low-bias $b_\rho = 1$ case. To illustrate that this is quite dependent upon the shape of the initial fluctuation spectrum, we also show the results for a model with enhanced large-scale power in Figure 4.3(c) which fits the high-luminosity data better than CDM does. The specific model (CDM+ν) shown has a 17 keV massive neutrino which decays with a 5 year lifetime added to a standard CDM model, which turns out to provide good agreement with large scale galaxy clustering data (Bond and Efstathiou 1991), whereas the standard CDM model does not. A very similar curve is obtained for a CDM model with non-zero cosmological constant ($\Omega_\Lambda = \Lambda/(3H_0^2) = 0.76$) and a lower CDM density, $\Omega_{cdm} = 0.19$, with $\Omega_B = 0.05$ and h=0.5). Both models not only have more power at larger scales, but flatter spectra than CDM does around the scales that are important for making the clusters. Kaiser (1991) and Evrard and Henry (1991) have also recently argued that flatter power-law spectra agree better with the Edge *et al.* data. All of these extra power models are close to (or exceed) primary microwave background anisotropy limits. We note that the conclusion that a CDM spectrum underproduces high-luminosity clusters is the result of just a few clusters with $L_X > 10^{45}$ erg s^{-1} in the Edge *et al.* data. More generally, we caution that larger homogeneous and complete X-ray cluster samples are essential to confidently constrain or rule out models.

Gas-dynamical effects can also improve the agreement between the CDM model and the data, and this is probably a more conservative path than the specific mechanisms discussed above for increasing the large-scale power in the density fluctuation spectrum.

278

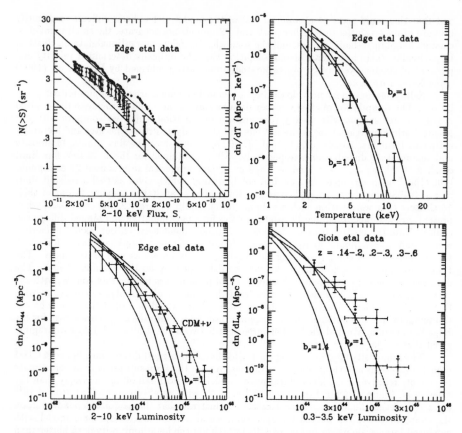

FIGURE 4.3 Predicted Flux counts, X-Ray Luminosity and Temperature functions versus observational data : (a) integral 2-10 keV Flux counts using analytic peak formulae for $b_\rho = 1$ and $b_\rho = 1.4$ CDM models (upper and lower solid curves), analytic Press-Schechter formulae (dashed curves), Monte-Carlo peaks simulation (circles), and Edge et al. (1990) X-Ray cluster data for $z \leq 0.108$ (circles with Poisson error bars); (b) differential Temperature counts for the same curves and data points (the upper solid $b_\rho = 1$ analytic peaks curves show the Poisson exclusion and no-exclusion results); (c) differential 2-10 keV Luminosity functions, curves as in (a), (b), plus an analytic peaks curve for a $b_\rho = 0.93$ CDM model that includes a decaying 17 keV massive neutrino (CDM+ν, short dashed curve) which shows how extra large-scale power can enhance the high-luminosity end of the function; (d) differential 0.3-3.5 keV Luminosity functions for the $b_\rho = 1, 1.4$ CDM model using the analytic peaks formula (solid curves), the $b_\rho = 1$ CDM model using the Press-Schechter formula (dashed), and the $b_\rho = 0.93$ CDM+ν model using the analytic peaks formula (short dashed), and the Gioia et al. 1990 EMSS cluster counts for redshift ranges $0.14 \leq z < 0.2$ (circles, solid error bars), $0.2 \leq z < 0.3$ (circles, no error bars), and $0.3 \leq z < 0.6$ (circles, dot-dash error bars). The CDM models show no discernible evolution over the redshift ranges of Gioia et al.

Since $L_X \propto n_B^2$, variation in n_B, whether because the average baryon abundance changes or the amount that finds its way into clusters changes, can dramatically alter the curves 4.2(a,b). The horizontal scale is uniformly shifted by $(\Omega_B/0.05)^2$ if we change the primordial value of Ω_B, as inferred from Big Bang nucleosynthesis studies, from the currently preferred value (*e.g.*, Olive *et al.* 1990), $\Omega_B = 0.05$, that we used for all of our figures. (Increasing Ω_B in this way would shift higher-bias models to the curves currently shown for low-bias models.) Since the luminosity and flux scaling depend upon an effective Ω_B, which includes only those baryons that have not cooled within the clusters and groups, the alteration of the horizontal axis is most likely to be nonuniform. For example, the cooled fraction in smaller clusters is likely to exceed that in more massive ones, which goes in the right direction to resolve the discrepancy. As well, if significant entropy generation occurs in the medium before the clusters form, *e.g.*, through gas shocking (Bond *et al.* 1984) or pre-heating (Evrard 1990, Kaiser 1991), the baryon density will be less than predicted by the primordial Ω_B and the decrease will be most marked for lower mass clusters and groups.

To assess the evolutionary aspects of the luminosity function, our theoretical models are compared in Figure 4.3(d) with the Einstein Medium Sensitivity Survey (EMSS) data of Gioia *et al.* (1990) for intermediate redshift clusters in three distance ranges. Since the results for the $b_\rho = 1$ CDM and $b_\rho = 0.93$ CDM+ν models exhibit only small evolution out to $z = 0.6$, we only show the results for the $0.14 \leq z < 0.2$ redshift bin. The Gioia *et al.* data do show a significant steepening with increasing redshift. As in panel (b), enhancing the large-scale power in the fluctuation spectrum flattens the luminosity function; a mass-dependent gas fraction can accomplish the same thing. Overall we conclude that high-bias CDM models are difficult to accommodate without high baryon densities ($\Omega_B > 0.2$). Our $b_\rho = 1$ and 1.4 CDM catalogues give temperature distribution functions which bracket that of Edge *et al.* , thus favouring bias factors in this range as well.

4.3 X-ray and SZ Maps, Radio Sources and OVRO

Because clusters are rare events defined by the tail of a Gaussian probability distribution, it is not surprising that the mass function will be highly sensitive to even the modest changes in the biasing factor used here. The most dramatic way to illustrate this is to show what various experiments would see as a function of biasing factor for fixed values of the other physical parameters such as core radius and Ω_B.

We used the cluster catalogue of Figure 4.1 to construct the $4° \times 4°$ X-ray maps shown in Figure 4.4. We passed the redshifted bremsstrahlung spectrum for each cluster through an approximate ROSAT filter ($0.1 - 2$ keV) to get the sky surface brightness of these sources. In Figure 4.4 panels (c) and (d) we give the $b_\rho = 1$ and 1.4 maps of the flux predicted for ROSAT's PSPC from these sources. The minimum contour is at $2.5 \times 10^{-15}(\Omega_B/0.05)^2$ erg cm^{-2} s^{-1}, and subsequent contours increase by a factor of 2 over previous ones. The value $\Omega_B = 0.05$ chosen for these maps is the preferred number from primordial nucleosynthesis, but the amplitude scales with Ω_B^2. However, the appropriate Ω_B to use is an effective value which excludes those baryons in the groups and clusters that are locked into stars, and therefore it may vary with group mass.

The 5σ sensitivity for the ROSAT all-sky survey was estimated to be $\sim 2 \times 10^{-13}$ before launch, and the highly sampled ecliptic pole region 5σ sensitivity was estimated to be $\sim 2 \times 10^{-14}$. Post-launch results indicate a slightly worse all-sky sensitivity. Deep observations in pointed mode look marginally promising for $b_\rho = 1$ and not very promising for $b_\rho = 1.4$. The $b_\rho = 1.7$ map is dismal. We have not applied any filter to the maps, although, in pointed mode, the resolution is about 30″. The ROSAT field-of-view is 2°, although only for the inner 1° is the resolution undegraded.

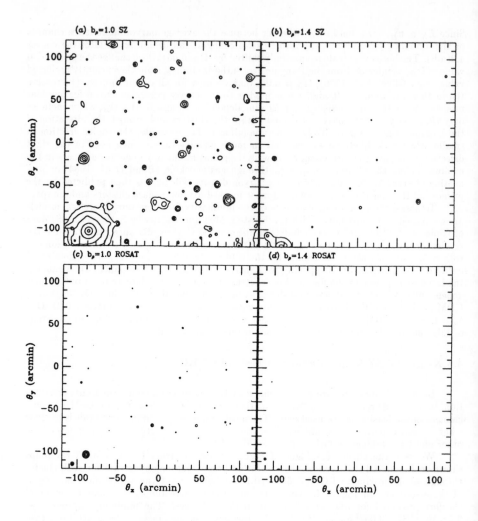

FIGURE 4.4 4° × 4° *contour maps of SZ sources smeared with a 1.8' beam from the (a)* $b_\rho = 1$ *and (b)* $b_\rho = 1.4$ *catalogues. The contour levels decrease by factors of two from* $-5 \times 10^{-6}(\Omega_B/0.05)$. *(c) and (d) give the corresponding* $b_\rho = 1$ *and 1.4 maps of X-ray emission from groups and clusters as would be seen by the ROSAT PSPC instrument. The contours increase by factors of two from a flux* $2.5 \times 10^{-15}(\Omega_B/0.05)^2$ *erg* $cm^{-2}\,s^{-1}$. *ROSAT's sensitivity in pointing mode is* $\sim 10^{-14}$.

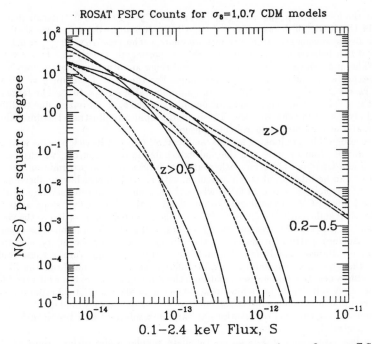

· ROSAT PSPC Counts for σ_8=1,0.7 CDM models

FIGURE 4.5 *The average number of counts per square degree above a flux cutoff S (in ergs* cm^{-2} sec^{-1}) *for the ROSAT PSPC instrument for all clusters* $(z > 0)$, *for clusters in the redshift bin 0.2–0.5 and for clusters with* $z > 0.5$. *The solid curve is the* $\sigma_8 = 1$ *CDM model, the short dashed curve is the* $\sigma_8 = 0.71$ *CDM model and the long dashed curve is the* $\sigma_8 = 0.71$ *CDM model with a 17 keV* ν *with a 5 year lifetime added to give extra large scale power. (The* b_8 *referred to in the figure is* σ_8^{-1}, *and is approximately* $1.07b_\rho$.) *These predictions are appropriate for the very deep fields that ROSAT will probe in pointing mode with varying exposures.*

These ROSAT maps indicate that it will be difficult to detect the influence of hot gas at $z \sim 0.5$ unless there is extra energy injected beyond the gravitational binding energy of groups and clusters. To further quantify the visual impression given by these maps, in Figure 4.5 we show the counts expected for the $b_\rho = 0.93$ ($\sigma_8 = 1$) and the $b_\rho = 1.3$ CDM models with the $r_{core} = 300$ kpc choice (unless r_{core} is smaller than r_{pk}, in which case we take the latter value). The all sky survey not only has a relatively high flux cutoff, but to register a cluster the flux smoothed over a Gaussian beam size 1.6′ within a square 'detect' cell of width 4.8′ must exceed this cutoff (Evrard and Henry 1991, and references therein). For the counts shown in Figure 4.5, which shows the integrated flux of the source rather than the smoothed flux within a detect cell, the primary effect is to truncate the counts at an increased flux cutoff.

The SZ contour maps corresponding to the X-ray maps (c) and (d) are shown in Figure 4.4 (a) and (b). The maps are smeared with the 1.8′ OVRO beam. The contours are negative, dropping by factors of two from the starting value of -5×10^{-6}. They scale as Ω_B.

For the $b_\rho = 1$ map, the average y-parameter is $\bar{y} = 1.0 \times 10^{-6}$. (Recall that $\overline{\Delta T/T}$ $= -2\bar{y}$ in the Rayleigh Jeans regime of the CMB spectrum. Hereafter, we shall assume Rayleigh Jeans wavelengths in the results we quote.) The rms is $\Delta T/T_{rms} = 6.3 \times 10^{-6}$, a factor of two below the sensitivity of current experiments, mostly due to a single large cluster. The minimum value is -1.9×10^{-4}, and there are other large excursions from the rms, as expected in such a non-Gaussian map. For the $b_\rho = 1.4$ map, $\bar{y} = 1.6 \times 10^{-7}$ and the rms is $\Delta T/T_{rms} = 9 \times 10^{-7}$, with minimum value -2.7×10^{-5}. The large cluster apparent in the $b_\rho = 1$ map is outside the boundary of the $b_\rho = 1.4$ image, due to differnent Zeldovich dynamics in the two cases. These maps graphically illustrate the unreliability of rms estimates as a guide to experimental design.

A severe problem in looking for this 'ambient' SZ effect at Rayleigh Jeans wavelengths is the confusion from discrete non-thermal extragalactic radio sources, which themselves generate a non-Gaussian radiation background. As a reasonable approximation, the spatial distribution can be taken as almost Poissonian, with a nearly power-law number-flux count relation. At frequencies above 15 GHz, where the attenuation of the galactic non-thermal background allows CMB experiments to be effective, the number-flux slope at faint flux density levels is sufficiently flat ($N \propto S^{-2.1}$) that the expected contribution to the rms of an experiment increases as the area surveyed is enlarged. This foreground must be superposed on the anisotropy pattern produced by the primary and other secondary backgrounds, and in some instances may in fact dominate. The statistical signature of these confusing sources, which are positive fluctuations, is different than the signal from the SZ effect, which produces negative fluctuations, and from the primary anisotropies, which tend toward a zero-mean. A sufficiently large low-noise dataset may allow statistical separation of the different components. A more direct and effective decontamination would involve coincident (and preferably concurrent) observations at varying resolutions and wavelengths to allow size and spectral discrimination.

The best available datasets on intermediate angular scales or less, those of OVRO, VLA and the South Pole experiments, suffer in varying degrees from this contamination. The South Pole experiment is at a sufficiently high frequency that the discrete-source contamination is likely to be negligible; however, interference by dust is certain to be a problem as the sensitivity level is increased. The single-dish measurements of the OVRO 8-field and 96-field are especially susceptible to the effects of radio sources, due to the relatively low frequencies and the high gain of the antenna. One of the 8 fields in the former experiment was deleted from the sample because of a 20 mJy source in the beam, and the 96 fields of the RING survey exhibit signs of significant infection. Although interferometric observations from the VLA are also subject to this problem, because of the information automatically provided over a range of angular scales they can be partially decontaminated.

The observations of Fomalont et al. (1988) with the VLA at 5 GHz, in addition to producing the limits discussed in §2, predict that the differential source counts follow $dN/dS = 22S^{-2.1}$ sr^{-1} at the OVRO frequency of 20 GHz (assuming an average source spectral index of -0.5). The composite $4° \times 4°$ image of the $b_\rho = 1$ SZ model (Figure 4.4a) with a Poisson source model using this extrapolation of the VLA counts is shown in Figure 4.6. Only the strong SZ source at the lower left edge of the map is visible above the numerous confusing sources.

For this map, the OVRO 8-field experiment would have a 40% chance of encountering at least one field contaminated at the 3σ level or worse (one was indeed found), as compared with only 1.0% for the $b_\rho = 1$ SZ model. The RING has a 79% chance of contamination compared with 0.99% for detection of SZ. The SZ probabilities are as large as they are due to the single large cluster in the map. Recent VLA data at 8.5 GHz have been reported by Windhorst to indicate that a population of flat spectrum radio sources may dominate at higher frequencies. Using his 8.5 GHz counts directly as an extreme extrapolation, we find

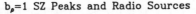

b_ρ=1 SZ Peaks and Radio Sources

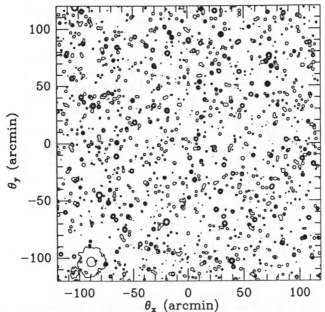

FIGURE 4.6 *Contour map of negative SZ sources (dotted) from the* $b_\rho = 1$ *catalogue with* $\Omega_B = 0.05$ *and positive radio sources (solid) drawn from faint counts extrapolated from the 5 GHz VLA data. The map is smoothed with the OVRO beam. The radio sources completely dominate the anisotropy signal. The lowest contour is* 2×10^{-5} *and the rest increase by factors of 4.*

a map which is even more packed with radio sources. The 3σ encounter probability rises to 73% and 98% for the 8-field and the RING, respectively.

Separate observations using the VLA at 8.5 GHz (Myers 1990) of 20 RING fields, 7 of which had measured temperatures above 3σ, show that 4 contain milliJansky-level radio sources, 3 of which are sufficiently bright to account for the observed fluctuations with extrapolated spectra. Three others remain unaccounted for by VLA sources, and follow-up observations at higher frequencies are underway.

We conclude that, unless the foreground can be removed accurately, a successful detection of the ambient SZ effect from unidentified groups and clusters is unlikely at these low frequencies. Experiments at millimetre and submillimetre wavelengths are likely to provide better prospects for observation of the ambient SZ effect.

4.4 Dust Anisotropies from Primeval Galaxies

BCH1 showed that if galaxies (even dwarfs) exist at $z \sim 10$ then they cover the sky, and if they are dust-laden then all energy from the near IR to the X gets absorbed and re-emitted in the far IR, with a peak wavelength $\lambda \sim 500 - 700\mu$ which is relatively model

284

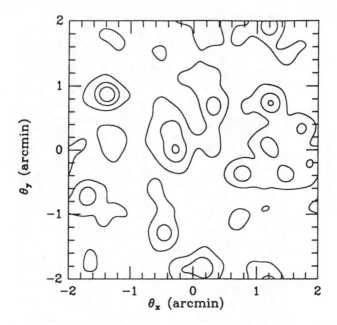

FIGURE 4.7 A 4' × 4' contour map for dust-emission from primeval galaxies at $z \sim 5$ convolved with an 18" beam appropriate for the Church et al. (1990) 800μm JCMT experiment. The lowest contour is 10^{-13} W cm^{-2} sr^{-1} and the rest increase linearly in equal steps.

insensitive. BCH1 and Bond (1988) showed that the intensity fluctuations in the sub-mm range would typically be $\Delta I_\nu / I_\nu \sim 0.01 - 0.1$ for a wide range of clustering models. BCH2 considerably extended the treatment of the emission and the expected anisotropy levels, using linear perturbation theory and shot noise models.

To focus the discussion, we consider BCH2 Models 7, 8 and 14 as typical examples of primeval galaxy dust models. For these, a burst of radiation generated at $z \gtrsim 5$ is absorbed by dust with a mass density Ω_{dust} that is located in primeval galaxies formed at $z = 5$ whose comoving number density is n_{G*}. The dust is assumed to be similar to typical Galactic dust. If n_{G*} is large, the depth across each individual galaxy is small, and the BCH2 linear perturbation treatment is appropriate. With $\Omega_{dust} = 10^{-5}$, which corresponds to Population I abundances in these galaxies, 86% of the energy is absorbed and re-radiated into a submillimetre background, while the remaining 14% is unabsorbed, contributing to a near infrared background; with $\Omega_{dust} = 10^{-6}$, only 18% is in the sub-mm and the rest is in the near IR. With a burst energy $\Omega_{RT}h^2 = 10^{-7}$, the average dust temperature is $T_d = 22$K, with the peak in $d\Omega_R/d\ln\lambda$ occurring at $\lambda \approx 500\mu$m. With the larger energy $\Omega_{RT}h^2 = 10^{-6}$, the dust is hotter, with average temperature 35K, and a spectral peak at $\lambda \approx 300\mu$m. However, since this peak emission band is shortward of where the current 6×10^{-7} FIRAS constraint comes from, this model is not yet ruled out. For this model, the IRAM and JCMT experiments, assuming their 95% upper limits are for *rms* anisotropies (non-Gaussian problems here, see below), give a source density limit of $n_{G*} \gtrsim 0.06(\,h^{-1}\mathrm{Mpc})^{-3}$, a factor of 3 above the bright galaxy abundance. If we assume that the primeval galaxies clustered with a $\gamma = 1.8$ power law correlation function, like galaxies that we see now do, then the limit on their clustering length (when they emit) is $r_o \lesssim 4\,h^{-1}\mathrm{Mpc}$, comparable to the current observed length. With the smaller energy, the constraints are a factor of 30 worse on source abundance and 7 worse on clustering length; *i.e.*, uninteresting.

For this same model with a $\Omega_{RT}h^2 = 10^{-6}$ energy release, we can predict *rms* levels in $\mathrm{W\,cm}^{-2}\,\mathrm{sr}^{-1}$ for various experiments, assuming a source abundance of $0.02(\,h^{-1}\mathrm{Mpc})^{-3}$, self-similar nonlinear clustering and biased linear clustering:

DIRBE (200μm)	SIRTF (200μm)	JCMT (800μm)	JCMT (500μm)
0.5×10^{-13}	4×10^{-13}	6×10^{-13}	30×10^{-13}

These estimates are obtained by multiplying the du_P and du_c power spectra in Figure 2.1b by the appropriate filters in Fig.2.1a (SIRTF's is somewhat similar to OVRO's), and integrating over $d\ln\ell$. The predicted levels are achievable with all of these instruments. However, the foregrounds will make background anisotropy determinations at 200μm and 500μm decidedly difficult, even though these are the bands where the largest signals are expected. Although the non-Gaussian aspects of the patterns make the use of *rms* values suspect, the relatively large beams of sub-mm experiments smooth the primeval galaxy images sufficiently that the pattern approaches Gaussian and there are few blank patches of sky (see Figure 4.7). It is easy to invent models with smaller anisotropies. Indeed, for a given Ω_{RT}, the JCMT estimates in the Table scale as $n_{G*}^{-1/2}$, and thus increasing the number of sources produces smaller relative anisotropies. As well, over the relatively short intergalactic distances gas dynamical processes are likely to be effective, and these could equally well wash out as enhance anisotropies. Another unresolved issue is whether gravitational lensing will effectively smear the anisotropies over small angular scales.

If the number n_{G*} is small, then the galaxies may be optically thick and perturbation calculations will not be appropriate. To adequately deal with such sources, physical modelling of the emission from each source is required. Here we give an example using the hierarchical peaks method for identifying primeval galaxies in a $b_\rho = 1.4$ CDM model. For illustration, we used just two filter scales, to identify primeval dwarf galaxies and primeval 'bright' galaxies. The objects had masses $6 \times 10^{10}\Omega_B\,\mathrm{M}_\odot$ and $8 \times 10^{11}\Omega_B\,\mathrm{M}_\odot$ in baryons and star-burst level luminosities $\sim 10^{44}$ erg/s and $\sim 10^{45}$ erg/s, respectively. The dust temperature was taken to be a constant 30K. Normal dust with a dust-to-baryon fraction in the galaxies of 0.2% was assumed. The scaling of the luminosity and of our Figure 4.7 map contours goes as $(T_d/30\mathrm{K})^5$ and linearly with the dust-to-baryon ratio. We assume these primevals were bursting between $z = 6$, before which few exist anyway, and an arbitrary cutoff redshift we chose to be $z = 4$. With these parameters, the energy in the background from these objects would be $\Omega_{RT}h^2 = 2.5 \times 10^{-7}$, 0.1% of the total energy in the CMB.

The $4' \times 4'$ map in Figure 4.7 is for the flux integrated over the entire dust emission spectrum. In that case, the lowest contour is 10^{-13} $\mathrm{W\,cm}^{-2}\,\mathrm{sr}^{-1}$, with the scaling given above. The radiation pattern is smoothed with an $18''$ beam, used in the chop and wobble experiment of Church *et al.* (1990), which had a throw of $40''$. The *rms* relative intensity fluctuations in the map are 0.5 and the maximum value is 4.1 times the average intensity distortion. We look forward to JCMT's SCUBA, which should be able to probe intensity levels like these of busting primeval galaxies. If so, then the non-Gaussian aspects of the radiation pattern would provide a powerful probe of the high redshift protogalactic medium including how galaxies are clustered at that time.

286

Acknowledgements: No attempt at complete references was made in this sightseeing tour. Thanks to JRB's collaborators, Shaun Cole, Hugh Couchman, George Efstathiou, Bernard Carr, Craig Hogan and Nick Kaiser, and STM's collaborators, Tony Readhead and Charles Lawrence. We also thank Hugh and Jens Villumsen for providing us with N-body simulations to compare with. Support of a Canadian Institute for Advanced Research Fellowship, a Steacie Fellowship and the NSERC of Canada is gratefully acknowledged by JRB.

6. REFERENCES

Bardeen, J.M., Bond, J.R., Kaiser, N. and Szalay, A.S. 1986, *Ap. J.*, **304**, 15.
Bardeen, J.M., Bond, J.R. and Efstathiou, G. 1987, *Ap. J.*, **321**, 28.
Bond, J.R., Centrella, J., Szalay, A.S. and Wilson, J.R. 1984, *M.N.R.A.S.*, **210**, 515.
Bond, J.R., Carr, B.J. and Hogan, C.J. 1986, *Ap. J.*, **306**, 428 [BCH1]; 1991, *Ap. J.*, **367**, 420 [BCH2].
Bond, J.R. and Efstathiou, G. 1987, *M.N.R.A.S.*, **226**, 655.
Bond, J.R. 1988, in *The Early Universe*, Proc. NATO Summer School, Vancouver Is., Aug. 1986, ed. Unruh, W.G. (Dordrecht:Reidel); 1989, in *Frontiers of Physics— From Colliders to Cosmology*, ed. Campbell, B. and Khanna, F. (Singapore: World Scientific); 1990, in *The Cosmic Microwave Background: 25 Years Later*, p. 45-65. ed. Mandolesi, N. and Vittorio, N. (Dordrecht: Kluwer).
Bond, J.R. and Couchman, H.M.P. 1987, in *Proceedings of the Second Canadian Conference on General Relativity and Relativistic Astrophysics*, eds. C. Dyer, A. Coley, (Singapore: World Scientific).
Bond, J.R., Efstathiou, G., Lubin, P.M. and Meinhold, P.R. 1991, *Phys. Rev. Lett.*, **66**, 2179.
Bond, J.R., Cole, S., Efstathiou, G. and Kaiser, N. 1991, *Ap. J.*, in press [BCEK].
Bond, J.R. and Efstathiou, G. 1991, submitted to *Phys. Lett. B*.
Boughn, S.P., Cheng, E.S., Cottingham, D.A. and Fixsen, D.J. 1990, preprint.
Cen, R.Y., Jameson, A., Liu, F. and Ostriker, J.P. 1990, *Ap. J. Lett.*, **362**, L41.
Church, S.E., Lasenby, A.N. and Hills, R.E. 1990, preprint.
Cole, S. and Kaiser, N. 1989, *M.N.R.A.S.*, **233**, 637.
Cunningham, C.R. and Gear, W.K. 1990, ROE preprint.
Davies, R.D., Lasenby, A.N., Watson, R.A., Daintree, E.J., Hopkins, J., Beckman, J., Sanchez-Almeida, J. and Rebolo, R. 1987, *Nature*, **326**, 462.
Edge, A.C., Stewart, G.C., Fabian, A.C. and Arnaud, K.A. 1990, *M.N.R.A.S.*, **245**, 559.
Efstathiou, G. and Bond, J.R. 1986, *M.N.R.A.S.* **218**, 103; 1987, *M.N.R.A.S.*, **227**, 33P.
Evrard, G. 1990, *Ap. J.*, **363**, 349.
Evrard, G. and Henry, P. 1991, preprint.
Fomalont, E.B., Kellerman, K.I., Anderson, M.C., Weistrop, D., Wall, J.V., Windhorst, R.A. and Kristian, J.A. 1988, *Astron. J.*, **96**, 1187.
Gioia, I.M., Henry, J.P., Maccacaro, T., Morris, S.L., Stocke, J.T., and Wolter, A. 1990, *Ap. J. Lett.*, **356**, L35.
Hogan, C.J. and Partridge, R.B. 1989, *Ap. J. Lett.*, **341**, L29.
Kaiser, N. 1991, preprint.
Klypin, A.A., Sazhin, M.V., Strukov, I.A. and Skulachev, D.P. 1987, *Soviet. Astr. Lett.*, **13**, 104; Strukov, I.A., Skulachev, D.P. and Klypin, A.A. 1987, in Proceedings I.A.U. Symposium 130, ed. Audouze, J. and Szalay, A.S. (Dordrecht: Reidel).
Klypin, A.A. and Kates, R.E. 1991, preprint.
Kreysa, E. and Chini, A. 1988, *Proc. 3rd ESO-CERN Symp. on Astronomy, Cosmology*

and Fundamental Particles, Bologna; 1989, *Proc. Particle Astrophysics Workshop, Berkeley*, (Singapore: World Scientific).

Martin, H.M. and Partridge, R.B. 1988, *Ap. J.*, **324**, 794.

Meinhold, P. and Lubin, P. 1991, *Ap. J. Lett.*, **370**, L11.

Meyer, S.S., Cheng, E.S. and Page, L.A. 1991, *Ap. J. Lett.*, **371**, L7.

Mushotsky, R. 1988, in *Hot Thin Plasmas in Astrophysics*, ed. R. Pallavicini (Dordrecht: Kluwer).

Myers, S.T. 1990, thesis, Caltech; Myers, S.T., Readhead, A.C.S. and Lawrence, C.R. 1991, in *After The First Three Minutes*, ed. C. Bennett, S. Holt and V. Trimble.

K.A. Olive, D.N. Schramm, G. Steigman and T.P. Walker 1990, *Phys. Lett.*, **B 236**, 454.

Partridge, R.B. 1991, , in *After The First Three Minutes*, ed. C. Bennett, S. Holt and V. Trimble.

Peebles, P.J.E. 1987, *Ap. J. Lett.*, **277**, L1.

Readhead, A.C.S., Lawrence, C.R., Myers, S.T., Sargent, W.L.W., Hardebeck, H.E. and Moffet, A.T. 1989, *Ap. J.*, **346**, 566.

Ryu, D., Vishniac, E.T. and Chiang, W.H. 1990, *Ap. J.*, **354**, 389.

Schaeffer, R. and Silk, J. 1988, *Ap. J.*, **333**, 509.

Smoot, G. *et al.* 1991, *Ap. J. Lett.*, **371**, L1.

Thomas, P. and Carlberg, R.G. 1990, *M.N.R.A.S.*, **240**, 1009.

Timbie, P.T. and Wilkinson, D.T. 1988, *Rev. Sci. Instrum.*, **59**, 914; 1990, *Ap. J.*, **353**, 140.

Vishniac, E.T. 1987, *Ap. J.*, **322**, 597.

Vittorio, N., Meinhold, P., Muciaccia, P.F., Lubin, P. and Silk, J. 1991, *Ap. J. Lett.*, **372**, L1.

MICROWAVE ANISOTROPIES FROM TEXTURE SEEDED STRUCTURE FORMATION

Ruth Durrer

Princeton University, Physics Department, Jadwin Hall, Princeton, NJ 08544

and

David N. Spergel

Princeton University, Observatory, Peyton Hall, Princeton, NJ 08544

Abstract

The microwave anisotropy in a scenario of large scale structure formation with cold dark matter and texture is discussed. We find that in this scenario small scale fluctuations (on angular scales of a few degrees and less) are very effectively damped by photon diffusion. On a scale of $10 - 20$ degrees we predict peaks with amplitudes of $\Delta T/T \approx (1 - 5) \cdot 10^{-5}$. About 5 to 10 such peaks are expected all over the sky. This result tells us that texture seeded structure formation is likely to be ruled out (or confirmed!) within the near future.

1 Introduction

The formation of large scale structure in the Universe is still one of the biggest unsolved problems in standard cosmology.

Recently the idea that inhomogeneities in the energy density of the Universe might be induced by seeds has attracted considerable interest. Especially since such seeds can be provided naturally by topological defects which form during phase transitions in the early Universe [1]. Possible candidates are cosmic strings (π_1–defects), global monopoles (π_2–defects) and textures (π_3–defects). An attracting feature of these defects is, that the only free parameter is the symmetry breaking scale η which we cast in the dimensionless quantity $\epsilon = 16\pi G\eta^2$.

While cosmic string induced structure formation is a difficult subject of much controversy [2, 3, 4, 5], texture and probably also global monopoles seem to lead to quite promising scenarios of structure formation [6, 7, 8, 9, 10, 11].

Here we consider texture. In contrary to global monopoles and cosmic strings, these π_3–defects are nonsingular, unstable field configurations. As they enter the horizon, they contract with the speed of light, eventually unwind building up a singularity for one instant of time, and then disappear [6]. As for each scenario of structure formation, also here the unexpected isotropy and homogeneity of the microwave background radiation (MBR) represents a bottle neck which has to be passed. A rough estimate of the MBR anisotropy, using an exact solution for the texture field in flat spacetime, has been calculated earlier [7, 8] and the anisotropy has been found to be on the order of ϵ. New calculations of texture seeded structure formation lead to a value of $\epsilon \approx 5 \cdot 10^{-4}$ to reconcile with observations [10]. With this value the above rough estimate already contradicts observations [12]. It is therefore important to calculate the MBR anisotropy in more detail and this attempt I shall describe in this talk.

We consider a flat universe with cold dark matter (CDM), baryons, photons and texture seeds. Numerical simulations and analytical estimates have already shown that such a scenario leads to early formation of small objects which are likely to reionize the universe as early as $z \approx 200$ [9, 10, 8]. For simplicity, we may therefore assume that recombination never takes place and photons and baryons remain coupled until they are too diluted to scatter effectively. This decoupling time is determined by the distance into the past, at which the optical depth becomes unity:

$$\tau(t_{dec}) = \int_{t_0}^{t_{dec}} \frac{dt}{\sigma_T n_e a} = 1$$

$$1 + z_{dec} \approx 37(10\Omega_b \cdot h_{50})^{-2/3} \, ,$$

where Ω_b denotes the density parameter of the baryons and h_{50} is Hubble's constant in units of $50 km/s/Mpc$. This last scattering surface is of course not as instantaneous as the recombination shell. The decoupling due to dilution is a rather gradual process.

In Section 2 we introduce our notation and present the linear perturbation equations which describe our system. In Section 3 we give a short account of the numerical techniques and in Section 4 we present our preliminary results. A more detailed description of the numerical techniques and especially a thorough interpretation of our results will be presented in a forthcoming paper [13].

2   Equations

Throughout this work we assumed a flat Friedman universe, $\Omega_{tot} = 1$ which is dominated (after the radiation epoch) by CDM. The baryonic contribution is parametrized by the density parameter of the baryons. In our numerical calculations we set $\Omega_b = 0.05$. We work with conformal time so that the background metric is given by

$$ds^2 = a^2(-dt^2 + dx^2) \, .$$

First we want to calculate the anisotropy induced in the radiation density by one spherically symmetric texture. From numerical simulations of texture collapse [9] we know that the assumption of spherical symmetry is quite justified. On the other hand, the concept of spherical symmetry is an acausal one and leads to accelerations outside the horizon. This acausality must be compensated, e.g. by subtracting off the mean of the perturbation [14, 15]. There are several different ways of doing this and we could not yet find convincing arguments to definitely prefer one compensation scheme. The difference of the results for different compensation schemes finally turns out to be the biggest uncertainty in our results.

Cosmological perturbations are usually split into scalar, vector and tensor contributions which do not couple in first order (Here the terms scalar vector and tensor refer to the transformation properties on the hypersurfaces of constant time.). The split into scalar, vector and tensor perturbations is nonlocal and thus acausal. But spherically symmetric perturbations are of course always of scalar type (Due to the adoption of spherical symmetry, we loose, e.g., all information about gravity waves produced during the collapse.). In this case it is thus very convenient to apply gauge–invariant cosmological perturbation theory which is adapted to a split into scalar, vector and tensor modes [16, 17, 18, 8].

The spherically symmetric ansatz for a texture knot which unwinds at a given time $t = t_c$ is [7]

$$\phi = \eta^2(\sin\chi\sin\theta\cos\varphi, \sin\chi\sin\theta\sin\varphi, \sin\chi\cos\theta, \cos\chi) , \tag{1}$$

where θ , φ are the usual polar angles and $\chi(r, t)$ has the following properties

$$\begin{aligned}
\chi(r = 0, t < t_c) &= 0 \\
\chi(r = 0, t > t_c) &= \pi \\
\chi(r = \infty, t) &= \pi .
\end{aligned}$$

Since the texture seed is already a perturbation, it can be evolved according to the background equation of motion which yields [6]

$$\partial_t^2\chi + 2(\dot{a}/a)\partial_t\chi - \partial_r^2\chi - \frac{2}{r}\partial_r\chi = -\frac{\sin 2\chi}{r^2} . \tag{2}$$

The energy momentum tensor of the texture field, $T_{\mu\nu} = \partial_\mu\phi \cdot \partial_\nu\phi - \frac{1}{2}g_{\mu\nu}\partial_\lambda\phi \cdot \partial_\lambda\phi$, can be written in the form

$$T_{00} = (M/l)^2 f_\rho , \quad T_{0i} = -(M^2/l)f_{v,i} ,$$

$$T_{ij} = M^2[(f_p/l^2 - \frac{1}{3}\triangle f_\pi)\delta_{ij} + f_{\pi,ij}] .$$

Setting $M^2 = 4\eta^2$ we find

$$f_\rho/l^2 = 1/8[(\partial_t\chi)^2 + (\partial_r\chi)^2 + \frac{2\sin^2\chi}{r^2}] , \tag{3}$$

$$f_p/l^2 = 1/8[(\partial_t\chi)^2 - \frac{1}{3}(\partial_r\chi)^2 - \frac{2\sin^2\chi}{3r^2}] , \tag{4}$$

$$f_v/l = -\frac{1}{4}\int_0^r (\partial_t\chi)(\partial_r\chi)dr , \tag{5}$$

$$\triangle f_\pi = \frac{1}{4}[(\partial_r\chi)^2 - \frac{\sin^2\chi}{r^2}] + \frac{3}{4}\int_\infty^r [(\partial_r\chi)^2 - \frac{\sin^2\chi}{r^2}]\frac{dr}{r} . \tag{6}$$

The length parameter l is introduced to keep the functions f_ρ to f_π dimensionless. A convenient choice is to set l equal the typical size of the perturbations, e.g., the horizonsize at texture collapse. Clearly $(M/l)^2 f_\rho$ and $(M/l)^2 f_p$ denote energy density and isotropic pressure of the texture field. $(M^2/l)f_v$ is the potential for the velocity field and f_π parametrizes anisotropic stresses.

The gravitational field produced by the texture can be given in terms of the gauge–invariant Bardeen potentials Φ_0, Ψ_0. For an explicit definition of the Bardeen potentials in terms of the metric perturbation see [16, 17, 18]. For perturbations which are much smaller than the size of the horizon, $l \ll l_H$, Ψ corresponds to the Newtonian potential and Φ is related to the perturbation of the 3–curvature of the hypersurfaces of constant time:

$$-(4/a^2)\triangle\Phi = \delta R^{(3)} + \mathcal{O}(l/l_H) .$$

Hence for subhorizon perturbations in the Newtonian limit for the matter ($|T^{00}| \gg |T^{ij}|$), we have

$$\triangle\Phi = -\triangle\Psi \ .$$

Einstein's equation for the Bardeen potentials induced by a texture yield [8]

$$\triangle\Phi_0 = -\epsilon(f_\rho/l^2 + 3(\dot{a}/a)f_v/l) \ , \tag{7}$$

$$\triangle(\Psi_0 + \Phi_0) = -2\epsilon\triangle f_\pi \ . \tag{8}$$

The gauge–invariant equation for the density perturbation induced in the CDM component by the texture field is [8]

$$\ddot{D} + (\dot{a}/a)\dot{D} - 4\pi G\rho_d a^2 D = \epsilon(f_\rho + 3f_p)/l^2 \ , \tag{9}$$

as one probably would have guessed naively. On large scales (i.e., when the size of the perturbation is comparable to the horizon size at the given time) the variable D denotes a combination of the density perturbation, ($\delta = (\rho - \rho_0)/\rho$,) the potential for the peculiar velocity, (lv ; $v_i = -lv_{,i}$) and geometric terms. For perturbations which are much smaller than the horizon, D reduces to the usual density contrast δ [1]. The Bardeen potentials induced by the dark matter are given by

$$\triangle\Psi_d = 4\pi Ga^2\rho_d D \ , \tag{10}$$

$$\triangle\Phi_d = -\triangle\Psi_d \ . \tag{11}$$

We shall neglect the perturbations in the geometry produced by baryons and photons. The total gravitational fields,

$$\Psi = \Psi_0 + \Psi_d \ , \quad \Phi = \Phi_0 + \Phi_m \ , \tag{12}$$

thus enter as source terms only into the equations for baryons and photons. To write down the gauge–invariant equations for the coupled baryon–photon system, we need two additional variables: A potential for the peculiar velocity of the baryons, lV ($lV_{,i} = -v_i + \mathcal{O}(l/l_H)$) and $\mathcal{M}(r,t,\gamma)$, which is a gauge–invariant version of the fractional perturbation of the energy integrated photon distribution function. The unit vector γ denotes the direction cosines of the photon momentum: If we define the distribution function of the photons by

$$f = f_0 + \delta f \quad \text{and} \quad \iota = \frac{4\pi}{\rho_r} \int dp \cdot p^3 \delta f \ ,$$

we have

$$\mathcal{M}(r,t,\gamma) = \iota(r,t,\gamma) + 4\mathcal{R}(r,t) + 4l\gamma^i\partial_i\sigma(r,t) \ ,$$

where $\mathcal{R}$ and $l\sigma$ are perturbations in the geometry induced by texture and dark matter. At this point we cannot say that on scales smaller than the horizon ι is the dominant contribution to $\mathcal{M}$, since the geometric terms $\mathcal{R}$ and σ are not induced by the perturbations in the radiation and thus not necessarily smaller than it. But soon after texture collapse, D grows like a dark matter perturbation

---

[1]For explicit definitions of all the gauge–invariant quantities used in this paper see, e.g. [8]. It is once more remarkable that the gauge–invariant result is much simpler and more intuitive than any specific choice of gauge! In synchronous gauge, e.g., we find $D = \delta + 3(\dot{a}/a)lv$, i.e., D also includes a kinetic energy of the perturbation. Equation (9) in terms of δ and lv is quite lengthy and not very illuminating.

without source term, $D \propto a$ leading to constant perturbations in the geometry. Therefore, $\mathcal{R}$ and σ stay more or less constant after texture collapse and never expand into regions much larger than t_c. On scales much larger than the horizon size at texture collapse, $r \gg t_c$ we thus expect $\mathcal{R}$ and σ to be negligible. This was also confirmed by numerical checks.

At times $t \leq t_{dec}$, the collisionless Boltzmann equation for the photons and the equation of motion for the baryons must be modified to take into account scattering. The dominant effect is nonrelativistic Thomson scattering by free electrons and in first order we may assume the differential cross section to be isotropic in the matter rest frame. The first order collision term can then be obtained in the following way: Denoting by $u = u^0(1, v)$ the four velocity of the baryon fluid, we find that the energy of a photon in the matter rest frame is given by

$$p' = u^0(p_0 + p_i \frac{dx^i}{dt}) = p(1 - \gamma_i(v^i - \beta^i)) ,$$

where β denotes the shift vector of the $\{t =$ constant$\}$ hypersurfaces. The brightness of the radiation scattered into the beam of photons moving in direction γ can now easily be calculated with the result

$$1 + \iota_+(\gamma) = 1 + \delta_r + 4\gamma_i(v^i - \beta^i) ,$$

where δ_r denotes the density contrast of the radiation field $\delta_r = \frac{1}{4\pi} \int \iota d\Omega$. This leads to a collision term

$$C(\gamma) = \delta_r + 4\gamma_i(v^i - \beta^i) - \iota .$$

(For a more detailed derivation of C in the the case $\beta = 0$ see [19] §92.)
Clearly this term must be gauge–invariant and inserting the definition of lV ($lV_{,i} = -(v_i - \beta_i + l\sigma_{,i})$), one finds

$$C = M - \mathcal{M} - 4l\gamma^i\partial_i V , \quad \text{with} \quad M = \frac{1}{4\pi} \int \mathcal{M} d\Omega . \tag{13}$$

This collision term has to be added to the right hand side of the collisionless Boltzmann equation (see [18, 20]) which then yields

$$\dot{\mathcal{M}} + \gamma^i\partial_i\mathcal{M} = 4\gamma^i\partial_i(\Phi - \Psi) + a\sigma_T n_e[M - \mathcal{M} - 4\gamma^i\partial_i V] , \tag{14}$$

where σ_T is the Thomson cross section, n_e the electron density and the factor a appears because we are working with conformal time.
The drag force due to Thomson drag of photons on the matter is given by

$$F_i = -\frac{a\sigma_T n_e \rho_r}{4\pi} \int C\gamma_i d\Omega = \frac{a\sigma_T n_e \rho_r}{3}(m_i + 4l\partial_i V) ,$$

with $m_i = (3/4\pi) \int \gamma_i \mathcal{M} d\Omega$. Including this drag force into the equation of motion for the matter one gets

$$l\partial_i\dot{V} + (\dot{a}/a)l\partial_i V = \partial_i\Psi - \frac{a\sigma_T n_e \rho_r}{3\rho_b}(4l\partial_i V + m_i) \quad \text{or}$$

$$l\Delta\dot{V} + (\dot{a}/a)l\Delta V = \Delta\Psi + a\sigma_T n_e \frac{\rho_r}{\rho_b}(\dot{M} - \frac{4}{3}l\Delta V) . \tag{15}$$

For the last equation we have used the zeroth moment of Boltzmann's equation, the continuity equation,

$$\dot{M} + (1/3)\partial_i m^i = 0 .$$

Using our assumption of spherical symmetry, (13) and (14) can be written in the form

$$M(r,t) = (1/2) \int_{-1}^{1} \mathcal{M}(r,t,\mu) d\mu \qquad \text{and} \tag{16}$$

$$\dot{\mathcal{M}} + \mu \partial_r \mathcal{M} = 4\mu \partial_r (\Phi - \Psi) + a\sigma_T n_e [M - \mathcal{M} - 4\mu l \partial_r V] , \tag{17}$$

where μ is the direction cosine of the momentum in the direction of $\mathbf{r}$.

The coupled system (2) to (12) and (15) to (17) is closed and can in principle be integrated numerically for a given initial configuration. But at very early times, when collisions are still extremely effective,

$$\frac{1}{a\sigma_T n_e} \equiv t_T \ll t ,$$

the system (15) to (17) is stiff and would require ridiculously small time steps. At very early times it is thus better to treat the baryons and photons as a single fluid. Then, (15) to (17) are replaced by

$$l\Delta \dot{V} + (\dot{a}/a) l\Delta V = \Delta \Psi \quad \text{and} \quad \dot{M} = \frac{4}{3} l\Delta V . \tag{18}$$

At later times, when collisions are still effective but equations (15) to (17) are not too stiff, we can truncate the distribution function at the second moments:

$$\mathcal{M} = M + \gamma_i m^i + \gamma_i \gamma_j \alpha^{ij} \qquad \text{with} \qquad \alpha_i^i = 0 . \tag{19}$$

The first and second moments of (14) then yield

$$\ddot{M} + a\sigma_T n_e \dot{M} = \frac{\Delta M}{3} + \frac{2}{15}\alpha - \frac{4}{3}\Delta(\Phi - \Psi) + \frac{4}{3}a\sigma_T n_e l\Delta V \qquad \text{and} \tag{20}$$

$$\dot{\alpha} + a\sigma_T n_e \alpha = 2\Delta \dot{M} \quad , \text{with} \quad \alpha = \partial_i \partial_j \alpha^{ij} . \tag{21}$$

Finally, at late times, $t \geq t_{dec}$ collisions are too sparse to prevent higher moments in the distribution function from building up and we have to integrate the full equations (15) to (17) which are then very loosely coupled.

3 Numerical Techniques

We integrate our system by using a modified staggered leapfrog scheme [21, 9]. The unstable damping terms, e.g. in (20) and (21) are absorbed by writing the equations in terms of the variables

$$\dot{M} \exp(\int a\sigma_T n_e dt) \quad \text{and} \quad \alpha \exp(\int a\sigma_T n_e dt)$$

respectively. For the late time evolution we evolve $\mathcal{M}$ at 10 different points $-1 \leq \mu_i \leq 1$ and then obtain M by a ten point Gauss Legendre integration. The comoving size of the texture, $r_c \approx 2t_c$ is initially about 90 grid points. The size of the total grid is always on the order of $r_{max} \approx \max\{t_{dec}, 3t_c\}$. The total number of grid points thus depends on the texture collapse time. In our routine it can vary between 128 for the late textures and 1024 for textures which collapse as early as at the time of equal matter and radiation density. Initially, we set in a texture who's size we adjust to make it collapse at a given time t_c. All other perturbations are set to zero. At very early times, $t_T \geq dt$ we use the

fluid approximation (18) to evolve the photon/baryon system. For $t_{dec} > t$ but $t_T < dt$ we use the truncated Boltzmann equation. For late textures, $t_c \geq t_{dec}/3$, where we have to continue the evolution well beyond t_{dec} we have to integrate the full Boltzmann equation at times $t > t_{dec}$.
A more detailed discussion of the numerical aspects of this work will be presented in [13].

The biggest uncertainty in the whole calculation comes in through the choice of a compensation scheme. As already mentioned in Section 2, spherical symmetry introduces acausalities and can lead to accelerations on super horizon scales. This would generally result in an over estimate of the final radiation anisotropy, $M(r,t)$ [15]. To prevent this, we use the following type of compensation: At each time t we choose a compensation radius r_{cp} and set $f_\rho = f_p = f_v = f_\pi = 0$ for $r > r_{cp}$. On small scales, $r < r_{cp}$ we subtract the mean of f_ρ and f_p, so that $\int_0^{r_{cp}} f_\rho = \int_0^{r_{cp}} f_p = 0$. The texture compensated in this way does not produce gravitational fields on scales larger than r_{cp}. In Fig. 1 we compare $M(r_{out}, t)$ with $r_{out} \approx 2t_{dec}$ for two different compensation schemes: For the solid line we used the most intuitive compensation scheme, $r_{cp} = l_H = t$. For the dashed line we used the most simple compensation scheme, $r_{cp} = \text{const.} = r_c$. The truth may lay somewhere between these two extremal cases, but let use just use this result as an estimate of the uncertainty in our calculation. Our results may thus be taken seriously only up to a factor of about 3.

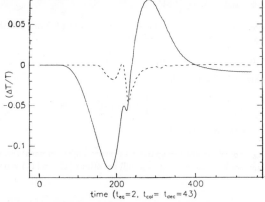

Figure 1: The photon density perturbation amplitude at $r_{out} \approx 2t_{dec}$ is compared for two different compensation schemes. Constant compensation (solid line) and horizon compensation (dashed line). $\Delta T/T$ is plotted in units of the parameter $\epsilon \approx 5 \cdot 10^{-4}$.

4 Results and Conclusions

We have numerically calculated the the radiation density variations, $M(r_{out}, t)$ with $r_{out} \approx t_{dec}$ and $0.01 \cdot t_c < t < \max\{5t_c, 2t_{dec}\} = t_{max}$ for a given texture collapsing at time t_c. Outside the simulation volume $r_{out} \times t_{max}$ the photons propagate freely, unaffected from the perturbations in the dark matter and without a significant amount of diffusion. To convert this result into a microwave anisotropy observed in the sky today, $t = t_0$, we must choose a distance r_0 between us and the texture. Clearly, if $r_0 > t_0$, the texture is outside the event horizon and we do not observe any signal from it. On the other hand, since the signal arrives at r_{out} roughly at time $t \approx t_{dec} \ll t_0$, if $r_0 < r_{out}$, the microwave signal arrives at r_0 long before t_0 and we do not expect to see anything of it today.

For the causal region outside the simulation volume, $r_{out} < r_0 < t_0$ the temporal result $M(t, r_{out}) \equiv M_{out}(t)$ can easily be converted into an angular anisotropy (see Fig. 2):

$$\frac{\Delta T}{T}(\theta) = M_{out}(t(\theta)) \quad \text{with} \quad \begin{aligned} t(\theta) &= t_0 - r(\theta) \quad \text{and} \\ r(\theta) &= r_0[\cos\theta - \sqrt{(r_{out}/r_0)^2 - \sin^2\theta}] \,, \end{aligned} \tag{22}$$

for $|\theta| \leq \theta_{max}$, where θ_{max} is determined by $\sin\theta_{max} = r_{out}/r_0$.

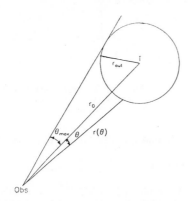

Figure 2: The conversion of a temporal signal at a given radius r_{out} into an angular signal: For $\theta \leq \theta_{max}$ a signal that appears to the observer at Obs. under an angle θ with respect to the center, was emmitted from the sphere with radius r_{out} at the time $t = t_0 - r(\theta)$.

Depending on the distance to the texture we may observe a cold spot, cold spot/hot spot or a hot spot. The distinct cold spot/hot spot feature obtained in the flat case [7, 8] has been somewhat distorted but it does not depend heavily on the compensation scheme (see Fig. 1). Two examples with collapse times $t_c \approx t_{dec}$ and $t_c \approx 0.2 \cdot t_{dec}$ are given in Figs. 3 and 4, where we have plotted the average between the two compensation schemes described above. A comparision of these two figures shows how effectively photon diffusion damps away the small scale perturbations in this scenario. For early textures $t_c \ll t_{dec}$, diffusion also causes a broadening of the signal to a width of the order of t_{dec} which corresponds to an angular scale of about $10°$ (for $h_{50} = 1$ and $\Omega_b = 0.05$), so that after the diffusion only the dominant cold spot survives (see Fig 4). A crude estimate of the damping factor for $t_c \ll t_{dec}$ is given below.

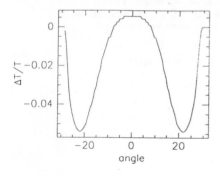 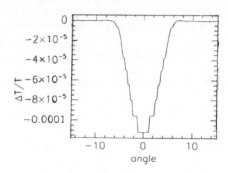

Figure 3: The angular anisotropy from a texture collapsing at $t_c = t_{dec}$ in units of the dimensionless parameter $\epsilon \approx 5 \cdot 10^{-4}$.

Figure 4: The angular anisotropy from a texture collapsing at $t_c = 0.2 t_{dec}$ in units of the dimensionless parameter $\epsilon \approx 5 \cdot 10^{-4}$.

The number of textures which collapse after a given time t_c and can be seen by us today is

$$N_>(t_c) = 4\pi t_0^2 \int_{t_c}^{t_0} \frac{dN}{dt} t \, dt \approx 4\nu (t_0/t_c)^2 = 4\nu(1 + z_c) \ .$$

$N(t)$ is the number density of textures forming at time t, $N(t) = \nu/(\frac{4\pi}{3} t^3)$, where ν is the probability for a texture to form at the horizon and is found analytically and numerically $\nu = 0.04$ [6, 9]. Since early textures are so effectively damped and late textures are so rare, the most prominent signals come from textures which collapse at the decoupling era, $t_c \approx t_{dec}$. For $z_{dec} \approx 60$, e.g., we expect to see about 10 such textures. Their angular size is

$$\theta_{dec} \approx 2 t_{dec}/t_0 = 2/\sqrt{1 + z_{dec}} \approx 14° \ .$$

The amplitude of the signal we expect from them is according to Fig. 3

$$\frac{\Delta T}{T}(\theta_{dec}) = 0.06\epsilon \approx 3 \cdot 10^{-5} \ ,$$

where we have set $\epsilon = 5 \cdot 10^{-4}$, which is required for successful large scale structure formation [10].

Our numerical calculations damp away the small textures completely. To get an idea of the order of magnitude of the damping, let us give a crude analytical estimate which is valid for $\theta \ll \theta_{dec}$

$$\frac{\Delta T}{T}(t_c) \approx \frac{\Delta T}{T}(\theta_{dec}) \exp(-\frac{\pi^2}{5}(t_c/t_{dec})^{3/4}) \ .$$

This exponential damping factor is obtained from the first order approximation for photon diffusion (given in [19] §92) for a perturbation with wavelength equal to twice the horizon size at texture collapse until $t_{end}(t_c)$ where the mean free path encompasses t_c, i.e., $t_T(t_{end}) = t_c$, and no damping after t_{end}.

This estimate has to be taken with a grain of salt, but it is save to conclude from our numerical results that a texture seeded CDM scenario of structure formation leads to a maximum amplitude of $\frac{\Delta T}{T} = (1-5) \cdot 10^{-5}$ on scales of about $14°$. Roughly 5–10 such patches are expected to be seen all over the sky. Since photon diffusion also moves all early perturbations up to about θ_{dec}, we expect much smaller anisotropies on small scales $\theta \leq 1° - 2°$.

This prediction is within the realm of present observational techniques [22]. It is thus possible to either rule out or confirm (!) this scenario.

A more thorough interpretation of these calculations, where we hopefully reduce our uncertainty to some extent, where we produce a whole sky map of the MBR, explicitly predict the magnitude of the quadrupole and of several interesting angular scales and where we also calculate the skewness of the anisotropies (i.e., their deviation from a Gaussian distribution) will be presented in a forthcoming paper [13].

References

[1] T.W.B. Kibble, *Phys. Rep.* **67**, 183 (1980).

[2] N. Turok and R. Brandenberger, *Phys. Rev.* **D33**, 2175 (1986).

[3] E. Bertschunger in: *Formation and Evolution of Cosmic Strings*, eds G. Gibbons, S. Hawking and T. Vachaspati (1990)

[4] A. Stebbins in: *Formation and Evolution of Cosmic Strings*, eds G. Gibbons, S. Hawking and T. Vachaspati (1990)

[5] R. Durrer in: *Formation and Evolution of Cosmic Strings*, eds G. Gibbons, S. Hawking and T. Vachaspati (1990)

[6] N. Turok, *Phys. Rev. Lett.* **63**, 2625 (1989).

[7] N. Turok and D.N. Spergel, *Phys. Rev. Lett.* **64**, 2736 (1990).

[8] R. Durrer, *Phys. Rev.* **D42**, 2533 (1990).

[9] D.N. Spergel, N. Turok, W.H. Press and B.S. Ryden, *Princeton University Preprint* PUPT–90–1182 (1990).

[10] A. Gooding, D.N. Spergel and N. Turok *Princeton University Preprint* PUPT–90–1207 (1990).

[11] D.P. Bennett and S.H. Rhie, Livermore National Laboratory Preprint (1990).

[12] E. Wright, private communication.

[13] R. Durrer and D.N. Spergel, in preparation (1991).

[14] J. Traschen, N. Turok and R. Brandenberger, *Phys. Rev.* **D34**, 919 (1986).

[15] S. Veeraraghavan and A. Stebbins, *Ap. J.* **365**, 37 (1990).

[16] J. Bardeen, *Phys. Rev.* **D22**, 1882 (1990).

298

[17] H. Kodama and M. Sasaki, *Prog. in Theor. Phys. Suppl.* **78** (1984).

[18] R. Durrer and N. Straumann, *H.P.A.* **61**, 1027 (1988).

[19] P.J.E. Peebles, *The Large Scale Structure of the Universe*, Princeton University Press (1980).

[20] R. Durrer, *Astron. and Astrophys.* *208*, 1 (1989).

[21] W.H. Press, B.P. Flannery, S.A. Teucholsky and W.T. Vetterling, *Numerical Recipies*, Cambridge University Press (1989).

[22] S.S. Meyer, E.S. Cheng and L.A.Page, *Ap. J. Lett.*, in press.

V. NEUTRINOS, COSMOLOGY AND NUCLEOSYNTHESIS

Tau Neutrinos and Cold Dark Matter*

G.F. Giudice

Theory Group
Department of Physics
University of Texas
Austin, Texas 78712

ABSTRACT: Neutrinos with standard weak interactions are possible *hot* dark matter candidates, but they cannot provide an explanation for a *cold* dark matter scenario, because of constraints from relic abundance. It is shown here, however, that a stable ν_τ with a mass in the range 1–35 MeV and a magnetic moment of 10^{-6} Bohr magnetons is a possible candidate for a *cold* dark matter particle. This hypothesis can be experimentally tested in the near future.

Of all the candidates proposed to solve the dark matter problem, neutrinos are the only particles known to exist. The relic abundance calculation shows however that a neutrino with mass in the range 100 eV $\lesssim m_\nu \lesssim$ 1 GeV would contribute to a present energy density larger than the critical density, thus leading to a too young Universe [1]. The heavy mass range, $m_\nu \gtrsim$ GeV, is experimentally excluded for the three known neutrinos and can be contemplated only for a heavy fourth generation neutrino. However, the possibility of a fourth generation neutrino (with mass up to 45 GeV) has been recently ruled out by the measurements of the Z^0 width. Therefore, we are led to conclude that, if the neutrino is responsible for the observed dark matter of the Universe, then its mass satisfies $m_\nu \lesssim$ 100 eV. In this mass range, neutrinos lead to the *hot* dark matter scenario, which is considered unsuccessful in predicting the large-scale structure of the Universe. Moreover, light neutrinos cannot account for the observed [2] halos in dwarf spheroidal galaxies, because of phase-space constraints [3].

*Research supported in part by the Robert A. Welch Foundation and NSF Grant PHY 9009850

A way to reconcile neutrinos with *cold* dark matter is to modify their predicted relic abundance or, ultimately, their annihilation rate in the early Universe. Assuming new interactions beyond the standard model, even neutrinos in the mass range 100 eV $\lesssim m_\nu \lesssim$ 1 GeV can have their present energy density sufficiently depleted to be consistent with constraints from the age of the Universe. Some authors [4] have suggested the existence of very light or massless scalar particles to open new channels for primordial neutrino annihilation. Here I will consider the possibility that the electromagnetic interactions due to a large neutrino magnetic moment enhance the annihilation rate of neutrinos in the early Universe.

As suggested in ref. [5], the τ neutrino is an acceptable cold dark matter candidate if it satisfies the following requirements: *(i)* it has a mass in the range 1 MeV $\lesssim m_{\nu_\tau} <$ 35 MeV; *(ii)* it is stable; *(iii)* it has a magnetic moment of about 10^{-6} Bohr magnetons. I will first briefly discuss these three conditions.

(i) The upper bound on m_{ν_τ} of 35 MeV is the direct experimental limit. The lower bound comes from the requirement that the $\nu_\tau - \bar{\nu}_\tau$ annihilation into an electron-positron pair is kinematically allowed. This annihilation channel will in fact turn out to be responsible for depleting the neutrino relic abundance. In the mass range 1-35 MeV, τ neutrinos are *cold* relics, since they are non-relativistic when their annihilations freeze-out.

(ii) Tau neutrinos must of course be stable (or quasi-stable in cosmological times) if assumed to form the galactic halos today. On the other hand, one would naturally expect a heavy neutrino to decay $\nu_\tau \to \nu_e e^+ e^-$ (via flavor mixing) or $\nu_\tau \to \nu\nu\nu$ (via flavor changing neutral currents) or $\nu_\tau \to \nu\gamma$ (via non-diagonal neutrino magnetic moments). All these decay modes must therefore be suppressed. This is realized, for instance, by simply assuming that the new interactions that produce the neutrino mass and magnetic moment have an approximate τ-lepton number conservation.

(iii) The presence of the neutrino magnetic moment provides new contributions to the neutrino annihilation rate, besides ordinary weak interactions. The dominant electromagnetic annihilation is $\bar{\nu}_\tau \nu_\tau \to e^+ e^-$ via a s-channel photon exchange, with a rate:

$$\sigma(\bar{\nu}_\tau \nu_\tau \to e^+ e^-) = \frac{\alpha\mu^2}{6} \sqrt{\frac{1 - 4m_e^2/s}{1 - 4m_\nu^2/s}} \left(1 + 8\frac{m_\nu^2}{s}\right) \left(1 + 2\frac{m_e^2}{s}\right), \quad (1)$$

where m_e, m_ν are the electron and neutrino mass, μ is the ν_τ magnetic moment and $\sqrt{s}$ is the total energy in the center of mass frame. The electromagnetic contributions to $\bar{\nu}_\tau\nu_\tau \rightarrow \gamma\gamma$ and $\bar{\nu}_\tau\nu_\tau \rightarrow \nu\nu$ can be neglected since they are suppressed with respect to eq.(1) by two extra powers of the magnetic moment. The standard relic abundance calculation yields a present neutrino energy density in units of the critical density (Ω_ν) [5]:

$$\Omega_\nu h^2 = 1.2 - 1.8 \left(\frac{\mu_0}{10^{-6}}\right)^{-2}, \qquad (2)$$

where μ_0 is the ν_τ magnetic moment in units of Bohr magnetons and h is the Hubble constant in units of 100 km sec^{-1} Mpc^{-1}. Therefore, a τ neutrino with $\mu_0 \simeq 10^{-6}$ can account for the missing mass of the Universe. This value of the magnetic moment is much larger than what can be predicted for a massive neutrino by ordinary electroweak interactions:

$$\mu_0 = \frac{3G_F m_e m_\nu}{4\sqrt{2}\pi^2} = 3 \cdot 10^{-12} \left(\frac{m_\nu}{10 \text{ MeV}}\right). \qquad (3)$$

New physics is required to explain the anomalously large value $\mu_0 \simeq 10^{-6}$. It is interesting to note a possible connection with the solution of the solar neutrino problem proposed in ref.[6], which requires an electron neutrino magnetic moment $\mu_0 \simeq 10^{-10} - 10^{-12}$, also much larger than predicted by weak interactions.

At present the best experimental limit on the τ neutrino magnetic moment comes from the study of $e^+e^- \rightarrow \bar{\nu}\nu\gamma$# [7]:

$$\mu_0 < 4 \cdot 10^{-6}, \qquad (4)$$

which is not too far from the value required to explain the dark matter as tau neutrinos.

The experimental bound in eq.(4) could however be improved in the near future. Searches at the Z^0 resonance are sensitive to a neutrino magnetic moment via the rare decay $Z^0 \rightarrow \bar{\nu}\nu\gamma$, which has a rate [5]:

$$\frac{d\Gamma}{dx}(Z^0 \rightarrow \bar{\nu}_\tau\nu_\tau\gamma) = \frac{\alpha\mu^2 M_Z^3}{96\pi^2 \sin^2\theta_W \cos^2\theta_W} x(1 - x + \frac{x^2}{12}), \qquad (5)$$

#The same analysis also yields a limit on the τ neutrino charge radius of $5 \cdot 10^{-16}$ cm [7]. Because of this limit, the contribution to τ neutrino annihilations coming from charge radius terms turns out to be always smaller than the weak contribution [8]. Therefore, a neutrino charge radius cannot sizably affect the calculation of Ω_ν.

where $x \equiv 2E_\gamma/M_Z$. Note that the distribution (5) is peaked at a relatively high photon energy, of about 25 GeV, allowing a clean signature at Z^0 factories. Integrating eq.(5), one obtains a Z^0 branching ratio of:

$$BR(Z^0 \to \bar{\nu}_\tau \nu_\tau \gamma) = 2 \cdot 10^{-7} \left(\frac{\mu_0}{10^{-6}}\right)^2. \tag{6}$$

Thus, $\mu_0 = 10^{-6}$ is at the border of observability for LEP.

A large neutrino magnetic moment can also enhance the decay of vector mesons (V) into a pair of neutrinos [8]. The contribution to this process from the neutrino magnetic moment grows with the vector meson mass M slower than the standard model weak contribution to the process, $\Gamma(V \to \bar{\nu}\nu)_\mu/\Gamma(V \to \bar{\nu}\nu)_{SM} \sim \mu^2\alpha(G_F M)^{-2}$. Therefore heavy meson systems are not suitable for detecting small neutrino magnetic moments. In fact, the decay rate for $\Upsilon \to \bar{\nu}\nu$ due to $\mu_0 = 10^{-6}$ is one order of magnitude smaller than the standard model rate. On the other hand, the decay $J/\Psi \to \bar{\nu}\nu$ due to $\mu_0 = 10^{-6}$ is ten times faster than in the standard model. In order to tag this invisible decay mode, one needs to consider cascade processes. In particular, the decay $\Psi(2S) \to J/\Psi + \pi\pi \to \bar{\nu}\nu\pi\pi$ has a substantial branching ratio in the presence of a large ν_τ magnetic moment [8]:

$$BR(\Psi(2S) \to J/\Psi + \pi\pi \to \bar{\nu}\nu\pi\pi) = 1.6 \cdot 10^{-7} \left(\frac{\mu_0}{10^{-6}}\right)^2. \tag{7}$$

Therefore, values of μ down to 10^{-7} Bohr magnetons could be probed at proposed tau-charm factories.

Future experiments can thus reach the sensitivity required to test the hypothesis of a ν_τ magnetic moment of 10^{-6} Bohr magnetons. However a different strategy to test the proposal of tau neutrino dark matter can be pursued by looking directly for signals from the halo particles. Unfortunately the tau neutrinos here considered are too light to be observed in presently operating detectors using nuclear recoil. Nevertheless, the occasional ν_τ annihilations in the halo can provide a characteristic monoenergetic positron flux. This flux have been estimated to be [8]:

$$j = 10^3 \left(\frac{\mu_0}{10^{-6}}\right)^2 \text{m}^{-2}\text{sec}^{-1}\text{sr}^{-1} \tag{8}$$

at an energy $E = m_{\nu_\tau}$. This is much larger than the experimental upper limit on the positron flux reported for energies $E > 20$ MeV [9]. It seems premature however to rule out the region $m_{\nu_\tau} > 20$ MeV, since solar wind modulation effects can deeply modify the flux prediction, eq.(8) [8].

In conclusion, a stable τ neutrino with mass in the range 1–35 MeV and magnetic moment 10^{-6} Bohr magnetons is a viable cold dark matter candidate and could constitute the observed galactic halos. Future experiments have the capability of testing this hypothesis.

References

[1] See, *e.g.*, E.W. Kolb and M.S. Turner, The Early Universe (Addison-Wesley, Redwood City, CA, 1989).

[2] S.M. Faber and D.N.C. Lin, Ap. J. **266** (1983) L17.

[3] S.D. Tremain and J.E. Gunn, Phys. Rev. Lett. **42** (1979) 407.

[4] P. Pal, Phys. Rev. **D30** (1984) 2100; E.D. Carlson and L.J. Hall, Phys. Rev. **D40** (1989) 3187.

[5] G.F. Giudice, Phys. Lett. **B251** (1990) 460.

[6] M.B. Voloshin, M.I. Vysotskii, and L.B. Okun, Zh. Eksp. Teor. Fiz. **91** (1986) 754 [Sov. Phys. JETP **64** (1986) 446].

[7] H. Grotch and R.W. Robinett, Z. Phys. **C39** (1988) 553.

[8] L. Bergstrom and H.R. Rubinstein, University of Stockholm preprint USITP-90-09 (1990).

[9] R.C. Hartman and C.J. Pellerin, Ap. J. **204** (1976) 927.

Testing Big Bang Nucleosynthesis

Robert A. Malaney

Institute of Geophysics and Planetary Physics, University of California, Lawrence Livermore National Laboratory, CA 94550, U.S.A.

1. Introduction

A great deal of effort has gone into the determination of the primordial abundances of the light isotopes as inferred from observation. The most recent of these efforts are reviewed in [Pag90]. Here we will discuss just two aspects of these observational studies. The first of these, the determination of the primordial ^{4}He abundance, tentatively suggests that the storm clouds may be gathering for the SBB (standard big bang) nucleosynthesis model.

2. Helium

The primordial ^{4}He abundance is often the most sensitive constraint on non-standard big bang models since its yield depends mostly on the universal expansion rate which in turn influences n/p ratio at the time of weak-reaction freeze-out.

In order to determine the primordial ^{4}He abundance one must correct for stellar processing. Such a correction is usually attempted by correlating the ^{4}He abundance with metallicity and extrapolating to zero metallicity. Assuming a linear correlation with metallicity, Z, we have

$$Y = Y_p + \frac{\Delta Y}{\Delta Z} Z \quad , \tag{1}$$

where Y is the observed ^{4}He mass fraction and Y_p is the actual primordial ^{4}He mass fraction.

The earlier observations were taken of highly-ionized H II regions in low-metallicity galaxies with oxygen as the metallicity tracer [Pei74]. However, oxygen may actually be a poor diagnostic of ^{4}He production (see [Ste89] and refs. therein) since ^{4}He is produced in stars of mass $M \gtrsim 2M_\odot$ whereas oxygen is produced only in stars of mass $M \gtrsim 12M_\odot$. Since it is expected that nitrogen and carbon are also produced in $M \gtrsim 2M_\odot$ stars, it is believed that these latter elements will represent an improved diagnostic of ^{4}He production.

The current maximum likelyhood regressions from the observed [Pag89] correlation of metallicity with helium abundance are,

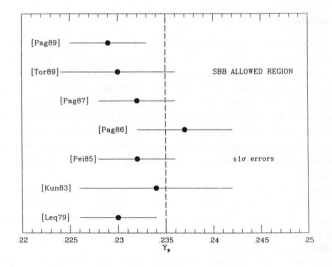

Fig. 1. Inferred values of Y_p and region allowed by SBB theory.

$$Y = 0.224(\pm 0.005) + 178(\pm 45)O/H \tag{2}$$

or

$$Y = 0.229(\pm 0.004) + 3120(\pm 780)N/H \ . \tag{3}$$

Chemical evolution corrections are important and deserve careful attention, especially since the observed data does not go below 0.01 of solar metallicity. It has been noted [Ful90] that if only low-metallicity objects ($Z < 1/4$ solar) are used in the linear regression to zero metallicity, a lower Y_p is obtained. The lack of very-low metallicity observations also leaves open the possibility that some process which produces ^{4}He, but no significant metals, early in the galactic history (perhaps massive stars) could lead to an overestimate of Y_p.

If it would be possible to truly determine that the upper limit to the primordial helium abundance were less than 23.5% then the SBB model would be inconsistent with the abundance limits of the other primordial isotopes. We note that the above primordial values already begin to hint at such a problem. This point is shown more graphically in figure 1 where it can be seen that the upper limit to Y_p has tended to decrease over the past decade or so. However, it is also clear that systematic and other uncertainties in the determination of Y_p do not yet allow for any robust invalidation of the standard model. Within the allowed uncertainties the SBB theory remains valid. Hopefully, continued effort in the observational determination of Y_p will resolve this very important issue.

3. Interstellar Lithium

Of the primordial isotopes, determination of the primordial abundance of ^{7}Li remains the most controversial. The problem has been that observations of the lithium abundance in the atmospheres of Pop I stars [Cay84,Hob88] result in a number fraction Li/H$\sim 10^{-9}$, whereas observations of Pop II stars [Spi82] result in Li/H$\sim 10^{-10}$ (all of the lithium observed is normally assumed to be in the form of ^{7}Li). The reason for the order of magnitude difference between the different sets of observations, and which measurement more closely represents the actual primordial abundance of lithium, has been the subject of much debate. The situation is further confused by galactic chemical and stellar evolution effects.

Due to the confusion involved in interpreting stellar lithium data it would be more advantageous to observe the lithium abundance in the halo or intergalactic medium. In this regard, observation of interstellar lithium toward SN 1987A seemed, in principle, a promising new approach to the determination of the primordial lithium abundance.

Four searches for interstellar lithium toward SN 1987A were undertaken, the results of which are given in table 1. The upper limits to equivalent widths and the corresponding upper limits to the primordial Li/H ratios are given. The early detection of interstellar lithium found in ref. [Vid87] was not verified by the later more detailed investigations. The much larger upper limit to the primordial Li/H ratio given by [Mal90] arises from a more conservative estimate of the present uncertainties involved in determining a lithium abundance from an interstellar Li I line toward SN1987A. These uncertainties primarily concern ionization corrections and depletion processes onto stellar grains.

Table 1.   Upper Li/H limits toward SN1987A.

Reference	Eq.Width(mÅ)	Interpretation
[Vid87]	0.2; 0.3; 0.5	detections?
[Baa88]	<0.15	Li/H$< 4 \times 10^{-10}$
[Sah88]	<0.13	Li/H$< 1.6 \times 10^{-10}$
[Mal90]	<0.25	Li/H$< 4.4 \times 10^{-9}$

Since the more conservative estimate of Li/H$\lesssim 4 \times 10^{-9}$ encompasses the Li/H ratio observed in both Pop I and Pop II stars, this is somewhat disappointing. The observational data of interstellar lithium cannot presently resolve the issue of the true primordial lithium abundance unless one is willing to believe that the ionization and grain depletion processes are currently much better understood than as discussed in [Mal90].

Recently [Baa91] have attempted to better constrain the allowed depletion of interstellar lithium. They show how much tighter constraints result *if* several assumptions are made regarding such depletion processes. They have also combined all the data of table 1 into to one spectra in an attempt to detect the interstellar lithium line. Although a detection remained elusive, a reduction in the [Mal90] upper limit by about a factor of two could be achieved.

The importance in determining the Li/H ratio toward SN1987A can be seen in the chemical evolution models of [Mat90]. Inhomogeneous non-standard nucleosynthesis models predict high primordial lithium. The chemical evolution models of [Mat90] can readily account for both a high and a low primordial lithium abundance. The key factor in discriminating between these models is the Li/H abundance in the Large Magellanic Cloud. Unfortunately, an upper limit of Li/H$\lesssim 2 \times 10^{-9}$ does not allow for any discrimination. If the Li/H ratio could be proven to be significantly below this value, however, the high-primordial lithium chemical-evolution models would be ruled out.

4. Inhomogeneous Nucleosynthesis

For a low value of Ω_b the inhomogeneous models arising from considerations of the QCD phase transition in the early universe (see [Mal91] and refs. therein) may produce substantially less ^{4}He than the SBB theory. If Y_p is found to be less than 23.5% then these inhomogeneous models could adequately account for the discrepancy. It is also interesting to note that an original failing of the inhomogeneous models was the large quantities of ^{7}Li they predicted [App87,Full88,Mal88]. However, for low values of Ω_b the lithium yield from these models is also reduced.

This work was performed under the auspices of the U.S. Department of Energy by the Lawrence Livermore National Laboratory under contract number W-7405-ENG-48.

References

[App87]   J. H. Applegate, C. J., Hogan, R. J., and Scherrer, R. J., Phys. Rev., D35 (1987) 1151.
          J. H. Applegate, C. J., Hogan, R. J., and Scherrer, R. J., Ap. J., 329 (1988) 572.
[Baa88]   D. Baade and P. Magain, Astron. Astrophys., 194, (1988) 237.
[Baa91]   D. Baade, S. Cristiani, T. Lanz, R. A. Malaney, K. C. Sahu and G. Vladilo, Astron. Astrophys., submitted (1991).
[Cay84]   Cayrel, R., Cayrel de Strobel, G., Campbell, B. and Dappen, W., Ap. J., 283 (1984) 205.
[Ful88]   G. M. Fuller, C. R., Alcock, and G. J. Mathews, Phys. Rev. D37 (1988) 1380.
[Ful90]   G. M. Fuller and D. Boyd, preprint (1990).
[Hob88]   L. M. Hobbs and C. Pilachowski, Ap. J., 334 (1988) 734.
[Kun83]   D. Kunth and W. Sargent, Ap. J. 273 (1983) 81.
[Leq79]   J. Lequeux,, M. Peimbert, J. F. Rayo, A. Serrano and S. Torres-Peimbert, Astron. Astrophys., 80 (1979) 155.
[Mal88]   R. A. Malaney and W. A. Fowler, Ap. J. 333 (1988) 14.
[Mal90]   R. A. Malaney and C. R. Alcock, Ap. J., 351 (1990) 31.
[Mal91]   R. A. Malaney and G. J. Mathews, Physics Reports, submitted 1991.
[Mat90]   G. J. Mathews, C. R. Alcock and G. M. Fuller, Ap. J., 349 (1990) 449.

[Pag86]   B. E. J. Pagel, R. J. Terlivich and J. Melnick, Pub. Astr. Soc. Pac., 98 (1986) 1005.

[Pag87]   B. E. J. Pagel in: "A Unified View of the Macro-and Micro-Cosmos" eds. A. DeRujula, D. V. Nanopoulos and P. A. Shaver (World Scientific, Singapore, 1988) p. 399.

[Pag89]   B. E. J. Pagel and E. A. Simonson, Rev. Mex. Astr. Astrofis. 18 (1989) 153.

[Pag90]   B. E. J. Pagel in Nobel Symposium 79- The Birth and Early Evolution of our Universe, ed. B. S. Skagerstam (Graftavallen; Sweden 1990).

[Pei74]   M. Peimbert and S. Torres-Peimbert, Ap. J. 193 (1974) 327.

[Pei85]   M. Peimbert, in Star Forming Dwarf Galaxies, ed. D. Kunth, T. X. Thuan and J. T. T. Van, p403 (Paris; Edition Frontiers 1985).

[Sah88]   K. C. Sahu, M. Sahu and S. R. Pottasch, Astron. Astrophys. Lett. 207 (1988) L1.

[Smi90]   V. V. Smith and D. L. Lambert, Ap. J. Lett., 345 (1989) L75.

[Spi82]   F. Spite and M. Spite, Astron. Astrophys., 115 (1982) 357.

[Ste89]   G. Steigman, J. Gallagher and D. N. Schramm, Comm. Astrophys. 14 (1989) 97.

[Tor89]   S. Torres-Peimbert, M. Peimbert and J. Fierro, Ap. J. (1989) submitted.

[Vid87]   A. Vidal-Madjar, P. Andreani, S. Cristiani, R. Ferlet, T. Lanz and G. Vladilo, Astron. Astrophys., 177 (1987) L17.

Prospects for relic neutrino detection

P F Smith

(Rutherford Appleton Laboratory, Chilton, Oxon., England)

1 Introduction

The standard big bang model predicts a universal background of relic neutrinos, comparable in number density to the background microwave photons [1]. Since the latter are observed, one can be confident that the neutrino background will also be present, and observations on this background could in principle provide information on neutrino masses and mixings (since all neutrino types would have been produced). In addition, if the heaviest neutrino has a mass ~ 25–50 eV it could both explain the Galactic dark matter and produce a closed universe.

This neutrino background is undetectable at the present time firstly because the neutrino energy is very low (10^{-4}–10^{-5} eV) resulting in a very low energy transfer to any conceivable detector, and secondly the low energy gives a low interaction cross section and hence a very low event rate per unit mass. These obstacles have so far precluded any realistic proposal for relic neutrino detection [1].

However, it is also difficult to accept that these neutrinos will never be detected, and it is therefore of interest to ask what technical developments would be needed to make this feasible, and whether any realistic experimental possibilities can be foreseen for the future. The aim of this paper is to illustrate the difficulties by summarizing six detection ideas which have been previously considered, indicating in each case the problems which have prevented the idea being developed into an experimental proposal. The most promising direction for further study would appear to be that of coherent interactions, from which a considerably increased cross section results for scattering from bulk matter, producing small macroscopic forces. So far, no investigations of this idea have resulted in a practical detection scheme, but in this paper one new variation is suggested which could in principle give an observable effect, if the necessary stringent experimental conditions could be created. It is suggested that this may become possible with the aid of foreseeable 21st century developments in nanotechnology.

2.. Basic neutrino numbers and energies

2.1 For the case of zero neutrino mass, the number density ρ_ν of relic neutrinos (which decouple from matter at $T \sim 1$ MeV, ~ 1s after the big bang) can be related [2] to the number density of relic photons (~ 400 cm^{-3}) giving $\rho_\nu \sim 100$ cm^{-3} ($\nu + \bar{\nu}$) for each of the three generations. The energy distribution would be spatially uniform and would have a red-shifted Fermi distribution with a present-day mean momentum $\sim 5.10^{-4}$ eV, and a flux $\sim 10^{12}$ cm^{-2}s^{-1}.

2.2 If one neutrino (for example the tau neutrino) has a mass $m_\nu \sim 20$–50 eV then it could be clustered in galaxies with a density similar to that estimated for the dark matter in our own Galaxy [3]. From considerations of phase space constraints, bound velocity and dark matter density ~ 0.4 GeV cm^{-3}, one obtains a "best fit" with $m_\nu \sim 25$ eV, characteristic momentum $p_\nu \sim 0.02$ eV, number density $\rho_\nu \sim 2.10^7$ cm^{-3}, and flux $\sim 5.10^{14}$ cm^{-2}s^{-1} [3,4,5].

2.3 For completeness a third possibility should be noted, in which the relic neutrinos have a non-zero mass but have not clustered in galaxies. In this case they would be accelerated into and through our Galaxy by its gravitational potential giving, in the region of the sun, a reduced number density $\rho_\nu \sim 1$ cm^{-3}, increased momentum $p_\nu \sim 0.1$ eV, and a relatively low momentum spread $\Delta p_\nu / p_\nu \sim 10^{-2}$ [6].

For studies of relic neutrino detection ideas, it is usual to assume the case of Galactic clustering, giving the set of numbers in paragraph 2.2 above.

3. Discussion of detection ideas

3.1 Single particle cross section. Electron scattering.

The lower curves in **Fig 1** show the momentum dependence of the neutrino cross section for elastic neutral current scattering on neutrons. They become constant at low momentum if the neutrino has a non-zero mass. Charged current interactions are a factor 4 higher (including the elastic scattering of an electron neutrino from an electron, which can also proceed via the W). The elastic scattering amplitude for a ν_μ or ν_τ on an electron, or any ν from a proton, is lower by the standard factor $(1 - 4\sin^2\theta_w)/2 \sim 0.04$.

Suppose we were to consider experiments based on a target of quasi-free electrons – for example in a metallic foil. The typical recoil energy from collision with a Galactic neutrino of momentum 2.10^{-2} eV would be $\sim 10^{-9}$ eV. At first sight this appears somewhat encouraging,

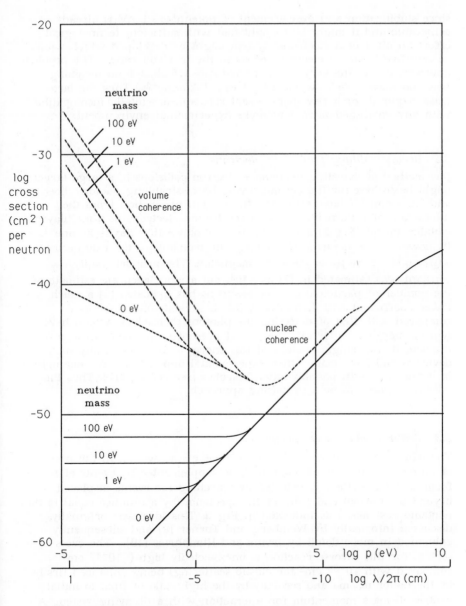

FIG 1 Neutrino elastic scattering cross section versus
momentum p and corresponding wavelength λ .
Full lines: single particle scattering.
Dashed lines: coherent scattering.

since stabilization and measurement of potentials < 1 nV is already achievable and it might be thought that with sufficient technological effort an ultra–cold electronic system might be developed which would be sensitive to single electron recoils in the nanovolt range. The problem, however, lies in the very low expected rate which, for an incoming neutrino mass ~ 30eV, would be ~ 1 day^{-1} ktonne^{-1}. Thus the target mass required for a reasonable event rate appears totally incompatible with any envisaged ultra–low noise experimental environment.

3.2 Bremsstrahlung from free electrons

One method of detecting neutrino + electron collisions in a large target might be to look for the accompanying bremsstrahlung photons. Loeb and Starkman [7] have estimated the photon production from the interaction of relic neutrinos with conduction electrons in a multilayer metallic target (**Fig 2**) and find a rate of only $\sim 10^{-3}$ day^{-1} ktonne^{-1}. If, however, the neutrino has a magnetic moment μ_ν the rate becomes $\propto (\mu_\nu / \mu_B)^2$ where μ_B is the Bohr magneton. The proportionality constant was estimated in [7] as $\sim 10^8$, but a more accurate estimate of the number of participating free electrons has indicated [8] a much lower coefficient $\sim 10^{-2}$. However, this is a relatively minor issue compared with that of detecting the photons themselves, which have typical energies $< 10^{-5}$ eV. There is at present no method foreseen for the detection of single photons of this energy – and indeed any such device would have many other experimental applications, for example in connection with possible axion searches [see ref 1 p217]. Thus this does not appear to be a promising approach.

3.3 Tritium beta decay spectrum

The tritium beta decay process $^3\text{H} \rightarrow {}^3\text{He} + e + \nu_e$ would, in the presence of an electron neutrino background, be supplemented by events of the form $\nu_e + {}^3\text{H} \rightarrow {}^3\text{He} + e$, detected by searching for events displaced beyond the normal end point of the spectrum (by a distance equal to the neutrino rest mass) as indicated in **Fig 3**. This idea was originally discussed informally by Weinberg and Turner [9] and subsequently examined in more detail by Irvine and Humphries [10]. The cross section for this inverse reaction is unexpectedly high ($\sim 10^{-43}$ cm^2 compared with 10^{-53} cm^2 for elastic scattering) being enhanced firstly by the factor m_e/m_ν and secondly by the high ratio of final to initial centre–of–mass momentum for interaction with a decaying system. A further enhancement ~ 100 was suggested in [10] due to Fermi momentum, but this is now believed to be incorrect. The outcome is a

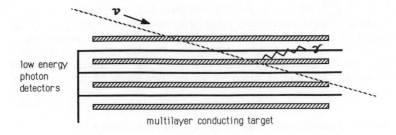

Fig 2   Suggested detection of bremsstrahlung photons
from scattering of relic neutrinos by free electrons.

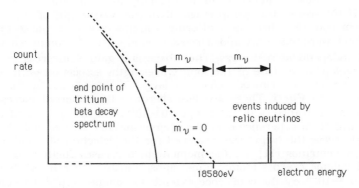

Fig 3   Suggested detection of Galactic electron neutrinos
by induced beta decay in tritium.

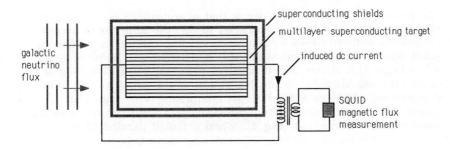

Fig 4   Hypothetical Galactic electron neutrino detector based on
coherent momentum transfer to superconducting electrons.

rate ~1 day^{-1} g^{-1}, at first sight encouraging but in practice not so, since existing tritium beta decay experiments operate with only ~1 µg of tritium, and have considerable difficulty in obtaining a resolution ~10 eV near the end point. In addition, the fact that this experiment would rely on the electron neutrino providing the dominant component of the Galactic dark matter discourages further study of its feasibility.

3.4 Coherent interactions: electrons in superconductors

The major hope for relic neutrino detection lies in the use of coherent scattering to increase the low energy cross section. The principle of coherent scattering is that if a number N of scattering centres are rigidly connected, ie the energy transfer is insufficient to excite the structure so that the group recoils as a whole, then the scattering amplitudes A must be added prior to squaring to obtain the cross section. If the size of the group is less than the wavelength of the momentum transfer, then the scattering amplitudes add in phase to give a total amplitude ~NA and a cross section $\propto$ (NA)^2, compared with NA^2 for independent scatterers. Thus there is a gain N in cross section, which in the case of the long wavelengths associated with relic neutrino interactions, can be extremely large. This is shown by the upper curves in **Fig 1**. There are two stages to the coherent increase. For momentum transfers in the range 10^4–10^6 eV there is a significant gain arising from coherence over the atomic nucleus, which could be used to decrease the target size required for the detection of solar or supernova neutrinos [11]. As the momentum decreases further the coherent volume reaches atomic dimensions, and for the relic neutrino region 10^{-2}–10^{-4} eV the coherence extends to volumes ~10^{-8}–10^{-2} cm^3, containing 10^{15}–10^{21} atoms and giving the very large cross section gains shown. The problem of applying this experimentally lies in the fact that the cross section now refers the recoil of a large assembly of particles, so that for a given momentum transfer the energy transfer (p^2/2m) becomes extremely small and difficult to observe.

The first application we consider is that of improving on the single electron scattering process considered in 3.1 by using the coherent interaction with electrons in a superconducting circuit, to produce a small but measurable current [12,13]. A detailed numerical analysis of this scheme [13] shows that it involves practical constraints which are at present unattainable. As an illustration of the problem, a typical requirement would be a 1 m^3 target comprising ~10^6 layers of superconductor each 10m x 1m, connected to a SQUID to measure the small changes in current (**Fig 4**). When oriented parallel to the

Galactic motion (resulting in a "neutrino wind") the coherent neutral current interaction would be equivalent to an electric field $\sim 10^{-36}$ V cm^{-1} and a total circuit voltage 10^{-27} V, giving a linear rise in current amounting to 10^{-18} A after 10 days. Thus unprecedented levels of noise and mechanical stability would be required, and to prevent spurious induced voltages from external sources it would be necessary to shield the system magnetically to changes $< 10^{-8}$ flux quanta $(\pm 10^{-19}$ gauss over $1 m^2)$ in 10 days.

Even if it is believed that these stringent requirements might be met with sufficient future technological effort, there remains a further problem of principle which has not been resolved – that a superconducting circuit is quantized, with a level spacing many orders of magnitude greater than the energy transfer from each neutrino scattering, so that it is not clear that the electrons will in fact recoil independently of the lattice material [12]. In addition, the idea suffers from the same disadvantage as that of *§3.3* – that it is applicable only to electron neutrinos, the cross section for muon or tau neutrinos being lower (see *§3.1* above) by a factor $\sim 10^{-3}$.

3.5 Coherent interactions: macroscopic first-order forces

The momentum transfer from a flux of low energy neutrinos by coherent scattering from a material target is equivalent to a small force on that target and we now consider the feasibility of measuring such forces. The easiest route to studying this is via neutrino optics, and a significant amount of work on this has been carried out in the past ten years. There is a close analogy with the more familiar case of neutron optics: for a low energy neutral particle passing through a region of matter the coherent interaction is equivalent to an average potential U (positive or negative) which changes the momentum p of the particle while in the material. This defines an effective refractive index $n = p_2/p_1$ which can then be used to calculate reflection and refraction effects in the medium, and any consequent forces on the material. The value of U is proportional to the weak coupling constant G_F, the number of atoms/cm^3, and a factor which depends on the neutrino type and the values of A and Z for the target material [4,15]. A typical value for U in a high density material is 2.10^{-13} eV (compared with 2.10^{-7} eV for neutrons). The refractive index n is related to U (and hence to G_F) by $n^2 = 1 - 2Um_\nu/p^2$, giving $(n-1)$ typically $\sim 10^{-8}$ for hypothetical Galactic dark matter neutrinos. This increases as p^{-2} until for that portion of the neutrino spectrum with $p < 10^{-6}$ eV total reflection would occur from a surface, as in the case of photons and sufficiently low energy neutrons.

An immediate point of interest is that one could in principle construct optical elements (from high density materials), for example Fresnel lenses and Bragg mirrors, which could provide a means of modifying or modulating the neutrino flux. One cannot, however, obtain a significant overall focusing or enhancement of the flux, in view of its largely isotropic nature (but with a superimposed directional component added by the Galactic motion of the solar system). The use of Bragg mirrors could provide selective reflection of a portion of the neutrino spectrum, but the total reflectivity cannot be increased by this means [6]. It is of interest that the size of neutrino optical systems would not be as great as the small value of $(n-1)$ would suggest. This is because the linear scale of an optical system increases only as $(n-1)^{-0.5}$ [6], indicating a typical size (eg focal length) $\sim 100-1000$m for neutrino optical elements. Thus the latter would be very substantial structures but not unreasonable in comparison with the scale of other modern particle physics or astrophysics projects.

Turning now to detection ideas, it was originally suggested that, since a prism deflects an incident beam through an angle $\propto (n-1)$, corresponding mechanical forces $\propto (n-1)$ could be produced in appropriately shaped targets (**Fig 5a**). Subsequent studies showed that for targets immersed in a uniform neutrino flux (**Fig 5b**) these first-order forces must always cancel, leaving only effects $\propto (n-1)^2$ [16]. However, it would be incorrect to conclude that no ideas can be created based on first order forces. What these proofs show is that the first order force is $\propto (n-1)\mathrm{grad}\rho_\nu$ where $\mathrm{grad}\rho_\nu$ is the number density gradient. However, although relic neutrinos would have negligible natural density gradients (eg arising from the gravitational potential gradient, or from random spatial fluctuations) it is nevertheless possible to envisage that significant distortions in ρ_ν could be produced by the above-mentioned neutrino lenses or mirrors, so that over a small spatial region (eg $10-100$cm) $\mathrm{grad}\rho_\nu$ would be sufficient to produce a force $\propto (n-1)$ which could be detected by a sensitive torsion balance (**Fig 5c**). Moreover, since alternating forces would be generally easier to detect than constant forces, one could envisage oscillating or rotating structures (perhaps in orbit) to produce the time varying gradients in ρ_ν. Although this general point was made many years ago [4] no attempt has been made to estimate the maximum value of $\mathrm{grad}\,\rho_\nu$ which might be achievable with neutrino optics. The magnitude of a first order force which might be produced artificially is thus still unknown, so although approximate (unpublished) calculations are somewhat discouraging, this idea cannot yet be excluded as a practical possibility.

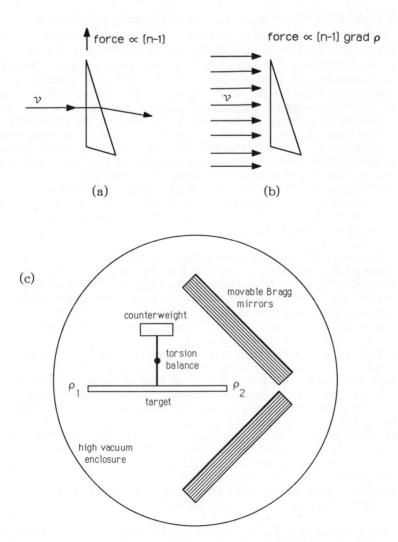

Fig 5   (a) Force proportional to n−1 resulting from deflection
             of neutrino by material prism.
         (b) Dependence of first order force on neutrino density
             gradient for prism immersed in neutrino flux.
         (c) Conceptual production of neutrino density gradient by
             Bragg mirror system, and detection with torsion balance.

3.6 Coherent interactions: macroscopic second-order forces

Coherent scattering of a uniform flux of neutrinos produces a force on the target $\propto (n-1)^2$ which is simply due to the optical reflection of the incident neutrinos. For example, at normal incidence the reflection coefficient from a single surface is $C = (n-1)^2/(n+1)^2$, with more complicated formulae for other angles and for a finite-thickness slab [4, 17]. By periodically varying the material density one obtains the equivalent of a Bragg mirror, which will increase the reflectivity for a restricted wavelength range, while the momentum transfer remains approximately constant (**Fig 6a**). The outcome of our detailed analysis of these effects [4-6] was that the magnitude of the force on a target due to reflection of a Galactic neutrino flux would not exceed $\sim 10^{-23}$ dyne g^{-1} (for a material target density $\sim 10^{-20}$ g cm^{-3}). Thus this case is more straightforward than that of $\S3.5$ in that an upper limit for the force per unit target mass due to second order forces can be rigorously established.

It is of some interest that measurement of a force of this order, although not technically feasible at the present time, appears possible in principle. For a freely suspended target (eg using a "drag-free" satellite enclosure to achieve the equivalent of zero-gravity [18]) the displacement would be $\sim 10^{-15}$ cm in a few hours, and $\sim 10^{-13}$ cm in a day, easily exceeding the displacements nowadays measurable in gravitational radiation detectors [18,19]. To reduce noise from gas collisions below this level requires a pressure $\sim 10^{-16}$ torr around the target – achievable by cooling to 50mK (at which temperature all vapour pressures fall below this level). Photon noise would be smaller than the gas collision noise, and noise from cosmic ray collisions could be reduced sufficiently by measuring the differential movement between the target and a surrounding (also freely floating) reference enclosure of lower density and hence reduced neutrino sensitivity (**Fig 6b**). There remains the problem of the constancy of the neutrino flux, making signal identification possible only through the correlation between the force direction and the Galactic orbit, but as indicated above it is also possible in principle to add a modulating component by an array of movable Bragg mirrors around the satellite.

Having established that a "physics design" is possible, it is nevertheless equally clear that such a system could present insurmountable difficulties in construction and operation, and this scheme is manifestly unrealistic both technologically and economically in the foreseeable future.

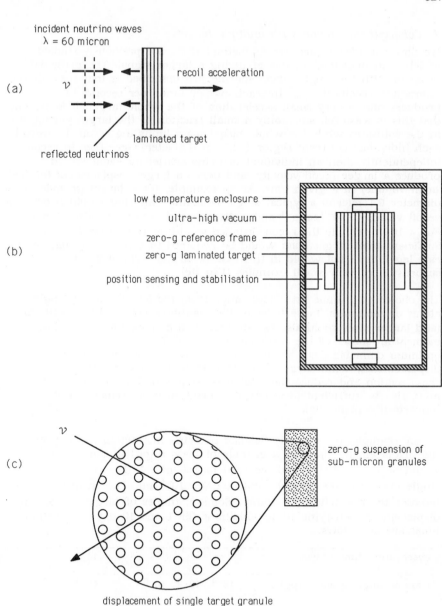

Fig 6 (a) Force proportional to $(n-1)^2$ from neutrino reflection.
(b) Conceptual zero-g experiment to detect second order force.
(c) Conceptual zero-g experiment with multiple granule target.

4 Coherent interaction with multiple targets

We now consider a variation of method *§3.6* , not previously studied, which is in principle capable of giving a larger signal. The principal problem with the single target scheme just described is that the coherent interaction results in $\sim 10^7$ interactions per tonne per second, but produces only a very small acceleration of the whole target. It is evident that this is wasteful, since only a small fraction of the target participates in the collisions while the whole target mass limits the recoil. If instead each individual coherent region (< 10 μm diameter) were free to recoil independently then an individual neutrino scattering event would produce a larger recoil velocity, and hence a larger displacement (of that small volume) in a given time. As an example, for a target granule of diameter 0.1 micron a typical single neutrino scattering would produce a recoil velocity $\sim 10^{-9}$ cm s^{-1} and a displacement $\sim 10^{-4}$ cm after 1 day. For a 100Å granule the recoil distance would be 1mm after 1 day. Sufficient total target mass would be required to give an adequate event rate (eg 1/day), but would in this case be in the form of a zero-g suspension of very small granules (**Fig 6c**).

The choice of granule size could range from the lower limit of single atoms or molecules, to particles of the maximum coherent size ~ 10 μm (and larger sizes could also be considered in conjunction with a lower momentum subset of the neutrino spectrum). There is no immediate optimum diameter since smaller granules would produce larger displacements, while larger granule sizes would improve the coherent cross section and reduce the total mass required. The displacement must also be sufficient to satisfy the quantum measurement limit (uncertainty principle).

As an example, for a single Galactic neutrino interaction in tungsten the displacement $d(\mu m)$ of a granule of mass m(pg) in time t(days) would be $d \approx 10^{-3}$ t/m, and the total target mass M(kg) required for a single event in time t(days) would be $M \approx 10^{-3}/mt$. The quantum limit imposes the restriction d > 10μm. Some examples of target masses and displacements satisfying these conditions for 1 and 10 day observation times are as follows:

observation time (days)	1			10		
mass of single granule (pg)	10^{-4}	10^{-5}	10^{-6}	10^{-4}	10^{-5}	10^{-6}
total target mass (kg for 1 event)	10	10^2	10^3	1	10	10^2
recoil distance (μm)	10	10^2	10^3	10^2	10^3	10^4

It can be seen that this scheme is of interest in allowing a reasonably small total target mass, and giving much larger (and in principle more easily observable) mechanical displacements than is possible in the case of a single target. It is equally clear that this type of scheme would be technically impossible at the present time since, in addition to the zero-g requirement, the system would require methods firstly for dissipating residual motion of the granules to provide an ultra-cold state with negligible initial velocity, and secondly for identifying individual displacements in the array (without using photons, which would cause larger recoils than the neutrinos). Other refinements could include the suspension of each granule within an outer shell as a means of discriminating against cosmic ray collisions.

It can, however, be envisaged that schemes based on this principle could become practicable with foreseeable 21st century technology. Todays technology, in which micro-engineering has reached the sub-micron level, could place substantial instrumentation and computing power within devices < 1cm in diameter. If in the future engineering becomes possible on the molecular scale it will then be similarly possible to place instrumentation within the sub-micron granules, allowing velocity changes relative to neighbouring members of the array to be recognised and recorded.

It might even be argued that this could already be the case, using selected molecules (or even biological cells) which would behave in a characteristic way on displacement and collision with neighbours. However, highly controlled engineering on the molecular scale will certainly begin to be possible within the next 20 years. The ability to lift and reposition single atoms and portions of molecules using the scanning tunneling microscope has already demonstrated[1] the beginnings of nanotechnology, which will lead to the design and construction molecular machines in turn capable of assembling other molecular machines (including self-replication) from atomic and molecular sub-constituents [20]. It is envisaged that some forms of these machines will act as programmed assemblers for the construction of micro-robots and macroscopic objects from a

---

[1] For example, nanometer-size words and patterns have been formed with single-atom line width by positioning Xe atoms, S atoms and CO molecules on a substrate (see New Scientist no.1712 (1990) 22, no.1753 (1991) 31, no.1757 (1991) 20).

reservoir of chemical constituents[2], with revolutionary consequences for medicine, engineering and artificial intelligence [21]. Of course, "conventional" microlithography and microelectronics have even today reached the 10–100 nm level, but it is likely that this "top–down" micro–machining approach will eventually be superseded by the "bottom–up" approach of atomic assembly.

These developments are still at an uncertain distance into the future, but (remembering the development speed of microelectronics and computer technology) are likely to happen faster than is generally appreciated, with major advances in the next 30 years. Thus, although not immediately applicable to neutrino physics, these developments are certainly relevant to questions of the form "will it ever be possible". The conclusions from the above discussion are thus (1) that coherent interaction with sub–micron targets could in principle give observable signals in the form of mechanical motion and (2) that complex structures and instrumentation within and around such sub–micron targets will eventually become possible, making it at least conceivable that suitable experiments could be proposed. It is also hoped that the above summary will stimulate further thought, and perhaps new ideas for relic neutrino detection.

Acknowledgements

The earlier studies of coherent detection were carried out in collaboration with J D Lewin, to whom I am grateful for continuing helpful discussions. I would also like to thank R W P Drever for discussions on the detectability of small forces.

---

[2] Just as plant seeds contain all of the information and molecular machinery necessary to extract and assemble C,H,O and N atoms into macroscopic edible objects or vertical cylinders of wood. This would be regarded as impressive technology by an alien robot community in which everything is manufactured (including themselves) and who would report back on these "programmed capsules" which automatically assemble food and structural materials. An alien transparency showing vegetables from planet earth was used to enliven the oral presentation, but permission was not received for its inclusion here !

References

1. For an earlier summary and additional references see §3 and §4 of P F Smith, J D Lewin, Phys. Rep. 187 (1990) 203.

2. See, for example, P D P Collins, A D Martin, E J Squires, "Particle Physics and Cosmology" (Wiley, 1989) 389; J Barrow, Fund. Cosmic Physics 8 (1983) 83; W J Marciano, Comments Nucl.Part.Phys. 9 (1981) 169.

3. S Tremaine, J E Gunn, Phys. Rev. Letters., 42 (1979) 407

4. P F Smith, J D Lewin, Phys. Lett. 127B (1983) 185

5 P F Smith, J D Lewin, Acta Phys.Polon., B16 (1985) 837.

6. P F Smith, J D Lewin, Acta Phys.Polon., B15 (1984) 1201.

7. A Loeb, G D Starkman, Princeton preprint IASSNS-AST 90/10 (1990).

8. W Hong, UCLA (personal communication)

9. See oral discussion in Proc AIP 72 (1981) 353.

10. J M Irvine, R Humphries, J Phys G 9 (1983) 847.

11. See, for example, A Drukier, L Stodolsky, Phys Rev D30 (1984) 2295; P F Smith, Z Phys C 31 (1986) 265.

12. J D Lewin, P F Smith, Astrophys. Lett 24 (1984) 59.

13. R Opher, Astrophys.J. 282 (1984) 398.

14. P F Smith, J D Lewin, Astrophys.J. 318 (1987) 738.

15. P F Smith, Nuovo Cimento 83A (1984) 263 (note misprint in eq 4.3 for which the square brackets should be positioned as in the corresponding eq 3.2).

16. N Cabibbo, L Maiani, Phys.Lett. B114 (1982) 115; P Langacker, J P Leveille, J Sheiman, Phys. Rev. D27 (1983) 1228.

17. The neutrino reflection formulae given in [4] contain a misprint in eq 13b, where the plus sign should be a minus sign as in eq 13a.

18. See, for example, V B Braginsky, A B Manukin, "Measurements of weak forces in physics experiments" (University of Chicago Press, 1977).

19. See, for example, V Braginski, Sov. Phys. Usp. 31 (1988) 836; and reviews by E Amaldi, Proc. 2nd Marcel Grossman Meeting on General Relativity Vol.1 (ed. R Ruffini, N. Holland 1982) 21; Proc 4th Winter School Hadronic Physics, Folgaria, Italy (ed. R Cherubini, World Scientific 1989) 423.

20. See, for example, review by C Schneiker in "Artificial Life", Proc Interdisciplinary Workshop on Synthesis of Living Systems (C Langton, ed., Los Alamos, 1987) 443.

21. For an account of future possibilities see ref [20] and K E Drexler, "Engines of Creation, the coming era of nanotechnology" (Anchor NY 1986).

NEUTRINOS FROM HELL

F. W. Stecker and C. Done

Laboratory for High Energy Astrophysics

NASA/Goddard Space Flight Center

Greenbelt, Maryland 20771

M. H. Salamon and P. Sommers

Physics Department

University of Utah

Salt Lake City, Utah 84112

Lasciate ogni speranza, voi ch'entrate!

(Abandon all hope, you who enter!)

- Dante, Inferno

ABSTRACT

We calculate the spectrum and high energy ν background flux from photomeson production in Active Galactic Nuclei (AGN), using the recent UV and X–ray observations to define the photon fields and an accretion disk shock acceleration model for producing high energy particles. Collectively, AGN produce the dominant isotropic ν background between 10^4 and 10^{10} GeV, detectable with current instruments. AGN ν's should produce a sphere of stellar disruption which may explain the 'broad line region' seen in AGN.

I. INTRODUCTION

Active Galactic Nuclei (AGN) have long been considered as potential sites for high energy ν production[1] as they are the most powerful emitters of radiation in the known Universe. They are presumably fueled by the gravitational energy of matter infalling onto a supermassive black hole at the AGN center, though the mechanism responsible for the efficient conversion of gravitational to observed luminous energy is not yet known. One class of models assumes that the infalling matter forms an accretion shock at some distance from the black hole[2]. First–order Fermi acceleration of charged particles at this shock then converts a significant fraction of the total gravitational energy into highly relativistic particles which have a characteristic power law spectrum[3]. Another possible energy generation mechanism involves electromagnetic acceleration in a black hole/accretion disk dynamo[4]. We will show that in the hellish conditions within AGN cores, ultrahigh energy cosmic rays (even neutrons) are trapped by the intense radiation fields. Thus, a necessary consequence of *any* acceleration mechanism in which protons reach energies of $\sim 10^6$ GeV is the production of high energy ν's through the collisions of these particles with the intense photon fields in AGN.

There are two main classes of AGN, radio–loud and radio–quiet. The radio–quiet AGN, called Seyfert 1's or quasars depending their luminosity, are thought to contain a similar "central engine" differing only in mass and luminosity[5]. However, in radio–loud objects, there is strong evidence that in addition to the black hole/accretion disk system there is also a relativistic jet[6]. If the jet is viewed end–on the it distorts both luminosity and size measurements because of relativistic boosting and time dilation, whereas if the jet is close to the plane of the sky large radio–lobes are seen. These are termed "core dominated" and "lobe dominated" sources, respectively.

Here we concentrate mainly on ν production from the cores of radio–quiet AGN, both because they are more numerous by a factor 10 than the radio–loud objects[7], and because their properties are better understood. However, in Section IX we also discuss previous work on radio–loud objects.

II. THE SPECTRUM OF A RADIO–QUIET AGN

A characteristic AGN spectrum is shown in Figure 1. There are comparable amounts of power in each logarithmic energy interval, but the spectrum is not completely flat. There is a thermal feature in the IR region, thought to be due to dust intercepting and reradiating the UV emission. Dust can only exist outside of the nucleus, as it is destroyed by sputtering, so the IR spectrum must arise at a large distance from the "central engine"[8]. The lack of variability in the IR spectrum also implies that the emission is from a large region[9]. The "UV–bump" is another thermal feature, due either to the release of energy through viscous processes in the accretion disk or to the accretion disk thermalising and reradiating the intense X–ray emission[10]. In either case the accretion disk is expected to be linked to the energy generation region, and so needs to be considered in any model of the central source. Its small size is confirmed by observations of variability on timescales of days[11]. Between the dust and disk emission, there may also be a further continuum component as the transition between the two thermal components is less sharp than prediced by standard dust and disk spectra[12]. However, this component also shows little variability on short timescales, so is not associated with the very central regions that we are considering here[13].

The most rapidly variable part of the spectrum is the X–ray region. Its emission is similar in luminosity to that of UV bump but its rapid time variability indicates that the X–rays are produced in a smaller region than the UV flux[14]. Thus, in the very central regions, only the UV and X–ray photon fields are of importance.

III. ENERGY LOSS MECHANISMS

The lack of strong X-ray absorption features in AGN spectra[15] implies that the secondary X-rays are produced in regions of low column density. This puts strict limits on the amount of target gas for pp interactions, and one finds that the very large photon density in the AGN core makes $p\gamma$ the dominant energy loss mechanism. This is a threshold interaction, and the dominant ν production channel at energies just above the threshold is $\gamma p \to \Delta \to n\pi^+$.[16]. The cross-section for this reaction peaks at photon energies of $\epsilon' = 0.35 m_p c^2$ in the proton rest frame. In the observers frame $\epsilon E_p = 0.35$ eV EeV with E_p in EeV and ϵ in eV. For UV bump photons, with a mean energy of 40 eV, this translates to a characteristic proton energy of $E_c \sim 10^7$ GeV.

The optical depth to photomeson production for protons of energy E_p can be approximated by $\tau(E_p) \sim \epsilon n(\epsilon) \sigma_o R$ where σ_o at the Δ resonance peak[17] $\approx 5 \times 10^{-28}$ cm^2 and $n(\epsilon)$ is the differential number density of photons at this energy in cm^{-3} eV^{-1}. A crude estimate for $n(\epsilon)$ from the accretion disk can be found by assuming that all sources run at some constant efficiency. This is quantified in terms of the the Eddington limit ($L_{edd} = 4\pi GM m_p c / \sigma_T$ where M is the mass of the black hole), the maximum steady state luminosity that can be produced before radiation pressure disrupts the accretion flow. Fitting accretion disk models to the UV spectrum[10] gives a typical luminosity of $(0.03 - 0.1)L_{edd}$ ergs s^{-1}, with $\sim 70\%$ of the luminosity emitted within 30 Schwarzschild radii ($R_s = 2GM/c^2$) of the AGN core. Hence, here we assume that a luminosity of $\sim 0.05L_{edd}$ is emitted isotropically within $R \sim 30R_s$. Given the weak dependence of the characteristic accretion disk temperature on black hole mass ($T \propto M^{-1/4}$) and the very similar UV bump shapes seen in the observational data, we can assume a luminosity independent 'generic' UV AGN spectrum. The AGN X–ray spectrum is also fairly universal, typically being an $\epsilon^{-1.7}$ power law above 2 keV[15], with a cutoff around 1–2 MeV[18]. The total X–ray luminosity is roughly the same as that in the UV bump i.e. $L_x \approx L_{UV} \approx 0.05L_{edd}$. While their more rapid variability suggests that the X–rays are emitted in a smaller region than that of the UV photons[14], here we calculate the mean X–ray photon density within the UV emission region. Normalizing so that $L \sim 4\pi R^2 c \int \epsilon n(\epsilon) d\epsilon = 0.1L_{edd}$, where $R = 30R_s$, we obtain

$$n(\epsilon) \approx \frac{10^{14}}{L_{45}} \text{cm}^{-3} \text{eV}^{-1} \begin{cases} \epsilon & \epsilon < 1 \text{ eV} \\ \epsilon^{-0.9} & 1 < \epsilon < 40 \text{ eV} \\ 3.25 \times 10^{-4} \epsilon^2 e^{-\epsilon/15} & 40 < \epsilon < 192 \text{ eV} \\ 2.45 \times 10^{-1} \epsilon^{-1.7} & 192 < \epsilon < 10^6 \text{ eV} \end{cases} \tag{1}$$

where L_{45} is the total UV luminosity in units of 10^{45} ergs s^{-1}. The photon spectrum, $n(\epsilon)$, is plotted in Figure 2 for $L_{45} = 1$. The $1/L$ scaling of $n(\epsilon)$ follows from the linear scaling of both R and L with M (i.e. with L_{edd} and R_s above.)

In our generic AGN core model, the X–rays are optically thick to the γ–rays owing to pair production, as the mean compactness parameter[19] is $\ell = L\sigma_T/(Rm_e c^3) \sim 20$. All photons of energy above 5 MeV are likely to be absorbed in the source. Such electromagnetic pair cascades can explain the shape of the X–ray spectrum[19], so here we have a self consistent model in which the proton energy loss gives rise to the secondaries which produce the X–ray spectrum.

IV. PROTON ACCELERATION

First-order Fermi acceleration of protons in strong shocks produces a power-law proton energy spectrum $\propto E_p^{-2}$ up to a maximum energy E_{max}. This maximum proton energy is determined by equating the $p\gamma$ lifetime $t_{p\gamma}(E_p) \approx (N_\gamma \sigma_{p\gamma} c \kappa)^{-1}$, where κ is the mean elasticity, with the proton acceleration time[2] $t_{acc}(E_p) \approx 2.2 \times 10^{-4} (R_{shock}/R_s)(E_p/m_p)B^{-1}$. The shock radius, R_{shock}, is fixed at $\sim 10R_s$ by our assumption that the X–ray luminosity is $0.05L_{edd}$. The magnetic field B is taken to be 10^3 G, which assumes approximate equipartition with a typical AGN UV luminosity. Setting the loss time equal to the acceleration time gives E_{max} as an implicit function of luminosity which is obtained numerically and plotted in Figure 3. For $L_{45} > 0.1$, $E_{max} \propto L$. Below that energy, the function levels off because the effective N_γ determining the loss time is smaller. The low maximum proton energy implies that not all of the photon spectrum is available for the $p\gamma$ collisions owing to the energy threshold of the interaction. We find an approximate fit

$$E_{max} = 10^9 GeV \begin{cases} 3 \times 10^{-3} & L_{45} < 2.4 \times 10^{-3} \\ 0.25L_{45}^{0.73} & 2.4 \times 10^{-3} < L_{45} < 1 \\ 0.25L_{45} & 1 < L_{45} < 100 \\ 25 & L_{45} > 100 \end{cases} \tag{2}$$

to our numerically generated curve. For $L_{45} > 100$, we have taken a constant E_{max} so as not to overcount the effect of the preponderance of high luminosity AGN at high redshift which may be operating at a larger fraction of their Eddington luminosity. For all luminosities, the maximum proton energy is limited by $p\gamma$ interactions rather than by the scale size of the shock region.

V. NEUTRINO PRODUCTION

The characteristic straight line escape optical depth of high energy nucleons to accretion disk UV photons is ~ 100 at 40 eV. The optical depth is independent of the luminosity since $\tau(E_p) \propto n(\epsilon)R$ and $n(\epsilon) \propto 1/L$ while $R \propto L$. The function $\tau(\epsilon)$ is shown in Figure 4. The high optical depth of UV photons implies that the secondary neutrons will not in general escape the shock volume, so that comparable amounts of power are generated via $n\gamma$ collisions. With roughly half of the energy loss going into π^0's and the other half into $\pi^{\pm}$'s, the luminosity for ν_μ $(\overline{\nu}_\mu)$ is $\sim 0.2L_x$, and for ν_e $(\overline{\nu}_e)$ is $\sim 0.1L_x$. As the AGN photon spectrum drops exponentially between 40 and 200 eV, it is overwhelmingly the UV bump photons which interact with the high-energy nucleons. We note that for this typical AGN spectrum, the X–rays are not optically thick. However, it is probable that there is a range of X–ray compactness in AGN, and hence a range of optical depths to the $p\gamma$ process. For more compact AGN, the maximum energy of the protons would be lower, with the $p\gamma$ interactions with X–rays determining the cutoff [20].

The mean pion energy is $< E_\pi > \sim 0.2E_p$, while the mean ν energy is $\sim E_\pi/4 \sim 0.05E_p$.[21] Thus the ν spectrum from proton interactions with the UV bump extends from $E_c/20$ to $E_{max}/20$ i.e. $5 \times 10^5 - 10^9$ GeV with a spectral index of ~ 2, mirroring that of the proton spectrum. Above 10^9 GeV the cutoff in the proton spectrum leads to a cutoff in the ν spectrum. Below $E_c/20$ relativistic kinematics give rise to a flat ν energy distribution[21]. This flattening of the ν spectrum below the threshold energy is a critical point that has been neglected in previous studies [22]. This is shown in Figure 5, where the ν spectrum of NGC4151, the brightest X–ray radio–quiet AGN, is calculated using Monte

Carlo techniques.

VI. THE DIFFUSE NEUTRINO BACKGROUND

The diffuse ν flux can be found by integrating over the luminosity function (LF) for

AGN. The most recent determinations of the X–ray LF come from the GINGA satellite[23].

Locally (*i.e.* at the present epoch) it is given by

$$\rho_o(L) = \begin{cases} n_1/L_1 & L_{min} < L < L_1 \\ n_1/L_1(L/L_1)^{-2.6} & L_1 < L < L_{max} \end{cases} \tag{3}$$

where $n_1 = 9.4 \times 10^{-78} L_1/(10^{42} \text{ergs}^{-1})$ cm^{-3}, $L_{min} = 3 \times 10^{38}$ erg s^{-1}, $L_1 = 3 \times 10^{42}$ ergs

s^{-1} and $L_{max} = 10^{46}$ ergs s^{-1}. These luminosities refer to the 2–10 keV X–ray flux so must

be multiplied by a factor 10 to transform them into a total X–ray flux. The luminosity

function at any epoch can be calculated by determining the effects of evolution, and is well

represented as a function of redshift, z, as

$$\rho(L, z) = \frac{g(z)}{f(z)} \rho_o\left(\frac{L}{f(z)}\right) \qquad 0 < z < z_{max} \tag{4}$$

where the luminosity and density evolution are given by the functions $f(z) = (1+z)^3$ and

$g(z) = (1+z)^{2.6}$ respectively[23]. We calculate the diffuse ν background assuming an epoch

of formation of AGN activity at $z = 2.2$ [24]. Figure 6 shows the results, obtained assuming

~ 30 % of the X–ray background is produced by AGN[23] and assuming that all of the

AGN X–ray flux is from the $p\gamma$ interaction and subsequent electromagnetic cascade. The

diffuse background from AGN can be seen above the atmospheric background in the range

$10^4 - 10^{10}$ GeV. Note that a comparison with Figure 5 implies that an angular resolution

of $\sim 3^o$ is sufficient for detection of NGC4151 over the diffuse AGN background. Such a

resolution is attainable in high energy neutrino detectors[25].

Galactic black hole candidates (such as Cyg X–1) may similarly produce high energy ν's. Their much higher X–ray (and hence potential ν) fluxes at Earth may well provide a better possibility for detection of an individual point source over the AGN diffuse background. This possibility will be discussed in a later paper.

VII. EVENT RATES

Event rates in the various ν detectors can be calculated from the spectrum. Fly's Eye with its energy threshold of 10^8 GeV would expect to see ~ 0.02 downward events per year. This is consistent with the present non–detection of an extragalactic ν background from the Fly's Eye. The next generation instrument, the HiRes Eye could see ~ 0.7 downward events per year. Above the HiRes energy threshold of $\sim 10^8$ GeV, upward ν's are severely attenuated by absorption in the earth[26]. At lower energies, covered by detectors[27] sensitive to both ν_μ and $\bar{\nu}_\mu$, the upward event rates at energies greater than 10^5 GeV are of order 10^2 yr^{-1} for IMB and MACRO, and of order 10^3 to 10^4 yr^{-1} for DUMAND II.

We note that $n\gamma$ and $\bar{p}\gamma$ interactions produce a significant flux of $\bar{\nu}_e$'s at 6.3 PeV, the Glashow resonance energy for the $\bar{\nu}_e e^- \rightarrow W^-$ process. An observation of an enhanced event rate at this energy either indicates that AGN are optically thick to neutrons (as indicated in this paper) or that there are anti–matter AGN in the Universe[28]. We estimate that there would be ~ 10 throughgoing resonance events per year in IMB and MACRO and that DUMAND II would see ~ 300 throughgoing and ~ 10 contained μ events per year from W^- resonance production.

VIII. THE BROAD LINE REGION

The "broad line region" (BLR), a characteristic feature of AGN, is thought to be a large number of dense, illuminated "clouds"[29]. The large line widths of ~ 5000 km s^{-1} can be explained by doppler shifts due to orbital motions. If the "clouds" have predominantly circular orbits (which is by no means certain!) then the velocity can be calculated as $v = \sqrt{GM/r}$, or $\beta = \sqrt{R_s/2r}$. For these high velocities the "clouds" have to be within $\sim 1700 R_s$, or 5×10^{16} cm for a $10^8 M_o$ AGN black hole.

One outstanding problem is the formation and stability of these "clouds". While the two phase instability, in which dense condensations can exist in pressure equilibrium with a hot, rarified gas is seen as the best and currently accepted model[30], there are many problems with this explanation[31]. Here we explore stellar disruption as an alternative explanation for the orgin of the BLR, one in which the problems of stability and formation are already answered.

The high energy ν luminosities predicted from AGN will have a profound effect on stars close to the center of the host galaxy, producing stellar winds, swelling the atmospheres and even causing their total disruption[32]. The column density for absorption of a $\sim 10^6$ GeV ν is $X = m_H/\sigma_{\nu N} \sim 2 \times 10^9$ g cm^{-2}, while total column density for a solar mass star $< \rho > R \sim 10^{12}$ g cm^{-2}. Using a conservative disruption criterion that the ν energy deposited in the star be greater than its nuclear energy generation we find a sphere of stellar disruption, R_{SSD}, within a radius of

$$R_{SSD} \approx 30 L_{45}^{1/2} \left(\frac{M_*}{M_O} \right)^{-1.1} \text{ light days} \tag{5}$$

where M_* and M_O are the masses of the irradiated star and the Sun respectively. This radius is the same as that inferred above for the BLR, and matches well with the current best observational determinations[33] and previous calculations have shown that outflowing material from stellar disruption provides, at least qualitatively, an environment similar to that in which the "Broad Lines" must arise[34].

IX. RADIO–LOUD AGN

The strong possibility of beaming in the radio–loud AGN implies that the intrinsic luminosities and sizes of the emission regions are more uncertain than in radio–quiet objects. However, if the above discussion of the origin of the BLR is correct, then the cores of radio–loud objects should similarly produce ν fluxes, since the radio–loud quasars (or QSO's) have a BLR similar to that of the radio–quiet objects. In this case the QSO's would contain the same sort of nucleus as the radio–quiet objects, but with the addition of jet emission from further out. This extra emission makes it difficult to scale the ν luminosity to the isotropic X–ray flux, since the X–rays may be dominated by anisotropic, beamed emission from the jet.

It is possible to estimate the isotropic X–ray flux in one of the nearby radio–loud objects, 3C273. Here the X–ray spectrum contains an iron line at 6.4 keV of equivalent width ~ 50 eV[35]. The currently accepted explanation for this line in AGN is fluorescence from the X–ray illuminated accretion disk[36], so much of the X–radiation is probably unbeamed. The isotropic X–ray emission in the 2–10 keV band is then $\sim 10^{46}$ ergs s^{-1}, giving a total X–ray flux of $\sim 10^{47}$ ergs s^{-1}. Our calculated ν flux is an order of magnitude lower than that of NGC4151 at energies below 1 PeV, but extends out to higher energies (see Figure

7), providing a possible point source of ultrahigh energy neutrinos. However, we stress that the uncertainties in the underlying assumptions are much higher for 3C273 than for NGC4151. The luminosity could be overestimated by orders of magnitude if the X–ray emission is in part from the relativistic jet, which is known to contribute strongly to the emission at lower energies[37].

The above calculation differs from previous work on radio–loud objects[38] in several aspects. Firstly, below ~ 1 PeV, the ν spectrum flattens owing to relativistic kinematics (see Section V). Secondly, only the X–ray and UV spectra are relevant in our scenario. For 3C273 there is much evidence that the radio, sub–mm and even IR spectra are from the base of a relativistic jet[37] and so should not be considered as part of the same emission region. This cuts down our estimate of the total neutrino luminosity by an order of magnitude.

SUMMARY

AGN should be copious emitters of ν's, providing an observable diffuse background which is much larger than the atmospheric ν background in the range 10^4 to 10^8 GeV. The high diffuse background masks all but the brightest individual AGN ν point sources. The diffuse background ν's should be observable with both present and planned detectors, providing an observational test of AGN acceleration mechanisms.

Such intense ν fluxes will have an effect on any stars close to the core. If the energy deposition drives winds from the disrupted stars then the resultant X–ray illumination could give rise to the emission lines which characterise the BLR in AGN.

ACKNOWLEGMENTS

The authors thank F. Halzen, D. Kazanas, J. Learned, R.F. Mushotzky, D. Seckel and T. Stanev for helpful discussions. CD and MHS acknowledge support from a National Research Council–NASA/GSFC Research Associateship and NASA grant NAGW-1999 respectively.

REFERENCES

[1] V. S. Berezinsky, *Proc. Neutrino 77 (Nauka Publ., Moscow)* 177 (1977); D. Eichler, *Astrophys. J.*, **232**, 106 (1979); R. Silberberg and M. M. Shapiro, *Proc. 16th Intl. Cosmic Ray Conf. (Kyoto)* **10**, 357 (1979); V. S. Berezinsky and V. L. Ginzburg, *Mon. Not. R. Astron. Soc.*, **194**, 3 (1981).

[2] D. Kazanas and D. C. Ellison *Astrophys. J.*, **304**, 178 (1986); D. Kazanas and D. C. Ellison, *Nature*, **319**, 380 (1986).

[3] W. I. Axford, E. Leer and G. Skadron *Proc. 15th Int. Cosmic Ray Conf.*, **11**, 132, Plovdiv (1977); G. F. Krimsky *Dokladay Acad. Nauk. SSR.*, **234** 1306 (1977); A. R. Bell, *Mon. Not. R. Astron. Soc.*, **182** 147 (1978); R. D. Blandford and J. P. Ostriker, *Astrophys. J. Lett.*, **221**, L29 (1987).

[4] P. Morrison, *Astrophys. J.*, **151**, L73 (1969); L.M. Ozernoi, *Astron. J. USSR*, **43**, 300 (1969); R.D. Blandford, *Mon. Not. R. Astron. Soc.*, **176**, 465 (1976); R.J.E. Lovelace, *Nature*, **262**, 649 (1976); R.D. Blandford and R.L. Znajek, *Mon. Not. R. Astron. Soc.*, **179**, 433 (1977).

[5] F. Cheng, L. Danese, G. DeZotti and A. Franceschini, *Mon. Not. R. Astron. Soc.*, **212**, 857 (1985).

[6] R.D. Blandford and A. Konigl, *Astrophys. J.*, **232**, 34 (1979); R.R.J. Antonucci and J.S. Ulvested, *Astrophys. J.*, **294**, 158 (1985).

[7] R. A. Sramek and D. W. Weedman, *Astrophys. J.*, **221**, 468 (1978); K. I. Kellerman, R. Sramek, M. Schmidt, D. B. Shaffer and R. F. Green, *Astron. Jour.* **98**, 1195 (1989).

[8] R. E. Barvainis, *Astrophys. J.*, **320**, 537 (1987).

[9] R.A. Edelson and M.A. Malkan, *Astrophys. J.*, **323**, 516 (1987).

[10] A. Laor, *Mon. Not. R. Astron. Soc.*, **246**, 369 (1990); A. Wandel and V. Petrosian, *Astrophys. J. Lett.***329**, L11 (1988); W. H. Sun and M. A. Malkan, *Astrophys. J.*, **346** 68 (1987); B. Czerney and M. S. Elvis, *Astrophys. J.*, **312**, 325, (1987); J.H. Krolik *et al.*, *Astrophys. J.*, in press (1991).

[11] J.H. Krolik *et al.*, ref 10.

[12] Z. Loska and B. Czerny, *Mon. Not. R. Astron. Soc.*, **244** 43 (1990); A. Laor, ref 10.

[13] C. Done *et al.*, *Mon. Not. R. Astron. Soc.*, **243**, 713 (1990).

[14] A. Lawrence, K. A. Pounds, M. G. Watson, and M. S. Elvis, *Nature*, **325**, 694 (1987); I. M. McHardy, *Mem. Soc. Ast. It.*, **59**, 239 (1988); I. M. McHardy and B. Czerny, *Nature*, **325**, 696 (1987).

[15] R. F. Mushotzky, *Astrophys. J.*, **256**, 92, (1982); T. J. Turner and K. A. Pounds, *Mon. Not. R. Astron. Soc.*, **240**, 833 (1989).

[16] F. W. Stecker, *Phys Rev. Lett*, **21** 1016 (1968).

[17] Review of Particle Properties, Particle Data Group, *Phys. Lett. B*, **239**, 1 (1990).

[18] R. E. Rothschild, R. F. Mushotzky, W. A. Baity, D. E. Gruber, J. L. Matteson, and L. E. Peterson, *Astrophys. J.*, **269**, 423 (1983).

[19] R. Svensson, In: *Proc. IAU Coll. No. 89, 'Radiation-Hydrodynamics in Stars and Compact Objects'*, p.325, edited by D. Mihalas and K.-H.A. Winkler, (Springer-Verlag, Berlin 1986), p. 325; A. A. Zdziarski, *Astrophys. J.*, **305**, 45 (1986); C. Done and A. C. Fabian *Mon. Not. R. Astron. Soc.*, **240**, 81 (1989).

[20] M. Sikora, J.G. Kirk, M.C. Begelman and P. Schneider, *Astrophys. J. Lett.*, **320**, L81 (1987).

[21] F. W. Stecker *Astrophys. J.*, **228**, 919 (1979).

[22] M. C. Begelman, B. Rudak, M. Sikora, *Astrophys. J.*, **362**, 38 (1990); P. L. Biermann and P. A. Strittmatter *Astrophys. J.*, **322**, 643 (1987).

[23] K. Morisawa and F. Takahara, *Pub. Astron. Soc. Jap.*, **41**, 873 (1989)

[24] J. S. Dunlop and J. A. Peacock, *Mon. Not. R. Astron. Soc.*, **247**, 19 (1990).

[25] R. Svoboda, *Nucl. Phys. B (Proc. Suppl.)* **14A**, 97 (1990).

[26] M. H. Reno and C. Quigg, *Phys. Rev.*, **D37**, 657 (1988).

[27] *Proc. Arkansas Gamma–Ray and Neutrino Workshop, Nucl. Phys. B (Proc. Suppl.)* **14A**, (1990); *DUMAND II proposal*, DUMAND collaboration (1988).

[28] Berezinsky and Ginzberg, ref 1; R. W. Brown and F. W. Stecker, *Phys. Rev.*, **D26**, 373 (1982).

[29] K. Davidson and H. Netzer, *Rev. Mod. Phys*, **51**, 715 (1979).

[30] J.H. Krolik, C.F. McKee and C.B. Tarter, *Astrophys. J.*, **249**, 422 (1981).

[31] W.G. Mathews, *Astrophys. J.*, **305**, 187 (1986); M. Elizur and G.J. Ferland, *Astrophys.*

J., **305**, 35 (1986).

[32] F. W. Stecker, A. K. Harding and J. J. Barnard, *Nature*, **316**,418 (1985); T. K. Gaisser, F. W. Stecker, A. K. Harding and J. J. Barnard, *Astrophys. J.*, **309**, 674 (1986).

[33] J. Clavel *et al.*, *Astrophys. J.*, **366**, 64 (1991); H. Netzer, *Comments Astrophys.*, **14**, 137 (1989).

[34] D. Kazanas, *Astrophys. J.*, **347**, 74 (1989).

[35] M.J. Turner *et al.*, *Mon. Not. R. Astron. Soc.*, **244**, 310 (1990).

[36] K.A. Pounds *et al.*, *Nature*, **344**, 132 (1990).

[37] A.P. Marscher and W.K. Gear, *Astrophys. J.*, **298**, 114 (1985); C.D. Impey, M.A. Malkan and S. Tapia, *Astrophys. J.*, **347**, 96 (1989).

[38] P. L. Biermann and P. A. Strittmatter, ref 22.

[39] D. B. Sanders, E. S. Phinney, G. Neugebauer, B. T. Soifer, and K. Matthews, *Astrophys. J.*, **347** 29 (1989).

[40] G. Piccinotti *et al.*, *Astrophys. J.*, **253**, 485 (1982).

[41] F.W. Stecker, ref 21; M. F. Crouch *et al.*, *Phys. Rev.*, **D18**, 2289 (1978); L. V. Volkova, *Sov. J. Nucl. Phys.*, **31**, 784 (1980).

[42] F.W. Stecker *et al.*, to be published (1991).

FIGURE CAPTIONS

Figure 1. A characteristic AGN spectrum for radio–quiet and radio–loud objects. Taken from Sanders *et al.*[39] with permission.

Figure 2. Our generic AGN photon spectrum for a total UV luminosity of $L_{45} = 1$.

Figure 3. Maximum proton energy versus total UV luminosity assuming the photon spectrum given in Equation 1.

Figure 4. The optical depth of the AGN at each point in the spectrum assumed spectrum. This is independant of the luminosity of the source.

Figure 5. The predicted ν_μ $(\overline{\nu}_\mu)$ flux from NGC4151, with $L_x = 3 \times 10^{43}$ ergs s^{-1} and a distance of 4.5×10^{25} cm[40]. The ν_e $(\overline{\nu}_e)$ flux is half that of the ν_μ $(\overline{\nu}_\mu)$ flux.

Figure 6. The integrated high energy ν_μ $(\overline{\nu}_\mu)$ neutrino background from AGN. Also shown is the horizontal ν_μ $(\overline{\nu}_\mu)$ flux from high energy cosmic rays interacting with the Earth's atmosphere[41] (ATM) and the background expected from photomeson production of the extragalactic high energy cosmic rays with the cosmic background radiation[42] (CBR).

Figure 7. The predicted ν_μ $(\overline{\nu}_\mu)$ flux from 3C273, with $L_x = 10^{47}$ ergs s^{-1} and a distance of 3×10^{27} cm[35]. The ν_e $(\overline{\nu}_e)$ flux is half that of the ν_μ $(\overline{\nu}_\mu)$ flux.

CONTINUUM ENERGY DISTRIBUTIONS OF QUASARS

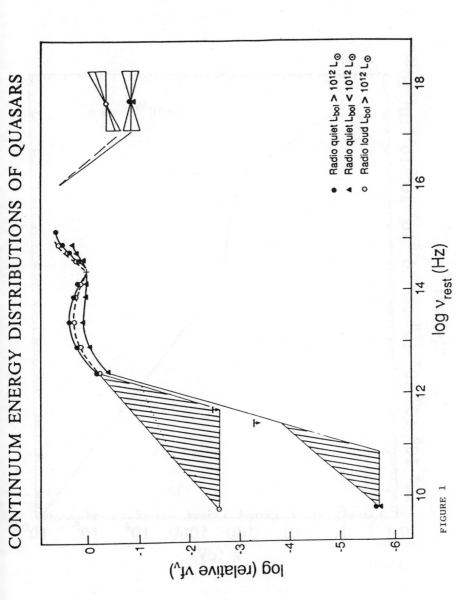

Radio quiet $L_{bol} > 10^{12} L_\odot$
Radio quiet $L_{bol} < 10^{12} L_\odot$
Radio loud $L_{bol} > 10^{12} L_\odot$

log ν_{rest} (Hz)

log (relative νf_ν)

FIGURE 1

344

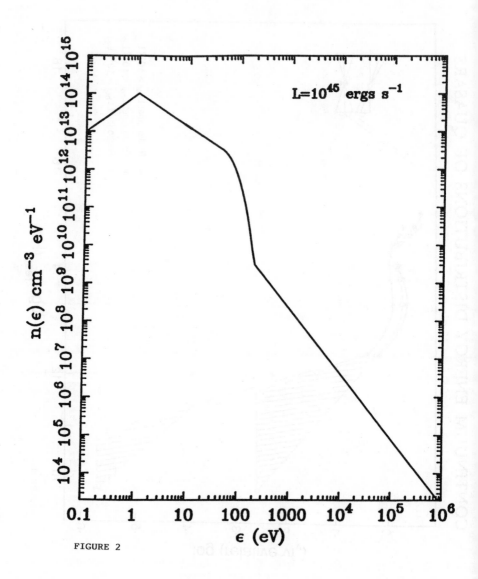

FIGURE 2

345

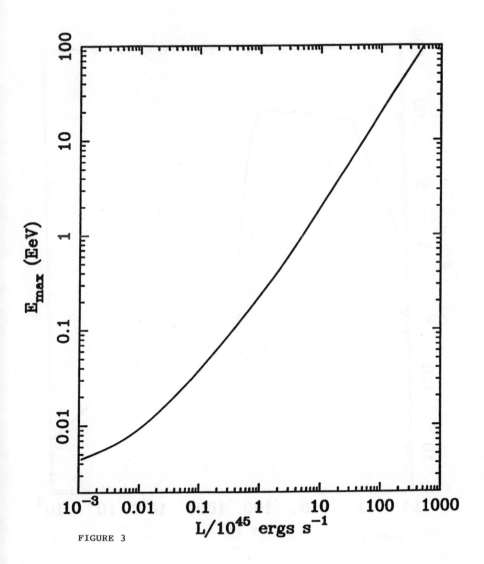

FIGURE 3

346

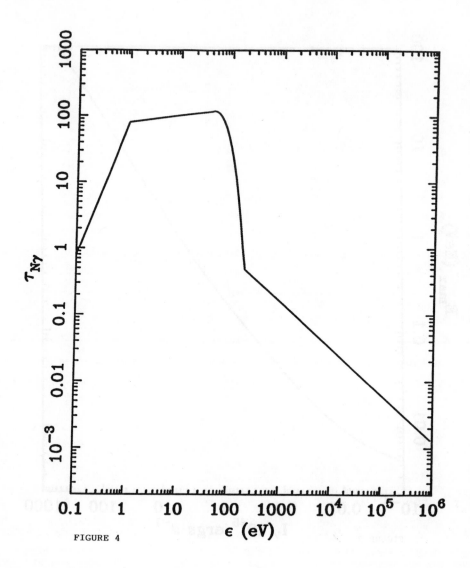

FIGURE 4

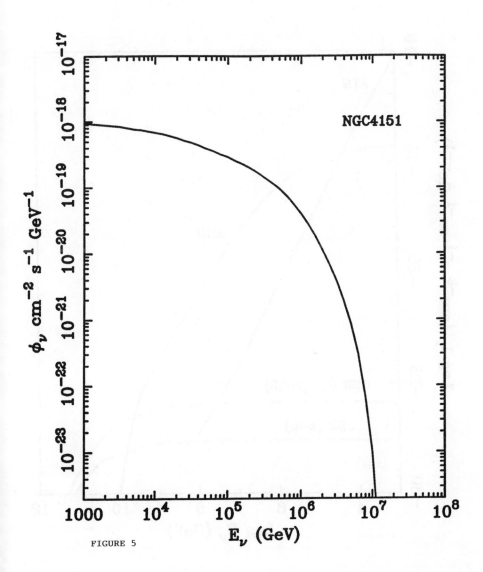

NGC4151

FIGURE 5

348

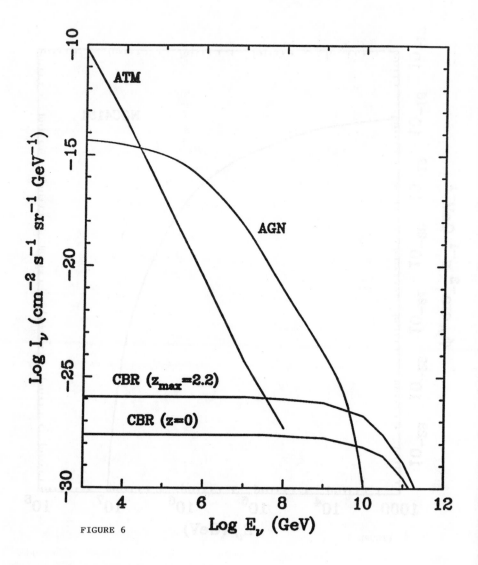

FIGURE 6

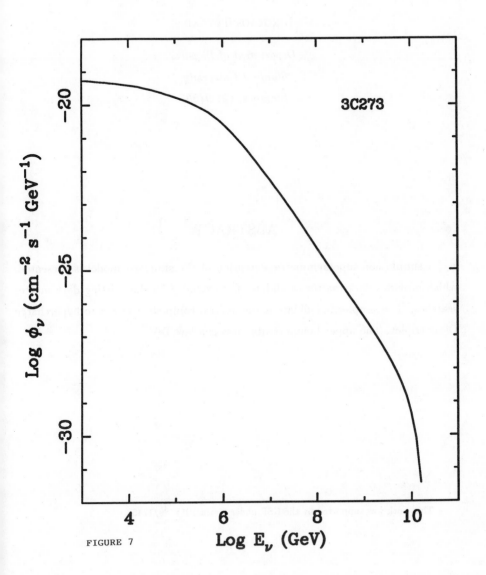

FIGURE 7

Can a Massive Neutrino Close the Universe?[*]

NIKOLAOS TETRADIS

Department of Physics,
Stanford University,
Stanford, CA 94305

ABSTRACT

A simple non-supersymmetric extension of the standard model is presented, which provides dark matter candidates not excluded by the existing dark matter searches. The simplest candidate is the neutral component of a zero hypercharge Dirac triplet. The upper bound on its mass is a few TeV.

---

[*] This work was supported by the NSF under Grant PHY 89/17438.

Introduction: For many years the simplest extension of the standard model which led to an $\Omega = 1$ universe involved the introduction of a fourth generation neutrino with a mass of a few GeV.[1] This had the advantage, over other possibilities such as the lightest superpartner, that the mass was tightly constrained. Recent measurements of the Z^o width at SLC[2] and LEP[3] have ruled out this possibility.[4] The question I would like to address in this report is whether simple non-supersymmetric extensions of the standard model exist that provide a viable dark matter candidate for which definite predictions can be made. My presentation will be based on work done in collaboration with Savas Dimopoulos, Rahim Esmailzadeh and Lawrence Hall.[5]

The mass for which one obtains closure density for the universe depends on the annihilation cross section of the heavy new particle into lighter ones. Therefore, we must classify the possible models according to the $SU(3) \times SU(2) \times U(1)$ quantum numbers of the left-handed and right-handed members of the extended sector. A good dark matter candidate cannot carry electrical charge.[6] Moreover, the experimental bounds on strongly interacting dark matter are very stringent and effectively exclude this possiblity.[7] This leaves us with the possibility of a neutral weakly interacting particle, to which we refer as massive neutrino. The neutrino with a mass in the GeV range is excluded, as has been mentioned already. However, it has been suggested that a neutrino with mass around a TeV would close the universe.[8] The electroweak cross-section responsible for the annihilation of neutrinos and anti-neutrinos has the proper value for closure for neutrino masses of a few GeV; the cross-section is too large for larger masses which means that the neutrinos do not close the universe. However, since the cross-section eventually drops like m^{-2} for particles heavier than the W, it is expected that for neutrino masses of the order of a few TeV, one can have closure again. Another issue which is crucial is the mechanism through which the neutrino obtains its mass. If it gets its mass throught the usual breaking of the $SU(2)$ group, its coupling to the Higgs particle must be strong. In other words the theory becomes strongly interacting and, contrary to recent claims,[9] one is unable to compute the annihilation cross section for the neutrino and therefore find the proper mass for closure. Griest and Kamionkowski[10] have shown that unitarity of the annihilation cross-section limits the mass to be less than 340 TeV, even if the Higgs annihilation channel is strongly coupled. In order to be able to perform analytic calculations, we need a model which remains perturbative in the region of neutrino masses that are interesting cosmologicaly.

The model: We consider the case of a Dirac neutrino, whose left and right handed components belong to the same representation of the electroweak group. In this case, we can introduce a bare mass, while the theory remains perturbative, allowing us to compute the annihilation cross section and therefore the freeze out abundance accurately. The lagrangian of our model is

$$L = L_{SM} + L_{gauge}^{R} + M_R R \bar{R} \tag{1}$$

L_{SM} is the lagrangian of the minimal standard model with three generations of quarks and leptons and a Higgs doublet. L_{gauge}^{R} includes the gauge interactions of a new multiplet which we denote by $R(C, W, Y)$. C, W and Y represent the transformation properties of this multiplet under $SU(3)$ of color, weak $SU(2)$ and $U(1)$ of hypercharge. We shall consider only color singlets ($C = \mathbf{1}$). However, we shall study general representations of $SU(2)$ and $U(1)$. We also impose a global $U(1)_R$ symmetry, which prevents interactions of the new multiplet with the standard three generations and the Higgs. Thus, the lightest member of this multiplet is stable. The theory is anomaly free and the bare mass term is allowed by the above symmetries. The new members have vector gauge interactions. For example, the interaction term with the Z^o and the photon, for a member ℓ with charge Q of the multiplet with hypercharge Y, is

$$g\bar{\ell}\gamma^{\mu}\ell\{\frac{1}{cos\theta_w}[Qcos^2\theta_w - \frac{Y}{2}]Z_{\mu} + Qsin\theta_w A_{\mu}\}. \tag{2}$$

After the electroweak symmetry is broken, the mass degeracy of the the various members of the multiplet is lifted by radiative corrections. In order to have a viable model we must ensure that the new multiplet includes a neutral member, which becomes the lightest. The mass difference between the neutral member and a member with charge Q of the multiplet with hypercharge Y is calculated to be:

$$\frac{\delta m}{MeV} \simeq 350 \times Q^2 + 170 \times (Y - Q)Q. \tag{3}$$

For a doublet $(Y = -1, 1)$, or a triplet $(Y = -2, 0, 2)$, the neutral member is the lightest. For higher reresentations this is not always true. For example, in a $d = 5$ representation, with hypercharge of (-2), the member with charge $+1$ is the lightest. Also in a $d = 4$ representation, for some values of Y and Q, δm is almost equal to zero and its sign is determined by higher loop corrections.

The upper bound on the neutrino mass: The present abundance of the lightest member of the multiplet is determined through the usual Lee-Weinberg analysis.[1] The details of this calculation are presented in ref. 5. We obtain an upper bound on the neutrino mass, by demanding that the universe is not overclosed. The neutrino mass for which closure density is obtained is plotted in the following figure for multiplets of various dimensionalities d and small hypercharge Y.

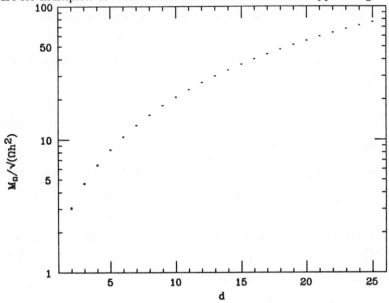

Experimental constraints: One basic idea of dark matter detection, is to detect the energy deposited by nuclei recoiling from elastic scattering with halo particles. Goodman and Witten[11] have calculated the coherent elastic scattering cross section between a weakly interacting particle and a nucleus. In the mass range of interest to us, this cross section is mass independent. For Germanium detectors, the relevant cross section is

$$\sigma \simeq 8 \times 10^{-32} Y^2 cm^2. \tag{4}$$

Any cross section larger than

$$\sigma_{Ge} \simeq 3 \times 10^{-32} (\frac{M}{10 \ TeV}) \ cm^2 \tag{5}$$

is excluded.[12] Therefore, for non-zero values of hypercharge, particles with masses up to $M_{Ge} = 25 \ Y^2$ TeV are ruled out as comprising the halo of our galaxy. This excludes all models with multiplets of dimension ten and less, non-zero hypercharge and stable neutral lightest member. In models with zero hypercharge, the neutral member does not have neutral current interactions, and hence, is very difficult to detect. Such models are not constrained by the experimental data. The representation with the smallest dimension, satisfying all of the above constraints is a triplet with hypercharge zero. The upper bound on the mass of the neutral member is $4.7\sqrt{\Omega h_o^2}$ TeV. This neutrino interacts with Germanium nuclei via two W-exchange. The effective coupling is estimated to be two orders of magnitude smaller than ordinary weak interactions, proceeding through Z^o exchange. Thus, the predicted cross section with Ge is at least four orders of magnitude smaller than the upper bound set by the experiment of reference 12.

Conclusion: There are simple extensions of the standard model, which provide dark matter candidates with mass in the TeV range, for which explicit calculations and predictions can be made. The simplest example is the weak vector triplet (R^+, R^o, R^-) whose lightest member R^o, can be the dark matter if it has a mass of $4.7\sqrt{\Omega h_o^2}$ TeV. The charged members of this triplet are a few hundred MeV heavier and can be produced at SSC or LHC.

REFERENCES

1. B. W. Lee and S. Weinberg, *Phys. Rev. Lett.* **39** (1977) 165,
   P.Hut, *Phys. Lett. B* **69** (1977) 85,
   M. I. Vysotskii, A.D. Dolgov and Ya. B. Zel'dovich *JETP Lett.* **26** (1977) 18.

2. Mark II Collaboration (Alan J. Weinstein, et al.), CALT-68-1591, Aug. 1989.

3. OPAL Collaboration (M.Z. Akrawy, et al.), *Phys. Lett. B* **231** (1989) 530.

4. K. Griest and J. Silk, *Nature* **343** (1990) 26,
   L. M. Krauss, *Phys. Rev. Lett.* **64** (1990) 999.

5. S. Dimopoulos, R. Esmailzadeh, L. Hall and N. Tetradis, *Nuc. Phys. B* **349** (1991) 714.

6. S. Dimopoulos, D. Eichler, R. Esmailzadeh and G. Starkman, *Phys. Rev. D* **41** (1990) 2388,
   R. S. Chivukula, A. G. Cohen, S. Dimopoulos and T. P. Walker, *Phys. Rev. Lett.* **65** (1990) 957,
   A. Gould, B. T. Draine, R. W. Romani and S. Nussinov, *Phys. Lett. B* **238** (1990) 337.

7. J. Rich, R. Rocchia and M. Spiro, *Phys. Lett. B* **194** (1987) 173,
   G. D. Starkman, A. Gould, R. Esmailzadeh and S. Dimopoulos, *Phys. Rev. D* **41** (1990) 3594.

8. A. D. Dolgov and Ya. B. Zeldovich, *Rev. of Mod. Phys.* **53** (1981) 1.

9. K. Enqvist, K. Kainulainen and J. Maalampi, *Nuc. Phys. B* **317** (1988) 647.

10. K. Griest and M. Kamionkowski, *Phys. Rev. Lett.* **64** (1990) 615.

11. M. Goodman and E. Witten, *Phys. Rev. D* **31** (1985) 3059.

12. D. O. Caldwell, et al., *Phys. Rev. Lett.* **61** (1988) 510; *Phys. Rev. Lett.* **65** (1990) 1305,
    D. O. Caldwell, *Mod. Phys. Lett. A* **5** (1990) 1543.

Non-MSW Solutions to the Solar Neutrino Problem

James Pantaleone

Physics Department, University of California, Riverside, CA 92521

Abstract

Recent experimental data have reaffirmed the solar neutrino problem. However the present ^{37}Cl, Kamiokande-II and ^{71}Ga data can be brought into agreement with the Standard Solar Model predictions by any of several different modifications of neutrino propagation. Some of the less discussed possibilities are reviewed here, along with possible experimental tests.

There is a long standing problem with the flux of neutrinos coming from the Sun. For two decades the ν_e flux observed by the ^{37}Cl experiment [1] has been consistently much less than the predicted flux [2]. This discrepancy has been confirmed recently be the Kamiokande-II (K-II) experiment [3] for the high energy ^{8}B neutrinos. Also, preliminary results from the SAGE [4] ^{71}Ga experiment indicate that there is a similar discrepancy in the lower energy pp ν_e flux. These results can be reconciled with the Standard Solar Model (SSM) if something happens to the ν_e as they propagate from the Sun to the Earth. There have been many suggestions along those lines, the most popular of which is the Mikheyev-Smirnov-Wolfenstein (MSW) solution [5-7]. Here, some of the other suggested solutions are reexamined in light of the recent experimental data. In addition, future experimental tests of the various proposals are discussed, emphasizing those tests which are independent of the SSM predictions.

Table 1: Experimental results on the average solar neutrino flux, given as a fraction of the SSM prediction.

Experiment	Process	$E_{Threshold}$	Exp./SSM
Davis et al.	$v_e + {}^{37}Cl \rightarrow v_e + {}^{37}Ar$	0.81 MeV	0.27 ± 0.04
K-II	$v_e + e \rightarrow v_e + e$	7.5 - 9 MeV	$0.46 \pm 0.05 \pm 0.06$
SAGE	$v_e + {}^{71}Ga \rightarrow v_e + {}^{71}Ge$	0.24 MeV	0.0 ± 0.5

Besides the above average flux values, there is some additional experimental information. The K-II experiment is somewhat sensitive to the neutrino spectrum and has ruled out drastic (step-function like) distortions of the high energy ^8B neutrino spectrum, independent of the SSM predictions [8]. Also, the ^{37}Cl and K-II experiments have been running for two decades and 3 years, respectively, and so the data can be examined for possible time variations. The ^{37}Cl data shows some evidence for an anticorrelation with sunspots [1] at a significance level of 1-C.L. $\approx 1\%$. However the real time K-II experiment has not seen this effect, and observes a constant flux on yearly, seasonal and day/night time scales [9]. These observations constraint theoretical interpretations.

I. Neutrino Mixing.

If the neutrinos have masses, then typically the mass basis will be different than the weak interaction basis with a unitary mixing matrix, U, connecting the two.

$$|v_i\rangle = U_{\alpha i} |v_\alpha\rangle \qquad (1)$$

where i = 1, 2, 3 labels the mass eigenstates and α = electron, muon, tau labels the interaction eigenstates (recent LEP and SLC experiments have confirmed that their are only three light neutrino species). Since the present solar neutrino experiments are all primarily sensitive to ν_e, the only kind of neutrino the SSM predicts the Sun to produce, mixing solves the solar neutrino problem by causing the ν_e to leak into other species which can not be seen by the present experiments [10]. The specific way that this happens can take many different forms, one of which is the MSW effect.

Neutrino masses and mixing are readily incorporated into the standard SU(3)xSU(2)xU(1) model, without any fine tuning (i.e. the see-saw mechanism [11]). In fact, they were suggested long before the solar neutrino problem existed, by analogy with the quark sector. Experimentally, neutrino masses and mixings have not been observed in laboratory experiments. However, because of the long Earth-Sun propagation distance, solar neutrinos are sensitive to neutrino masses many orders of magnitude smaller than can be probed in any purely terrestrial experiment.

I.A. Large vacuum mixing.

One way to solve the solar neutrino problem with neutrino mixing is to have large mixing among all three neutrino species. The average ν_e survival probability, neglecting matter effects, is given by

$$P(\nu_e\text{->}\nu_e) = \sum_{i=1,3} |U_{ei}|^2 |U_{ei}|^2 \tag{2}$$

This is valid for $m_3^2\text{-}m_1^2$ and $m_2^2\text{-}m_1^2$ values anywhere in the ranges

$$10^{-11} \text{ eV}^2 < \ < 10^{-8} \text{ eV}^2 \tag{3}$$

$$10^{-4} \text{ eV}^2 < \ < 10^{-1} \text{ eV}^2 \tag{4}$$

Below 10^{-11} eV^2 the oscillation wavelength is longer than the Earth-Sun distance (see solution I.B), and above 10^{-1} eV^2 laboratory experiments are relevant. The intermediate region between (3) and (4) is where the MSW effect operates. The minimum value for Eq. (2) is $P(\nu_e \to \nu_e) \approx 1/3$ which occurs when $|U_{e1}|^2 \approx |U_{e2}|^2 \approx 1/3$ (N.B. $|U_{e3}|^2 \approx 1 - |U_{e1}|^2 - |U_{e2}|^2$ by unitarity). This value for $P(\nu_e \to \nu_e)$ agrees with that flux reduction given in Table 1 for the ^{37}Cl (roughly) and ^{71}Ga experiments. For the K-II experiment, this solution yields a flux reduction = $P + (1-P) \times (1/7) = 0.43$ since ν_μ's or ν_τ's have an electron scattering cross section smaller by about 1/7. Thus this solution can explain all of the present average flux measurements.

This solution can be tested in many ways by continuing and next generation experiments. The continuing SAGE and starting GALLEX [12] ^{71}Ga experiments should measure 1/3 of the predicted SSM flux. Also, the upper mass range, Eq. (3), can be probed independently of SSM predictions by (1) measurements of the atmospheric neutrino flux (preliminary results here already show some evidence for neutrino mixings [13]) and (2) proposed long-baseline accelerator neutrino experiments [14] as has been suggest using a FERMILAB ν beam to either the DUMAND, IMB or SOUDAN detector.

This solution, and in fact any neutrino mixing solution, can also be tested by a comparison of charged and neutral current measurements of the flux. A neutral current measurement is insensitive to the neutrino flavor and should measure a flux equal to the SSM value. Neutral current measurements may be possible in the planned SNO [15] detector and in the discussed BOREX detector [16].

The lower mass range, Eq. (4), can be probed independently of SSM predictions by observing the ^{7}Be solar neutrino line [17]. Because of the narrow neutrino spectrum, neutrino oscillation would remain coherent for long distances and would be directly observable using the annual 3.5% variation in the Earth-Sun distance. Thus a small neutrino mass in the range of Eq. (4) would produce time variations of 1 week to 6 months in the ^{7}Be flux. A high statistics, real time measurement of the ^{7}Be flux may be possible in the proposed BOREX prototype, BOREXINO [18] detector.

Fig. 1 shows the neutrino oscillation parameter regions that can be probed by the above techniques. Also shown is the 90% C.L. allowed region for the MSW solution. Note that a large fraction of this region is predicted to produce observable day/night variations in the ^{7}Be flux [17]. Thus a high statistics measurement of this neutrino line, as may be possible in the BOREXINO detector, would also provide crucial information on the MSW effect.

I.B. Neutrino Oscillations.

Another way to solve the solar neutrino problem with neutrino mixing is to exploit the possibility of flavor oscillations [19-20]. Neglecting the third neutrino flavor, the two-flavor survival probability can be written as

$$P(\nu_e \rightarrow \nu_e) = 1 - \sin^2 2\theta \ [\ 1 - \cos 2\pi R/\lambda\ ] \ ; \ \lambda = 4\pi E/ m_2^2 - m_1^2 \quad (5)$$

where θ is the mixing angle and the quantum interference term has been included with λ the oscillation wavelength and R the distance propagated. If the Earth-Sun distance corresponds to half of an oscillation wavelength, then large reductions in the flux will occur. The ^{37}Cl and K-II detectors are primarily sensitive to neutrino energies of about 10 MeV and so substituting this into the expression for the wavelength yields a mass squared difference of 10^{-10} eV^2. The full allowed parameter region, using the results in Table 1 and the K-II spectral information, is shown at

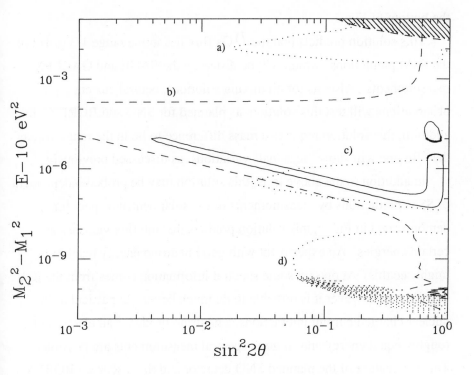

Fig. 1 Solid lines show 90% C.L. limits for two-flavor solutions to the solar neutrino problem using data from Table 1 and K-II spectral information; the upper large region is the MSW solution, bottom small region is the just-so solution. The upper shaded region is excluded by the Gosgen reactor data. The dot and dash regions can be probed by next generation experiments independent of SSM predictions: a) upper dotted region by a long-baseline accelerator neutrino experiment, b) the dashed region by a SNO or BOREX neutral current measurement, and c) d) the lower dotted regions by a measurement of the ^{7}Be line, as proposed for BOREXINO, where c) is from day/night and d) from "seasonal" time variations.

the bottom of Fig. 1.

This solution predicts that the ^{71}Ga flux lies in the range 1/3 to 1/2 of the SSM predictions--which will be tested by the SAGE and GALLEX measurements. Also, as for all mixing solutions, neutral current observations will test this solution, as planned for SNO and BOREX. In addition, this solution requires a mass difference to lie in the range where the ^{7}Be line will show "seasonal" variations, as discussed previously.

In addition to the above tests, this solution may be probed independent of SSM predictions by measurements of the solar neutrino spectrum. As demonstrated in Fig. 2, this solution predicts that the flux vanishes at certain energies. An experiment with good neutrino energy resolution might see this. At present some spectral information comes from the K-II experiment, however it is possible to do much better. In particular, in nuclear charged current interactions, a neutrino produces an electron of roughly equal energy and so good spectral measurements are possible. This is a feature of the planned SNO detector and the discussed BOREX and ICARUS [21] detectors.

II. Neutrino Decay.

One previously proposed solution to the solar neutrino problem is neutrino decay [22]. Ignoring mixing, the electron-neutrino survival probability has the form

$$P(\nu_e \text{->} \nu_e) \approx \exp(- (t/480s) (m/E)) \tag{6}$$

where 480 seconds is the Sun-Earth propagation time, t is the neutrino lifetime, m the neutrino mass and E the neutrino energy. This form predicts the rough observation that suppression of the solar neutrino flux is largest at low energies, in agreement with Table 1 (see Fig. 2). The decay must take the form $\nu_2 \text{->} \overline{\nu}_1 + \phi$ so that similar decays of

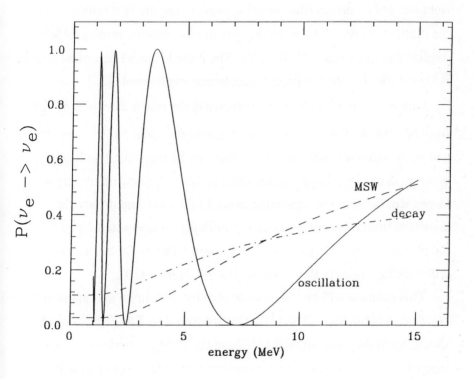

Fig. 2 Comparison of some typical solar electron-neutrino survival probabilities, as a function of energy; MSW, $\sin^2 2\theta = 0.1$, $(m_2^2\text{-}m_1^2) = 4\times10^{-7}$ eV^2 (dash); vacuum oscillation, $\sin^2 2\theta = 1$ $(m_2^2\text{-}m_1^2) = 5.2\times10^{-11}$ eV^2 (solid); and neutrino decay, $\sin^2 2\theta = 0.9$, $\tau m/10$ MeV = 700 sec (dot dash).

charged leptons are forbidden. This can be incorporated into the standard model, for example consider v_2 to have a Dirac mass, v_1 and the scalar ϕ to be light, and ϕ to have a lepton number of 2 with coupling only to the right-handed neutrinos (this model appears to require fine tuning). Then the lifetime is given by $\tau = 16\pi/(g^2 m)$ and the solar neutrino problem implies that gm $\approx (6 \times 10^{-4})(0.1 \text{ eV})$. The particle model described above is very difficult to test in present accelerator experiments [23].

One problem with the above scenario is the recent detection of $\overline{v}_e$'s from SN1987A. Here the neutrinos traveled for about 5×10^{12} seconds and yet a signal was still observed. This can be reconciled with solar neutrino decay by adding neutrino mixing [24]. If neutrino mixing is large then, because the supernova emits all flavors of neutrinos, the depletion from neutrino decay can be partially compensated for. All of the present supernova data and solar neutrino data in Table 1 can be fit with mixing angles in the range of 40 to 60 degrees [19].

This solution will be tested somewhat by continuing and planned solar neutrino experiments. The ^{71}Ga flux measured by SAGE and GALLEX should lie in the range of 0.08 to 0.20 of the SSM predictions. Also, this solution is different than pure neutrino mixing solutions in that neutral current flux measurements should also show substantial reductions from the SSM values, between 50% and 70%.

This solution does require neutrino masses near the present laboratory bounds and so may be observed in future terrestrial experiments. For example, large oscillations in the atmospheric neutrino flux are predicted (and may already have been observed[13]). These large masses may also be probed by long baseline accelerator experiments [14] as has been suggested using either the DUMAND, IMB or SOUDAN detectors (see Fig. (1)).

This solutions also predicts that there should be a substantial flux of low energy $\overline{\nu}_e$'s coming from the Sun. Such neutrinos could be observed in a low background, low threshold experiment such as BOREXINO.

III. Neutrino Magnet Moment.

This solution has been well discussed in the literature [25]. Briefly, the neutrino precesses as passing through the solar magnetic field from a left-handed state to a "sterile" or "semi-sterile" right-handed state. The angle of precession if roughly given by

$$\begin{matrix} \text{spin} \\ \text{precession} \end{matrix} \approx (\frac{\mu}{10^{-10}\mu_B})(\frac{B}{1kg})(\frac{L}{R}) \qquad (7)$$

where μ is the neutrino magnetic moment, B the solar field strength and L the dimensions of the field. Because the solar magnetic field is poorly constrained, the minimum size of μ for this solutions to be viable is somewhat uncertain. A reasonable lower limit appears to be $10^{-11} \mu_B < \mu$. Present laboratory bounds are a factor of 40 larger than this.

This solution has several problems. It's most attractive feature is that it might explain the anticorrelation between the ^{37}Cl data and sunspots. However this effect has not been observed in K-II (although this might be explained by a value for the magnetic moment at, or slightly larger than, the present laboratory bound [26]). Also, the necessary value for the magnetic moment is several order of magnitude larger than predicted by the standard model although it might be accomadated in extensions thereof. Finally, their are astrophysical bounds [27] on the magnetic moment which potentially could rule out this solution.

One way to further constrain this solution would be to improve the laboratory bounds on the neutrino magnetic moment. This can be done using the ^{7}Be solar neutrinos in BOREXINO [17]! By looking for distortion in the electron spectrum, the present laboratory bounds could be improved by an order of magnitude, independent of the SSM predictions. This is because the ν-e scattering via a magnetic moment is singular at low neutrino energies. A direct test of this solution is possible if the neutrino is majorana and mixings are large, then a substantial flux of $\overline{\nu}_e$'s may come from the Sun and could be observed in next generation low background experiments like BOREXINO.

IV. Summary.

A new experiment, Kamiokande-II, has confirmed the ^{37}Cl result that there is a significant difference between the predicted and measured flux of neutrinos coming from the ^{8}B reaction in the Sun. In addition, preliminary evidence from the SAGE ^{71}Ga experiment indicates that a similar discrepancy is also present for the lower energy neutrinos from the pp reaction. These discrepancies can be resolved if the neutrino has some new property which manifests itself during propagation from the Sun to the Earth. There are many possibilities for this new neutrino physics that can explain all of the present solar neutrino data and still be in agreement with known laboratory results. A list of the qualitatively distinct possibilities mentioned herein is given below (however note that any of these possibilities can be combined with any other to form a "heterotic" solution to the solar neutrino problem).

$$MSW, n_f = 2$$

$$\text{Neutrino Oscillation, } n_f = 2$$

$$\text{Large vacuum mixing, } n_f = 3$$

$$\text{Neutrino Decay, } n_f = 2$$

$$\text{Neutrino magnetic moment, } n_f = 2.$$

Each type of solution above can be tested directly, independent of SSM predictions, by continuing and next generation solar neutrino experiments.

I would like to thank Andy Acker and Sandip Pakvasa for a useful collaboration [19] relevant to much of the above work.

References

1. K. Lande et al., talk presented at NEUTRINO'90 (CERN). R. Davis, Jr., D.S. Harmer and K.C. Hoffman, Phys. Rev. Lett. 20, 1205 (1968).
2. J.N. Bahcall and R.K. Ulrich, Rev. Mod. Phys. 60, 297 (1988).
3. K.S. Hirata et al., Phys. Rev. Lett. 63, 16 (1989); 65, 1297 (1990).
4. M. Cherry, these proceedings.
5. L. Wolfenstein, Phys. Rev. D17, 2369 (1978); D20, 2634 (1979).
6. S.P. Mikheyev and A.Yu. Smirnov, Nuovo Cimento C9, 17 (1986); Sov. J. Nucl. Phys. 42, 913 (1985).
7. For a recent review, see T.K. Kuo and J. Pantaleone, Rev. Mod. Phys. 61, 937 (1989).
8. K.S. Hirata et al., Phys. Rev. Lett. 65, 1301 (1990).
9. K.S. Hirata et al., ICRR-222-90-15.
10. B. Gribov and B. Pontecorvo, Phys. Lett. B28, 493 (1969); J.N. Bahcall and S.C. Frautschi, Phys. Lett. B29, 623 (1969).
11. For a discussion, see P. Langacker, Phys. Rep. 72, 158 (1981).
12. T. Kirsten et al., Talk Presented at NEUTRINO'90 (CERN).
13. K. Hirata et al., Phys. Lett. B205, 416 (1988).
14. J. B. Learned and V.Z. Peterson, HDC-5-89; R. Becker-Szandy et al. UCI-Neut. No. 89-22; W. Allison et al., Soudan 2 letter of Intent PDK-444; J. Pantaleone Phys. Lett. B264, 245 (1990).
15. G.T. Ewan et al., Sudbury Neutrino Observatory Proposal, October

(1987).

16. The Borex Collaboration, Letter of Intent to Gran Sasso Laboratory, Sept. 1988.

17. S. Pakvasa and J. Pantaleone, Phys. Rev. Lett. 65, 2479 (1990).

18. R.S. Raghaven, Presented at the 25th ICHEP, Singapore Aug. 1990.

19. A. Acker, S. Pakvasa and J. Pantaleone, Phys. Rev. D (in press).

20. V. Barger, R.J.N. Phillips and K. Whisnant, MAD-PH-579.

21. J. Park, these proceedings.

22. J.N.Bahcall, N. Cabibbo and A. Yahil, Phys. Rev. Lett. 28, 316(1972); S. Pakvasa and K. Tennakone, Phys. Rev. Lett. 28, 316 (1972).

23. S. Pakvasa and J. Pantaleone, in preparation.

24. J.A. Frieman, H.E. Haber and K. Freese, Phys. Lett. B200, 115(1988).

25. A. Cisneros, Astrophys. Space Sci. 10, 87 (1981); L.B. Okun, M.B. Voloshin, M.I. Vysotsky, Sov. J. Nucl. Phys. 44, 440(1986).

26. A. Suzuki et al., KEK Preprint 90-51.

27. J. Bernstein, M. Rudermanm and G. Feinberg, Phys. Rev. 132, 1227 (1963); J.M. Latimer and J. Cooperstein, Phys. Rev. Lett. 61, 23 (1988); R. Barbieri and R.N. Mohapatra, Phys. Rev. Lett. 61, 27 (1988); G. Raffelt, Phys. Rev. Lett. 64,2856 (1990).

"Long-term" Neutrino Flux Integrations

Wick C. Haxton

Department of Physics, University of Washington, Seattle, Washington 98195

Abstract

The standard solar model predicts that the sun's luminosity has increased by 40% over the past 5 Gyrs of main-sequence burning, reflecting the evolving chemistry of the solar core. This increase is accompanied by an exponential growth in the ^{8}B neutrino flux, with a doubling time of 0.85 Gyr. I describe an unusual nuclear system that, in principle, could yield a quantitative terrestrial record of these past changes, and discuss some of the practical obstacles to reading this record.

I also argue that there exists a "twin" of the ^{37}Cl solar neutrino experiment that could be mounted with modest effort. The experiment, in which neutrinos incident on an iodine-bearing liquid produce the noble gas ^{127}Xe, should enjoy a number of advantages over ^{37}Cl: a substantially larger counting rate, a potentially greater relative sensitivity to ^{7}Be neutrinos, lower cosmic ray backgrounds, and a cleaner signal for the ^{127}Xe decay. I discuss the physics that could be learned from such a detector, emphasizing in particular the virtues of a passive, high counting rate experiment that could be operated cheaply over the long times that might pass before the next galactic supernova.

In this talk I would like to discuss two ideas for novel solar or supernova neutrino experiments. The first [1] is the more speculative, but addresses an issue that has not been raised previously: is it possible to perform a measurement that tests solar behavior over time scales characteristic of main-sequence burning? If such a measurement were possible, it would directly check the evolutionary predictions of our standard theory of stellar structure. In the second half of this talk I discuss the possibility of mounting a radiochemical experiment that is remarkably similar to the present ^{37}Cl detector, but has advantages from the perspectives of counting rate, backgrounds, and the ease of detecting the final product [2]. Such an experiment would be of great interest as a conventional solar neutrino measurement: the high statistics and low backgrounds would yield tight constraints on any short-term variations in the solar flux, and the combination of this experiment and the present Homestake effort would hopefully permit one to separate the ^{7}Be and ^{8}B neutrino fluxes. This separation could help establish that some portion of the Kamioka II counting rate arises from heavy flavor neutrinos. Simple, passive detectors with high statistics are also significant from the perspective of the supernova watch, as they can be operated with little effort over very long times. Thus I will argue that lifetimes of such experiments might be compatible with another evolutionary scale, the rate of stellar collapse in our galaxy.

One of the more interesting results of the standard solar model is the predicted monotonic increase in the photon luminosity of approximately 40% during the past five billion years of main-sequence burning [3]. Geophysicists have had great difficulty in reconciling this prediction with the known continuity of the Earth's biology and geology. While one billion years ago the standard model luminosity was 8% below today's value, paleoclimatic evidence leads to the conclusion that the Earth could not have been cooler than today [4]. Indeed, numerical models of climate predict that a 5% luminosity decrease is sufficient to trigger a major glacial epoch, an event that could not have been overlooked in the geophysical record [5]. This may indicate that such models are seriously flawed, perhaps failing to account for alternative energy sources or feedback mechanisms that contribute to climatic stability. However, it is also possible that this discrepancy, along with the solar neutrino puzzle [5], indicates a failure of stellar evolution models. Phenomena outside the standard solar model, such as mixing of the core or mechanisms for early solar mass loss, could in principle lead to a softer long-term luminosity growth. Such changes in the physics of the standard solar model would likely have broad implications for stellar evolution, chronology, and galactic chemistry.

The solar luminosity profile in time tracks the rising core temperature that accompanies helium synthesis and the associated opacity increase. The solar neutrino flux also depends critically on the core temperature. In particular, the flux of high energy neutrinos emitted in the β decay of ^{8}B varies as

$$\phi(t) = \phi_0 \, e^{-t/\tau} \tag{1}$$

where t is the time before present (so that the beginning of main sequence burning corresponds to $t = 4.6$ Gyr), $\tau = 1.22$ Gyr, and $\phi_0 = 5.76 \cdot 10^6/\text{cm}^2\text{sec}$ is the present standard model ^{8}B flux. (This formula is a fit, accurate to about 5%, to the standard model results of Bahcall and Ulrich [3].) Thus this flux is our most sensitive monitor of the evolving chemistry of the sun, doubling in 0.85 billion years, a time scale that opens possibilities for experimental tests.

There have been serious efforts to develop geochemical solar neutrino measurements exploiting the parent isotopes ^{97}Mo/^{98}Mo, ^{81}Br, and ^{205}Tl. The daughter isotopes are unstable with lifetimes $\lesssim 10^7$ year. Thus these experiments, in the context of the standard solar model, are contemporary flux measurements. In the context of nonstandard models, the ^{97}Mo/^{98}Mo experiment is also interesting as a possible test of solar variability on the Kelvin timescale. In contrast, a test of the core temperature over the much longer nuclear burning time scales of interest here would have the virtue of probing a predicted variation in the *standard* solar model, directly probing the evolutionary aspects of this model. However, the prospects for such a geochemical measurement seem extraordinarily bleak when one considers the unique properties of ^{98}Tc that make the ^{98}Mo experiment so attractive [6]. No stable isotopes of Tc exist, so that the ambient concentration of both ^{98}Tc and neighboring Tc isotopes is extraordinarily low. Thus the mass spectrometric isolation of ^{98}Tc is feasible, and the usually fatal neutral particle backgrounds from natural radioactivity (such as (n, γ)) are absent. Furthermore the ^{98}Tc half life is short compared to most geologic timescales. Thus the integration time is known, given an ore body with a geologic age clearly much longer than the half life.

The time scale for the standard model ^{8}B neutrino luminosity change is about one billion years. All unstable isotopes with comparable lifetimes have appreciable abundances, presumably the residual abundances from the time these elements were synthesized in the galaxy. Thus any geochemical experiment must involve a daughter isotope that is effectively or actually stable, with the neutrino signal an isotopic anomaly. Presumably this anomaly would be extraordinarily small in all but the most exceptional cases. Furthermore, even if a measurement of the solar neutrino-induced isotopic excess proved feasible, the integration time would be highly uncertain: the effective integration time may or may not correspond to the geologic age of the ore body, and even that age probably could not be determined with certainty.

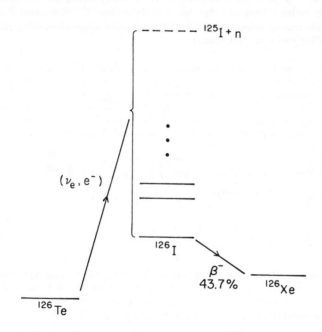

Fig. 1. Level scheme illustrating the production of ^{126}Xe by neutrino reactions on ^{126}Te.

One case, and perhaps the only case, where these powerful general arguments are circumvented is illustrated in Fig. 1 [1]. Charge current neutrino reactions off ^{126}Te (abundance 18.95%) produce ^{126}I, which then β-decays to ^{126}Xe with a branching ratio of 0.437. The threshold for ^{126}Te$(\nu, e^-)^{126}$I is 2.15 MeV, so only the ^{8}B neutrinos (with an endpoint energy of approximately 15 MeV) contribute. Xenon is a heavy rare gas, and thus its abundance in rock is extraordinarily low because of the efficient fractionation of rare gases that occurred when the earth's crust was formed. The isotope of interest, ^{126}Xe, is rarest of the stable xenon isotopes, with an abundance of 0.089%. This isotope is shielded: ^{127}Xe and ^{125}Xe are short-lived, so that (n, γ) and similar neutral particle backgrounds are absent. The stable isotopes two mass units away, ^{124}Xe and ^{128}Xe, have relatively low abundances of 0.10% and 1.91%, respectively, permitting excellent mass spectrometric discrimination for ^{126}Xe.

I will argue below that the production of ^{126}Xe by solar neutrinos can, in suitable deep telluride ores, lead to an isotopic excess of ^{126}Xe that approaches 0.1%. It also appears that this production will exceed that from natural radioactivity backgrounds. Geochemists have succeeded in measuring isotopic anomalies as small as $\sim 10^{-5}$ in ideal systems [7]. In the case of xenon, the best present techniques have a sensitivity of about 10^{-3} for the principal isotopes [8]. These experiments involve small samples ($\lesssim 1$g) carefully isolated from atmospheric contamination. The low natural abundance of ^{126}Xe requires one to achieve this level of sensitivity while processing much larger quantities of telluride ($\sim$ one kilogram).

If the isotopic excess proves measurable, the difficulty of accurately determining the xenon retention age of the ore would appear to remain. However tellurium contains two remarkable natural clocks: the other principal isotopes of tellurium, ^{130}Te (33.8%) and ^{128}Te (31.7%), double beta decay to ^{130}Xe and ^{128}Xe, respectively, producing measurable isotopic excesses in these mass numbers proportional to the xenon gas retention age of the ore. The isotopic anomalies Δ_{126}, Δ_{128}, and Δ_{130} are

$$\Delta_{126} = \frac{1}{R} \left(\frac{< \phi\sigma >}{10 SNU} \right) \left(\frac{\tau}{10^9 y} \right) \left(1 - e^{-t_0/\tau} \right) 2.93 \cdot 10^{-17}$$

$$\Delta_{128} = \frac{1}{R} \left(\frac{10^{24} y}{\tau_{1/2}(128)} \right) \left(\frac{t_0}{10^9 y} \right) 1.15 \cdot 10^{-14}$$

$$\Delta_{130} = \frac{1}{R} \left(\frac{10^{21} y}{\tau_{1/2}(130)} \right) \left(\frac{t_0}{10^9 y} \right) 5.71 \cdot 10^{-12} \tag{2}$$

where t_0 is the xenon retention age of the ore (either the ore formation age or time of last recrystalization), and R is the initial ratio of trapped natural xenon to natural tellurium in the ore. Studies of double beta decay, typically performed with $\lesssim 1$g samples, have established incidental limits on Δ_{126} of a few parts per hundred [9].

A reasonable value [1] for $< \sigma\phi >$ in the standard solar model is ~ 12 SNU, and a favorable value for R would be that typical of Kalgoorlie ore, $R \sim 0.46 \cdot 10^{-13}$ [9]. Using the Kalgoorlie ore age of $2.46 \cdot 10^9 y$ one obtains a standard solar model $\Delta_{126}^{SSM} = 0.08\%$. The requirement that cosmic ray and natural radioactivity backgrounds contribute no more than 10% of this signal yields limits on the ore depth of $\gtrsim 1400$m (a geologic average), on the uranium content of $\lesssim 4.5$ppm, and on the thorium content of $\lesssim 160$ppb [1]. The U and Th limits are relatively easy to satisfy, while the ore depth requirement eliminates a number of tellurides.

The standard solar model predicts that the quantity $\log (\Delta(126/\Delta(132)$ will vary linearly with the $\beta\beta$ decay excess $\Delta(130)$. A definitive geochemical or laboratory determination of the ^{130}Xe $\beta\beta$ decay half life would permit one to extract τ from the slope of such a graph, thereby providing a direct check on standard model evolution over ~ 1

Gyr and its prediction of significant long-term luminosity growth. The scale of this experiment, involving ~ 1kg of ore, is well beyond the capabilities of any existing facility. However it is not obviously beyond what is technically possible. The prospect that one might be able to measure a standard-model evolutionary parameter τ may encourage experimentalists to explore the feasibility of this very difficult measurement.

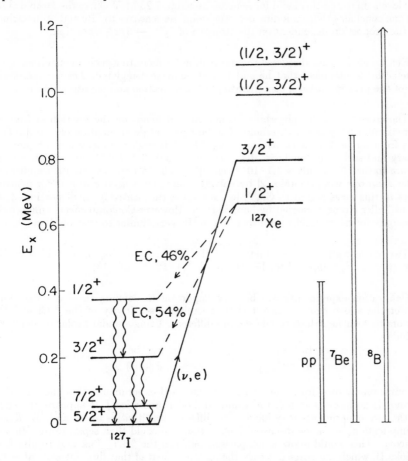

Fig. 2. Level scheme showing relevant weak and electromagnetic transitions in ^{127}I and ^{127}Xe.

Let us now turn to an experiment that very likely can be performed with existing techniques. Three years ago it was noted [2] that the reaction ^{127}I$(\nu_e, e^-)^{127}$Xe ($\tau_{1/2}$ = 36.4d), producing a noble gas daughter nucleus, is strikingly similar to the reaction

$^{37}\text{Cl}(\nu_e, e^-)^{37}\text{Ar}(\tau_{1/2} = 35.0\text{d})$ that Davis and his collaborators have exploited in the Homestake experiment. A level scheme illustrating some of the relevant nuclear physics is given in Fig. 2. As the transition from the ground state of ^{127}I to the first excited state is very weak, ^7Be solar neutrinos are absorbed only through a single transition, $5/2^+ \rightarrow 3/2^+$ (125 keV). The high-energy ^8B neutrinos can produce ^{127}Xe by exciting many levels up to the threshold for neutron breakup, 7.23 MeV above the ground state. Thus one concludes that an iodine detector would be sensitive to ^7Be and ^8B neutrinos, with the proportion dependent on the strength of $5/2^+ \rightarrow 3/2^+$ transition.

Ken Lande [10] has discussed the motivation for mounting such an experiment and the detector development that he and Ray Davis have completed. Let me summarize some of the principal issues, concentrating on cross section and counting rate issues.

The cross section for absorbing ^8B neutrinos depends on the fraction of Gamow-Teller strength lying below threshold. The best present determination of that strength comes from forward-angle (p, n) scattering [11]. Although the measurements are preliminary and still under analysis, they suggest $\sigma(^8\text{B}) \sim 3.2 \cdot 10^{-42}$ cm^2, which one can compare to the ^{37}Cl result $\sim 1.1 \cdot 10^{-42}$ cm^2 [12]. The ^{127}I cross section is significantly smaller than one would obtain [2] by naively scaling cross sections for other neutrino targets in this mass region ($^{71}\text{Ga}, ^{98}\text{Mo}$), reflecting the relatively small portion of the Gamow-Teller strength carried by bound states. However Coulomb effects enhance the phase space for heavy nuclei, resulting in a $\sigma(^8\text{B})$ very similar to that of ^{71}Ga.

Two nuclear structure calculations of the ^7Be cross section have yielded $\sigma(^7\text{Be})$ $\sim 2.0 \cdot 10^{-45}$ cm^2, compared to the ^{37}Cl result $2.4 \cdot 10^{-46}$ cm^2 [13].

Calibration experiments will be very important in defining these cross sections. However the above results suggest that the relative sensitivity of the ^{37}Cl and ^{127}I detectors to solar neutrinos could be quite different. Using standard solar model fluxes one finds

$$\frac{< \sigma\phi(^7\text{Be}) >}{< \sigma\phi(^8\text{B}) >}\Bigg|_{37\text{Cl}} = 0.17 \qquad \frac{< \sigma\phi(^7\text{Be}) >}{< \sigma\phi(^8\text{B}) >} = 0.50$$

(Of course one knows, from the ^{37}Cl and Kamioka II experiments, that the flux of ^8B neutrinos is no more than half the standard model prediction.) It appears, therefore, that the two experiments may have quite different sensitivities to ^8B and ^7Be fluxes. The implication, as Lande discussed [10], is that one could then separate the ^7Be and ^8B fluxes. This would allow a comparison between the ^8B ν_e flux and results from Kamioka II, which measures not only the ν_e component of that flux but also, at $\sim 1/6$ the relative efficiency, any ν_μ or ν_τ component that might arise from flavor-changing neutrino oscillations.

Another advantage of the ^{127}I experiment is that only one stable iodine isotope exists, while ^{37}Cl has an abundance of only 24.2%. To illustrate the net effect of cross sections and abundance on counting rate, we take a methylene iodide target of the same volume as the current Homestake experiment. The comparative rates for detectors of this volume are

$$R(^{127}\text{Xe}) = (9.0/d)\left(\frac{\sigma(^8\text{B})}{3.2\cdot10^{-42}\text{cm}^2}\right)\tilde{\phi}(^8\text{B})$$

$$+ (4.5/d)\left(\frac{\sigma(^7\text{Be})}{2.0\cdot10^{-45}\text{cm}^2}\right)[\tilde{\phi}(^7\text{Be}) + 0.12\tilde{\phi}(pep)$$

$$+ 0.08\tilde{\phi}(^{13}\text{N}) + 0.21\tilde{\phi}(^{15}\text{O})]$$

$$R(^{37}\text{Cl}) = (1.22/d)\tilde{\phi}(^8\text{B}) + (0.21/d)[\tilde{\phi}(^7\text{Be}) + 0.20\tilde{\phi}(pep)$$

$$+ 0.09\tilde{\phi}(^{13}\text{N}) + 0.31\tilde{\phi}(^{15}\text{O})]$$

where $\tilde{\phi}(^8\text{B})$ denotes the ^8B neutrino flux in units of the standard-model flux, etc. Also only the $5/2^+ \to 3/2^+$ transition has been included in the ^{127}I CNO and pep estimates. One concludes that, for detectors of equal volume, the ^{127}I is seven times more sensitive to ^8B neutrinos and 22 times more sensitive to ^7Be neutrinos than ^{37}Cl. The total counting rate, for standard solar model fluxes, would be in the ratio $(15.3/d):(1.55/d)$. Tentatively it appears that the counting rate for an ^{127}I detector would be about an order of magnitude greater than that for the Homestake experiment, which itself has the highest counting rate of any functioning detector.

Calibration of the detector involves two separate experiments. It is crucial to measure the $5/2^+ \to 3/2^+$ transition that governs ^7Be neutrino absorption. I believe that there is only one method of performing such a calibration, production of an intense ($\sim$ 100 kCi) ^{37}Ar source [14]. While ^{37}Ar can, in principle, be produced by $^{36}\text{Ar}(n,\gamma)$, the cost of enriched ^{36}Ar ($\sim$ 90%) is considerable. There are also substantial engineering problems associated with irradiating a compressed gas in a reactor where the product ^{37}Ar has a large burnup cross section (5.2b). The more attractive solution is the irradiation of a large volume of ^{40}Ca, an inexpensive target, with fast ($\sim$ few MeV) neutrons: the cross section for $^{40}\text{Ca}(n,\alpha)^{37}\text{Ar}$ is 60-200 mb in the 3-8 MeV range. An attempt will soon be made to produce such a source at the LAMPF beamstop.

The ^8B cross section requires a knowledge of the Gamow-Teller distribution from threshold up to the continuum. The forward-angle (p,n) measurement provides this profile, but there is no obvious strong, well-separated transition of known strength that can be used to normalize the profile. We think the best solution to this problem is to measure the ^{127}Xe production rate for an iodine detector placed in LAMPF beamstop stopped-muon ν_e spectrum [15]. Although this cross section is very closely related to that for ^8B neutrinos, first-forbidden transitions and other corrections to the long-wavelength limit will be of some importance at LAMPF neutrino energies ($< E_\nu > \sim$ 40 MeV). Such "nuclear form factor" corrections, which we anticipate will be on the order of 10-20%, can be estimated in nuclear model calculations. This theory input together with the LAMPF cross section then, in effect, determines the normalization of the Gamow-Teller distribution. Therefore one can then predict the ^8B cross section, given an accurate LAMPF cross section. A proposal to perform the LAMPF calibration for both ^{37}Cl and ^{127}I detectors has been approved. Since the ^{37}Cl cross section is independently known from ^{37}Ca β-decay studies [12], we can test the above procedure for this case [15].

There are a number of very pleasing aspects to the proposed ^{127}I experiment in addition to the favorable counting rate and sensitivity to ^{7}Be neutrinos. As discussed in [2], the higher Z of ^{127}I and unfavorable thresholds suppress many neutron-and α-induced background reactions, including ^{127}I$(p,n)^{127}$Xe induced by energetic protons from cosmic ray muon reactions. It appears very likely that background production of ^{127}Xe will be quite low. The counting of ^{127}Xe can be done very cleanly because the decay proceeds only to excited states in ^{127}I (see Fig. 2). Thus one can exploit the coincidence between the nuclear γ-ray and the Auger radiation [2,10]. In contrast, the only signal available in the ^{37}Cl and ^{71}Ga experiments is the atomic rearrangement energy released by the daughter decay. Davis and Lande have shown that the counting of ^{127}Xe can be performed with background rates $\lesssim 1$/year [10]. Finally, there is a possibility that iodine can be obtained from the U.S. stockpile of ~ 3 kilotons [2].

There are several strong physics motivations for pursuing this experiment. As discussed previously, the prospect of combining the ^{37}Cl and ^{127}I results to determine the separate ^{7}Be and ^{8}B ν_e fluxes is very exciting. Such a separation would allow one to use the Kamioka II results, or perhaps more accurate results from a future Super Kamioka, to test the flavor of the high-energy solar neutrino spectrum. Speculations about time variations [16] in the Homestake signal would be put to a rigorous test by operating a second, higher statistics experiment in phase with the ^{37}Cl detector. Finally, while highly instrumented detectors like Kamioka II and SNO can provide wonderful data, the prospect of continuing these experiments for a time comparable to the interval between galactic supernovae (~ 100 years) is daunting. In contrast, an ^{127}I detector could be maintained as a ν_e observatory with little additional expense and only periodic maintenance. Even if operated in a passive mode (so that the detector has reached solar neutrino saturation), a supernova at the galactic center (8 kpc distant) would produce ~ 100 ^{127}Xe atoms, a signal $\sim 4\sigma$ above the solar background [2].

This work was supported in part by the U.S. Department of Energy and by NASA (grant #ATP-90-103.)

References

[1] W.C. Haxton, Phys. Rev. Lett. 65, 809 (1990).

[2] W.C. Haxton, Phys. Rev. Lett. 60, 768 (1988).

[3] J.N. Bahcall and R.K. Ulrich, Rev. Mod. Phys. 60, 297 (1988).

[4] E.N. Parker, in *Basic Mechanisms of Solar Activity*, edited by V. Bumba and J. Kleczek (Reidel, Dordrecht, 1976), p. 31.

[5] E.J. Öpik, Contrib. Armagh. Obs. No. 9 (1953); E. Erikson, Meteorol. Monogr. 8, 68 (1968); W.D. Sellers, J. Appl. Meteorol. 8, 392 (1969).

[6] G.A. Cowan and W.C. Haxton, Science 216, 51 (1982); K. Wolfsberg *et al.*, in *Solar Neutrinos and Neutrino Astronomy* (Ref. 5), p. 196.

[7] T. Kirsten, in *Science Underground*, edited by M.M. Nieto *et al.*, AIP Conference Proceedings No. 96 (American Institute of Physics, New York, 1983), p. 396.

[8] C. Hohenberg (private communication).

[9] O.K. Manuel, in *Nuclear Beta Decay and Neutrinos*, edited by T. Kotani *et al.* (World Scientific, Singapore, 1986), p. 71; E.W. Hennecke, Phys. Rev. C 17, 1168 (1978); E.C. Alexander *et al.*, Earth Planet Sci. Lett. 5, 478 (1969); B. Srinivasan *et al.*, J. Inorg. Nucl. Chem 34, 2381 (1972); N. Takaoka and K. Ogata, Z. Naturforsch. 21, 84 (1966); T. Kirsten in *Nuclear Beta Decay and Neutrinos*, (Ref. 12), p. 81, and references therein; T. Kirsten *et al.*, Z. Naturforsch. 22a, 1783 (1967).

[10] K. Lande, talk presented at the 1990 UCLA Supernova Watch Workshop.

[11] J. Rapaport, private communication.

[12] A. García *et al.*, submitted to Phys. Rev. Lett.

[13] F. Dellagioacoma and F. Iachello, Phys. Lett. 218B, 399 (1989); J. Engel, S. Pittel, and P. Vogel, Bartol Research Inst. preprint BA-91-03 (1991).

[14] W.C. Haxton, Phys. Rev. C 38, 2474 (1988).

[15] LAMPF experiment E1213, Ken Lande, spokesman.

[16] J.K. Rowley, B.T. Cleveland, and R. Davis, in *Solar Neutrinos and Neutrino Astronomy*, ed. M.L. Cherry, K. Lande, and W.A. Fowler (AIP Conf. Cosmic Ray Conf., Adelaide, 1990 (to be published); G.A. Bazilevskaya *et al.*, Yad. Fiz. 39, 856 (1984) [Sov. J. Nucl. Phys. 39, 543 (1984)].

Seesaw Model Predictions for the τ-Neutrino Mass

Sidney A. Bludman[1,2], D.C. Kennedy[2], and P.G. Langacker[2]

[1] Center for Particle Astrophysics, University of California, Berkeley, CA 94720
[2] University of Pennslyvania, Philadelphia, PA 19104

ABSTRACT: The small neutrino masses and flavor-mixing apparently observed in the Sun are most naturally interpreted in terms of grand unification theories incorporating the seesaw mechanism. In two such theories, SO(10) GUT and SUSY GUT, that are consistent with all laboratory data, the neutrino mixing is like up-quark (CKM) and the neutrino masses are proportional to the squares of the up-quark masses. For the SO(10) GUTS model, the symmetry breaking scale is intermediate and the μ-neutrino mass is close to that observed in solar neutrino oscillations. Although the seesaw model mass predictions are less reliable than the mixing angle predictions, the τ-neutrino mass may lie in the cosmologically important range (4-28) eV and be accessible to laboratory neutrino oscillation experiments or to observation in a nearby supernova.

A National Science Foundation
Science & Technology Research Center

1. The Deficit of Low-energy Solar Neutrinos

Of the solutions of the solar neutrino problem considered, only matter- amplified neutrino (MSW) oscillations occur naturally for a range of neutrino parameters. In this interpretation of the combined Homestake, Kamiokande II and Sage data [1], either semiadiabatic or large-mixing adiabatic oscillations are taking place in the Sun because of a neutrino mass-squared difference $\delta m^2 =$0.05-20 meV^2 and vacuum mixing $\sin\theta$ >0.03. We believe, from laboratory limits and from theoretical expectations in the next section, that the neutrino mixing matrix is similar to the quark (CKM) mixing matrix. The oscillations in the Sun are therefore most likely $\nu_e \to \nu_\mu$ with the μ-neutrino mass $m_{\nu_\mu} =$0.2-4 meV. For Cabbibo mixing, the solar neutrino oscillations are semiadiabatic with $m_{\nu_\mu} \simeq$ 0.45 meV, but m_{ν_μ} could be as much as ten times larger if the neutrino mixing is much smaller or much larger than Cabbibo mixing so that $\delta m^2 =$20 eV^2 in the Sun.

2. Seesaw Model for Small Neutrino Masses

The MSW solution requires new physics beyond the Standard Model (SM) of electroweak unification. Unless one wants to invoke a new symmetry, the most natural and minimal extension of the SM that explains the smallness of ordinary ν_L masses is the Seesaw Model. This model [2] invokes a superheavy right-handed Majorana neutrino N_R, which can form a (Majorana) mass, $M_N \overline{N}_L^c N_R$, with itself and a (Dirac) mass, $m_D \overline{\nu}_L N_R$ with the ν_{iL} of the Standard Model i.e. a mass term

$$m_D \overline{\nu}_L N_R + M_N \overline{N}_L^c N_R = \frac{1}{2} \left(\overline{\nu}_L \quad \overline{N}_L^c \right) \begin{pmatrix} 0 & m_D \\ m_D^T & M_N \end{pmatrix} \begin{pmatrix} \nu_R^c \\ N_R \end{pmatrix}$$

in the Lagrangian. Allowing for three flavors for both the R- and L- neutrinos, m_D and M_N are each 3×3 matrixes. Diagonalizing the Lagrangian leads to three (unobserved) superheavy neutrinos and three light neutrinos whose masses are $m_l = m_D M^{-1} m_D^T$. Quarks and charged leptons are expected to have masses of order m_D. All of these masses are evaluated at the GUTS scale X, so that we finally have for each family $m_{\nu_i}(X) = m_D^2(X)/M_{Ni}(X)$. These masses now have to be run down to the low-energy scale by renormalization group calculations that are model-dependent [1].

We discuss two theoretical models [3] that are each consistent with all present experimental constraints and induce a quadratic seesaw with up-quark masses and neutrino flavor mixings approximately those of the quark (CKM) mixings i.e. $m_e : m_{\nu_\mu} : m_{\nu_\tau} = m_u^2 : m_c^2 : m_t^2$ and $\sin^2 2\theta_{e\mu} =$0.18,$\sin^2 2\theta_{\mu\tau} =$(0.004-0.014),$\sin^2 2\theta_{e\tau} =$

(4×10^{-6}-1×10^{-4}. The first model, SO(10) GUT, predicts intermediate scale symmetry breaking and a μ-neutrino mass near that derived from solar neutrino oscillations and a τ-neutrino mass that may be cosmologically important.

In the SO(10) GUT, a first symmetry breaking down to a left-right symmetric model takes place at a GUT scale $M_X = (3 - 30) \times 10^{16}$ GeV, followed by a second symmetry breaking down to the SM at an intermediate scale $M_R = (0.7-3)\times10^{10}$ GeV. Such an intermediate scale may arise naturally from the breakdown of Peccei-Quinn symmetry at a scale $\sim 10^{11} - 10^{12}$ GeV needed to close the Universe by invisible axions or may arise from the breakdown of hidden-sector supergravity symmetry-breaking at $\sqrt{m_{3/2}M_{Planck}} \sim 10^{11}$ GeV, where $m_{3/2} \sim$ TeV is the gravitino mass .

A reasonable Yukawa coupling of Higgs to right-handed neutrinos, then gives these right-handed neutrinos a mass $M_N \approx (0.01 - 1)M_R = (0.7-3)\times10^{11\pm1}$ GeV and the quadratic seesaw formulas

$$m_{\nu_e} = (0.05)\frac{m_u^2}{M_N} \quad ,$$

$$m_{\nu_\mu} = (0.07)\frac{m_c^2}{M_N} \quad ,$$

$$m_{\nu_\tau} = (0.18)\frac{m_t^2}{M_N} \quad .$$

The SO(10) model predicts neutrino masses $m_{\nu_e} < 2 \times 10^{-5}, m_{\nu_\mu} = (.0055 - 2.2), m_{\nu_\tau} = (60 - 10^5)$ eV. At its lower limit, the μ-neutrino mass is only five times larger than that observed in solar neutrino oscillations for Cabbibo angle mixing. At its lower limit, the τ-neutrino mass is almost within the cosmological upper limit of 28 eV. Smaller neutrino masses would be obtained if the SO(10) model were supersymmetrized or the intermediate scale Higgs content altered so as to realize the intermediate scale symmetry breaking at a somewhat larger value of M_R.

In the minimum supersymmetric GUT (MSSM), starting from the unification scale $M_X = (0.6 - 4.0) \times 10^{16}$ GeV a single symmetry breaking takes place at the SUSY scale. A right-neutrino mass $M_N = (0.06 - 40) \times 10^{15}$ GeV appears near the GUT scale. The radiatively corrected MSSM predictions for the neutrino masses are:

$$m_{\nu_e} = (0.05)\frac{m_u^2}{M_N} \quad ,$$

$$m_{\nu_\mu} = (0.09)\frac{m_c^2}{M_N} \quad ,$$

$$m_{\nu_\tau} = (0.38)\frac{m_t^2}{M_N} \quad .$$

The MSSM predicts neutrino masses $m_{\nu_e} < 2 \times 10^{-11}$, $m_{\nu_\mu} = 6 \times 10^{-9} - 4 \times 10^{-6}$,

m_{ν_τ}=(0.011-0.87) eV. The MSSM model gives a μ-neutrino too small a mass for $\nu_e \to \nu_\mu$ to appear in solar neutrino oscillations. If we instead use the small $\nu_e \to \nu_\tau$ mixing, we obtain a δm^2 and mixing at the upper left corner of the MSW triangle, close to but somewhat outside the range needed to solve the solar neutrino problem. The neutrino mass contribution to the cosmological matter density is then small, less than the baryon mass density.

A mass 17 keV neutrino appearing with 1% mixing in some β-decay experiments cannot arise by the seesaw mechanism. Such a bump in the β-spectrum appears in some crystalline detector experiments [4], but not in magnetic spectrometer experiments [5]. If it exists, a 17 keV neutrino needs to be a Dirac neutrino, in order to evade double β-decay constraints, and needs to decay invisibly and fast, in order to evade cosmological and astrophysical constraints [6]. Non-seesaw theoretical models can be contrived for this [7], but small neutrino masses do not occur naturally and require a small (TeV) symmetry- breaking scale, much smaller than the intermediate scale or GUT scale obtained above for our seesaw models.

3. Cosmological and Astrophysical Constraints on the τ-Neutrino Mass

We have just seen that at least some theoretical models imply the quadratic seesaw relation with quark masses m_{ν_τ}=(2-4)$(m_t/m_c)^2 m_{\nu_\mu}$=(1-3)$\times 10^4$. If $\nu_e \to \nu_\mu$ are taking place in the Sun, m_{ν_μ}=(0.2-4) meV, so that m_{ν_τ} lies in the cosmologically important range (2-100) eV. Irrespective of theoretical motivation, this mass range will be pursued in terrestial and Galactic experiments.

The deceleration and dynamical age of the Universe together constrain the present mass density of a Friedman universe $\Omega_o h^2$=(0.05-0.3) and 0.4<h<0.7 [6]. Since $m_{\nu_\tau} = 92\Omega_\nu h^2$ eV, the mass density in neutrinos will equal that in baryons $(\Omega_\nu h^2 \simeq 0.01)$ if $m_{\nu_\tau} \simeq 1$ eV. The Universe can be closed with, neutrinos comprising 80% of the mass density if, $m_{\nu_\tau} = 74 h^2$=14-22 eV, for h=0.44-0.54. In that case, massive neutrinos provide a natural source of hot dark matter. Together with non-Gaussian primordial fluctuations, massive neutrinos could then be responsible for large-scale structure, especially if the canonical cold dark matter scenario proves unable to account for the very large cosmological structures now observed.

Present laboratory limits on $\nu_\mu \to \nu_\tau$ oscillations, from the E531 emulsion experiment at Fermilab [8], are $\sin^2 2\theta < sin^2 2\theta_c$ for $\delta m^2 > 2$ eV and $\sin^2 2\theta$ <0.004 for $\delta m^2 > 20$ eV, at 90% C.L. These experiments apparently already exclude τ-neutrinos massive enough to close the universe if, as we expect from the CKM matrix, $\sin^2 2\theta \approx 0.008$. These experiments deserve repitition and refinement.

Atmospheric neutrino oscillation experiments [9] require $\sin^2 2\theta < 0.4$ for $\delta m^2 > \times 10^{-4\pm 0.7}$ eV^2. These experiments could give terrestial confirmation of solar neutrino oscillations only if δm^2 is quite large in the Sun and their sensitivity could be a thousandfold improved.

4. Conclusions

At least two theoretical models consistent with all laboratory data lead to neutrino mixing like up-quark mixing and a quadratic seesaw formula for neutrino masses. For the SO(10) GUTS model, the symmetry breaking scale is intermediate and the μ-neutrino mass is close to that observed in solar neutrino oscillations. Although the seesaw model mass predictions are less reliable than the mixing angle predictions, the τ-neutrino mass may lie in the cosmologically important range (4-28) eV and be accessible to laboratory neutrino oscillation experiments or to observation in a nearby supernova.

References

[1] S.A. Bludman,D. Kennedy and P. G. Langacker, Pennsylvania preprint UPR 0443T and Center for Particle Astrophysics preprint CfPA-TH-91-002

[2] M. Gell-Mann, P. Ramond, R. Slansky, in *Supergravity*, F. van Nieuwenhuizen and D. Freedman, eds. (Amsterdam: North Holland, 1979) p. 315; T. Yanagida, *Prog. Theo. Phys.* **B135**, 66 (1978).

[3] P. Langacker *et al., Nucl. Phys.* **B282**, 589 (1987).ref. 21 of [11a].

[4] A. Hime, N. A. Jelley, Oxford preprint OUNP-91-01 (1991); B. Sur *et al,* LBL Preprint (1990); J.J. Simpson, Phys. Lett.—bf174B, 113, 1986.

[5] J. Markey and F. Boehm, Phys. Rev. C32, 2215, 1985; D.W. Hetherington *et al,* Phys. Rev. C **36**, 1504, 1987.

[6] S. A. Bludman, Center for Particle Astrophysics preprint CfPA-TH-91-004.

[7] S. L. Glashow, Harvard preprint HUTP-90/AO75; K. S. Babu and R. N. Mohapatra, Maryland preprint UMD-PP-91-186.

[8] N. Ushida *et al., Phys. Rev. Lett.* **57**, 2897 (1986).

[9] K. S. Hirata *et al., Phys. Lett.* **B205**, 416 (1988); Ch. Berger *et al., Phys. Lett.* **B245**, 305 (1990);

Constraints on Heavy Particles decaying into Neutrinos

PAOLO GONDOLO

Department of Physics, University of California, Los Angeles, CA 90024, USA

ABSTRACT

We consider constraints on the lifetime, the abundance and the mass of a generic very heavy $(1 - 10^{14}\,\text{TeV})$ unstable particle decaying into neutrinos at cosmological epochs. We obtain mass dependent lower limits of order 10^{23}–10^{25} s on the lifetime of those long lived particles that could be the dark matter today. For lifetimes shorter than the age of the universe, the extension of the excluded regions is determined mainly by the loss of the lower energy neutrinos due to experimental thresholds or to the neutrino opacity of the universe.

1. Introduction

Upper bounds on the flux of high energy neutrinos, obtained from proton decay experiments (IMB, Fréjus, Kamiokande) or cosmic rays observatories (such as Fly's Eye, DUMAND, or the Lake Baikal experiment), constrain the cosmic abundance of heavy unstable particles that decay into neutrinos. If these neutrinos are produced with a typical energy E_e equal to a fraction f of the parent mass m_X, $E_e = f m_X$, then, given m_X and f, upper bounds can be obtained on the product of the abundance n_{X0} of the heavy parent particles and of their branching ratio into neutrinos B_ν as a function of their lifetime τ_X. n_{X0} is the number density the parent particles would have at present if they had not decayed and coincides with their actual present number density for lifetimes longer than the age of the universe.

We present here the bounds on m_X, B_ν, n_{X0}, τ_X imposed by the non-observation of a signal above the one expected from atmospheric neutrinos in proton decay experiments and in Fly's Eye. Earlier attempts to set these bounds (see [1,2]) were made before any experimental data were available. We take into account the experimental thresholds, the neutrino opacity of the early universe, the neutrino absorption in the Earth and the proper cross sections at high energies. A full exposition will be given elsewhere [3].

2. Cosmological neutrino absorption

High energy neutrinos can be absorbed in interactions with the cosmic neutrino background and with matter at cosmological redshifts. The predominant processes are the annihilation of the high energy neutrinos (or antineutrinos) with a background antineutrino (or neutrino respectively) and the inelastic scattering off nucleons. The latter process turns out to be dominant at redshifts $z \lesssim 10^3$ for neutrino energies $E_\nu \lesssim 10\,\text{TeV}$, gradually loosing importance for higher neutrino energies and becoming negligible at all redshifts for $E_\nu \gtrsim 10^6\,\text{TeV}$. The inelastic scattering off nucleons affects only marginally the value $z_a(E_e)$ at which the neutrino opacity of the universe r_ν is unity for a neutrino of initial energy E_e. Less than a fraction $1/e$ of the neutrinos emitted at redshifts larger then $z_a(E_e)$ survive until today.

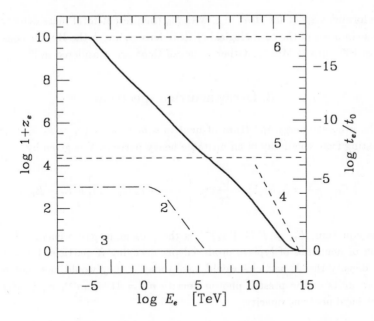

Fig. 1. The absorption redshift z_a as a function of the neutrino energy at emission E_e (line 1). The other lines are: (2) separation between the two regions where annihilation and scattering absorption dominate; (3) the present time; (4) the Z boson pole; (5) the matter-radiation equality; (6) the time of neutrino decoupling.

The absorption redshift is shown as line 1 in the E_e - t_e (or E_e - z_e) plane in fig. 1. Lines 3 and 6 mark the present time and the time of neutrino decoupling respectively. Also indicated is the matter-radiation equality redshift (line 5) and the location of the Z boson pole (line 4) in the $\nu\bar{\nu} \rightarrow f\bar{f}$ annihilation cross section. The bending at $E_e \gtrsim 10^{10}$ TeV is due to neutrino-nucleon scattering. The dash-dotted line (line 2) separates the two regions where annihilation and scattering absorption dominate respectively.

Approximate expressions for the absorption redshift $z_a(E_e)$ are [3]

$$1 + z_a(E_e) = \begin{cases} 4.37 \times 10^6 \left(\frac{E_e}{\text{TeV}}\right)^{-2/5}, & E_e \gtrsim 5.2 \times 10^5 \text{ TeV}, \\ 1.31 \times 10^7 \left(\frac{E_e}{\text{TeV}}\right)^{-1/2}, & E_e \lesssim 5.2 \times 10^5 \text{ TeV}. \end{cases} \qquad (1)$$

These formulas apply for $\Omega_0 h^2 = 1$, with Ω_0 the present total mass density of the universe in units of the cosmological critical density, and h the Hubble constant in units of $100 \text{ km s}^{-1} \text{ Mpc}^{-1}$. Other values of $\Omega_0 h^2$ are considered in [3].

3. Decay neutrino spectrum

The present energy spectrum of neutrinos of type ν_i ($\nu_i = \nu_e, \bar{\nu}_e, \nu_\mu, \bar{\nu}_\mu, \ldots$) originating from the decay of an unstable heavy particle X is given by

$$E_{\nu_i} \frac{\mathrm{d}\phi_{\nu_i}}{\mathrm{d}E_{\nu_i}} = \phi_\gamma B_{\nu_i} Y_X \, \kappa \frac{t_e}{\tau_X} \exp\left[-\frac{t_e}{\tau_X} - r_{\nu_i}(t_e, E_e) \right] \theta(E_e - E_{\nu_i}). \tag{2}$$

In this equation, $E_{\nu_i} = E_e(1 + z_e)^{-1}$ is the present neutrino energy, B_{ν_i} is the number of neutrinos of type ν_i produced per decaying X particle, $Y_X = n_{X0}/n_\gamma$ is the density the X particles would have today if they would have not decayed given in units of the present photon density $n_\gamma = 412.5 \, \text{cm}^{-3}$, $r_{\nu_i}(t_e, E_e)$ is the cosmological neutrino opacity,

$$t_e = t_0 \left(\frac{E_{\nu_i}}{E_e} \right)^\kappa \tag{3}$$

is the neutrino emission time, $\kappa = 2$ for $E_{\nu_i}/E_e < 4.44 \times 10^{-5}$, $\kappa = \frac{3}{2}$ for $E_{\nu_i}/E_e > 4.44 \times 10^{-5}$, and

$$\phi_\gamma = \frac{cn_\gamma}{4\pi} = 0.98 \times 10^{12} \, \text{cm}^{-2} \, \text{s}^{-1} \, \text{sr}^{-1} \tag{4}$$

is the present background photon flux per unit solid angle. The neutrino flux is plotted in fig. 2 by means of solid curves for $\tau_X = 10^{-9} t_0$ and $E_e = \frac{1}{3}, \frac{1}{3} \times 10^3$ and $\frac{1}{3} \times 10^5 \, \text{TeV}$ (lines 1, 2 and 3 respectively). In order to show the effect of the cosmological neutrino absorption, the dotted lines represent what the neutrino flux (2) would have been without absorption, i.e. $r_{\nu_i} = 0$. These three curves are translations one of the other, because the differential flux (2) with $r_{\nu_i} = 0$ depends on the ratio E_{ν_i}/E_e only.

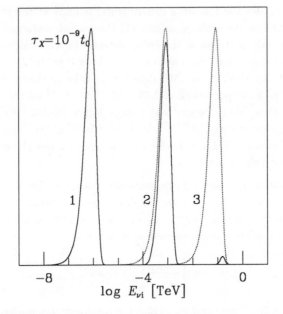

Fig. 2. The present decay neutrino energy spectrum for $\tau_X = 10^{-9} t_0$ and $E_e = \frac{1}{3}$ TeV (line 1), $E_e = \frac{1}{3} \times 10^3$ TeV (line 2) and $E_e = \frac{1}{3} \times 10^5$ TeV (line 3).

4. Present constraints

At present, the best means to detect a diffuse high energy neutrino background is through the production of an energetic charged lepton in the collision of such a neutrino with a nucleon. According to the location of the nucleon, we consider three types of signal: first, a so-called (vertex) contained event occurs when the struck nucleon lies inside an underground detector, such as in the Fréjus, IMB and Kamiokande experiments; second, a flux of (through-going) muons is registered for collisions occuring in the material surrounding the detector (rock in the underground experiments and water in the underwater experiments, such as DUMAND and Lake Baikal's); third, if the nucleon belongs to the Earth atmosphere, then an extensive air shower (EAS) is originated, which could be detected in cosmic rays observatories, such as Fly's Eye, DUMAND and Lake Baikal's.

We will consider the following experimental bounds at the 90% C.L. on the neutrino flux of non-atmospheric origin: (1) the rate of contained events induced by non-atmospheric neutrinos at the Fréjus detector [4] with electron and/or muon energies greater than 3 GeV does not exceed 19 $kton^{-1} yr^{-1}$; (2) the same rate but with threshold 140 MeV at the IMB detector [5] is lower than 3 $kton^{-1} yr^{-1}$; (3) IMB obtained [6] an upper bound of 2.65×10^{-13} $cm^{-2} s^{-1}$ on the flux of up-going muons of non-atmospheric origin with energy larger than 2 GeV; (4) Fly's Eye published [7] upper limits of $10^{-46} s^{-1} sr^{-1}$, $3.8 \times 10^{-47} s^{-1} sr^{-1}$ and $10^{-47} s^{-1} sr^{-1}$ on the rate of neutrino-induced EAS's for neutrino energies higher than 10^5 TeV, 10^6 TeV and 10^7 TeV respectively.

We have taken into account the neutrino absorption by the Earth (if any), because the Earth is opaque to very high energy neutrinos. Using a simplified Earth model with uniform density the flux of neutrinos coming from below the horizon at zenith angle ϑ is suppressed by a factor of

$$\exp\left[-\frac{\sigma_{\nu_i N}(E_{\nu_i})}{\sigma_\oplus}|\cos\vartheta|\right], \tag{5}$$

with $\sigma_\oplus = 2.4 \times 10^{-34}$ cm^2 and $\sigma_{\nu_i N}(E_{\nu_i})$ the total neutrino-nucleon cross section. For the latter, we consider only standard model physics. For the neutrino energies we are considering, $\sigma_{\nu_i N}(E_{\nu_i})$ includes only the charged current cross section $\sigma_{\nu_i N}^{CC}(E_{\nu_i})$, because the energy and momentum fractions transferred to the nucleon in a neutral current process are negligible at these energies.

The charged current cross section $\sigma_{\nu_i N}^{CC}(E_{\nu_i})$ is well-known for $E_{\nu_i} \lesssim 10$ TeV, but at higher energies it becomes more and more uncertain, even by a factor of 10 at $E_{\nu_i} \simeq 10^9$ TeV, because of the poor knowledge of the nucleon structure functions at small arguments [8]. We use the results of ref. [9], in particular the differential charged current cross section in their fig. 3 up to $E_{\nu_i} = 10^7$ TeV. At higher energies, we have matched the asymptotic form of the cross section in ref. [10] to the results of ref. [9].

We make the assumptions that the same amounts of neutrinos and antineutrinos of each type are produced in the decay of the particles X, $B_{\nu_e} = B_{\bar{\nu}_e} = B_{\nu_\mu} = B_{\bar{\nu}_\mu} = \ldots \equiv B_\nu$, and that their production energy is always $E_e = \frac{1}{3}m_X$.

The computation of the expected signals is described in detail in [3]. In fig. 3 we present the results in terms of the X lifetime τ_X, its mass m_X and the combination

$B_\nu Y_X m_X$. Notice that for $\tau_X \gtrsim t_0$ one has $Y_X m_X = \Omega_{X0} h^2 25.5\,\mathrm{eV}$, where Ω_{X0} is the actual present X particle mass density in units of the cosmological critical density. For sake of comparison with other results, we set $B_\nu = 1$. The shaded regions are excluded by the present experimental data. The solid lines refer to the IMB up-going muons, the dotted and short-dashed to the Fréjus and IMB contained events respectively, and the long-dashed to the Fly's Eye EAS's. For $B_\nu \neq 1$, these lines are to be appropriately shifted upward or downward. The short-dashed–dotted line corresponds to a present mass density $\Omega_0 h^2 = 1$ either in X particles (for $\tau_X \gtrsim t_0$) or in decay products (for $\tau_X \lesssim t_0$). The region $\Omega_0 h^2 > 1$ is excluded. For comparison, the long-dashed–dotted line in the lower left quadrant indicates the upper limit on $Y_X m_X$ for photon-producing unstable particles obtained in [11] requiring a non-excessive photodissociation or photoproduction of primordial light elements.

Figs. 3a, b, c and d correspond to $m_X = 1, 10^5, 10^6$ and 10^{10} TeV respectively. Increasing the X mass, at first the bounds at $\tau_X \lesssim t_0$ shift to the left as the neutrinos become more and more energetic and the signal energy moves more and more above the experimental thresholds. Then the bounds shift to the right because the redshift of neutrino absorption in the cosmic background decreases with increasing neutrino energy and the neutrinos produced by relatively short-lived particles do not survive in appreciable quantities until the present. For $\tau_X \gtrsim t_0$, the best constraints come from the bounds of IMB on up-going muons at $m_X \lesssim 5 \times 10^5$ TeV and from those of Fly's Eys on EAS's at $m_X \gtrsim 5 \times 10^5$ TeV. In both cases, the lower bound on τ_X at fixed X density decreases with increasing X mass. In particular, when the X particles could constitute the dark matter today, i.e. $\Omega_{X0} h^2 \simeq 1$, the lower bound on their lifetime varies from $\tau_X \gtrsim 10^7 t_0$ at $m_X = 1\,\mathrm{TeV}$ to $\tau_X \gtrsim 10^6 t_0$ at $m_X = 10^5$ TeV. It becomes $\tau_X \gtrsim 10^8 t_0$ when the Fly's Eye threshold is reached at $m_X = 10^6$ TeV and decreases thereafter, being $\tau_X \gtrsim 10^7 t_0$ at $m_X = 10^{10}$ TeV and disappearing at $m_X \gtrsim 10^{14}$ TeV, when the present universe is opaque to high energy neutrinos.

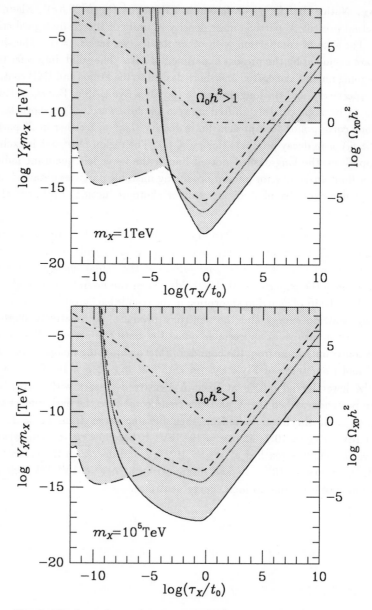

Fig. 3. The bounds on the X particle lifetime τ_X, its mass m_X and its abundance in the combination $B_\nu Y_X m_X$. The shaded regions are excluded by the present experimental data. We have set $B_\nu = 1$. The different lines refer to: IMB up-going muons (solid lines),

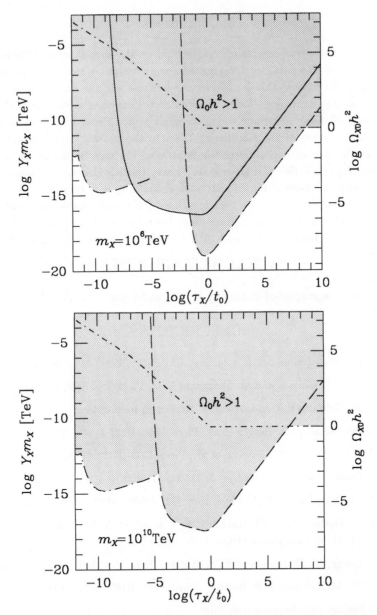

Fig. 3. (*cont.*) Fréjus and IMB contained events (dotted and short-dashed lines resp.), Fly's Eye EAS's (long-dashed lines). Also indicated are the line $\Omega_0 h^2 = 1$ (short-dashed–dotted line) and the upper limit from primordial light elements abundances (long-dashed–dotted line).

5. Conclusions

We have considered constraints on the lifetime, the abundance and the mass of very heavy unstable particles decaying into neutrinos at cosmological epochs. We have taken into account that both the early universe and the Earth are opaque to very high energy neutrinos. We have evaluated the expected signals from a diffuse background of decay neutrinos at underground proton decay experiments and at cosmic rays observatories. Comparing them to the present experimental upper limits on the flux of non-atmospheric neutrinos, we have been able to severely constrain the properties of a generic very heavy particle decaying into neutrinos. In particular, we have obtained that such a particle, if it generates photons in addition to neutrinos, must be very long-lived (effectively stable) in order to be the dark matter today. The precise lower bound on its lifetime depends on its mass, and varies between $10^6 t_0$ and $10^8 t_0$ for masses between $1\,\text{TeV}$ and $10^{10}\,\text{TeV}$.

ACKNOWLEDGEMENTS

This work is supported in part by D.O.E. under grant No. AM03-76SF00010.

REFERENCES

[1] P. H. Frampton and S. L. Glashow, Phys. Rev. Lett. 44 (1980) 1481

[2] J. Ellis, T. Gaisser and G. Steigman, Nucl. Phys. B177 (1981) 427

[3] G. B. Gelmini, P. Gondolo and S. Sarkar, in preparation

[4] Fréjus Collab., Ch. Berger et al., Phys. Lett. B227 (1989) 489

[5] IMB Collab., J. M. Lo Secco et al., Phys. Lett. B188 (1987) 388

[6] R. Svoboda et al., Astrophys. J. 315 (1987) 420

[7] R. M. Baltrusaitis et al., Phys. Rev. D31 (1985) 2192

[8] A. V. Butkevich, A. B. Kaidalov, P. I. Krastev, A. V. Leonov-Vendrovski and I. M. Zheleznykh, Z. Phys. C39 (1988) 241

[9] C. Quigg, M. H. Reno and T. P. Walker, Phys. Rev. Lett. 57 (1986) 774

[10] D. W. McKay and J. P. Ralston, Phys. Lett. B167 (1986) 103

[11] S. Sarkar, private communication

STUDY OF MIXING OF SOLAR NEUTRINOS
WITH A 1000 TON ICARUS DETECTOR

M. CHENG, D. CLINE, J. PARK and M. ZHOU

Department of Physics, University of California at Los Angeles,
405 Hilgard Avenue, Los Angeles,
California 90024-1547

ABSTRACT

We review recent measurements of Solar neutrinos with the view towards the goals of 1000 ton Liquid Argon detector for the Gran Sasso laboratory (ICARUS). The technique allows the study of both elastic scattering on electrons and inverse β decay and is thus independent of calculations of solar neutrino flux. We show how the flux of neutrinos (ν_μ, ν_τ) that could arise from neutrino mixing can be inferred from these two measurements for a large range of parameters that are allowed by current measurements.

1. Introduction

The prospects for detection neutrino oscillation, and hence a finite neutrino mass, using Solar neutrinos remain high. Current experiments suggest the existence of neutrino oscillations but do not prove it. These experiments typically measure a single scattering reaction such as $\nu_e + e^- \Rightarrow \nu_e + e^-$ and thus must compare their results with calculations of the Solar neutrino flux. In contrast a new generation of Solar neutrino telescope is being constructed, and that can be used to measure two reactions simultaneously. This avoids the necessity to use the Solar neutrino flux calculations. In the ICARUS detector the reactions that are measured are

$$\nu_{e,\mu,\tau} + e^- \Rightarrow \nu_{e,\mu,\tau} + e^- \qquad (1)$$

and

$$\nu_e + Ar \Rightarrow e^- + K^* \Rightarrow e^- + K + \gamma . \qquad (2)$$

In this note we show how a 1000 ton ICARUS detector can be used to conclusively prove the existence of neutrino oscillations (whether vacuum or MSW induced) and to measure the critical parameters of the neutrino oscillations $[(\sin^2 2\theta$ and $(m_1^2 - m_2^2)]$.

2. Matter Oscillation

The neutrino states, $|\nu_e\rangle, |\nu_\mu\rangle, |\nu_\tau\rangle$, that are produced in weak-interaction decays are the flavor eigenstates. The flavor eigenstates are linear combinations of the mass eigenstates, $|\nu_1\rangle, |\nu_2\rangle, |\nu_3\rangle$ that diagonalize the free Hamiltonian. Since different mass eigenstates move with different speeds, what was initially a pure electron neutrino will become an admixture of the three flavor states (vacuum oscillation). However, vacuum oscillations have some difficulty to explain the solar neutrino problems because it need maximal mixing to achieve a minimum of the survival probability of 1/3. In addition, these maximal mixing angles are much larger than the known quark mixing angles.

Mikheyev and Smirnov[1] showed that under resonance condition electron neutrinos may convert into μ neutrinos on their way out through the sun. Wolfenstein[2] (1979) pointed out earlier that in matter the masses of neutrinos are changed as a result of the weak interaction. The neutral weak currents act equally all neutrino flavors. But ν_e will exchange W with the electrons in matter. Consequently, ν_e will have a different phase from the other neutrinos in its passage through the sun. This exchange adds the effective matter Hamiltonian to the vacuum oscillation one which leads to resonant neutrino oscillation, the MSW effect. At the resonant point, which depends on the density of matter, the state vector, which was originally in the direction of $|\nu_e\rangle$ turns to the direction $|\nu_\mu\rangle$. In the MSW effect, the mixing angles and mass differences can each vary by orders of magnitude and still remain within the parameter domain that solves the solar neutrino problem since no matter how small the vacuum mixing angle may be, there is always a density for which the neutrino will oscillate with maximum mixing. According to Wolfenstein the time development of vacuum and matter oscillation is described by the differential equation

$$
i \frac{d}{dt} \begin{pmatrix} \nu_e \\ \nu_\mu \end{pmatrix} = \begin{pmatrix} A & B \\ B & D \end{pmatrix} \begin{pmatrix} \nu_e \\ \nu_\mu \end{pmatrix} \tag{3}
$$

in the basis of eigenstates $|\nu_e\rangle$ and $|\nu_\mu\rangle$. The elements of the Hamiltonian matrix are

$$
\begin{aligned}
A &= \frac{m_1^2 c^2 + m_2^2 s^2}{2E} + \sqrt{2}\, G_F N_c \\
B &= \frac{\Delta m^2}{2E} \, cs \\
D &= \frac{m_1^2 s^2 + m_2^2 c^2}{2E}
\end{aligned} \tag{4}
$$

where m_1 and m_2 are the masses of neutrino mass eigenstates $|\nu_1\rangle$ and $|\nu_2\rangle$, respectively, and $\Delta m^2 = m_2^2 - m_1^2$ is measured in units of eV^2. The vacuum mixing angle θ_ν is defined as the angle between mass eigenstates and flavor eigenstate, and G_F is the Fermi constant, N_e is density of electron in the sun, and E is the neutrino energy measured in MeV.

When density of electrons is constant, the probability that a neutrino born as a ν_e remains a ν_e at distance L considering the MSW effect can be written as follows:

$$P(\nu_e, \nu_e; R) = 1 - \sin^2 2\theta_M \sin^2(\pi R/L_M) \tag{5}$$

and the oscillation parameters are obtained by diagonalizing the Eq (3)

$$\sin^2 2\theta_M = \frac{\sin^2 2\theta}{\sin^2 2\theta + (L_\nu/L_e - \cos 2\theta)^2} \tag{6}$$

$$L_M = \frac{L_\nu}{[\sin^2 2\theta + (L_\nu/L_e - \cos 2\theta)^2]^{1/2}} \tag{7}$$

$$L_\nu = \frac{4\pi\rho}{m_1^2 - m_2^2} \tag{8}$$

$$L_e = \frac{2\pi}{2^{1/2}G_F N_c} . \tag{9}$$

There are two regions where the MSW mixing is almost negligible.
(A) If the vacuum oscillation length (L_ν) is much greater than the electron neutrino interaction length (L_e), in other words, electron density is high, then the mixing angle in matter is much smaller than that in vacuum and oscillations are suppressed.
(B) If the vacuum oscillation length (L_ν) is much less than the neutrino electron interaction length (L_e), in other words, electron density is low, then the mixing angle in matter is reduced to vacuum oscillation.

For the MSW resonance case, the electron density ρ has to be smaller than the central density of the sun so that the following relation should be satisfied.

$$L_\nu/L_e = \cos 2\theta_\nu . \tag{10}$$

There are two approximations of allowed regions on the MSW oscillation (adiabatic approximation, nonadiabatic approximation).
(A) The adiabatic approximation holds when the width of the resonance layer is much larger than the oscillation length (L_M) in matter. It means that the eigenvectors of the equation of motion (Eq. 3) change slowly as the neutrino travels through the sun.
(B) In the nonadiabatic regime, situation is reversed. The width of the resonance layer should be much smaller than L_M. The eigenvectors change rapidly in a small region.

We proceed to use these relationship for the simulation of the detection of neutrino oscillations by a 1000 Ton ICARUS detector.

3. Status of current measurement of solar neutrino

Recently, three experiments are running to detect the solar neutrino. They are the ^{37}Cl experiment by J. Davis, et al., water cherenkov experiment by Kamiokanda II and the ^{71}Ga experiment by SAGE.

With its twenty year operating data, ^{37}Cl solar neutrino experiment gives out the solar neutrino capture rate 2.1 ± 0.3 SNU (solar neutrino unit, 1 SNU $= 10^{-36}$/(target atom) sec), which is less than 1/3 of the expected capture rate of SSM (standard solar model).[3] This is so-called solar neutrino puzzle. The experiment also suggested the possible observation of neutrino events associated with large solar flares.[4] A possible explanation is the large magnetic moment of electron neutrino. Unfortunately, this experiment cannot provide further information about solar neutrino such as the neutrino spectrum and day-night effect of the earth by the limit of detecting principle.

Kamiokanda II, a real time water Cherenkov detector which can give out direction and energy information of neutrino-induced electron, has run since 1987. The detector has an energy threshold of 7 MeV that is in the energy range of ^{8}B neutrinos. The recent measurement of the solar neutrino flux (three year statistics) relative to the SSM flux by Bahcall and Ulrich[5] is $0.46\pm0.05(\text{stat})\pm 0.06(\text{syst})$.[6] Taking the advantage of its recoil electron energy measurement, Kamiokanda II claimed that the adiabatic solution with $\Delta m^2 = 1.3{}^*10^{-4}(\text{ev}^2)$ and $6.3{}^*10^{-4} < \sin^2(2\theta) < 2.2{}^*10^{-2}$ is disfavoured at 90% C.L. if the difference of real data and the SSM expected result is explained by the MSW effect. On the other hand, the measurement[7] of neutrino event ($E_\nu > 50\text{MeV}$) in correlated with the great solar flare activity in March 1989 showed no evidence of the excess of the rate, which is opposed to the observation of the ^{37}Cl experiment.

The SAGE experiment claimed zero counts with its short running period and set an upper limit of $(\text{Exp.}/\text{SSM} = 0.0\pm0.4)$[8] for low energy neutrino flux. Combining the results of solar neutrino flux from above three experiments, T. K. Guo et al.[9] showed the adiabatic solution is excluded with more than 99% C.L.

4. Simulation for the ICARUS 1000 Ton detector[10,11]

We apply numerical methods to solve the equation of motion with two neutrino flavors. Electron density in the Sun is approximated with an exponential function of distance from the core of the Sun. We calculated the surviving probability of ν_e at the surface of the Sun using fourth order Runge-Kuta method. This value is used to generate the ν_e spectrum. A Monte Carlo program is written to simulate the recoil electron. At 5 Mev cutoff energy all the elastic scattering events are confined to less than 20^0. This is very important in the event selection since solar neutrino should point toward the Sun.

Table (I) Rates for Events in ICARUS (1000 ton) for $E_e > 5$ MeV per Year.

Solution	Parameters	Absorption	Elastic	Sum
	No Oscillation	832	918	1740
Case I	$\sin^2 2\theta = 10^{-3}$	136	564	700
	$\Delta m^2 = 10^{-4} eV^2$			
Case II	$\sin^2 2\theta = 1 \times 10^{-0.5}$	276	442	718
	$\Delta m^2 = 5 \times 10^{-5} eV^2$			
Case III	$\sin^2 2\theta = 1 \times 10^{-1.5}$	526	546	1072
	$\Delta m^2 = 1.1 \times 10^{-6} eV^2$			

Three pairs of parameters are used to typify the wide range of solutions to the solar neutrino puzzle.[12] In figure 1 through 3 we show the spectrum of ^8B ν_e as well as the recoil electron. In each case we also show the event rate for absorption and elastic scattering, (Figure 4a–c). In case I where high energy neutrinos are converted, we see a drastic change in absorption rate. This is because absorption channel is sensitive only to neutrinos above 5.85 MeV threshold. In figure 4d we show the ratio of absorption event and elastic scattering event for each of these cases. In table I we show that the event rate for the SSM prediction and three cases we studied with a 1000 ton ICARUS detector. In case I and II we see that ratio are drastically different from SSM prediction. It is obvious that different solutions can be distinguished if both interaction modes are used.[12]

5. Conclusion

We demonstrate in this note that a exclusive conclusion can be drawn for solar neutrino problem if two independent modes are used to measure the neutrino events. Until now only one mode is employed in experiments currently under operation, and the result is dependent on neutrino flux calculation. Instead a new generation of neutrino detector is under development which may take the advantage of using both absorption and elastic scattering modes and therefore free of any flux calculation discrepancies.

Acknowledgement

We thank L. Fortson and W. Hong for helpful discussions.

References

(1) S. P. Mikheyev and A. Yu. Smirnov, Nuovo. Cim **9C** (1986) 17.
(2) L. Wolfenstein, Phys. Rev. **D20** (1979) 2634.
(3) Davis, R., Jr. et al., in Neutrino'88, eds. J. Schneps et al. (1988) 518.
(4) Davis, R., Jr. 1986, in proc. seventh workshop on Grand Unification/ICOBAN'86, Toyama, Japan, ed. J. Arafune (Singapore: World Scientific), p. 237. Davis, R., Jr. and Evans, J. C. 1973, in Proc. 13th international Cosmic Ray Conference, Denver (university of Denver), Vol. 3, p. 2001.
(5) J. N. Bahcall and R. K. Ulrich, Rev. Mod. Phys., **60**, 297 (1988).
(6) "Constraints on neutrino Oscillation parameters from the Kamiokanda II Solar Neutrino Data", ICRR-Report-214-90-7
(7) "Search for Neutrino Events in the Kamiokanda II Detector in Correlation with the Solar flare Activity in March 1989", ICR Report-201-89-18
(8) V. N. Gavrin et al., Talk presented at 25th International Conference on High Energy Physics (Singapore) and at Neutrino'90 (CERN).
(9) Kuo, T. K. and Pantaleone, James, UCR-HEP-T-60 preprint.
(10) ICARUS proposal, CERN, Harvard, Milano, Padova, Tokyo, Wisconsin Collaboration, INFN/AE-85/7 Frascati (1985).
(11) ICARUS I; An Optimized, Real Time Detector of Solar Neutrinos, LNF-89/005(R).
(12) J. N. Bahcall, M. Baldo-Ceolin, D. B. Cline and C. Rubbia, Physics Lett. B178 (1986) 324.

Figure Captions

(1a) ^{8}B neutrino spectrum from the Sun. $\Delta m^2 = 10^{-4}eV^2$, $\sin^2\theta = 10^{-3}$.

(1b) Recoil electron energy spectrum with v.e. scattering at 5 Mev cutoff.

(1c) Recoil electron angular spectrum with v.e. scattering at 5 Mev cutoff.

(1d) Recoil electron energy spectrum with absorption at 5 Mev cutoff.

(2a) ^{8}B neutrino spectrum from the Sun. $\Delta m^2 = 5 \times 10^{-5}eV^2$, $\sin^2 2\theta = 1 \times 10^{-0.5}$.

(2b) Recoil electron energy spectrum with v.e. scattering at 5 Mev cutoff.

(2c) Recoil electron angular spectrum with v.e. scattering at 5 Mev cutoff.

(2d) Recoil electron energy spectrum with absorption at 5 Mev cutoff.

(3a) ^{8}B neutrino spectrum from the Sun. $\Delta m^2 = 1.1 \times 10^{-6}eV^2$, $\sin^2 2\theta = 1 \times 10^{-1.5}$.

(3b) Recoil electron energy spectrum with v.e. scattering at 5 Mev cutoff.

(3c) Recoil electron angular spectrum with v.e. scattering at 5 Mev cutoff.

(3d) Recoil electron energy spectrum with absorption at 5 Mev cutoff.

(4a) Event rate per year as a function of cutoff energy of recoil electron at $\Delta m^2 = 10^{-4}eV^2, \sin^2 2\theta = 10^{-3}$.

(4b) Event rate per year as a function of cutoff energy of recoil electron at $\Delta m^2 = 5 \times 10^{-5}eV^2$, $\sin^2 2\theta = 1 \times 10^{-0.5}$.

(4c) Event rate per year as a function of cutoff energy of recoil electron at $\Delta m^2 = 1.1 \times 10^{-6}eV^2$, $\sin^2 2\theta = 1 \times 10^{-0.5}$.

(4d) Ratio of absorption and elastic scattering event as a function of cutoff energy. Case I is $\Delta m^2 = 10^{-4}eV^2$, $\sin^2 2\theta = 10^{-3}$. Case II is $\Delta m^2 = 5 \times 10^{-5}eV^2$, $\sin^2 2\theta = 1 \times 10^{-0.5}$. Case III is $\Delta m^2 = 1.1 \times 10^{-6}eV^2$, $\sin^2 2\theta = 1 \times 10^{-1.5}$.

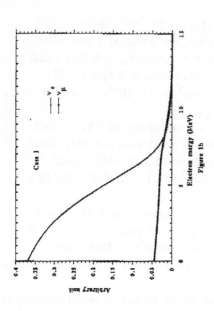

Figure 1a

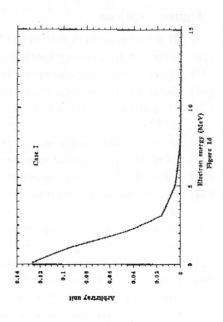

Figure 1b

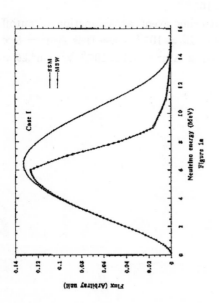

Figure 1c

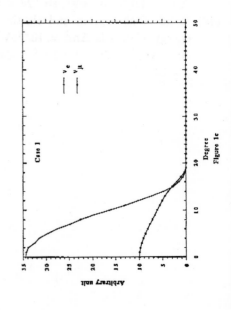

Figure 1d

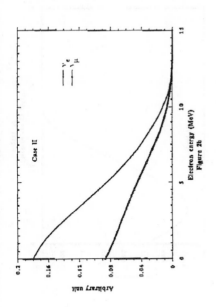

Case II

Flux (Arbitrary unit)

Neutrino energy (MeV)

Figure 2a

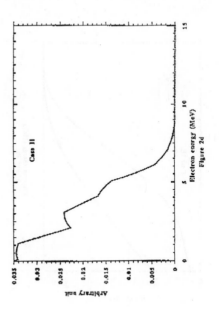

Case II

Arbitrary unit

Electron energy (MeV)

Figure 2b

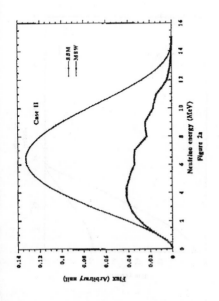

Case II

Arbitrary unit

Degree

Figure2c

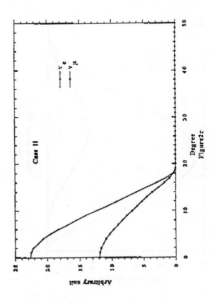

Case II

Arbitrary unit

Electron energy (MeV)

Figure 2d

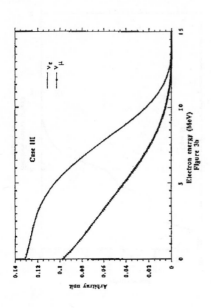

Figure 3a

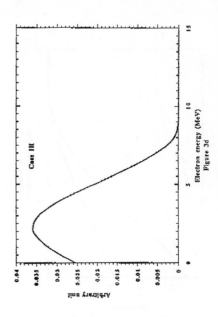

Figure 3b

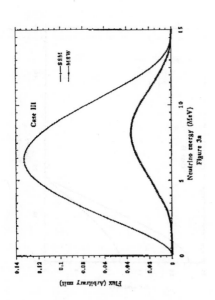

Figure 3c

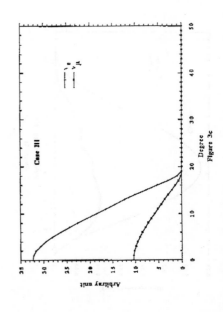

Figure 3d

DIFFERENCE CROSSSECTION FOR ELASTIC SCATTERING

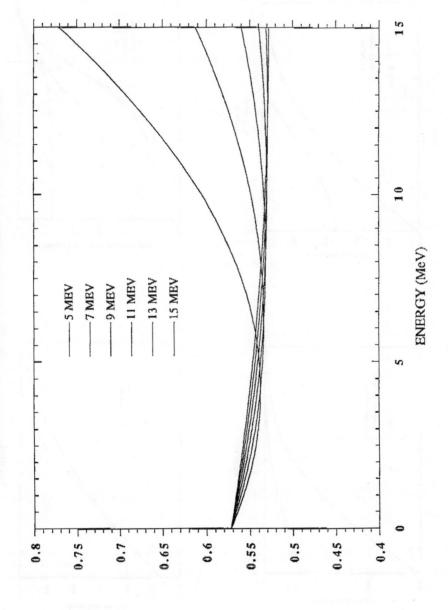

404

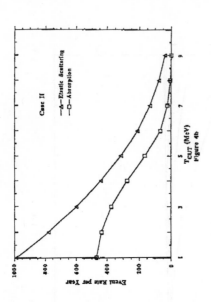

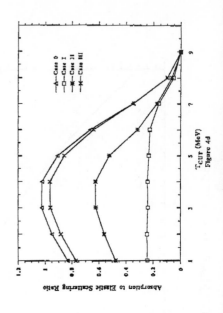

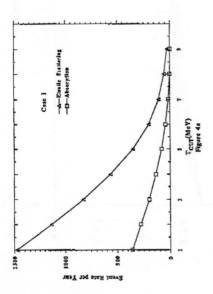

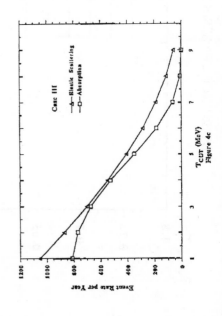

Light Neutralino Dark Matter is Still OK

Kim Griest

Center for Particle Astrophysics, University of California, Berkeley, CA 94720

Abstract

We dispute recent claims that neutralinos lighter than 20-30 GeV have been ruled out as dark matter by the recent LEP accelerator experiments.

The recent results from LEP[1] have had an important impact on those of us interested in particle dark matter (DM) candidates. The LEP experiments had (and still have) great potential to discover these neutral and very weakly interacting particles, and their failure to find them has given us our strongest limits yet. In brief, LEP tells us that new particles, with masses under 45 GeV which couple with neutrino strength to the Z boson, do not exist. So popular particle dark matter candidates such a new massive Majorana or Dirac neutrino have been ruled out.[2] Other candidates, such as the sneutrino and several types of cosmion, while previously under pressure from various experimental results, have now also been firmly disallowed by the LEP results.[3,4,5] The spectacular LEP results have also led to claims that one of the leading particle dark matter candidates, the neutralino, must be heavier than 20-30 GeV,[6,3,7,8] and even to claims by some that the neutralino is no longer a viable dark matter candidate. I would like to take exception to these last two claims.

In minimal supersymmetry (SUSY), the neutralinos are linear combinations of the SUSY partners of the photon, Z, and two neutral Higgs bosons. (Alternatively we can use W^3-ino and B-ino instead of the photino and Z-ino as basis states). So there are four neutralinos, the lightest of which $(\tilde{\chi}_1)$ is typically stable and constitutes the dark matter candidate. The masses and couplings of these four particles are determined, in the minimal model, by 4 parameters: M_1, M_2, μ, and $\tan\beta$, where the first two parameters are soft SUSY breaking parameters, μ is the Higgsino mass, and $\tan\beta = v_2/v_1$ is the ratio of Higgs vacuum expectation values. In addition we have as free parameters, the squark and sfermion masses ($\widetilde{M}_{sq}$ and $\widetilde{M}_{sf}$), the gluino mass ($\widetilde{M}_g$), and one Higgs mass (m_{H2}). (There are actually five Higgs bosons in these models, but all are determined by the above parameters.)

We can check the viability of neutralino dark matter by exploring the above parameter space. To do this we pick values for a set of parameters and calculate the neutralino masses and couplings. We then find the neutralino relic abundance Ωh^2, which is calculated[9] from the neutralino annihilation cross section $\sigma(\tilde{\chi}_1 \tilde{\chi}_1 \to \text{all})$. To be the dark matter we require $0.025 \leq \Omega h^2 \leq 1$, where $h \gtrsim .5$ parameterizes ignorance of the Hubble parameter. Finally, we calculate various cross sections and masses for which experimental constraints exist. In particular, we consider:

1. The chargino masses: $\tilde{m}_{\tilde{\chi}}^+ > \sim 45$ GeV (from LEP) and $\tilde{m}_{\tilde{\chi}}^+ > m_\chi$ (from consistency). (m_χ is the mass of $\tilde{\chi}_1$.)

2. The ASP cross section $\sigma_{ASP}(e^+e^- \to \tilde{\chi}_1 \gamma\gamma) < .03$ pb (from the ASP[10] experiment).

3. The squark and gluino masses: For simplicity consider all squarks degenerate and all sfermions degenerate. Then $\widetilde{M}_{sq} > \sim 100$ GeV (from CDF), and $\widetilde{M}_{sf} > \sim 60$ GeV (from ASP).

4. The width of the Z boson into neutralinos: $Z \to \tilde{\chi}_1 \tilde{\chi}_i$, where $\tilde{\chi}_i$, $i > 1$ are the more massive neutralino eigenstates. These limits (from LEP[1]) are very roughly approximated by $Z_{13}^2 - Z_{14}^2 < \sim .03$, $Z_{13}Z_{23} - Z_{14}Z_{24} < \sim .04$, with similar relations for the other $Z_{13}Z_{i3} - Z_{14}Z_{i4}$ couplings.[11] The Z_{ij} are the elements of the matrix which diagonalizes the neutralino mass matrix. The Z–neutralino couplings are given by the above factors.

5. The Higgs boson masses and mixings: $\tan\beta > \sim 1.3$ and $m_{H2} > \sim 25$ GeV. (from LEP).[1]

6. Other constraints from LEP such as the total Z width, and the Z width into invisible channels.[1]

Can all these constraints and conditions be met simultaneously? The answer is yes. How low can m_χ be and there still be a solution? Without even a thorough search I can find masses below 10 GeV. For example (with all masses in GeV) with $\mu = -14$, $M_2 = 10$, $M_1 = \frac{5}{3}\tan^2\theta_W M_2 \approx 5$, $\tan\beta = 1.5$, $m_{H2} = 35$, $\widetilde{M}_{sf} = 70$, and $\widetilde{M}_{sq} = 200$, we find $m_\chi = 6.1$ GeV and $\Omega h^2 = 0.3$. (This would give $\Omega = 1$ for $h = .55$.) So neutralino dark matter candidates with $m_\chi < 20\text{-}30$ GeV certainly exist. Is this true in only a tiny region of parameter space? In one sense yes, but in another no. Is is important to keep in mind; however, that this is not a well-posed question, since there is no meaningful measure on parameter space. But roughly, for $-50 < \mu < -10$ GeV and $10 < M_2 < 13$ GeV we find neutralino DM candidates with masses below 10 GeV. In evaluating the "size" of parameter space, it is important to remember that the constraint that the neutralinos make up the dark matter is very limiting. For example, for neutrino dark matter, the constraint $\Omega = 1$ determines the neutrino mass precisely; that is, the allowed parameter space is a set of measure 0. Also we are working on a small hypersurface of parameter space. The interesting region in (μ, M_2) space changes drastically as other parameters such as $\widetilde{M}_{sf}$, $\tan \beta$, or M_1 are varied. For example, by taking $M_1 = .15 M_2$, instead of $M_1 \approx .5 M_2$, the above region becomes roughly $-400 < \mu < -100$ GeV, and $15 < M_2 < 35$ GeV, a much "larger" area.

The inability of LEP to rule out light neutralino DM is not surprising. The neutralino states can be written

$$\tilde{\chi}_i = Z_{i1} \tilde{B} + Z_{i2} \tilde{W}_3 + Z_{i3} \tilde{H}_1 + Z_{i4} \tilde{H}_2. \tag{1}$$

The decay which LEP measures,[1] $Z^\circ \to \tilde{\chi}_1 \tilde{\chi}_i$ depends only upon the combination $Z_{13} Z_{i3} - Z_{14} Z_{i4}$ and so LEP can only constrain this combination. In particular, LEP cannot constrain the $\tilde{B}$ or $\tilde{W}_3$ (i.e. the photino or Z-ino) components. So LEP cannot touch two of the four components which make up the neutralinos. The ASP experiment does constrain the photino component, and in fact the combination of LEP + ASP is very powerful. However, ASP says nothing if the selectron is heavier than around 60 GeV.

So why do my conclusions differ from those of other workers[7,8,3,6] who claim neutralinos lighter than 20-30 GeV are ruled out? The main difference is in the meaning of a "minimal" SUSY model. When I talk about the "minimal" SUSY model, I mean a supersymmetric extension of the standard model with minimal particle content and arbitrary soft-SUSY breaking parameters. However, "minimal" is also sometimes used to mean minimal particle content *plus* some grand unification and renormalization group assumptions. The relations

$$\begin{aligned} M_1 &\approx \frac{5}{3} \tan^2 \theta_W M_2 &\text{(a)} \\ M_3 &\approx \frac{\alpha_s}{\alpha} \sin^2 \theta_W M_2 &\text{(b)} \end{aligned} \tag{2}$$

are often imposed. Here M_3 is basically the "gluino" mass, so the relations become $\widetilde{M}_g \approx 4 M_2$, and $M_1 \approx \frac{1}{2} M_2$. These relations are found using the renormalization group, where M_1, M_2, and M_3 are all set equal at the unification scale. In fact, to reduce the number of free parameters, I used Eq. (2)(a) in the example above. However, I did not use Eq. (2)(b). Eq. (2)(b) is crucial. If this relation is imposed, the CDF limit on the gluino mass ($\widetilde{M}_g > \sim 100$ GeV) is turned into a limit on M_2. The parameter M_2 determines the light photino eigenstate mass. It is this relation, in conjunction with the LEP limits which eliminate light neutralinos. Note that the CDF limit concerns only strong interaction physics, while LEP is probing only weak interactions. Thus Eq. (2)(b) is a powerful assumption.

Is Eq. (2)(b) necessary? No! It has been know for a long time, that while the GUT scale relation $M_1 = M_2 = M_3$ is elegant and simple, it is not necessary. As shown in reference 15, one simple way to have this relation not hold, even within the framework of a minimal Supergravity model, is to slightly complicate the kinetic terms of the super Lagrangian. For example, the standard Lagrangian contains kinetic terms such as

$$\mathcal{L}/e \supset -\frac{1}{4}f_{ab}\bar{\lambda}^a D^\mu \gamma_\mu \lambda_b - \frac{1}{4}f_{ab}F^a_{\mu\nu}F^{b\mu\nu} + \cdots, \tag{3}$$

where f_{ab} is a function of the chiral superfields which transforms like a product of two adjoint representations. The function f_{ab} is normally taken to be the Kronecker delta function δ_{ab}, in which case Eq. (2) results. When f_{ab} is allowed to have off-diagonal terms, new terms must be added to the Lagrangian, and as a result the normal GUT relation $M_1 = M_2 = M_3$ does not obtain. The values of M_1, M_2, and M_3 become arbitrary.[15] Thus even within the framework of a standard GUT supergravity theory, Eqs. (2)(a) and (2)(b) are far from being necessary. With Eq. (2) gone, the CDF limit has no effect on the neutralino search and light neutralino dark matter is allowed. If, in addition, one relaxes Eq. (2)(a), which I imposed only to simplify the search through parameter space, one has much "more" parameter space for light neutralino DM. In this case, which has not been carefully considered in the literature, the standard μ vs M_2 mass and mixing contours are greatly changed. The resulting changes this makes on neutralino dark matter are being studied by Leszek Roszkowski and myself.[12]

Some philosophy: When exploring parameter space in search of an interesting phenomena (such as a light DM candidate), one imposes constraints such as Eq. (2)(a), in order to speed and simplify the search. If one finds an interesting result, the result will not go away when the constraint is relaxed; one will just have a larger parameter space which contains the interesting result. However, attempting to use experimental results to rule out a particular model is quite a different thing. If one rules out the hypersurface in parameter space which lies along a constraint equation, one can not correctly claim to have ruled out the model. In this case, one *must* allow all the free parameters to vary over their natural ranges, before claiming to rule out the model. So, in my opinion, light neutralino DM is still viable within the minimal SUSY model, LEP results notwithstanding.

Finally, three points: In my opinion the biggest threat to minimal SUSY dark matter does not come from the neutralino searches. As mentioned, the Z does not couple to half of the neutralino basis states. The best chance of ruling out (or discovering) the neutralino comes from the Higgs searches. If the top quark has a mass under ~ 150 GeV,[13] then the minimal SUSY models have a neutral Higgs boson under ~ 90 GeV. LEP or LEP 200 should be able to discover or rule out such a Higgs boson. Already LEP has searched much of this parameter space. If the top quark is very heavy; however, the minimum mass of the Higgs can be larger.[13] Note that there is no such minimum mass for the neutralino, and a neutralino substantially heavier than the Z, makes a fine dark matter candidate and is out of the reach of currently planned accelerators.[9,14] The possibility of very massive neutralino dark matter exists whether or not the GUT relations Eq. (2)are imposed, and so popular claims that LEP has ruled out WIMP dark matter are patently false.

Next, I think it is important to note that while light neutralinos within the minimal model still make excellent DM candidates, this model is becoming increasingly constrained. However, by moving just slightly beyond the minimal model, many of the constraints disappear. For example, by extending the low-energy gauge group to include an extra U(1), as in the popular E6 models, one loses many constraints, since in this case the neutralino is a linear combination of five states. While not "minimal" in the senses I used earlier, these are attractive models which contain a light neutralino dark matter candidate.[16,17] In fact, they are in many respects just as attractive as the well-studied "minimal" model, especially since we really have very little information concerning the ultimate unification gauge group (or even whether there is grand unification via a simple gauge group). So even if the minimal model is ruled-out in the future, supersymmetric particles will still make excellent dark matter candidates, with the neutralino or its counterpart probably playing a leading role.

Third, does LEP tell us anything about the detectability of neutralino DM, if, in fact, it is the DM? In direct detection experiments, one typically one looks for elastic $\tilde{\chi}_1$–nucleus scattering in a kilogram size crystal. If the neutralino is under 45 GeV, LEP tells us it couples at most very weakly to the Z. Thus we do not expect Z exchange to play much of a role, unless the dark matter is heavier than 45 GeV. The CDF limits on squark masses imply that squark exchange will also be small. While the sfermions can play a role in determining the relic abundance, sfermions exchange typically will not contribute to elastic scattering off nuclei. So one expects the elastic interaction to be dominated by the exchange of the two neutral Higgs bosons. Once again, it is the Higgs sector which is crucial for light WIMP dark matter. Finally note that several of the points made in this talk have also been made by Drees and Tata.[17]

REFERENCES

1. For example, F. Dydak, plenary talk at 25th Intl. Conf. on High Energy Physics, Singapore, Aug. 1990; D. Decamp, *et al.,Phys. Lett.* **244B,** 541 (1990); *ibid.* **235B,** 399 (1990); *ibid.* **237B,** 291 (1990).

2. K. Griest and J. Silk, *Nature,* **343,** 26 (1990).

3. L. M. Krauss, *Phys. Rev. Lett.* **64,** 999 (1990).

4. G. F. Guidice, G. Ridolfi, and E. Roulet, *Phys. Lett.* **211B,**370, 1988.

5. S. Raby and G. West, *Phys. Lett.* **194B,** 194, 1987.

6. L. M. Krauss, talk at *Z* Symposium, Madison, 1990.

7. J. Ellis, D. V. Nanopoulos, L. Roszkowski, and D. N. Schramm, *Phys. Lett.* **245B,** 251, 1990.

8. L. Roszkowski, *Phys. Lett.* **252B,** 471, (1990).

9. K. Griest, *Phys. Rev.* **D38,** 2357 (1988); K. Griest, M. Kamionkowski, and M. S. Turner, *Phys. Rev.* **D41,** 3565 (1990).

10. C. Hearty, *et al., Phys. Rev. Lett.* **58,** 1711 (1987).

11. We thank Leszek Roszkowski for pointing out that the 3rd and 4th lightest neutralinos can give an important contribution to the *Z* width and change the allowed parameter space.

12. K. Griest and L. Roszkowski, in progress.

13. Y. Okada, M. Yamaguchi, and T. Yanagida, Tokyo University preprint TU-363 (1990).

14. K. A. Olive and M. Srednicki, *Phys. Lett.* **230B,** 78 (1989).

15. J. Ellis, K. Enqvist, D. V. Nanopoulos, and K. Tamvakis, *Phys. Lett.* **155B,** 381, 1985; M. Drees, *Phys. Lett.* **158B,** 409, 1985; M. Drees, *Phys. Rev.* **D33,** 1486, 1986.

16. R. Flores, K.A. Olive, and D. Thomas, *Phys. Lett.* **245B,** 509, 1990.

17. M. Drees and X. Tata, CERN-TH-5770/90 (1990).

VI. NEW DETECTORS FOR DARK MATTER AND OTHER NEW PARTICLES

Greenland '90: A First Step Toward Using the Polar Ice Cap as a Cherenkov Detector

S. W. Barwick[1], F. Halzen[2], D. Lowder[3], T. Miller[3], R. Morse[2], P.B. Price[3], and A. Westphal[3]

[1]Department of Physics, Univ. of California, Irvine, CA 92717
[2]Department of Physics, Univ. of Wisconsin, Madison, WI 53706
[3]Department of Physics, Univ. of California, Berkeley, CA 94720

Introduction

When Galileo first peered into the depths of space using his newly created optical telescope, he glimpsed a heavens that appeared far different than the one explored by the unaided eye. His observations, such as the moons of Jupiter and phases of Venus, were to revolutionize astronomy. Today, by looking within our Universe with telescopes designed to collect neutrinos instead of photons, we have the same opportunity as Galileo to discover entirely new objects which have no counterpart in the optical sky. Neutrinos, like photons, carry no charge and thus are not affected by magnetic fields which pervade our galaxy. However, one difference between neutrinos and photons is that high energy photons *cannot* escape from cosmic sites shielded by more than a few hundred grams of surrounding matter, whereas neutrinos can escape with ease. Therefore, the intensity and time distribution of high energy neutrinos can reveal the nature of the central engine powering these sources. It is important to understand the nature of these engines because they may be the cosmic accelerators of high energy ($> 10^{15}$eV) cosmic rays. Furthermore, a variety of models involving neutron stars or black holes in binary systems have suggested that these systems are likely to be powerful emitters of high energy neutrinos. A new class of detector is required to significantly constrain these models.

Enormous detectors with unprecedented sensitivity will be required to search for astrophysical sources of high energy neutrinos. A new idea has been introduced which suggests that optically transparent ice found deep under the Antarctic ice cap may be used in much the same way that the DUMAND facility uses ocean water as an integral part of a large volume particle detector. Photomultiplier tubes embedded within the clear ice would sense the Cherenkov radiation emitted by charged particles traversing the active volume of the detector. The feasibility and cost effectiveness critically depend on the optical transparency of the in-situ polar ice. Barwick and Halzen[1] have pointed out that if the *in situ* ice properties are similar to those measured in the laboratory, then an ice-sheet detector is very likely to be the most cost effective to reach the detector volumes required to search for astrophysical point sources of high energy neutrinos, diffuse sources from objects such as Active Galactic Nuclei, neutrino oscillations, or WIMP annihilation within the sun.

The experimental challenge of high energy neutrino astrophysics is to conceive detectors of large sensitivity in an environment shielded from the severe cosmic ray backgrounds at the surface of the earth. We have begun the AMANDA (Antarctic Muon And Neutrino Detector Array) project to test the feasibility of using polar ice as a Cherenkov detector. We report on the initial investigation of the *in situ* transparency of Greenland ice. These tests gave the encouraging result that the mean optical attenuation length is consistent with laboratory measurements.

The Polar Ice Coring Office (PICO) has implemented a number of different technologies to drill boreholes in the ice. One relatively inexpensive method of drilling (compared to the alternatives) uses hot water to melt through the ice. Surface snow is collected, heated to near boiling in a pressure tank, and

414

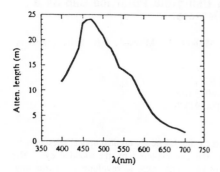

Figure 1

distributed via a high pressure hose. The hot water melts the ice within a cylinder whose radius is determined by the rate of descent. Numerous holes of modest diameter (6"-7" diameter) have been drilled to depths in excess of one kilometer and a large diameter hole (14" diameter) has been drilled to 500m.[2] After a borehole is drilled, a string of photomultiplier tubes is lowered to depth.

Gow and his colleagues[3] have measured the distribution of bubble diameters and bubble density as a function of depth in Antarctic ice. At depths between 800-1100 meters, trapped air is forced into the ice lattice forming nitrogen and oxygen hydrates (molecules of $N_2O_2 * 6H_2O$). Approximately 0.06% of the ice changes from the usual hexagonal structure into a cubic crystal, called a clathrate hydrate, to accommodate the hydrate molecules. The remaining ice retains its structure and becomes bubble-free.[4] The optical attenuation length, shown in figure 1,

has been measured in the laboratory by Grenfell and Perovich[5] who found that it is comparable to the measured attenuation length of distilled water. Until last summer, when we tested Greenland ice (GISP site) at a depth of 200m, there was no information regarding *in situ* ice. However, the Greenland tests demonstrated that the attenuation length of shallow ice is similar to that obtained in laboratory measurements .

Why Antarctica? Physics advantages

It is no secret that Antarctica, located on the far end of the globe, has a rather unforgiving environment. Experimental campaigns must be carefully planned to accommodate the formidable logistical requirements. We consider an Antarctic-based experiment because the AMANDA concept has several compelling science advantages:

(i) instrumenting polar ice with photomultiplier tubes allows the experiment to follow science. The detector can be gradually increased in resolution or (alternatively) expanded in coverage over the essentially unlimited expanse of the stable icefields;

(ii) ice is a quiet, sterile medium. No bioluminescence is expected, but this phenomenon will be investigated in the initial series of optical studies. Studies of Antarctic ice cores have shown that the concentration of radioactive β-emitters such a ^{40}K is nearly four orders of magnitude below concentrations in seawater. In addition to the lack of sources of background light, intrinsic dark noise is reduced by a factor of 50-100 in the constant -55 C temperatures. Since the background rates are small, extremely simple trigger schemes can be implemented with off-the-shelf, inexpensive electronics;

(iii) the ice is expected to be transparent to UV radiation (the attenuation length of lab ice is 5 meters at a wavelength of 300nm); it remains to be seen if this will be useful in practice;

(iv) a cosmic source is observed *continuously* and remains at fixed angle, making the subtraction of comic ray backgrounds extremely simple. Although only the northern half of the celestial sphere can be studied from a detector placed at the South Pole, the best studied candidate sources of neutrinos are located in the northern hemisphere. The unique opportunity provided by the uninterrupted observation of sources should be emphasized because they are expected to be impulsive and episodic at TeV energy scales;

(v) A large fraction of the detector electronics can be located at the surface.

(vi) The existing Bartol-Leeds South Pole air shower array (SPASE[6]) can be used to determine the absolute pointing accuracy of AMANDA by looking at events which simultaneously trigger both detectors (although the hemispherical PMTs are pointed down, they have some sensitivity to down-going muons).

(vii) Most new air shower arrays are located in the northern hemisphere (CASA, CYGNUS II, AKENO, etc.). Thus, AMANDA can monitor the neutrino signal and air shower arrays can monitor gamma emission from any northern source *at the same time* . This capability is especially important for sporadic emission since short bursts are difficult to confirm using air shower arrays alone (due to the fact that most arrays are separated by many hours in right ascension).

In addition, there are a number of practical advantages to considering South Pole Station as a base for AMANDA. It is manned throughout the year, providing food and lodging. As much as 10 kilowatts of power are available to experimenters which is adequate for AMANDA. There are communication links via satellite which are capable of transmitting ~5 MBytes of data per day. While these rates do not permit real time transmission of data, they do permit routine monitoring of the detector.

Greenland Campaign of August 1990: Observation of Muons Using Ice as a Particle Detector

Our initial study of ice transparency was conducted at the GISP II site in Greenland (72N 38W) during 13-16 August 1990. The test consisted of lowering a string of three Thorn EMI (model 9870) photomultiplier tubes (PMTs), 5" in diameter, to a depth of ~217 meters which is about 100 meters below the firn ice layer. The three PMTs were arranged vertically, with a 1 meter separation between PMTs. The hemispherical windows of the PMTs were oriented facing upwards. High voltage was supplied from the surface and the PMT analog output was routed via coaxial cable (RG180) directly to electronics on the surface that were specially designed by Phillips Scientific (Cable-Aid restorer) to compensate for the attenuation and dispersion introduced by the cable. Using the Cable-Aid restorer, we observed only a small attenuation (less than 25%) in the amplitude of the PMT output which has a characteristic risetime of 10ns. Discriminator thresholds were set to 30 mV, resulting in sensitivity to signals greater than 1 photoelectron (p.e.) for the top PMT, 1 p.e. for the middle PMT, and 2 p.e. for the bottom PMT.

At a depth of 217 meters, the effective overburden is about 170 meters of water equivalent and the ice temperature at this depth is -31.4 C. Only muons (other shower particles are effectively removed in the first 10 meters of water equivalent).with an energy of 30 GeV or greater can penetrate to the string of photomultiplier tubes. The vertical flux at this depth is 1 muon/m^2/s/sr, decreasing roughly as cos$^2(\theta)$ for zenith angle θ. Each PMT with its associated electronics was set to produce a digital pulse when the PMT output exceeded a threshold of 30 mV. Noise rates varied between 10 kHz and 20 kHz. The event trigger required that all three PMTs produce a digital pulse in coincidence. The particle rates were measured for coincidence windows of 200ns, 100ns, 50ns, and 30ns. The observed coincidence rate stayed constant at 1.8 Hz for all windows which demonstrates that the events were cosmic ray muons. The random coincidence rate was also measured by inserting additional time delays into the signal paths of the PMTs to destroy the real coincidence window. The random coincidence rate was less than 15% of the observed event rate which provides additional evidence that muons were observed at a rate of (1.-0.15)(1.8 Hz) = 1.6 Hz. Lastly, the pulse height distribution from each PMT was much broader than would be expected if the distribution had been generated by background light or PMT dark noise.

The timing and single photon resolution of the PMTs were established by pulsing green LEDs mounted on each PMT. Attempts were made to optically isolate the PMTs from each other, although they did see each other through the reflective and diffusive walls of the borehole. Despite the larger than expected singles rates (a small amount of surface light leaked down the borehole), we were able to establish a muon signal using the triple coincidence trigger. From the measured rate of cosmic ray muons, we can determine the effective acceptance of the apparatus using the calculated cosmic ray flux and the modeled solid angle of the 3 PMTs. We account for the 1.6 Hz coincidence rate if we assume that the muons are observable within a radius of 1.0 m, centered on the PMT. This would give a sensitive volume in the shape of a cylinder, with a radius of 1 meter and a length of 2 meters (Fig 2.). The geometrical aperture of this configuration is 1.6 m²sr which is what is required to produce a 1.6 Hz coincidence rate. *This experiment explicitly shows that a PMT placed at a depth of ~200m in polar ice is capable of sensing muons to a radius of 1 meter* Monte Carlo studies (see next section) show that if all three PMTs were indeed operated at single photoelectron sensitivity, then the effective radius would increase from 1.0m to 1.5m.

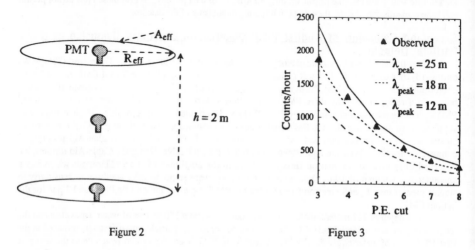

Figure 2   Figure 3

Interpretation of results: photon diffusion by random walk in Greenland ice

At a depth of 200m, ice cores retrieved from the hole are milky and translucent due to a uniform distribution of bubbles. Davis[7] has calculated the intensity distribution of light scattered by air bubbles in water, which is a good approximation to spherical air bubbles trapped in ice. Monte Carlo studies of photon propagation were used to generate radial and angular distributions as a function of the number of scatters, N. For the case where N>20, a numerical fit to radial distribution gives $d=(1 + 3N^{1/2})\lambda$, where d is the radial distance, and λ is the mean free path between scatters = $1/(\sigma\rho)$, where σ is the cross-sectional area of a bubble and ρ is the number density of bubbles. Analysis of cores taken previously from this depth in Greenland indicate that there are 200 bubbles/cm³ of mean radius 0.017cm, giving λ ~5.5 cm. These results were incorporated into a second Monte Carlo program to predict muon rates and pulse height distributions for each PMT. The program takes into account the hemispherical geometry of the PMT, the small decrease in PMT gain at low temperatures (~15%), photoelectron collection efficiency and photon quantum efficiency. The cosmic ray muon flux and angular distribution were based on the work of Miyake[8]. The frequency-dependent attenuation length was scaled as a fraction of the laboratory value. Figure 3 compares the predicted muon rate with the observed rate as a function of the number of photoelectrons collected by each PMT. The upper curve assumes that the attenuation length is identical to the lab values which have a maximum attenuation length of 25 meters. The lower dashed line assumes that the maximum attenuation length is 12 meters. Good agreement between the predicted flux and the observed flux is achieved only if the attenuation of the ice between the wavelengths of 300-700 nm is greater than 80% of the laboratory values.

This measurement established the technique of detecting cosmic ray muons via Cherekov radiation emitted in ice. The attenuation length of the Greenland ice is consistent with the value observed for polycrystalline ice grown in the laboratory. Since the values of ionic purity and particulate concentration of the ice at the GISP site are comparable to those of distilled water[9], it is not surprising that the optical properties of Greenland ice are comparable to those measured for distilled water[10]. Trodahl, Buckley, and Brown[11] have studied the diffusive transport of light through bubbly *sea* ice and conclude that the optical attenuation length at a wavelength of 500nm must be at least 10m. Grenfell[12] has found that the spectral albedo of both blue ice and sea ice below the eutectic point can be modelled accurately if the optical attenuation length was assumed to be the same as measured in the laboratory.

Future Plans

We are currently constructing a string of PMTs designed to evaluate the usefulness of deep Antarctic ice as an integral component in a large-volume particle detector. The proposed string of photomultiplier tubes will measure the optical transparency of ice located approximately one kilometer beneath the South Pole, a depth where bubbles trapped within the ice are known to be absent. Additional goals of the protostring include the measurement of up/down discrimination and timing resolution. The string will provide experience with gain and timing calibration techniques, reliability assessment, quality control, remote monitoring and data acquisition. These tests will be performed during the next Antarctic campaign (Nov. 91 through Jan 92).

[1] F. Halzen and S. W. Barwick, Proc. Summer Study on High Energy Physics(Snowmass ,CO,1990).

[2] J. Sonderup, director of PICO, private communication.

[3] A.J. Gow, J. Geophy. Res. 76(1971)2533.
   A.J. Gow and T. Williamson, Cold Regions Research and Engineering Laboratory, Research Report 339 (Oct. 1975).

[4] S. Miller, Science, 165 (1969)489.

[5] T. C. Grenfell and D.K. Perovich, J. Geophys. Res.86 (1981)7447.

[6] N.J. T. Smith, et al., Nucl. Inst. and Meth. A276 (1989)622.

[7] G. E. Davis, J. Opt. Soc. Amer. 45 (1955)572.

[8] S. Miyake, Proceedings of the 13th International Cosmic Ray Conference, Vol. 5, (Denver, CO,1973),3638.

[9] A. J. Gow, private communication.

[10] C. Bower, Indiana University Thesis (May, 1988).

[11] H. J. Trodahl, R.G. Buckley, and S. Brown, App. Opt., 26(1987)3005.

[12] T. C. Grenfell, J. Geophys. Res., 88 (1983)9723.

A Double Beta Decay Experiment Using Bolometric Detectors

A. Alessandrello[*], C. Brofferio, D. Camin, O. Cremonesi, E. Fiorini,
G. Gervasio, A. Giuliani, F. Passoni, M. Pavan, G. Pessina, E. Previtali, L. Zanotti

Istituto Nazionale di Fisica Nucleare and Dipartimento di Fisica dell'Università di Milano, Italy
* I.N.F.N. Laboratori Nazionali del Gran Sasso

Abstract

A new way to search for double beta decay and heavy dark matter candidates consists in using large mass (100 g or more) cryogenic particle detectors. The operation of such large detectors with good energy resolution requires the optimization of the system at very low temperatures, 20 - 30 mK typically, in order to achieve low heat capacity.

We report on several tests that have been performed on double beta candidate materials, with the aim of selecting the most promising ones to detect this rare process. Among them the best results have been obtained with a 6 g crystal of TeO_2. Preliminary background measurements with crystals of this compound will take place in the Gran Sasso Laboratory, with an especially built low–activity cryostat.

Introduction

In the search of double beta decay, the best experimental results have been so far obtained with the nucleus ^{76}Ge [1][2]. Thanks to the excellent characteristics of Germanium as a semiconductor, Germanium detectors are in fact almost ideal instruments for this search, having both excellent energy resolution and high detection efficiency, with the whole detector as active material. Theoretical and experimental interest on double beta decay, however, is not limited to ^{76}Ge; there exist in fact many nuclei, as ^{82}Se, ^{100}Mo, ^{130}Te and ^{136}Xe that could have detectable half lives, and perhaps, in some case, shorter than ^{76}Ge, owing to their higher transition energy or larger nuclear matrix elements.

In particular, double beta decay with 2–neutrino emission has been detected for the nuclei ^{76}Ge, ^{82}Se and ^{100}Mo, while extensive searches are in progress on ^{136}Xe[3]. For ^{130}Te and ^{128}Te there is geochemical evidence[3], but no direct measurement, of double beta decay. A high sensitivity experiment on the 0–neutrino decay of ^{130}Te could give stringent limits on the Majorana mass of neutrino and the structure of the weak interactions.

The bolometric technique provides a unique way of realizing homogeneous particle detectors from a large number of materials different from Germanium.

The technique

The principle of operation of bolometric particle detectors is straightforward: the absorber material which is traversed by the particle is placed in loose thermal contact with a low temperature heat sink, which is the mixing chamber of a dilution refrigerator (fig. 1). If one assumes that the energy released in the absorber is totally thermalized in a negligible time, then a temperature rise given by

$$\Delta T(t) = \frac{E}{C} e^{-t/\frac{C}{G}}$$

is observed, where G is the thermal conductance between the absorber and the heat sink, and C is the heat capacity of the absorber. In order to obtain a sizable temperature rise, the heat capacity must be kept at very low values. This is obtained by using a dielectric pure crystal absorber, for which C is proportional to $(T/\theta_d)^3$, where θ_d is the Debye temperature.

The ultimate energy resolution of such a detector should be given by the statistical thermal fluctuations in the absorber, that is[4]

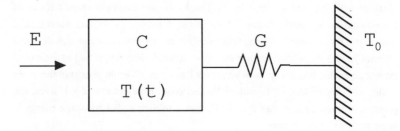

Fig. 1:   Thermal scheme of the detector

In our detectors,the temperature rise, which carries the information on the energy deposited in the absorber, is converted into a voltage signal by a thermistor glued to the absorber, and the signal is fed to a low–noise GaAs preamplifier which is located inside the cryostat, in thermal contact with the main helium bath at 4.2 K, in order to minimize the stray capacitance of the connections and the pick–up noise. The amplified signal is finally processed with a digital optimum filter and its amplitude is stored in a pulse height spectrum.

Since in this approximate picture the parameter of importance is the heat capacity of the detector, not its mass, one could easily operate large detectors, and compensate the mass increase with a relatively small decrease of the heat sink temperature T_0. In practice, below 50 mK several difficulties arise. In particular it is observed that the energy resolution performance of the detector is not limited by thermal fluctuations in the absorber, or the noise in the preamplifier, but by microphonic noise due to mechanical vibrations of the sample and the electrical wires.

This low–frequency noise contribution is difficult to filter, since thermal detectors are intrinsically slow (rise time of the pulse is of the order of a few milliseconds), and there is a large overlap between the power spectra of noise and signal. In the worst cases, the energy released by microphonic noise can be so large to prevent the crystal from reaching the lowest temperature.

Result of the tests

Several tests have been performed on materials of interest for double beta decay. all detectors have been irradiated with γ ray sources of ^{232}Th, ^{60}Co and ^{22}Na

Crystal	Mass	Resolution (FWHM)	T
Ge	11 g	13 keV	35 mK
Ge	190 g	220 keV	45 mK
Te	21 g	48 keV	44 mK
CaF$_2$	5.6 g	80 keV	47 mK
TeO$_2$	6 g	32 keV	26 mK

Energy resolution, as expected, was found to be limited only by microphonic noise and by stability of the working point. Not surprisingly the stability problem was worst in the case of TeO$_2$, which operated at the lowest temperature. In this case, during long runs, a deterioration of the resolution above the quoted value and proportional to the energy was observed.

The pure Tellurium sample provided satisfactory results at the temperature quoted above, but in that case the fragility of the material prevented the use of a holder stiff enough to keep the microphonic noise at an acceptable level.

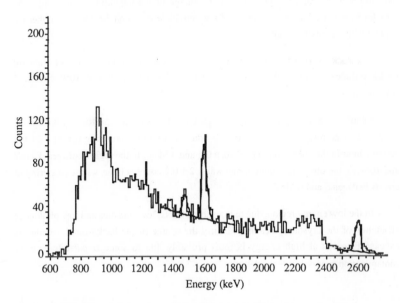

Fig 2. TeO$_2$ spectrum with ^{232}Th source and ^{40}K background peaks

Low Background Cryostat

A high sensitivity experiment on the double beta decay requires maximum suppression of the background contributes in the energy region where the events are expected, at the value of the transition energy for the 0-ν decay mode, and from 0 to this value for the 2-ν mode. This interval is fully within the range of the natural radioactivity background, thus special care must be devoted to the selection of the materials to be used in proximity of the detector, to the shielding against the surrounding γ radioactivity and, finally, to the intrinsic activity of the detector itself.

For this purpose, a special cryostat (fig. 3) has been built in the Laboratori Nazionali del Gran Sasso, with materials selected for their low radioactivity contamination.

All the components of the cryostat have been checked with γ-ray germanium spectrometers; the maximum intrinsic contamination allowed was 0.01 Bq/kg for all the major components, and 0.1 Bq/kg for all the materials present in an amount less than 1 kg.

The whole cryostat is surrounded by a Faraday cage, to ensure protection against electromagnetic disturbances, and by a 10 cm lead shield in all directions.

The cooling power of the cryostat is 1 mW at 100 mK, and the volume available for the sample is 21 dm^3. We plan to take advantage of the capability of cooling large samples by placing a 3 cm inner shield of low activity lead around the detector, in thermal contact with the mixing chamber.

The rock overburden of the Gran Sasso Laboratory (3500 hg cm^{-2} of standard rock) reduces muon and neutron fluxes to 3×10^{-8} and 4×10^{-6} cm^{-2} s^{-1} respectively[5][6].

Results of a very preliminary background measurement, performed with a Ge crystal of 11 grams, for a live time of 145 hours, gave a total counting rate of 2.2± 0.2 counts / hour in the energy region between 0.5 and 3 MeV. In the region between 3 MeV and 10 MeV the observed counting rate was 1.2 ± 0.2 counts / hour, with a clustering of counts between 5 and 6 MeV.

In the low energy region, due to threshold effects, low statistics and low photopeak efficiency of the crystal, we cannot identify the source of the background, while the excess of counts at high energy is most probably due to some α-particle emitter contamination near or inside the detector.

A more sensitive measurement, with the 6 g TeO₂ detector, will be the next step of our background analysis. From this test we expect to extract more informations on the γ background below 3 MeV and on the radiopurity of the detector itself.

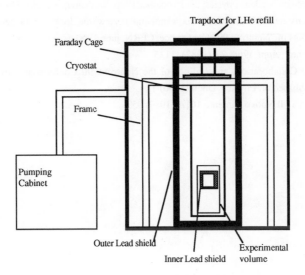

Fig 3 low activity cryostat

Conclusions

Results presented above show that the bolometric technique allows the realization of particle detectors of significant mass with a variety of substances different from germanium. In particular tellurium dioxide has been found to give excellent performances. This makes TeO₂ a first choice material for the search of double beta decay with this method.

The next stages to realize a double beta experiment on [130]Te will be the realization of a larger detector (50-100 g mass) and a more detailed analysis of the background and of the long time stability with the 6 g detector. The ultimate sensitivity of the experiment will be of course dependent on the final background and energy resolution obtained. Taking conservative values of 100 keV for the energy resolution and 10^{-4} counts keV^{-1} hour^{-1} for the background rate, a sensitivity of $\approx 10^{21}$ y in one year of measurement is expected for the 0-ν mode.

References

1] D. O. Caldwell et. al., *Nucl. Phys.* B, proceedings, suppl., 13, (1990), 547

2] I. V. Kirpitchnikov et al., F.T. Avignone III et al. Proceedings of the XIV EPS Conference on Nuclear Physics, 22-26 october 1990, Bratislavia, Czecho-Slovakia

3] see for instance M. K. Moe, "Experimental review on double beta decay", Proceedings of "Neutrino 90" Conference, CERN, june 1990

4] J. C. Mather , *Appl. Opt,.* 21 (1982), 1125

5] MACRO Collaboration, proceedings of the Neutrino '90 Conference, Geneva, 1990, published on Physica, D666 (1990) 1

6] P. Belli et al. Il Nuovo Cimento 101A (1989) 959

A Cosmic Axion Experiment*

C. Hagmann† P. Sikivie, N. S. Sullivan, and D. B. Tanner
University of Florida, Gainesville, Florida 32611

Abstract

We report on the status of a pilot experiment to search for cosmic axions. The detector utilizes the electromagnetic coupling of the axion by stimulating the decay of halo axions into photons inside a microwave cavity. A narrow range of axion masses m_a was explored by sweeping the cavity resonance frequency with a tuning rod. The negative result of the search puts a constraint on the axion-photon coupling assuming our halo is made of axions.

1 Introduction

The axion is a hypothetical particle proposed to solve the strong CP problem. A new global U(1) symmetry is introduced into the QCD Lagrangian along with a complex scalar field σ[1]. The symmetry is spontaneously broken at a scale f_a and the resulting Goldstone boson is the axion[2]. Massless at the classical level, the axion acquires a small mass due to instanton effects. The mass was originally thought to lie in the keV range, making it accessible to accelerator searches and beam dump experiments[3]. However, these particular searches were unsucessful, and a new kind of 'invisible' axion[4] was

---

*This Research was supported by Department of Energy Grant No. DOE-F605-86ER-40272.

†Present address: University of California, Department of Physics, Berkeley, CA 94720

proposed. This 'invisible' axion has a much smaller mass and is very weakly coupled. The energy scale f_a and the axion mass m_a are related by

$$m_a \simeq 0.6 \times 10^{-5} \, \text{eV} \left(\frac{10^{12} \text{GeV}}{f_a} \right) \tag{1}$$

and with axion couplings to ordinary matter proportional to m_a.

Invisible axions can have important implications for astrophysics and cosmology[5]. Thermally produced axions can freely stream out from the interior of stars and contribute to their radiative energy loss. In order not to cool stars excessively, the axion couplings need to be sufficiently small, and this constraint places an upper limit of a few eV on the axion mass. An even better limit comes from the recent observation of supernova SN1987a. Axion emission can dominate neutrino emission for a certain axion mass range. Yet the duration of the observed neutrino pulse implies that all of the energy released during core collapse was emitted in form of neutrinos. This excludes the axion mass range $10^{-3}\text{eV} < m_a < 1\text{eV}$.

Cosmological considerations provide a lower limit on m_a. This lower limit follows from the requirement that the axion cosmological energy density does not overclose the universe. There are two contributions to the axion cosmological energy density, the first from initial vacuum misalignment[6], the second from cosmic axion strings[7]. The second vanishes if there is inflation after the U(1) symmetry breaking transition. The initial misalignment mechanism produces $\Omega_a \simeq O(m_a/\mu\text{eV})^{-7/6}$ whether or not inflation occurs. The size of the contribution from cosmic axion strings have been subject of some controversy in the past. The problem is that it has proved impossible thus far to derive analytically the energy spectrum of axions radiated by decaying axion strings. However that spectrum has been recently obtained by means of computer simulations. The results support the view that cosmological axion strings contribute to the cosmological axion density an amount of the same order of magnitude as the contribution due to initial vacuum misalignment. In that case whether of not inflation occurs after the U(1) symmetry breaking transition, the critical axion mass which yielda $\Omega_a = 1$ is in the μeV range, Axions in this mass range provide a significant fraction of the energy density and would be all or part of the dark matter which is known to exist in our universe. In particular, our own galaxy has a dark halo of unknown composition which could conceivably be axionic. Our detector

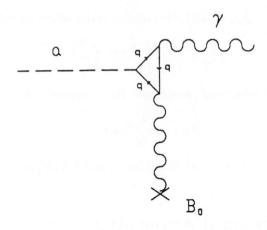

Fig.1: Feynman diagram of $a \rightarrow \gamma$ conversion in background field B_0.

uses the axion-photon coupling

$$L_{a\gamma\gamma} = g_{a\gamma\gamma}\, a\, \mathbf{E} \cdot \mathbf{B} \qquad (2)$$

where $g_{a\gamma\gamma}$ in the DFSZ model is given by

$$g_{a\gamma\gamma} = 1.38 \times 10^{-15}\mathrm{GeV}^{-1} \left(\frac{m_a}{10^{-5}\mathrm{eV}} \right) . \qquad (3)$$

This term allows for axion-photon transitions in a background field $B_0(E_0)$[8]. B_0 provides a virtual photon (see Fig.1) and the axion energy is carried away by a single real photon of frequency $\omega = m_a(1 + O(10^{-6}))$ where the broadening is due to the kinetic energy distribution of halo axions. The conversion efficiency is enhanced in a microwave cavity tuned to the axion mass. The power delivered this way is mode dependent and given by

$$P_{nlp} \simeq 4 \times 10^{-23}\,\mathrm{W} \left(\frac{V}{10\ell} \right) \left(\frac{B_0}{8\mathrm{T}} \right)^2 C_{nlp} \left(\frac{\rho_a}{0.3\,\mathrm{GeV/cm^3}} \right) \left(\frac{m_a}{2\pi(3\mathrm{GHz})} \right) \left(\frac{Q_{nlp}}{10^5} \right)$$

$$(4)$$

where the quantity Q_{nlp} is the loaded quality factor of the cavity and C_{nlp} is a form factor

$$C_{nlp} = \frac{\left(\int_V \mathbf{B}_0 \cdot \mathbf{E}_{nlp} \, d^3x\right)^2}{V \int_V B_0^2 \, \epsilon \, E_{nlp}^2 \, d^3x} \cdot \tag{5}$$

For TM modes of cylindrical cavities and $\vec{\mathbf{B}}_0 = B_0 \hat{z}$ one has

$$C_{nlp} = \frac{4}{\lambda_{0n}^2} \delta_{l0} \delta_{p0} \tag{6}$$

where λ_{0n} is the nth zero of the Bessel function $J_0(x)$ and $\epsilon=1$.

2 Experimental Apparatus

Our axion detector (see Fig.2) has three major components: a strong magnet, a tunable cavity, and a sensitive microwave receiver. The superconducting solenoid provides a field of about 8 T and is operated in persistent mode. Two cylindrical microwave cavities were built out of OFHC copper parts, which were electropolished and clamped together. To maximize the form factor C_{nlp}, we operated the cavities in their TM$_{010}$ mode. We have achieved unloaded Q's of around 1.6×10^5 at $T = 2$ K for a resonance frequency of 1.5 GHz and a volume of 7ℓ, which is a factor 1.5 lower than the expected value for copper in the anomalous skin depth regime. The cavities were coupled to coax cables using coupling loops or probes. The output port of the cavity was strongly coupled with $Q_{hole} = Q_{wall}/3$ to maximize the search rate of the experiment. The input port was very weakly coupled and used only for determining resonance frequency and Q in transmission. Tuning was accomplished by moving dielectric or metal rods inside the cavity. Our dielectric material was the low loss ceramic Magnesium Titanate/Magnesium Aluminate with $\epsilon \simeq 10$. Each cavity contained a large tuning rod out of metal or dielectric for coarse tuning and a smaller, stepper motor controlled, fine tuning rod. The large rods extended over the full cavity length to avoid longitudinal mode localization[9] and were moved sideways. Figures 3 and 4 show tuning curves for a metal and a dielectric tuning rod respectively. Also shown in the figures are the TE modes in the vicinity of the TM$_{010}$ mode. At a few frequencies, they cross the TM$_{010}$ mode, leading to mixing and holes

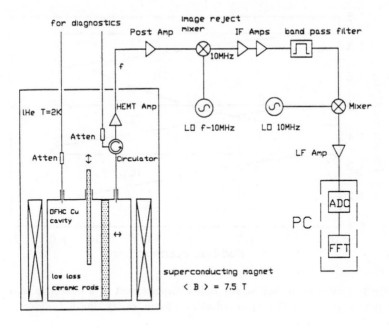

Fig.2: Block diagram of experimental setup.

in the tuned frequency band. For our present cavities this reduces the useful band by less than one percent.

We selected cryogenic HEMT preamplifiers[10] for the first amplification stage. They have excellent noise characteristics and are easy to use. Figure 5 shows noise temperature versus frequency for an L-band HEMT. A cryogenic circulator was inserted between cavity and preamp to improve impedance matching. It also allows direct reflection measurements of the cavity.

The room temperature part of the electronics included further amplification and mixing stages. The cavity bandwidth $f/Q \simeq 30\,\text{kHz}$ is larger than the expected axion linewidth of $\simeq 1\,\text{kHz}$ and it is advantageous to use a multichannel receiver with a bin width of roughly 1kHz. The audio signal was fed to a 16 bit A/D converter, oversampled at 70 kHz and fourier analyzed by a TMS320C25 digital signal processor in real time. The time for one 32 point power spectrum was 0.9 ms.

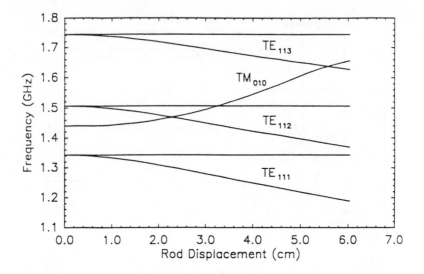

Fig.3: Cavity frequency versus radial displacement of a dielectric rod for cavity radius R=6.83cm, rod radius r=0.64cm, cavity length L=38cm, and ϵ=9.5.

3 The Search

We conducted five cavity scans from fall 1989 to summer 1990[11]. Using two different cavities, we covered the frequency range of 1.32–1.63 GHz and 1.80–1.83 GHz. Data taking was largely automated. The fine tuning rod had a range of roughly 25 MHz and was stepped under computer control. For each rod setting, initially 10^5 spectra were averaged in about 90 seconds and searched for candidate peaks due to axion-photon conversion. For each bin, the deviation from the mean was calculated according to

$$\#\mathrm{sigma} = \frac{P - \bar{P}}{\bar{P}}\sqrt{N} \tag{7}$$

and the program searched for 2σ peaks. If a peak appeared, another spectrum was taken and checked for coincidence peaks. This process was repeated up to 5 times and the peak was flagged for later rechecks if it proved statistically significant. A typical spectrum obtained during a scan is shown in Fig.6. The candiate peaks were reeaxmined with the magnetic field turned off. None of them survived this test, $i.e.$ all persisted without field. Most of them originated from electronic equipment nearby.

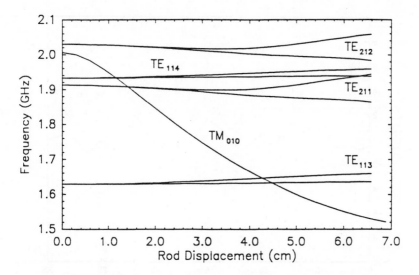

Fig.4: Cavity frequency versus radial displacement of a metal rod for R=6.83cm, r=0.62cm, and L=38cm. To avoid cluttering, the higher TM modes as well as the TEM modes are not shown.

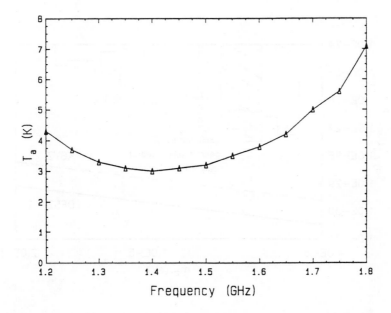

Fig.5: Noise temperature of Berkshire L-1.5-30H at a physical temperature of 4K.

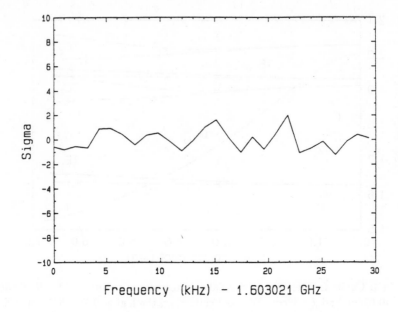

Fig.6: Sample spectrum of cavity output.

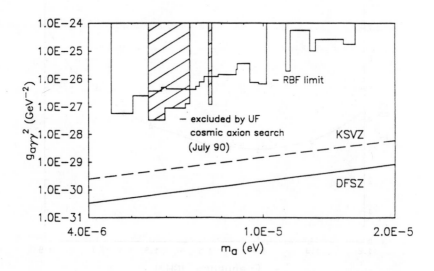

Fig.7: Experimental limits on the axion-photon coupling $g_{a\gamma\gamma}$.

Since no peak with the signature of an axion signal was found, it is possible to put an upper limit on the electromagnetic coupling of the axion, provided the galactic halo is axionic. The noise fluctuations in the detector have a Gaussian distribution, and one can calculate the probability for a given signal to be detectable. Our detector is looking for 2σ peaks above noise. The probability for a negative fluctuation of $> 2\sigma$ is 0.025 or in other words, a 4σ signal will be seen at a 97.5% confidence level. In terms of the system noise temperature T_n, bin bandwidth B and # of averages N, the minimum detectable signal power is (at 97.5% C.L.)

$$P_{\min} = 4k_B T_n B/\sqrt{N} \tag{8}$$

where

$$T_n = T_{bath} + T_{amp} . \tag{9}$$

For our experimental parameters, $P_{\min} \simeq 10^{-21}\,\text{W}$. Figure 7 shows the upper limit on $g_{a\gamma\gamma}^2$ obtained from our search. Also shown is the limit set by the RBF collaboration in a previous experiment[12]. Our presently detectable power level P_{min} is about a factor 1000 too large for the detection of DFSZ axions in our halo. Because of the $\sim 2K$ noise limits of currently available amplifiers, this factor can only be gained by increasing $P_{a\to\gamma}$, and most straightforwardly by increasing the detector volume. A follow up experiment along these lines has been proposed and is described in this volume[13].

References

[1] R.D. Peccei and H. Quinn, *Phys.Rev.Lett.* **38**, 1440 (1977); *Phys.Rev.* **D16**, 1791 (1977).

[2] S. Weinberg, *Phys.Rev.Lett.* **40**, 223 (1978); F. Wilczek, *ibid.* **40**, 279 (1978).

[3] For a review of axion properties and experiments see: J.E. Kim, *Phys.Rep.***150**, 1 (1987); H.Y. Cheng, *ibid.* **158**, 1 (1988); R.D. Peccei, in *CP Violation*, C. Jarlskog, ed. (World Scientific, Singapore, 1989).

[4] M. Dine, W. Fischler, and M. Srednicki, *Phys.Lett.* **104B**, 199; A.P. Zhitnitskii, *Sov.J.Nucl.Phys.* **B166**, 260 (1980); J.E. Kim, *Phys.Rev.Lett.* **43**, 103 (1979); M.A. Shifman, A.I. Vainshtein, and V.I. Zakharov, *Nucl.Phys.* **B166**, 493 (1980).

[5] Recent reviews include: M.S. Turner, *Phys.Rep.* **197**, 67 (1990); G.G. Raffelt, *ibid.* **198**, 1 (1990).

[6] J. Preskill, M. Wise and F. Wilczek, *Phys.Lett.* **120B**, 127 (1983); L. Abbott and P. Sikivie, *ibid.* **120B**, 133 (1983); M. Dine and W. Fischler, *ibid.* **120B**, 137 (1983).

[7] R. Davis, *Phys.Lett.* **180B**, 225 (1986); R. Davis and R.P.S. Shellard, *Nucl.Phys.* **B324**, 157 (1989); D. Harari and P. Sikivie, *Phys.Lett.* **195B**, 361 (1987); A. Dabholkar and J.M. Quashnock, *Nucl.Phys.* **B333**, 815 (1990); C. Hagmann and P. Sikivie, submitted to *Nucl.Phys.*

[8] P. Sikivie, *Phys.Rev.Lett.* **51**, 1415 (1983); *Phys.Rev.* **D32**, 2988 (1985); L. Krauss, J. Moody, F. Wilczek, and D. Morris, *Phys.Rev.Lett.* **55**, 1797 (1985).

[9] C. Hagmann, P. Sikivie, N.S. Sullivan, and D.B. Tanner, *Rev.Sci.Instrum.* **61**, 1076 (1990).

[10] Berkshire Technologies Inc., Oakland, CA 94609.

[11] C. Hagmann, P. Sikivie, N.S. Sullivan, and D.B. Tanner, *Phys.Rev.* **D42**, RC, 1297 (1990).

[12] S. DePanfilis, A.C. Melissinos, J.T. Rogers, Y.K. Semertzidis, W.U. Wuensch, H.J. Halama, A.G. Prodell, W.B. Fowler, and F.A. Nezrick, *Phys.Rev.Lett.* **59**, 839 (1987); *Phys.Rev.* **D40**, 3153 (1989).

[13] K. van Bibber, these proceedings.

AN OBSERVATIONAL SEARCH FOR AXIONS

M. Ted Ressell

Department of Astronomy & Astrophysics
Enrico Fermi Institute
The University of Chicago
Chicago, IL 60637-1433

NASA/Fermilab Astrophysics Center
Fermi National Accelerator Laboratory
Batavia, IL 60510-0500

Abstract. An axion with a mass between $3\,\mathrm{eV}$ and $8\,\mathrm{eV}$ would have a cosmologically interesting abundance. These axions should produce a detectable signature in the night sky via their decay to two photons. I describe an unsuccesful telescopic search for this signature in clusters of galaxies which effectively closes this window of axion mass.

Peccei-Quinn (PQ) symmetry and its resulting pseudo-Goldstone boson, the axion, are among the best motivated of the many extensions to the standard model of particle physics currently being investigated.[1] PQ symmetry provides a simple and elegant solution to the strong-CP problem which plagues Quantum Chromodynamics (QCD). Additionally, the PQ scheme arises naturally in theories of supersymmetry and superstrings. The axion's couplings to ordinary matter are inversely proportional to the energy scale of PQ symmetry breaking, f_a which also determines the axion mass

$$\frac{m_a}{\mathrm{eV}} \simeq \frac{6 \times 10^6 \, \mathrm{GeV}}{f_a/N},$$

where N is the color anomaly of PQ symmetry.[2] Like the π^0, the axion decays into two photons with a lifetime of

$$\tau_a(a \rightarrow 2\gamma) = 6.8 \times 10^{24} \sec \left(\frac{m_a}{\mathrm{eV}}\right)^{-5} \zeta^{-2}$$

where $\zeta = (E/N - 1.95)/0.72$ and E is the electromagnetic anomaly of PQ symmetry. In the simplest models $E/N = 8/3$ and $\zeta = 1$, but different values, such as $E/N = 2$, are not excluded; such values result in a significant increase in the axion lifetime. This model dependance of τ_a, as embodied in ζ, can be very important when discussing astrophysical and cosmological constraints on the axion mass.

There is currently very little theoretical guidance as to the mass of the axion. Its mass might lie anywhere in the range $10^{-11}\,\mathrm{eV} \lesssim m_a \lesssim 10^5\,\mathrm{eV}$. Fortunately, laboratory experiments and a host of astrophysical and cosmological arguments have narrowed this region to two "windows" $10^{-6} \lesssim m_a \lesssim 10^{-3}$ and $3\,\mathrm{eV} \lesssim m_a \lesssim 8\,\mathrm{eV}$.[2,3] A search for axions in this latter window is the subject of this talk. Axions with masses of a few eV are coupled strongly enough so that they come into thermal equilibrium in the early Universe. They decouple when the temperature is slightly less than 100 MeV and today should have an abundance of about $n_a \simeq 50\,\mathrm{cm}^{-3}$, comparable to that of neutrinos or microwave photons. Knowing this, it is simple to calculate their contribution to the critical density

$\Omega_a \simeq m_a/(49h_{50}^2 \,\mathrm{eV})$ (the Hubble constant is $H_0 = 50h_{50}\,\mathrm{km\,sec^{-1}\,Mpc^{-1}}$). These axions cannot be *the* dark matter but, as I will show, this abundance is sufficient to allow their detection via their decay to two photons, as was first pointed out by the authors of reference 4.

First, let me motivate the existence of the multi-eV window of axion mass. The lower limit of 3 eV is fairly robust and comes from considering the effects of axion emission upon the observed neutrino burst of Supernova 1987A.[5] This analysis is fairly free of ambiguity as the axion couplings to matter are known to within factors of order unity and as the entire cooling history of the neutron star is available; there is no need to appeal to statistical arguments. The upper limit of the window, $m_a \lesssim 8\,\mathrm{eV}$, is rather more uncertain. By appealing to the effect of axions upon the evolution of red giant (RG) and horizontal branch (HB) stars, Raffelt and Dearborn[6] have found the limit $m_a \lesssim 2 \cdot \zeta^{-1}\,\mathrm{eV}$, seemingly closing the window. In spite of this there are two good reasons to consider this window: the first is the model dependence of the relevant axion-matter coupling (for $E/N = 2$ the limit becomes $m_a \lesssim 29\,\mathrm{eV}$); the second is the fact that the authors must appeal to the statistics of small numbers. Since it is impossible to observe the entire life cycle of a RG-HB star, the argument is based on the number of stars on the HB (5) in the cluster M67; perhaps we are witnessing a fluctuation from what would be expected. While neither of these concerns may in the end be relevant, they justify giving the window a closer look.

If, for the moment, I forget the limit imposed by the RG-HB star evolution, what other limits may be placed upon axions in this mass range? One limit is to consider the effects of decaying relic axions upon the diffuse extragalactic background radiation (DEBRA).[7] With the assumptions that axions are uniformly distributed throughout space and that $\Omega_{\mathrm{TOT}} = 1$ it can be shown that they produce a glow in the night sky of[4]

$$I_a = \frac{n_a m_a c^3}{4\pi\tau_a H_0 \lambda_a}\left(\frac{\lambda_a}{\lambda}\right)^{7/2} = 1.8\times 10^{-23}\left(\frac{m_a}{\mathrm{eV}}\right)^7 \frac{\zeta^2}{h_{50}}\left(\frac{\lambda_a}{\lambda}\right)^{7/2} \mathrm{erg\,cm^{-2}\,arcsec^{-2}\,\mathring{A}^{-1}\,s^{-1}},$$

where $\lambda_a = 24800\,\mathring{A}(\,\mathrm{eV}/m_a)$ and λ is the observed wavelength. By insisting that the glow from axion decay not exceed the observed DEBRA flux, I find that for $E/N = 8/3$, $m_a \lesssim 3.8\,\mathrm{eV}$. Unfortunately, this suffers from the same model dependence as the RG-HB star limit. For $E/N = 2$ the limit moves to $m_a \lesssim 8\,\mathrm{eV}$; a significant improvement, and the claimed upper bound of the window.[8] As a final note, the above expression is only valid for $m_a \lesssim 25\,\mathrm{eV}$, axions heavier than this decay in a time less than the age of the Universe (for $\zeta = 1$).

The assumption that axions remain unclustered is a conservative one. Relic thermal axions will, in accord with the equivalence principle, fall with baryons (and any other particles) into the various potential wells that develop in the Universe, and will be found in extended structures around galaxies (halos) and in clusters of galaxies, as they cannot dissipate energy and condense further. Their decays will produce a line at the wavelength λ_a, which is Doppler-broadened by the velocities that axions have in these objects, and for distant objects the line is also redshifted $\lambda = (1 + z)\lambda_a$. The most promising case for detection is that of cluster axions where the expected line intensity is $I_a \sim 10^{-17}\zeta^2\,(m_a/3\,\mathrm{eV})^7\,\mathrm{erg\,cm^{-2}\,arcsec^{-2}\,\mathring{A}^{-1}\,s^{-1}}$.[4] The background with which this must compete is the "night sky," which is dominated by the glow of the atmosphere, includes many strong lines, and has a continuum intensity of roughly $10^{-17}\,\mathrm{erg\,cm^{-2}\,arcsec^{-2}\,\mathring{A}^{-1}\,s^{-1}}$.[9] For the remainder of this talk I will discuss an unsuccessful search for an intracluster axion decay line carried out by Matthew Bershady, Michael Turner, and myself at Kitt Peak National Observatory (KPNO) which effectively closes the 3 eV to 8 eV "window."[8,10]

We selected three well studied, rich clusters of galaxies:[11] Abell 1413 ($z = 0.143$), 2218 ($z = 0.171$), and 2256 ($z = 0.0601$).[11] The red shifts of these clusters are large enough so that their angular sizes are comparable to our field of view (5 arcmin), and small enough so that the axion line is not shifted too far into the red or diminished too much in strength. Each of the clusters has been the subject of extensive dynamical studies and x-ray observations, which aids in modeling the axion line expected.

Our strategy for dealing with the night sky is simple. Since axions share the same potential well as the galaxies and x-ray emitting gas, the axion-line intensity should decrease going from the cluster center outward. We take a spectrum from near the cluster center and subtract from it a spectrum from the outer reaches of the cluster. In principle, this procedure should completely remove the night sky with minimal effect to the axion line. Since the night sky brightness varies both spatially and temporally, it is important that the "off cluster" data and the "on cluster" data be taken as close in space and time as possible. Even so, the bright emission lines and the OH bands in the red are still troublesome. The solid line in the figure is typical of the subtractions we were able to achieve. Subtraction works quite well except where there are bright night-sky lines, and sensitivities of a few percent of the continuum night-sky brightness are typical. Because the axion line is red shifted by different amounts in the different clusters while the night-sky lines remain fixed, the spectral regions where the subtraction is poor correspond to different axion masses in the different clusters.

How can we recognize an axion line? First, the line should be seen at different wavelengths in the different clusters, because the red shifts of the clusters are different. Next, the line should have the predicted strength and should be Doppler-broadened to a gaussian of width $\Delta\lambda/\lambda = 2\sqrt{\ln 4}\,\sigma/c$, where σ is the one-dimensional velocity dispersion of the cluster. Finally, since cluster axions behave as a gas of collisionless particles they should have a density profile described by an isothermal sphere,[12] and the spatial variation of the line strength provides another check. While it is impossible to make the case with certainty, e.g., an intracluster atomic-emission line could satisfy all these criteria, we have enough redundancy to make a strong case which could then be followed up by other methods.

In modeling the axion line, we used the analytic King model approximation[11,12] to an isothermal sphere, which has a density run $\rho(r) = \rho_c / \left(1 + r^2/a^2\right)^{3/2}$, where a is the core radius, r is the radial distance from cluster center, and the central density is $\rho_c = 9\sigma^2/4\pi G a^2$. Relic axions only make up a fraction of the cluster mass. In objects that have not undergone significant dissipation—including clusters—the ratio of the axions to baryons should remain constant. This means that the axion mass density in the cluster should be related to that of baryons: $\rho_a \simeq (\Omega_a/\Omega_B)\rho_B$. Of course, the mass of the cluster may not be primarily baryonic—it could be particle dark matter. In that case, $\rho_a \simeq (\Omega_a/\Omega_B)(\Omega_B/\Omega_{TOT})\rho$. We shall use the most conservative assumption: $\rho_a = \Omega_a\rho \simeq (m_a/49h_{50}^2\,\mathrm{eV})\rho$, which follows by assuming $\Omega_{TOT} = 1$ and $\Omega_{TOT} \gg \Omega_B$.

Having related ρ_a to the cluster density ρ, it is now straightforward to write down the predicted axion-line intensity in the *cluster rest frame*:

$$I_a(R, \lambda) = \frac{\Sigma_a(R)c^3}{4\pi\sqrt{2\pi}\sigma\lambda_a\tau_a} \exp\left(\frac{-(\lambda - \lambda_a)^2}{\lambda_a^2}\frac{c^2}{2\sigma^2}\right),$$

$$= 1.0 \times 10^{-20} \left(\frac{m_a}{\mathrm{eV}}\right)^7 \left(\frac{\sigma_3}{h_{50}a_{250}}\right)\zeta^2 \exp\left(\frac{-(\lambda - \lambda_a)^2}{\lambda_a^2}\frac{c^2}{2\sigma^2}\right)\Big/(1 + R^2/a^2),$$

in units of $\mathrm{erg\,cm^{-2}\,arcsec^{-2}\,\mathring{A}^{-1}\,s^{-1}}$, where $\Sigma_a = 9\sigma^2\Omega_a/2\pi G a^2(1 + R^2/a^2)$ is the axion surface density, R is the projected radial distance from cluster center, $a = a_{250}\,250\,\mathrm{kpc}$, and $\sigma = \sigma_3\,1000\,\mathrm{km\,s^{-1}}$. In our rest frame the line is red shifted and I_a is "dimmed" by $(1 + z)^{-4}$. The key features of the axion line are: extreme sensitivity to the axion mass ($\propto m_a^7$); varies as R^{-2} for $R \gg a$; and axion-model dependence ($\propto \zeta^2$).

Our observations were made on the nights of 24 and 25 May 1990 at KPNO using the 2.1 meter telescope and Gold Camera CCD spectrograph at a spectral resolution of about

10 Å (the axion-line width is about 250 Å · eV/m_a). On the first night we took multiple exposures of all three clusters in the wavelength range 4762 Å − 8441 Å (axion masses from 6.1 eV to 3.2 eV) by "walking" the slit from cluster core outward along an E-W axis. We did this to cover a large enough portion of the cluster to determine the spatial profile of the axion line (spatial coverage of 3 to 10 core radii). On the second night, we changed the grating to cover the spectral range from 3737 Å − 7606 Å (axion masses from 7.8 eV to 3.7 eV), and took single exposures of the two smaller clusters. Our exposure times ranged from 30 to 75 minutes. The data were then reduced using the Image Reduction and Analysis Facility (IRAF) software.

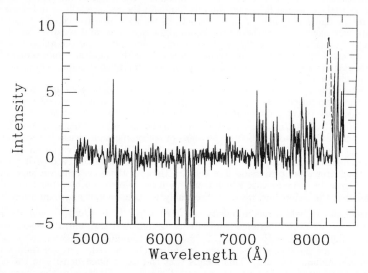

In the figure I show the "on cluster" spectrum (near cluster center) minus the "off cluster" spectrum (about three core radii from center) for A2256. The Intensity units are 10^{-18} erg cm^{-2} arcsec^{-2} Å^{-1} s^{-1}. The dashed line shows how the spectrum would change if a 3.2 eV axions were present in the cluster. There is no obvious candidate line in the "on − off" spectrum (or in any of our "on − off" spectra), even as weak as that expected for a 3.2 eV line. Since $I_a \propto m_a^7$, the line expected for $m_a > 3.2$ eV should be much more prominent. On this basis, we exclude the existence of an axion line for masses from 3.2 eV to 7.8 eV, the full range of our search.

Given the uncertainties in predicting the strength of the axion line and the fact that our search may have other relevance, we felt it useful to derive a quantitative flux limit to the existence of any intracluster line. By Fourier cross-correlating our "on − off" spectra with a noiseless template, we were able to determine that, at the 2-σ level, there are no line features with a line-flux of more than $I_a \simeq (1.0 \pm 0.5) \times 10^{-18}$ erg cm^{-2} arcsec^{-2} Å^{-1} s^{-1}, where the errors represent the variation throughout the spectra and not the formal statistical error. This value is significantly smaller than the predicted value of any intracluster axion line. For the details of this procedure see references 8 and 10.

I have described the results of an unsuccessful telescope search for line radiation from the decay of cluster axions of mass 3.2 eV to 7.8 eV. While our results seem to close this window, I remind the reader that there are astrophysical and particle physics uncertainties in the predicted line strength. We have been very conservative in treating the astrophysical uncertainties, perhaps by even a factor of 30. A remaining worry is that axions could be more smoothly distributed than we have assumed so that in our "on – off" subtractions we have eliminated the axion line too. To address this we have subtracted the "on cluster" spectrum of A2218 from that of A2256; the axion signature is one negative line and one positive line displaced in wavelength by a factor of $(1 + z_{2218})/(1 + z_{2256})$. While the night-sky subtraction was typically a factor of four worse, it was adequate and we found no evidence for the axion signature.

The most significant axion model dependence is embodied in ζ. In the simplest axion models $\zeta = 1$; but if E/N is very close to two, ζ could be very small. Fortunately, because the axion flux is proportional to m_a^7, the ζ dependence is not of great importance. Allowing $E/N = 2$ ($\zeta = 0.07$) only moves our mass limit to $m_a \lesssim 4\,\mathrm{eV}$.[8]

Finally, I note that our flux limits are applicable to any relic species that resides in clusters and whose decays produce mono-energetic photons of energy $1.6\,\mathrm{eV}$ to $3.9\,\mathrm{eV}$. The expected line strength can be generalized to another relic X by multiplying I_a by a factor of $(n_\gamma/2)(\Omega_X/\Omega_a)(\tau_a/\tau_X)$, where $n_\gamma = 1$ or 2 is the number of photons produced per decay. For example, as applied to the decay of relic neutrinos, $\nu \to \nu' + \gamma$,

$$I_\nu(R, \lambda) = 4.8\times10^{-20} \left(\frac{m_\nu}{\mathrm{eV}}\right)^2 \left(\frac{\sigma_3}{h_{50}a_{250}}\right) \left(\frac{10^{24}\,\mathrm{sec}}{\tau_\nu/B_\gamma}\right) \exp\left(\frac{-(\lambda - \lambda_a)^2}{\lambda_a^2}\frac{c^2}{2\sigma^2}\right) \bigg/ (1+R^2/a^2),$$

in units of $\mathrm{erg\,cm}^{-2}\,\mathrm{arcsec}^{-2}\,\text{Å}^{-1}\,\mathrm{s}^{-1}$ where $\tau_{24} = \tau/10^{24}\,\mathrm{yr}$ and B_γ is the radiative branching ratio. Using our line-flux limits we find $\tau_{24}/B_\gamma \simeq \mathcal{O}(1)$ throughout the relevant range of mass.

This work was supported by NASA, through grant NAGW-1340 (at FNAL) and a GSRP fellowship, and the DOE at Chicago. I thank my collaborators, Matthew Bershady and Michael Turner, for many useful discussions.

References

1. R.D. Peccei and H.R. Quinn, *Phys. Rev. Lett.* **38**, 1440 (1977); F. Wilczek, *Phys. Rev. Lett.* **40**, 279 (1978); S. Weinberg, *Phys. Rev. Lett.* **48**, 223 (1978). For a recent review of the axion see R.D. Peccei, in *CP Violation*, edited by C. Jarlskog (WSPC, Singapore, 1989). There are two types of axion: one which couples to both quarks and leptons, the DFSZ axion, and one which has no tree-level couplings to leptons, the hadronic axion.
2. M.S. Turner, *Phys. Rep.*, **197**, 67 (1990).
3. G.G. Raffelt, *Phys. Rep.*, **198**, 1 (1991).
4. T. Kephart and T. Weiler, *Phys. Rev. Lett.* **58**, 171 (1987); M.S. Turner, *Phys. Rev. Lett.* **59**, 2489. (1987).
5. A. Burrows, M.T. Ressell, and M.S. Turner, *Phys. Rev.* **D 42**, 3297 (1990).
6. G.G. Raffelt and D.S.P. Dearborn, *Phys. Rev.* **D 36**, 2211 (1987).
7. M.T. Ressell and M.S. Turner, *Comments on Astrophysics*, **14**, 323 (1990).
8. M.T. Ressell, to be submitted to *Phys. Rev.* **D** (1991).
9. P. Massey et al., *Pub. Astron. Soc. Pac.* **102**, 1046 (1990).
10. M.A. Bershady, M.T. Ressell, and M.S. Turner, to appear in *Phys. Rev. Lett.* March 1991.
11. C.L. Sarazin, *Rev. Mod. Phys.* **58**, 1 (1986); A. Dressler, *Astrophys. J.* **223**, 765 (1978); A. Dressler, *ibid* **226**, 55 (1978); T.C. Beers and J.L. Tonry, *ibid* **300**, 557 (1986); L.A. Thompson, *ibid* **306**, 384 (1986); S.M. Lea and J.P. Henry, *ibid* **332**, 81 (1988); P.E. Boynton, et al., *ibid* **257**, 473 (1982); D.G. Fabricant et al., *ibid* **336**, 77 (1989); C. Jones and W. Forman, *ibid* **276**, C. Jones et al., *ibid* **234**, L21 (1979); A. Dressler, *Astrophys. J. Supp.* **42**, 565 (1980).
12. J. Binney and S. Tremaine, *Galactic Dynamics* (Princeton Univ. Press, Princeton, 1987), Chs. 1, 2, and 4.

440

Searching for WIMPs with Mica

D. P. Snowden-Ifft, Y. D. He, and P. B. Price

Department of Physics, University of California at Berkeley, Berkeley, CA 94720, USA

Abstract

We propose two new ideas which use mica to search for WIMPs in our galaxy. The first sets an upper limit on WIMP elastic scattering cross sections by utilizing the long integration time associated with natural mica. The second, using annealed mica, not only limits WIMP cross sections but also provides a strong signature for WIMPs, allowing the possibility of detection. Both ideas can, potentially, set better limits than existing experiments.

Introduction

Weakly interacting massive particles (WIMP)s are one of the most thoroughly studied and well motivated dark matter candidates available today. Since WIMPs interact only weakly with matter they are extremely elusive and to date only a few of the many WIMP candidates have been completely ruled out. The detection and identification of a WIMP would have tremendous implications for cosmology and particle physics. As such, the need for better and more sensitive WIMP detectors is extremely important. In this paper we present two new ideas which use the solid state nuclear track detector (SSNTD) mica to search for WIMPs.

Two experiments have attempted to rule out WIMPs (Ahlen et al. 1987 and Caldwell et al. 1988a) by applying the techniques and detectors originally used in searches for double β decays. In principle these experiments can not only rule out WIMPs with various masses and cross sections, but might actually detect WIMPs. The signature of WIMPs is an annual variation in the rate of recoil events with time. The cause of this variation is the earth's motion around the sun. This motion is such that the rate of recoils in June, S_{June}, is higher than the rate of recoils in December, $S_{December}$. The signal is $S_{June}/S_{December} - 1 > 0$. In practice, however, the noise rate in these experiments, due to cosmic ray induced spallation products which undergo β decay with energies in the interesting keV regime, is high while the signal is weak ~2.5% (Primack, Seckel, and Sadoulet 1988a). At present no experiment could actually see WIMPs if they existed.

Mica as a charged particle detector

Mica is a SSNTD which is capable of recording the ~keV/amu recoils produced in WIMP-nucleus collisions (Borg et al. 1982). As with any SSNTD the response one obtains from mica is s. In the case of mica, s is defined as,

$$s = \frac{v_T}{v_\perp} \tag{1}$$

where v_T is the etch rate along a charged particle's path and $v_\perp$ is the etch rate perpendicular to the cleavage plane of the mica. Price and Salamon (1986a) have studied the response of mica. It is their conclusion that,

$$s = C[(dE/dx)_e + (dE/dx)_n]$$ (2)

for low velocity ($\beta < 10^{-2}$) ions. C is an important number. For muscovite mica $C \approx 1/(2$ GeV/g/cm^2). Since s must be at least 1 for a track to register, $1/C$ represents a minimum value of dE/dx required for a track to register. 2 GeV/g/cm^2 is a very large cutoff. It is so large, in fact, that electrons of any energy will never record in mica. This is the first advantage of using mica to detect WIMPs. It is insensitive to the primary background of most other dark matter detectors. Exploiting this and other advantages, we will discuss two ideas for using mica as a WIMP detector.

Using old mica to detect WIMPs

The first idea exploits the fact that mica is old (1 billion years old in some cases). Every WIMP recoil produced within the mica is recorded and the latent track damage is stored for future study. Furthermore, mica found well underground is well shielded from cosmic ray neutrons, protons, and muons which can mimic WIMPs. Since the mica is essentially randomly oriented during its lifetime, it is impossible to obtain a definitive signature of WIMPs from its motion through the WIMP halo. Nevertheless, the long integration time and low background make it possible to set very good limits on the mass and cross section of WIMPs in our Galaxy. We now investigate this idea in detail.

Fission track dating
Having obtained a sample of mica for investigation, we must determine the time interval during which the temperature of the mica was low enough for WIMP recoil tracks to be retained. This can be done using the fission track dating method (Fleischer, Price, and Walker 1975a) which measures this time interval. This method has been in use since 1963 and is very accurate.

Track fading
Tracks, when heated, can fade. Track-fading certainly occurs during the formation of mica since it is created at high temperatures from molten magma. Of concern to an experimentalist using mica is whether any track-fading has gone on since the mica was formed and cooled. Fortunately there is a method for determining the extent of track-fading in a piece of mica since formation. As outlined by Price and Salamon (1986b) the method relies on the fact that the fading rate is a function of dE/dx along the track. The higher dE/dx is, the lower the fading rate of the damaged region. Price and Salamon used spontaneous fission tracks which have a very high dE/dx (see Table 1). For comparison they used tracks produced when energetic alphas (from radioactive decays within the mica) collide with magnesium or silicon and form a compound nucleus. Alphas alone do not have a sufficiently high dE/dx to record etchable tracks in mica. These α-interaction tracks have very low dE/dx but do produce etchable tracks (see Table 1). Since both the spontaneous fission tracks and the α-interaction tracks stem from the same source, radioactivity

Table 1 - Background tracks in 1 billion year old mica.					
Track Type	Z	β	(dE/dx) (GeV/g/cm^2)	Range (μm)	Track density (cm^{-2})
Spontaneous Fission	~38 and ~52	~0.031 - ~0.046	~35 - ~55	~5 - ~15	~35
α-interaction	14 - 15	0.006	3	0.5	~5
α-recoil	82 - 90	0.0013	~15	~0.05	~3x10^5

within the mica, their ratio is in some fixed proportion which can be calculated. This ratio can be measured by etching the mica and determining the surface density of each kind of track. If the ratio of the surface density of α-interaction tracks to spontaneous fission tracks is greater than 0.1, no thermal fading has taken place in the mica for tracks with stopping powers greater than α-interaction tracks. As we shall see, the WIMP recoil tracks will easily satisfy this requirement.

Constraints on target atoms

The idea then is to choose an old mica in which no fading of the α-interaction tracks has occurred and look for tracks caused by WIMP-nucleus recoils. Mica is used as both the target and the detector. Table 2 shows the makeup of a typical mica. The chemical composition of any particular

Table 2 - Elements in a typical mica.			
Element	mole fraction	$R(\beta = 2\beta_c)$ (μm)	d_{max} (μm)
O	0.61	0.20	0
F	0.04	0.23	0
Al	0.11	0.26	0.026
Si	0.19	0.25	0.043
K	0.05	0.22	0.11
Fe	0.01-0.1	0.26	0.16
Rb	0-0.001	0.27	0.20
Cs	0-0.001	0.30	0.24
U	$< 10^{-9}$	0.30	0.27

mica may be measured using an electron microprobe analyzer. A graph of dE/dx vs velocity ($\beta < 2\beta_c$=maximum recoil velocity) for the constituents of mica is shown in Figure 1. β_c is the galactic escape velocity. dE/dx was calculated using the TRIM90 code (Ziegler, Biersack, and Littmark 1985). The threshold dE/dx is also shown. As can be seen in this graph, only recoils heavier than fluorine will record in mica. Thus less than 35% of the constituents of mica can be considered as targets.

There is an additional constraint on the atoms to be used as targets. As can be seen in Table 2 the range of recoils with $\beta = 2\beta_c = 0.004$ is less than 0.3 μm. Spontaneous fission and α-interaction tracks therefore do not constitute a background to a search for WIMP recoil tracks because they can easily be distinguished by their range. Unfortunately there is a third type of track caused by the decay of ^{238}U. It is the recoil of the decay products of ^{238}U which emit an α particle. These α-recoils produce tracks which are very short and in very high concentration, see Table 1. If the WIMP-recoil tracks are indistinguishable from the α-recoil tracks no useful limits (i.e. better than existing limits) can be set. A way must be found to distinguish the WIMP-recoil tracks from the α-recoil tracks. At first sight this would appear relatively simple since the range of the WIMP-recoils is far greater than the range of the α-recoils. The simplest parameter to measure, however, is the depth, d, of the etched tracks. The maximum depth is $d_{max} = R_{max}(1 - 1/s)$ where R_{max} is the range obtained by a $\beta = 2\beta_c$ recoil. These maximum etched depths are shown in Table 2. As can be seen there is not much hope left for using aluminum or silicon as target atoms. The depth of these tracks is too close to the depth of the α-recoil tracks. As there is not enough uranium in most micas to be of much use, we are left with

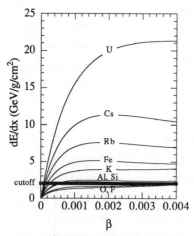

Fig. 1) dE/dx vs β for various ions in mica.

potassium, iron, rubidium and cesium as possible target atoms.

Backgrounds

Unfortunately WIMPs are not the only particles which can cause target atoms to recoil with keV/amu energies. Cosmic ray muons and radioactive products can also cause such recoils. We estimate the background of tracks which could mimic WIMP recoils from neutrons from the decay of radioactive elements in the rock surrounding the mica to be $1/20$ cm^2. The background from muons will vary with the depth of the mica. If mica can be found at depths of 200 m.w.e, the background from muons will be $1/$cm^2. These calculations are crude and almost certainly are overestimates.

Example

We now consider a concrete example of such an experiment. The first step is to collect as many samples of mica as possible from deep mines. The second is to determine the age and track fading of each sample. The chemical composition of each sample must then be measured. From these samples one is selected. For the purposes of clarity, let us assume we can obtain a mica which has the following characteristics,

1) $t_{mica} = 1$ billion years old
2) no fading of the WIMP recoil tracks
3) 0.1% cesium (mole fraction)

Assuming we can achieve an experimental limit of $1/$cm^2, at the 90% confidence level, Figure 2 shows the limits (dashed curve) which would be placed on WIMP masses and cross sections. to do this calculation we assumed WIMPs have a mass density of $\rho_w = 0.3$ GeV/cm^3, the measured dark matter density near the Earth (Primack, Seckel, and Sadoulet 1988b). We also assumed they have a Maxwellian velocity distribution with a velocity dispersion of $\beta_0 = 0.0007$ truncated at $\beta_c = 0.002$, the local escape velocity from the galactic gravitational potential (Primack, Seckel, and Sadoulet 1988c).

This plot is a little misleading because the two experiments shown on this graph use different targets (germanium for the Caldwell et al. experiment and cesium for our experiment). As such, WIMP candidates will produce different curves on this graph depending on which target is under consideration. Let us first consider candidates with spin independent interactions. These candidates will interact with cross sections proportional to A^2c where A is the atomic number of the nucleus and c is the coherence factor (Griest 1988). For masses between 500 GeV and 10^4 GeV, where we can set good limits, this factor differs by no more than 10% between germanium and cesium. Thus for WIMPs which couple with spin independent interactions and masses between 500 GeV and 10^4 GeV our experiment is ~5 times better than the Caldwell et al. experiment at high masses. Let us next consider spin dependent interactions. The germanium (Ge^{73}) which is represented by the spin dependent curve in Figure 2 has a nuclear spin of 9/2. Cs^{133} which constitutes 100% of natural cesium has a nuclear spin of 7/2. Thus spin dependent candidates for cesium will have cross sections which are 60% ($7^2/9^2$) lower than spin dependent candidates for germanium with 100% Ge^{73}. Even with this taken into account, our experiment is 30 times better than the Caldwell et al. experiment for high masses. Furthermore, our background calculations are conservative. We may find backgrounds much lower than predicted and will be able to set correspondingly better limits.

Advantages and disadvantages

There are several advantages to this experiment. The first is that the exposure is done. All we have to do is analyze the mica. The second is that very good limits on WIMP masses and cross sections can be attained. Moreover, Griest (1990) has pointed out that mica this old has been around the center of the Galaxy four times since its formation. This allows it to average its exposure over the entire dark matter distribution, unlike any other WIMP detector.

There are several disadvantages as well. Muons may be a limiting factor if mica cannot be found sufficiently deep. Moreover, we are at the mercy of geology when it comes to choosing a target atom. We must take what we find in the mica. And finally, perhaps most importantly, we do not, as yet, have a signature of WIMPs. Concern over these difficulties led us to a second idea for using mica to detect WIMPs.

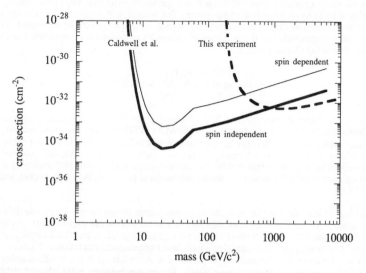

Fig. 2) Mass vs cross section exclusion plot for the Caldwell et al. (1988) experiment and the limits attainable using ancient (10^9 yrs) mica with cesium as a target. See text for a discussion of spin dependent vs spin independent interactions

Using annealed mica to detect WIMPs

The second idea relies on the fact that mica can be annealed. When properly annealed essentially all background tracks over large areas of mica can be removed without decomposing the mica. After annealing, the basic idea is to place it deep underground in contact with a suitable target. WIMPs will then cause target atoms to recoil into the mica (see Figure 3). As we will see, with sufficiently large amounts of mica and exposure time, this idea can set better limits than the existing experiments. Furthermore, this idea has already solved three of the major disadvantages associated with our first idea. First, this detector can be located sufficiently deep underground that muon induced recoils are not a concern. Second, we can choose our target material. And third, there is also a very strong WIMP signal associated with this detector.

Forward/backward asymmetry

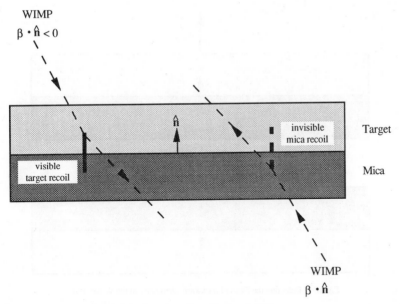

Fig. 3) Detection scheme for active experiment

This signal comes from a directionality inherent to this detector. Consider Figure 3 again. Recoils can be produced in either the mica or the target material. Remember, though, the conclusion we inferred from Figure 1. Less than 35% of the atoms in mica will leave an etchable track. This can certainly be optimized so that essentially no mica recoils will record in the mica. A target may then be chosen with sufficiently high Z that target recoils will record in the mica. Target recoils, which enter the mica, can only be produced by WIMPs travelling in one direction. To be more precise, imagine a unit vector $\hat{n}$, which points normal to the surface of the mica and towards the target (see Figure 3). Only WIMPs with velocities such that $\hat{n}\cdot\beta < 0$, where β is the velocity vector of the WIMP, will produce recoils which can record in the mica. This detector, therefore, "sees" only 1/2 of the incident WIMPs. This directionality and the ease with which it is obtained are extremely important.

Because the typical WIMP velocity, $\beta_0 = 0.0007$, is close to the velocity of the sun through the Galaxy, $\beta_{sun} = 0.000767$ (Caldwell et al. 1988b), not many recoils are produced which travel parallel to β_{sun}. Many more are produced anti-parallel to β_{sun}. This forward/backward asymmetry, $S_{forward}/S_{backward} - 1$, is therefore large. To detect this forward/backward asymmetry one places two of these detectors back to back ($\hat{n}$ vectors antiparallel) and orients one of the $\hat{n}$ vectors parallel to the velocity of the sun through the Galaxy, β_{sun} (see Figure 4). If WIMPs there be, the forward detector, $\beta_{sun} \parallel \hat{n}$, will accumulate roughly 100 times as many recoils as the backward detector. This signal is ~4000 times larger than the June/December signal associated with existing detectors. Furthermore, any background will uniformly produce tracks on both surfaces. Since the signal from our detector is the ratio of the number of etched tracks on the forward detector to the number of etched tracks on the backward detector minus one, the background will, within statistical limits, cancel out. With this simple design one simultaneously solves two of the outstanding problems associated with WIMP detection, namely directionality and background subtraction.

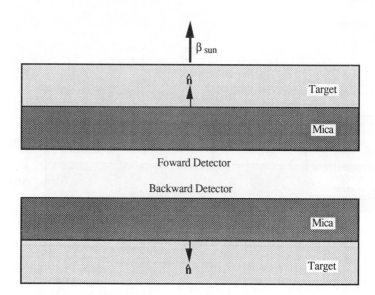

Figure 4) Detecting the forward/backward asymmetry in the WIMP flux.

Backgrounds

We can locate this detector far enough underground that muons will not significantly contribute to the production of background tracks. We also estimate that neutrons from the decay of radioactive elements in the rock will not affect us. If they do, the experiment can be shielded from fast neutrons. Our main worry is α-recoils inside the mica. Fortunately mica typically has a very low concentration (<<ppb) of radioactive elements in it. Moreover, it may be possible to distinguish α-recoils from WIMP recoils.

Example

Let us again ground these ideas with a proposal for an experiment. We will use germanium as an example of a target atom. Once again we would select a number of micas. We do not require that they come from deep mines. We do, however, require a large amount of undeformed mica of uniform clarity and in single large crystals. For definiteness let us assume that each detector, forward and backward, will have 2 m^2 of mica. Each mica would then be tested for uranium. Uranium at the <<ppb level can be measured using the methods described earlier to measure the age. Those micas which have << 1 ppb uranium would then move on to a second phase of testing. This second set of tests would make sure that the constituents of the mica do not record while the target atoms record with high efficiency. Let us assume that a mica can be found which is insensitive to constituent recoils while still retaining a sensitivity to germanium recoils. Germanium is a good choice because it can readily be obtained, is easily handled, and has << ppb uranium. The annealing of this mica then needs to be studied to assure that large areas of the mica can be cleared of all tracks. We would then cleave 4 m^2 of mica, to obtain fresh surfaces. This would then be cut to a standard size, 12.5 cm × 12.5 cm (to fit on a microscope stage), and annealed. Typically all tracks can be annealed from a piece of muscovite mica by raising the temperature to ~600 °C for ~1 hour (Fleischer, Price, and Walker 1975b). A mask, with regularly spaced holes, would then be placed over each piece of mica and exposed to a beam of germanium atoms. Those atoms passing through the holes in the mask act to calibrate the mica. The mica would then be moved underground to ~50 m.w.e for a 1 year exposure. During this move the mica would be separated from the germanium covered plates, turning the detector off. Once underground, the germanium would be placed next to the mica in an evacuated container. The evacuation serves two purposes. First, the range of recoils in air is only ~100 μm. Because we cannot be sure that the distance between the mica and germanium is much less than this we need to remove the air. Second it removes all radon from the detector, which reduces background. This chamber would then be placed on a telescope mount and pointed in the proper direction.

After a one year exposure, the mica is returned to the surface and etched. We envisage scanning with a phase contrast microscope. It is possible this process can be automated. Since we know the approximate number and location of the calibration tracks we will be able to locally measure the efficiency of our detector on ~1 cm scales.

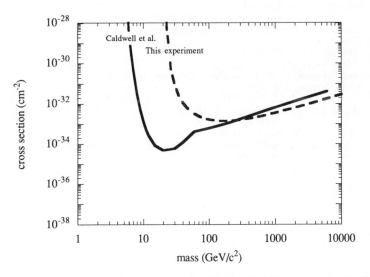

Fig. 5) Mass vs cross section exclusion plot for the Caldwell et al. (1988) experiment and the limits attainable using annealed mica with a germanium target.

Assuming no events are observed on the forward detector, we can place an experimental limit of $1/m^2$ at the 90% C.L. and set the limits shown in Figure 5. There is an important difference between the two curves. The Caldwell et al. experiment is background limited so there is no way, with the current design, to improve the limits. Building a bigger detector or integrating longer is pointless. We, however, can always improve our limits by building a bigger detector or integrating longer. If no background is encountered, our limits will grow with the size of the detector and exposure time. If at some point we encounter background problems, our limits will grow only as the square root of the size of the detector and exposure time, but they will still grow.

Conclusion

We have investigated two new ideas for using mica as a WIMP detector. The first idea utilizes the long integration time (10^9 yrs) of natural mica. The second utilizes the fact that mica can be annealed to remove all background tracks. Both ideas should be able to set better limits on the mass and cross section of WIMPs than existing dark matter detectors. The second can detect the forward/backward asymmetry associated with our motion through the WIMP halo providing a strong signature for WIMPs.

References

Ahlen, S. P., Avignone III, F. T., Brodzinski, R. L., Drukier, A. K., Gelmini, G., and Spergel, D. N. 1987, Phys. Lett. B., **195**, 603.

Borg, J., Dran, J. C., Langevin, Y., Maurette, M., and Petit, J. C. 1982, Rad. Eff., **65**, 133.

Caldwell, D. O., Eisberg, R. M., Grumm, D. M., Witherell, M. S., Sadoulet, B., Goulding, F. S., and Smith, P. F. 1988a, Phys. Rev. Lett., **61**, 510.

_____ 1988b, Phys. Rev. Lett., **61**, 512.

Fleischer, R. L., Price, P. B., and Walker, R. M. 1975a, *Nuclear Tracks in Solids: Principles and Applications*, (University of California Press: Berkeley), p. 163.

_____ 1975b, ibid., p. 82.

Greist, K. 1988, Phys. Rev. D, **38**, 2366.

Griest, K. 1990, private communication.

Price, P. B. and Salamon, M. H. 1986a, Phys. Rev. Lett., **56**, 1227.

_____ 1986b, Nature, **320**, 425.

Price, P. B. and Walker, R. M. 1963, J Geophys. Res., **68**, 4860.

Primack, J. R., Seckel, D., and Sadoulet, B. 1988a, *Ann. Rev. Nucl. Part. Sci.*, **38**, 764.

_____ 1988b, *Ann. Rev. Nucl. Part. Sci.*, **38**, 767.

_____ 1988c, *Ann. Rev. Nucl. Part. Sci.*, **38**, 767.

Ziegler, J. F., Biersack, J. P., and Littmark, U. 1985, in *The Stopping and Range of Ions in Solids*, (New York: Pergamon Press).

Current Statues of the PASS Superheated Superconducting Granule Detector

M. Le Gros, G. Meagher, A. Kotlicki, B.G. Turrell

Department of Physics, University of British Columbia,

Vancouver, B.C., V6T 1Z1, Canada

and A.K. Drukier

Applied Research Corp., Landover, Maryland, 20785 U.S.A.

We recently reported[1] a new method of fabricating planar arrays for the superheated superconducting granule detector which was suggested some time ago[2] as a very sensitive detector for radiation and particles.

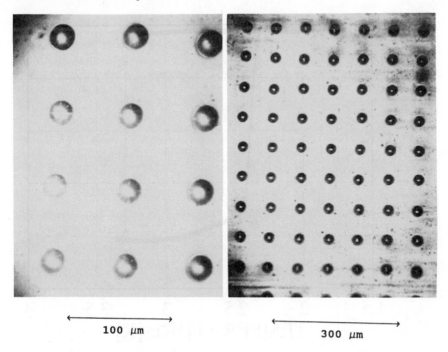

Fig. 1. Photographs of the array of spherical indium granules
(two magnificiations).

In the version suggested by our group[1] the detector consists of planar arrays of spherical granules of type I superconductor. The applied magnetic field, B_0, and the temperature, T, are chosen to put the granules in the superheated superconducting state very close to the phase transition line. An incoming particle or radiation deposits energy which heats the granule and causes it to "flip" to the normal state. The induced magnet flux change is then detected by a SQUID magnetometer.

The results we obtained previously were for a 20 × 30 granule array obtained by a fairly rudimentary photolithographic technique in which an initial arrays of squares was produced and then melted in the presence of a wetting agent to obtain spheres. This planar array of superheated superconductors (PASS) exhibited a spread in individual transition temperatures in a fixed field which was an order of magnitude narrower than ones obtained for colloidal samples.

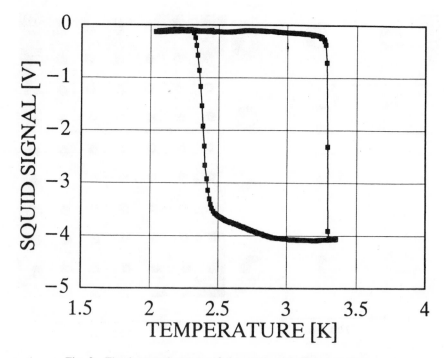

Fig. 2. The hysteresis curve of the array of indium granules for $B_0 = 2.3\ mT$. Both superconducting superheated to normal and normal supercooled to superconducting transitions are shown.

We report here results obtained for arrays consisting of 10^4 indium granules. To improve the quality of the arrays we used a custom–made mask with several different grid configurations including a 100×100 grid of 40×40 μm squares separated by 30 μm. This grid was used to produce photolithographically a 100×100 grid of indium squares of 3.5 μm thick. These squares were then melted into spheres of radius 12 μm (Fig. 1).

The quality of the new arrays is very good. An imaging technique[3] allowed us to determine that standard deviations in both the radius and the distance distribution are smaller than 4%.

A cryogenic test on the arrays was performed to observe the transitions for a range of B_0 from 0.05 to 16 mT. The full hysteresis curve for $B_0 = 2.3$ mT is shown in Fig. 2. The superheated superconducting to normal transition in $B_0 = 4$ mT is shown in Fig. 3 and it should be noted that the spread in transition temperatures for the 10^4 grains is only 20 mK.

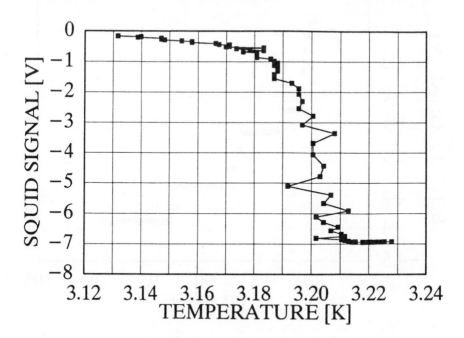

Fig. 3. The superheated superconducting to normal transition
in $B_0 = 4$ mT with expanded temperature scale.

The individual grain flips in a slow temperature sweep for $B_0 = 16$ mT is shown in Fig. 4. The larger steps correspond to two or three granules flipping at the same time. This represents only 40 s of a complete sweep which took a total of 4 hours. A numerical analysis of the SQUID output signal allowed us to obtain the distribution of signal size for individual flips which is shown in Fig. 5. This distribution clearly shows that "multiflips" (2 or more granules flipping simultaneously) are easily distinguished from single ones which is important for some future applications of the detector. More than 9000 flips were recorded and we assume the missing 1000 flips occurred during the SQUID resets. This slight reduction in detector efficiency would only be important for relatively high counting rates which are not expected in WIMP or neutrino detection. The effects can, in any case, be eliminated by suitable electronic design for other applications.

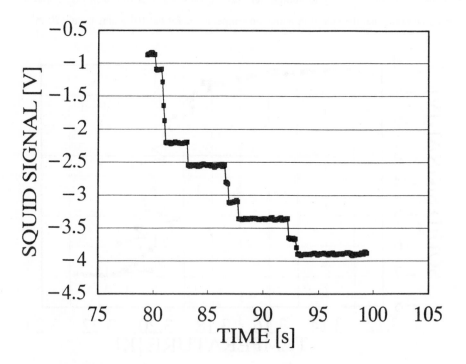

Fig. 4. Single, double or triple granules flip in the planar array.

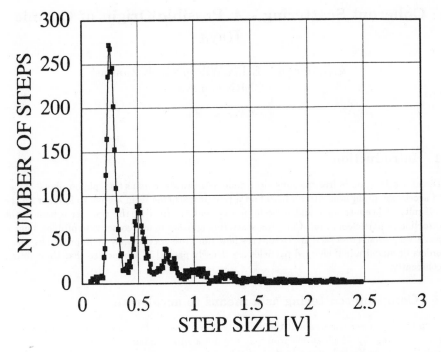

Fig. 5. Distribution of flip sizes. The peaks at 0.25, 0.5 and 0.75 V correspond to single, double and triple flips respectively.

This work is supported by the Science Council of B.C. through an STDF-AGAR grant, EBCO Industries Inc., and NSERCC.

REFERENCES

1. M. Le Gros, A. Da Silva, B.G. Turrell, A. Kotlicki and A.K. Drukier, Appl. Phys. Lett. 56, 2234, 1990.

2. See e.g. A.K. Drukier and L. Stodolsky, Phys. Rev. D. 30, 2295, 1984.

3. M. Le Gros, G. Meagher, A. Kotlicki, B.G. Turrell and A.K. Drukier, Proceedings of the 2nd London Conference on Position-Sensitive Detectors, Sept. 4–7 1990, Imperial College, London, to be published in NIM.

Coherent Scattering - A Possible Origin of Cosmic Rays

A.G.Morgan*, E.J.N.Wilson and K.Zioutas[†]

CERN,Geneva

2 August 1991

1 Introduction

Direct acceleration in free space by electromagnetic waves remains the "philosophers' stone" of acceleration physics. Attempts to find a practical scheme seem always to founder on the difficulty of keeping wave and particle in step and in directing the transverse field in the direction of particle motion. Compton scattering appears to overcome these problems but it has a very small cross section. We discuss how this cross section may be improved by many orders of magnitude if charged particles are densely packed in a bunch so that they absorb coherently.

2 Compton scattering as a means to accelerate

In spite of our reservations about the difficulty of acceleration in-vacuo, a mechanism, *Compton scattering*, by which an electron may acquire a small amount of energy from a photon striking it from behind, seems to do just this.

To accelerate electrons by Compton scattering would be a very laborious and inefficient process for the energy transferred as a photon finds an electron is at most a few eV. Supposing a high power laser could deliver a pulse of 1000 Joules in a mm diameter spot. The cross section an electron presents to this beam of 10^{22} photons is $\sigma_c = 0.6 \times 10^{-28} m^2$. Only about 1 of the photons will interact with the electron and the energy each will deposit will be only a small fraction of that of the photon. Although this paper treats the acceleration of electrons, the idea of coherent response could equally be applied to protons or other charged particles. In fact more massive particles may acquire more energy before the γ dependent terms in the expressions for acceleration limit the rate of acceleration. Cosmic rays are, of course, mainly these massive particles.

3 Acceleration by Coherent Scattering

In the coherent regime the scattered waves can reinforce each other coherently and we can add up the amplitudes of the N scattering centres within a box of one wavelength dimensions *before* squaring them to find the scattered intensity. The scattering probability then rises as N^2 where N is the number of centres participating. If the density is high enough for a complete layer, one third to one half of a wavelength thick, to scatter with unit probability, the system becomes opaque. The backward scattered wave returning from a smooth layer of such scatterers will be reflected as from a mirror.

Consider N charged particles in a unit cube illuminated with an electromagnetic beam of unit cross section. If the cross section for the incoherent Compton scattering is σ_c the probability of an interaction will be $N\sigma_c$ while the probability of an interaction if all the particles interact coherently is $N^2\sigma_c$.

Now take N particles in a cube of edge $\lambda/2$ illuminated by a beam whose dimensions match the face of the cube. The probabilities of incoherent and coherent interaction will be:

$$P_{inc} = \frac{4N\sigma_c}{\lambda^2} \tag{1}$$

and

$$P_{coh} = \frac{4N^2\sigma_c}{\lambda^2} \tag{2}$$

respectively.

The number density of electrons within the box is

$$\rho = \frac{8N}{\lambda^3} \tag{3}$$

so we may substitute $2N/\lambda = \rho\lambda^2/4$ to obtain the density, ρ', at which the probability for the coherent effect is one

$$1 = \left(\frac{\rho'\lambda^2}{4}\right)^2 \sigma_c.$$

The density required becomes smaller as the wavelength increases.

Conversely, the frequency below which light will not penetrate is:

$$\omega' = \frac{2\pi c}{\lambda'} = \pi c \rho^{\frac{1}{2}} \sigma_c^{\frac{1}{4}} = \pi \left(\frac{\sigma_c}{\mu_0^2}\right)^{\frac{1}{4}} \sqrt{\frac{\rho}{\epsilon_0}}.$$

However, the frequency of incident radiation below which the electrons of a plasma are totally reflecting is well known and is called *plasma frequency* , given by the expression,

$$\omega_p = \sqrt{\frac{q^2\rho}{m_0\epsilon_0}} \tag{4}$$

In *MKS* units for electrons $(q = e)$ we compare the two expressions,

$$\omega' \approx 83\,\rho^{\frac{1}{2}} \quad \text{and} \quad \omega_p \approx 56\,\rho^{\frac{1}{2}}$$

the difference in the constant is simply due to the fact that we made the rough approximation of using $\lambda/2$ for the side of the effective coherent cube.

4 Lorentz Transformation of the Reflecting Layer

In the rest frame of a bunch moving away from source, Expression 4 holds but the wave is seen as red shifted. In the Lab frame the observed plasma frequency is such that,

$$\frac{\omega_{p,L}}{\omega_p} = (1 + \beta)\gamma, \tag{5}$$

i.e., $\omega_{p,L}$ is red-shifted to ω_p. Moreover, the Lab observer sees the moving plasma as more dense than a co-moving observer ($\rho_L = \gamma\rho$). Taking both factors into consideration the *observed* plasma frequency of a relativistically moving ($\beta \longrightarrow 1$) plasma is,

$$\omega_{p,L} \approx 2\sqrt{\frac{q^2\gamma\rho_L}{m_0\epsilon_0}}. \tag{6}$$

The depth of the reflecting layer (the coherence length) to an observer in the lab is $\delta \sim \lambda_p/2$ which may be written in terms of observed quantities in the Lab frame,

$$\delta = \frac{\lambda_{p,L}}{2} = \frac{\pi c}{\omega_{p,L}} = \frac{\pi c}{2}\sqrt{\frac{m_0\epsilon_0}{q^2\gamma\rho_L}}. \tag{7}$$

5 A linear accelerator

In this section we we treat the bunch as a perfect mirror. There is a problem associated with the rate of transfer of energy to a relativistic particle bunch. The distance that the lab observer sees the bunch has travelled, as it is overtaken by a pulse of electromagnetic radiation, increases rapidly as the bunch becomes relativistic. Consider a laser that produces a pulse of duration τ measured in the laser lab frame. We take small element ($d\tau$) of this pulse and imagine that its leading edge is just incident on the surface of the bunch. Before the trailing edge has overtaken the surface plane of the bunch the bunch is observed to move a distance, ds, in the laser lab frame. The space-time coordinates of the second event are obtained from the intersection of the world line of the bunch, $\beta ct = x$, and that of the trailing edge of the element, $c(t - d\tau) = x$. For a bunch moving away from the laser at speed βc,

$$ds = \frac{\beta}{1 - \beta}c\,d\tau \equiv \frac{(1 + \beta)\beta}{1 - \beta^2}c\,d\tau \tag{8}$$

which at high speeds ($\beta \longrightarrow 1$) gives

$$ds \approx 2\gamma^2 c\,d\tau. \tag{9}$$

Such a pulse carries with it a small increment of energy from the laser, $dE = \Phi A\,d\tau$ (Φ is the laser flux, and A the cross sectional area of the beam), which is transferred to the bunch and due to the coherence described above has an accelerating effect.

Now at any velocity the energy of the bunch is given by,

$$E = Nm_0c^2\gamma \tag{10}$$

Laser of...	today	tomorrow	units
Power Output, ΦA	10^{14}	10^{17}	W
Pulse length, τ	10^{-14}	10^{-14}	s
Initial γ,	100	1000	γ_I
Final energy, E_F	0.6	624	GeV
Linac length, S	3.8	30×10^3	m
Mean accel.force	151	0.28	$MeV\,m^{-1}$

Table 1: Parameters of Terrestrial Accelerators (for a bunch of 10^{10} electrons whose height is $2\,mm$ and whose length $\sim 300\,nm$ - the assumed wavelength of the laser.)

It follows that,

$$\frac{d\gamma}{ds} = \frac{d\gamma}{dE}\frac{dE}{d\tau}\frac{d\tau}{ds} \approx \frac{\Phi A}{2Nm_0c^3\gamma^2}. \tag{11}$$

Thus the distance covered by the bunch in being accelerated from γ_I to γ_F is,

$$S = \frac{2Nm_0c^3}{3\left(\frac{dE}{d\tau}\right)}\left[\gamma_F^3 - \gamma_I^3\right].$$

We know that the energy gained by the bunch must eventually become that of the whole laser pulse $\Phi A\tau$ and so we may substitute for γ_F to obtain,

$$S = \frac{2Nm_0c^3}{3\Phi A}\left[\left(\gamma_I + \frac{\Phi A\tau}{Nm_0c^2}\right)^3 - \gamma_I^3\right].$$

Conversely we can consider a bunch in a continuous stream of radiation, here,

$$E_F = m_0c^2\gamma_F = \sqrt[3]{E_I^3 + \frac{3m_0^2c^3\Phi AS}{2N}}.$$

We see that the energy gain per unit length decreases as the particles' energy increases and in particular $E_F \sim S^{\frac{1}{3}}$.

Table 1 contains the parameters for two linac's with lasers at and just beyond the forefront of current technology. (Note: we have chosen γ_I to avoid problems with space-charge effects.)

At the present time (or in the near future) we do not envisage the application of this technique for accelerating electrons to high ($\sim TeV$) energies. The acceleration rate one can expect from today's laser is only just comparable to an S-band linac up to 0.6 GeV. Above this energy the S-band linac is better. Tomorrow's laser is comparable with the expected performance of CLIC.

However, the key parameter in this acceleration scheme is τ—how quickly you can transfer the energy: the shorter the pulse the (very much) shorter the Linac.

6 A Stellar accelerator

The origin of *cosmic rays* and the mechanisms by which they reach high energies apparently in free space are not understood. Many suggestions have been made so far, starting historically with Fermi's acceleration proposal[1].

The most distinguished feature of the cosmic ray high energy spectrum is the $\sim E^{-2}$ power-law. Any proposed acceleration scenario must explain:

1. the spectrum shape (E^{-2} law)

2. the highest energies observed ($\leq 10^{20}\,[eV]$).

If a coherent mechanism, such as the one presented here, were to exist in the Universe then it would occur in regions of both high charged particle density and high radiation flux.

It is apparent that the sun accelerates particles to high energies during solar flares, which are accompanied by enhanced radio emissions [1].

The radiation energy density in Quasars, and Active Galactic Nuclei is $\sim 10^{18}\,eV\,cm^{-3}$ for a luminosity $\sim 10^{39}\,W$[2]. The corresponding average energy density at near the pulsar surface ($R \sim 10 - 100\,km$) is $\sim 10^{28}eV\,cm^{-3}$; for a luminosity of $\sim 10^{33}\,W$. For comparison consider a laser with 1 J/fs and a spot of 1 mm diameter. The maximum energy density is of the order of $10^{25}eV\,cm^{-3}$. We find the observed Radio Pulsar luminosities range from $\sim 10^{22}\,W$(for the Crab Nebula) to as low as $\sim 10^{19}\,W$. Whereas X-ray pulsar luminosities reach $\sim 10^{32}\,W$[3, 4] in agreement with the expected Eddington limit (which does not take into account coherent modes of interaction).

We consider a simple model of a neutron star (of luminosity L). Over the surface of the star, of radius r_0, we find a thin shell of charged particles. If the thickness of the shell is δ of relation (7) then the star's radiation pressure will continuously expand the spherical envelope—accelerating the particles... The total energy, E, of all the particles in this spherical shell is given by relation (10). We can however substitute for N; expressing E in terms of the lab-observed particle density at a radius r,

$$E = 4\pi m_0 c^2 \gamma r^2 \delta \rho_r.$$

Substituting for δ we have,

$$E = \frac{2\pi^2 m_0 c^3}{q}\sqrt{m_0\epsilon_0}\,r^2\sqrt{\rho_r\gamma}.$$

Now it is not unreasonable to expect continuity of flow which demands that when its radius is r the particle flux represented by the expanding shell must obey the relation,

$$\beta_r r^2 \rho_r = \beta_s r_0^2 \rho_s$$

(the suffix s indicates measured at the surface), and since the particles we are considering are *relativistic* we can write,

$$\rho_r = \rho_s \left(\frac{r_s}{r}\right)^2. \tag{12}$$

Substituting this result into E and differentiating with respect to γ we obtain,

$$\frac{dE}{d\gamma} = \frac{\alpha r}{\sqrt{\gamma}}$$

where

$$\alpha \equiv \left(\frac{\pi^2 c^3 r_0}{q}\sqrt{m_0^3\epsilon_0\rho_s}\right)$$

Thus in the relativistic limit of relation (9) and following the reasoning of relation (11) we find

$$\frac{d\gamma}{dr} = \frac{L}{2c\alpha}\gamma^{-\frac{3}{2}}r^{-1}$$

with s taken as the distance from the centre of the star ($s \equiv r$) which when integrated gives the energy of a plasma particle at a radius r, as,

$$E_F = \left(E_I^{\frac{5}{2}} + \frac{5Lm_0cq}{4\pi^2r_0\sqrt{\epsilon_0\rho_s}}\ln\left(\frac{r}{r_0}\right)\right)^{\frac{2}{5}}. \tag{13}$$

The implications of equation (13) become apparent when it is *fed* with the parameters of a *typical* pulsar: $r_0 = 10^4\,m$; $L = 6 \times 10^{32}\,W$; $\rho_s = 10^{18}\,m^{-3}$. If we assume that the surface plasma, of electrons and positrons, is energetic (in an outward direction) with $\gamma_I > 1$ and consider the energy of particles reaching a distance $r = 10^5\,m$, we have

$$E_F \approx 4 \times 10^{12}\,eV.$$

We note that the frequency of the radiation that finally accelerates the shell is $\nu \approx 10^{13}\,Hz$— a radio frequency with a wavelength of $\lambda \approx 3 \times 10^{-5}\,m$.

7 Discussion

Finding that the acceleration rate to be expected from a man made device is typically only 6 GeV in 250 m and therefore no better than an S-Band Linac is a disappointment. So too is the prediction that acceleration rates get worse at higher energy. The celestial mechanism is even more intriguing. The numerical estimate above suggests a few TeV can be reached before an electron escapes from a pulsars influence. The obvious extension is to repeat the theory for protons to see if they reach higher energies.

References

[1] J.P.Wefel and W. H. Sorrel, *"Genesis and Propogation of Cosmic rays"*, Mathematical and Phys. Sciences Vol.220 (1987) pp 1 and 227 Edit.M.M.Shapiro and J.P.Wefel D. Reidel Publ. Company, (1987).

[2] G.F.Bignami and W.Hermsen, *"Annual Review of Astronomy and Astrophysics"* , Vol.21,(1983),p.67

[3] , S.L.Shapiro and S.A. Teukolsky, *"Black holes, White Dwarfs and Neutron Stars"*, J.Wiley and Sons, New York (1983) p.289, J.H.Taylor and D.R.Stinebring, *"Annual Review of Astronomy and Astrophysics"* Vol.24 (1986) pp 285,453.

[4] K.R.Lang , *"Astrophysical Formula, 2nd. Edition"* Springer Verlag, Berlin (1980) pp. 52,58

VII. THE SUPERNOVA WATCH

SUPERNOVA NEUTRINOS: LIFE AFTER SN1987A

A. BURROWS

Department of Physics and Astronomy
University of Arizona
Tucson, AZ 85721 USA

I. INTRODUCTION

The detection of neutrinos from SN1987A[1,2] was a glorious event in supernova science. However, its epoch is past and both theory and experiment must now concern themselves with the future: the detection of a galactic neutrino burst. Although there is ambiguity (and, at times, pessimism) concerning when we legitimately can expect the next collapse, this needn't concern us here. As this conference amply demonstrates, we may well be ready. IMB3, Kamiokande II (Super?), SNO, MACRO, LVD, and BAKSAN are either now on line, or under construction. This already impressive international network may be joined in the future by LSND, ICARUS, SNBO, a Homestake ^{127}I detector, BOREXINO, the CalTech scintillator, JULIA, and various low-temperatures devices. I refer the reader to the proceedings of this workshop for a full discussion of each of the above projects and will concentrate in this paper solely on the theory of neutrino bursts. I will attempt to provide a useful overview and preliminary answers to the following questions: What are the distinctive and diagnostic features of the neutrino emissions? What might we learn about supernova physics and the supernova mechanism from an abundance of events? Of what theoretical issues should the experimental teams be aware? This excursion into theory will perforce be brief. A more complete discussion of supernova and protoneutron star neutrino signatures can be found in Burrows[3,4,5], Myra and Burrows[6], Bruenn[7], Arnett et al.[8], Mayle and Wilson[9,10], Woosley and Weaver[11], Mayle, Wilson, and Schramm[12], Blinnikov et al.[13], and Bethe[14], and in references cited therein.

II. SUPERNOVA NEUTRINOS: THE BIG PICTURE

Stellar collapse is arguably the most violent event in stellar science. The neutrino emissions that accompany the death of a massive star, a Type II or Type Ib supernova, and the birth of a neutron star (or stellar mass black hole?) are of such a magnitude that a sober researcher could at one time easily have been charged by the sceptical with wild exaggeration. SN1987A changed all that. A multiple of 10^{53} ergs in $\sim 10^{58}$ neutrinos of all species radiated over several seconds in a burst that rivals the optical luminosity of the observable universe causes far fewer miscarriages today. Nevertheless, that this 0.1 to $0.2\,M_\odot$ of neutrino mass energy is released every second in the universe, that such bursts have occurred 10^8 times in our galaxy alone and have involved the mass equivalent of 10 giant globular clusters, and that, at any time, 10—100 $M_\odot$ of burst neutrinos in thin expanding shells reside in the galaxy can still quicken the pulse.

The physics of collapse involves nuclear physics, statistical mechanics, hydrodynamics, general relativity, and particle physics in a heady mix stirred over the last thirty years by many, many theorists (e.g. Burbidge *et al.*[15], Colgate and Johnson[16], Arnett[17], Colgate and White[18], Wilson[19], Mazurek[20], Sato[21], Van Riper[22], Lamb *et al.*[23], Imshennik and Nadyozhin[24], Bethe *et al.*[25], Burrows *et al.*[26], Bowers and Wilson[27], Hillibrandt[28], Burrows and Lattimer[29,30], Baron *et al.*[31], Bruenn[32], to list only a few). Figure 1 depicts, in storyboard form, the rough sequence divined. After 10–20 million years of quiescent thermonuclear evolution, the core ($\sim$1.5 $M_\odot$, $R \sim 3 \times 10^8$ cm) of a massive star ($> 8 M_\odot$, $R \sim 10^{11-14}$ cm) reaches the critical Chandrasekhar mass and collapses. In less than a second, it achieves densities near 10^9 tonnes per cubic centimeter, energy densities of 1 Megatonne of TNT equivalent per cubic micron, and temperatures near 10 MeV ($\sim 10^{11}$ K). Some of the matter experiences speeds near one-fourth the speed of light and accelerations of $\sim 10^{11}$ $g_\oplus$'s.

Upon reaching nuclear density, the core stiffens, bounces, and generates an out-going shock wave that either directly, or after hesitating for about a second[33], within hours ejects the stellar envelope. The protoneutron star that remains is hot and electron-rich. It is via copious neutrino radiation over seconds that it cools and neutronizes to become the standard "neutron" star. So dense and hot is the protoneutron star that it is opaque to all neutrino types during the first minute(s) of its life.

There is a bit of understatement in the above description, but no exaggeration. It was such a neutrino burst from the supernova explosion, SN1987A, of the blue supergiant Sanduleak –69° 202 in the Large Magellanic Cloud that was identified on Earth by the IMB and KII collatorations 50 kiloparsecs distant. I will not dwell long on what was concluded from these data in the multitude of papers that, for lack of space and with respect for the reader, I will not cite. Given these sparse data (see Table 1) and the nature of small-number statistics, the now standard model of neutrino bursts stood up rather well (however, see Lo Secco[34]). We could, in fact, conclude the following:

1) Neutrino bursts do at times accompany supernovae.

2) Most, if not all, of the events in water Čerenkov detectors were $\bar{\nu}_e$'s.

3) The duration of the emission was many *seconds*, not milliseconds.

4) $\langle \epsilon_{\bar{\nu}_e} \rangle \sim$ 10–15 MeV (some cooling noted)

5) $E_{\bar{\nu}_e} \sim 3 \times 10^{52}$ ergs ($\times 6 \simeq 2 \times 10^{53}$ ergs)

6) $\langle$ Radius $\rangle \leq$ 50 kilometers

7) $1.2 M_\odot \leq M_G \leq 1.6 M_\odot$.

This list is comfortingly consistent with our pre–SN1987A expectations, but is doubly significant for what it does not contain.

The half kilogram of neutrino mass energy that no doubt coursed through the Earth on February 23rd 1987 at 7:35:41.37 (UT) and the full nanogram that sped through KII and IMB were not adequate to fully diagnose "supernova" theory. What was the explosion mechanism? Was a bounce-shock formed? What were the magnitude of the ν_e, $\bar{\nu}_e$ ν_μ, $\bar{\nu}_\mu$, ν_τ, and $\bar{\nu}_\tau$ emissions? Was there a ν_e flash? What is the equation of state of nuclear matter? How does rotation factor into collapse? Are hydrodynamic instabilities of the protoneutron star important? What is the spectrum of the emitted neutrinos? What are the mass and mass accretion history of the core? Are neutrino oscillations or neutrino masses factors? Are there any non-neutrino emissions? SN1987A was mute on all these questions. It is the prospect that a galactic supernova neutrino burst might be detected by extant or planned neutrino telescopes and thereby that some of these questions might be answered that brings us together for this workshop.

III. PREDICTED FEATURES OF NEUTRINO BURSTS

In such a short space, all the theoretical reasons for the burst structure to be described below can not be stated. I will attempt only a bit more than a catalog of the predicted energies, luminosities, and timescales during each phase in supernova and protoneutron star evolution.

a) Infall and Break-Out: The Early Phase

Figure 2 is an embellished version of Figure 1 in Myra and Burrows[6] (hereafter MB). It depicts the luminosities of the various neutrino species versus time for approximately the first hundred milliseconds after bounce. This is the earliest, most luminous, most "featureful" phase of the neutrino emissions. MB's calculation followed the collapse of the small 1.18 $M_\odot$ iron core of Nomoto and Hashimoto's[35] 13 $M_\odot$ star. Cores with other masses and density profiles will give quantitatively, but not qualitatively, different results. For instance, Bruenn[32] has performed a similar calculation, but with a fatter core, and has discovered remarkably similar features. His results are depicted in Figure 3. Though, as expected for the more extended core he employed, Bruenn's corresponding timescales are longer ($\times 2$–4), the similarities between Figures 2 and 3 are quite striking.

During collapse, the accelerating rate of electron capture provides a ramp in the electron neutrino (ν_e) luminosity, but in that of no other species. This ν_e ramp is the first neutrino signal of collapse. Its spectrum gradually hardens and is of a non-thermal, capture shape with an average ν_e energy that grows from 12 MeV to 16 MeV. However, the trapping of the ν_e's on infall ensures that only $\sim 10^{51}$ ergs is radiated before bounce.[20,21] The ramp-up in ν_e luminosity is quasi-exponential with a time constant of ~ 4.5 milliseconds in the calculation of MB. Any significant deviation from this overall timescale and low radiated energy (for instance, $\times 10$), would indicate new and interesting physics at odds with the standard model.

The capture ramp in the ν_e luminosity does not continue indefinitely, but reaches $\sim 3 \times 10^{53}$ ergs/s, approximately one millisecond after bounce at nuclear densities. As suggested in §II, the bounce creates a strong shock wave deep ($R \sim 10$–20 kilometers) in the neutrino-opaque core. This shock begins life at speeds near 7×10^9 cm/s (70 km/ms $\sim \frac{c}{4}$) and, hence, will not long be buried or obscured. At such early speeds, within only one millisecond, the shock reaches the neutrinospheres ("optical" depth ~ 1) that lie between 50 and 100 kilometers for all the various neutrino species (ν_e, $\overline{\nu}_e$, ($\nu_\mu, \overline{\nu}_\mu, \nu_\tau, \overline{\nu}_\tau$) $\equiv$ "ν_μ"). Shock break-out is accompanied both by a ν_e flash, as the ν_e's newly liberated by electron capture behind the front are released[36] and by the rapid ($<$0.5 ms) turn-on of the $\overline{\nu}_e$ and ν_μ radiations made possible by the high temperatures of the shock. The short timescale of the $\overline{\nu}_e$ and ν_μ turn-on can be derived by convolving the light-travel time across the neutrinospheres (100 km/(300 km/ms) ~ 0.3 ms) with the time it takes the shock to traverse approximately one mean-free-path near the neutrinospheres (e.g. $\sim$10 km/(50 km/ms) ~ 0.2 ms) and is a robust prediction of the standard model.

The break-out ν_e flash is quickly superposed on the earlier capture ramp and a ν_e luminosity of $\sim 6 \times 10^{53}$ ergs/s is achieved (see Figure 2). This spike may decay over 3–5 milliseconds and contain some multiple of 10^{51} ergs (a weak spike is required if the prompt supernova mechanism obtains), but does *not* represent significant neutronization. The electron-lepton number is still trapped and will take many seconds of diffusion through the entire opaque core to emerge. The ν_e flash signals the neutronization of only the low-density periphery of the protoneutron star.

The $\overline{\nu}_e$ and ν_μ luminosities gradually ($\sim$20 ms) increase as heat from the shocked mantle reaches the neutrinosphere. The $\overline{\nu}_e$ component is initially suppressed by the large early lepton asymmetry in the mantle, but within a few score milliseconds of neutronization, the ν_e and $\overline{\nu}_e$ luminosities merge. From this time onward, the luminosities of the six neutrino species are roughly the same, though during the first $\sim$100 ms there is a slight preponderance of ν_e's and $\overline{\nu}_e$'s. However, the ν_e and $\overline{\nu}_e$ spectra are interestingly different. Immediately after break-out, the ν_e spectrum makes a radical transition to a more thermal distribution, with an average energy of $\sim$10 MeV. Since the $\overline{\nu}_e$ neutrinosphere is a tad deeper in the core, its spectrum is harder, with an average energy of $\sim$13 MeV. The rapid spectral change of the ν_e's, the break-out ν_e flash, the rapid turn-on of the $\overline{\nu}_e$'s and ν_μ's, and the quick decay of the ν_e spike are all important diagnostics of shock dynamics in the standard model.

b) The First Hundreds of Milliseconds After Bounce: The Middle Phase

The ν_e, $\overline{\nu}_e$, and ν_μ spectra taken from MB 43 milliseconds after bounce are shown in Figure 4. The solid lines are the theoretical spectra and the dashed lines are thermal fits with non-zero pseudo-chemical potentials. It is seen that the spectra are "pinched" (they have a deficit at both high and low energies relative to a zero-temperature Fermi-Dirac distribution with the same average energy), but that they can be fit with a $(T, \eta = \frac{\mu}{T})$ pair of (2.4 MeV, 3.2), (3.1 MeV, 3.0), and (5.1 MeV, 4.1) for ν_e's $\overline{\nu}_e$'s and ν_μ's, respectively. Notice that the ν_μ's start with a high average energy of $\sim$24 MeV, since the lower ν_μ cross sections allow us to see still deeper into the hot core. The predicted deficit at high ϵ_ν's should have consequences for the long-term mechanism[37], neutrino nucleosynthesis[38], and the gross detectability of the signal and would be exciting to verify.[39]

In the calculations of both MB and Bruenn[32], due to the debilitating combination of nuclear dissociation and neutrino losses, the bounce shock fizzled into an accretion shock, not a supernova. Wilson[33] has suggested that neutrino heating of the shocked envelope between ~100 and ~500 kilometers over periods from hundreds of milliseconds to seconds will reenergize the shock into a supernova. Without going into the details here, some variant of this "long-term", delayed mechanism seems compelling.[37,40,41,42] Notice that if the prompt mechanism were viable, accretion onto the protoneutron star would be drastically truncated (the envelope is then exploding, not imploding). Since accretion can supply a large fraction of the total neutrino luminosity ($L_A \sim \frac{GM}{R}\dot{M}$) in the early phases, the initial decay of the neutrino fluxes would be steeper in the first 100 milliseconds after bounce if the prompt mechanism works than if the delayed mechanism works. Furthermore, the explosion clears away mass that contributes to the neutrino opacities. This would enable one to see deeper into the core to higher temperatures for all of the neutrino species and would thereby harden all of the spectra. This quick spectral hardening is diagnostic of explosion, be it prompt or delayed. In addition, it seems that a core that "rings" after bounce[6,32,41] modulates the ν_e and $\bar{\nu}_e$ luminosities with amplitudes approaching 50% and periods between 10 and 50 milliseconds if the prompt mechanism aborts, but the long-term mechanism succeeds, but not if the prompt mechanism succeeds (see Figures 2 and 3). Therefore, the supernova mechanism is stamped onto the neutrino signatures. If the delayed mechanism obtains, the luminosities after bounce stay higher, longer, the ν_e and $\bar{\nu}_e$ luminosities are modulated for hundreds of milliseconds to seconds until explosion, the explosion is accompanied by an abrupt but "late" spectral hardening, and the ν_e flash is more energetic. If the direct mechanism obtains, the initial luminosities decay more quickly and are not appreciably modulated, there is a rapid, early spectral hardening, and the ν_e flash is weaker. Furthermore, if the prompt mechanism obtains, less mass accumulates in the protoneutron star. This implies that the gravitational binding energy of the young neutron star, (and, hence, the total integrated neutrino loss) is smaller, all else being equal, in the prompt explosion scenario, than in the delayed scenario.

However, it should be noted that even if the prompt shock fizzles, the subsequent accretion may gradually decay before the delayed explosion commences. This itself leads not only to a gradual decay in the luminosity, but to a *gradual* hardening in the neutrino spectra for the reasons cited above. Bruenn[32] has noted, and MB confirms, that during the first hundreds of milliseconds after shock fizzle the average neutrino energies can grow ~50% while the envelope is "waiting" to explode. This gradual hardening would be interesting to identify in future data, but should be distinguishable from the abrupt hardening predicted at "explosion."

c) Protoneutron Star Cooling: The Late Phase

Though the neutrino emissions during the first hundred milliseconds after bounce are rich with diagnostic features, only 10–20% of the neutron star's binding energy has been radiated. The energy sources for the middle phase that includes and continues the first 100 ms are not only the gravitational binding energy of the outer mantle of the protoneutron star, which due to the loss of entropy and lepton number collapses quasi-statically, and the accretion energy, but, quite possibly, also the "convective" overturn of the mantle. The latter is driven by the unstable entropy and lepton gradients dynamically impressed on the protoneutron star periphery by a stalling shock. The nature and neutrino signatures of these instabilities are not yet known, but are under active investigation and may prove to be crucial to the supernova mechanism itself.[42,43,44]

However, after the accretion/mantle collapse/overturn phase and after the supernova is truly launched, a more staid protoneutron star cooling phase commences. Though the first second of protoneutron star life described above may have been vigorous and luminous, the long-term phase that follows may account for most of the energy emitted. Figure 5 from Burrows[3] depicts the $\bar{\nu}_e$ luminosity versus time during this late phase for various core sizes and accretion regimes (models 52–57, 72). (The reader is referred to that paper for a fuller explanation.) Generically, the emissions break up into the two phases expected: the middle accretion/mantle cooling phase with a decay constant near one second and the long-term core cooling phase with a much longer duration ($\tau \gtrsim 4$ seconds) that smoothly matches onto the middle phase. In fact, the luminosities during the long-term phase do not decay exponentially, but as power laws (index $\sim 1 \pm 0.5$) that indirectly reflect the nonlinear nature of neutrino transport. The long duration of the late core cooling phase is a consequence of the high densities and neutrino energies ($\to$ high opacities) in the protoneutron star interior that is now the source of the neutrino emissions. Though KII and IMB detected neutrino events during the first ~ 10 seconds, we expect that the thousand or so events anticipated from a galactic collapse will be spread out over many *tens* of seconds to a minute. It should be remembered though that during the long-term phase the neutrino spectra gradually soften and that a larger and larger fraction of the emitted neutrinos are shunted below the detector thresholds of ~ 5–10 MeV.

The protoneutron star cooling phase ends when the core becomes transparent. At that stage, both the luminosities and average neutrino energies plummet. This nose-dive is a diagnostic of cooling physics and may be observable with a galactic event. Since the "ν_μ" opacities are lower than the ν_e and $\bar{\nu}_e$ opacities, we expect that the ν_μ emissions should fall off first. Calculations indicate that this may happen after a *minute* or more.[30] The precise time will depend on, among other things, the softness of the nuclear equation of state, the protoneutron star mass, and, importantly, the opacity of matter at supranuclear densities. The latter must be corrected for Fermi-liquid and Pauli-blocking effects[3,30,45] which can decrease them by as much as a factor of five. Indeed, uncorrected opacities have forced some to continue accretion for as much as ten seconds to fit the KII data at late times.[46] The core opacities in those calculations were too high to release appreciable energy on the observed timescale. This resulted in excessive core masses and binding energies that, if our suspicion and opacity calculations are correct, is unnecessary. Data at late times will indirectly probe the exotic neutrino transport at supranuclear densities.

It is possible that a fat iron core with a thick envelope will experience a large accretion rate after bounce that smothers the supernova. If this situation obtains, the $\bar{\nu}_e$ luminosity may behave as in model 72 of Figure 5. In model 72, a high accretion rate resulted in a high luminosity over more than a second that abruptly (≤ 0.5 ms) turned off as a *black hole* formed. The maximum neutron star mass was exceeded and the core became general-relativistically unstable. This mechanism for stellar-mass black hole formation involves a protoneutron star intermediary with a quite distinctive neutrino signature. The characteristic short (1-2 seconds) duration, quick turn-off, and high luminosity are accompanied by a softer spectrum, since the thick, accreting envelope down-scatters the emerging neutrinos more than does the more tenuous accreting envelope in the standard neutron star formation scenario. If such an exotic neutrino burst is detected, it may indicate that a stellar-mass black hole was formed in a massive star and that the supernova was aborted (or a different mechanism was employed?). Since we know that the supernova and neutron star birth rates are roughly comparable to the massive O/B star death rates[47], only a small fraction (at most) of collapses should lead to stellar-mass black holes. Whether this intriguing black hole scenario is realistic or merely a theoretical fantasy is unknown, but might be resolved with a galactic neutrino burst and a lot of luck.

As this paper has sought to demonstrate, the neutrino emissions that attend core collapse, supernova explosions, and neutron star formation are elaborately diagnostic of the internal dynamics and evolution of this most exotic and violent of astrophysical events. A galactic burst, captured by the international network of neutrino-telescopes that is being established, should be fabulously rich in the answers to the many questions that remain in supernova theory. SN1987A was just a foretaste.

The author expresses his thanks to Jim Lattimer, Dave Arnett, Jim Wilson, Eric Myra, Steve Bruenn, Ron Mayle, Stan Woosley, and the penitude of others too numerous to cite individually for stimulating conversations on neutrino bursts and supernovae over the last few years. Special thanks are extended to David Cline for convening such a productive workshop. This work was supported through the NSF and NASA with grants AST 89-14346 and NAGW-2145, respectively.

REFERENCES

1. Hirata, K. *et al.*, 1987, *Phys. Rev. Lett.*, **58**, 1490 (Kamiokande, KII).

2. Bionta, R. M. *et al.*, 1979, *Phys. Rev. Lett.*, **58**, 1494 (IMB).

3. Burrows, A., 1988, *Ap. J.*, **334**, 891.

4. Burrows, A., 1990a, *Ann. Rev. Nucl. Part. Sci.*, **40**, 181.

5. Burrows, A., 1990b, in *Supernovae*, ed. A. Petschek (Springer: New York), p. 143.

6. Myra, E., and Burrows, A., 1990, *Ap. J.*, **364**, 222 (MB).

7. Bruenn, S., 1987, *Phys. Rev. Lett.*, **59**, 938.

8. Arnett, W. D., Bahcall, J. N., Kirshner, R. P., and Woosley, S. E., 1989, *Ann. Rev. Astr. Ap.*, **7**, 629.

9. Mayle, R., and Wilson, J. R., 1988, *Ap. J.*, **334**, 909.

10. Mayle, R., and Wilson, J. R., 1990, in *Supernovae*, ed. S. E. Woosley (Springer-Verlag: New York), p. **333**.

11. Woosley, S. E., and Weaver, T. A., 1986, *Ann. Rev. Astr. Ap.*, **24**, 205.

12. Mayle, R., Wilson, J. R., and Schramm, D., 1987, *Ap. J.*, **318**, 288.

13. Blinnikov, S. I., Imshennik, V. S., and Nadyozhin, D. K., 1988, in *Neutrino-88*, eds. J. Schneps, T. Kafka, W. A. Mann and Pran Nath (Singapore: World Sci.), p. 165.

14. Bethe, H. A., 1988, *Ann. Rev. Nucl. Part. Sci.*, **38**, 1.

15. Burbidge, E., Burbidge, G., Fowler, W. A., and Hoyle, F., 1957, *Rev. Mod. Phys.*, **29**, 547.

16. Colgate, S. A., and Johnson, H. J., 1960, *Phys. Rev. Lett.*, **51**, 235.

17. Arnett, W. D., 1966, *Can. J. Phys.*, **44**, 2553.

18. Colgate, S. A., and White, R. H., 1966, *Ap. J.*, **143**, 626.

19. Wilson, J. R., 1971, *Ap. J.*, **163**, 290.

20. Mazurek, T. J., 1974, *Nature*, **252**, 287.

21. Sato, K., 1975, *Prog. Theor. Phys.*, **54**, 1325.

22. Van Riper, K. A., 1978, *Ap. J.*, **221**, 304.

23. Lamb, D. Q., Lattimer, J. M., Pethick, C. J., and Ravenhall, G., 1978, *Phys. Rev. Lett.*, **41**, 1623.

24. Imshennik, V. S., and Nadyozhin, D. K., 1979, *Astrophys. Space Sci.*, **62**, 309.

25. Bethe, H. A., Brown, C. E., Applegate, J., and Lattimer, J. M., 1979, *Nucl. Phys.*, **A324**, 487.

26. Burrows, A., Mazurek, T. J., and Lattimer, J. M., 1981, *Ap. J.*, **251**, 325.

27. Bowers, R. B., and Wilson, J. R., 1982, *Ap. J.*, **263**, 366.

28. Hillebrandt, W., 1982, *Astr. Ap.*, **110**, L3.

29. Burrows, A., and Lattimer, J. M., 1985, *Ap. J. (Letters)*, **299**, L19.

30. Burrows, A., and Lattimer, J. M., 1986, *Ap. J.*, **307**, 178.

31. Baron, E. A., Cooperstein, J., and Kahana, S., 1985, *Phys. Rev. Lett.*, **55**, 126.

32. Bruenn, S. W., 1989, *Ap. J.*, **341**, 385.

33. Wilson, J. R., 1985, in *Numerical Astrophysics*, eds. J. Centrella, J. LeBlanc, and R. Bowers (Boston: Jones and Bartlett), p. 422.

34. LoSecco, J. M., 1989, *Phys. Rev.*, **D39**, 1013.

35. Nomoto, K., and Hashimoto, M., 1988, *Phys. Repts.*, **163**, 13.

36. Burrows, A., and Mazurek, T. J., 1983, *Nature*, **301**, 315.

37. Bethe, H. A., and Wilson, J. R., 1985, *Ap. J.*, **295**, 14.

38. Woosley, S. E., Hartmann, D. H., Hoffman, R. D., and Haxton, W. C., 1990, *Ap. J.*, **356**, 272.

39. Janka, H.-T., and Hillebrandt, W., 1989, *Astr. Ap.*, **224**, 49.

40. Lattimer, J. M., and Burrows, A., 1984, in *Problems of Collapse and Numerical Relativity*, eds. D. Bancel and M. Signore (Dordrecht: Reidel), p. 147.

41. Mayle, R., 1985, Ph.D. thesis, U. C. Berkeley (UCRL preprint no. 53713).

42. Burrows, A., 1987, *Ap. J. (Letters)*, **318**, L57.

43. Burrows, A. and Lattimer, J. M., 1988, *Phys. Repts.*, **163**, 51.

44. Wilson, J. R., and Mayle, R., 1988, *Phys. Repts.*, **163**, 63.

45. Goodwin, B. T., and Pethick, C. J., 1982, *Ap. J.*, **253**, 816.

46. Mayle, R., 1987, in *Proceedings of the Minnesota Workshop on SN 1987 A*, eds. T. Walsh and K. Olive (Minneapolis: Independence Press), p. 53.

47. van den Bergh, S., and Tammann, G. A., 1991, *Ann. Rev. Astr. Ap.*, in press.

FIGURE CAPTIONS

Figure 1:   A storyboard of the evolution of the core of a massive ($>8\,M_\odot$) star into a young neutron star via stellar collapse, supernova explosion, and copious neutrino emission.

Figure 2:   The luminosities of ν_e's, $\overline{\nu}_e$'s, and "ν_μ"'s (in 10^{51} ergs/s) versus time in milliseconds during the first 100 milliseconds after bounce from Myra and Burrows.[6] Bounce was at 112 milliseconds. Note the infall ν_e ramp, the break-out ν_e flash, the abrupt turn-on of the $\overline{\nu}_e$'s and ν_μ's, and the 10–20 millisecond oscillations of the ν_e and $\overline{\nu}_e$ emissions.

Figure 3:   Luminosities (in 10^{52} ergs/s) of the ν_e's, $\overline{\nu}_e$'s, and ν_μ's versus time after bounce in seconds taken from Bruenn.[32] Notice the qualitative similarity with Figure 2.

Figure 4:   The spectra (in 10^{51} ergs MeV^{-1} s^{-1}) of the ν_e's, $\overline{\nu}_e$'s, and ν_μ's versus neutrino energy (in MeV) at 43 milliseconds after bounce from Myra and Burrows.[6] The solid lines are results of the multi-group calculation and the dashed lines are fits with Fermi-Dirac spectra with non-zero chemical potential. See text for details.

Figure 5:   The long-term luminosity (in ergs/s) versus time (in seconds) for various protoneutron star models taken from Burrows.[3] A stiff EOS, as described in Burrows and Lattimer[30], was employed. Two phases of neutrino emission are clearly discerned. Notice model 72 in which a black hole (BH) formed after $\lesssim 2$ seconds of accretion.

Table 1 (from Burrows 1990a)

SN1987A NEUTRINO DATA

Detector	Event #	Time[a] (seconds)	Electron energy (MeV)	Angle wrt LMC (degrees)	Comment
IMB	1	7:35:41.374(UT)	38 ± 7	80 ± 10	±50 ms
	2	0.411	37 ± 7	44 ± 15	
	3	0.650	28 ± 6	56 ± 20	
	4	1.141	39 ± 7	65 ± 20	
	5	1.562	36 ± 9	33 ± 15	
	6	2.683	36 ± 6	52 ± 10	
	7	5.010	19 ± 5	42 ± 20	
	8	5.581	22 ± 5	104 ± 20	
KII	1	7:35:35(UT)	20.0 ± 2.9	18 ± 18	±1 min.
	2	0.107	13.5 ± 3.2	40 ± 27	
	3	0.302	7.5 ± 2.0	108 ± 32	
	4	0.323	9.2 ± 2.7	70 ± 30	
	5	0.507	12.8 ± 2.9	135 ± 23	
	6	0.685	6.3 ± 1.7	68 ± 77	background
	7	1.540	35.4 ± 8.0	32 ± 16	
	8	1.728	21.0 ± 4.2	30 ± 18	
	9	1.915	19.8 ± 3.2	38 ± 22	
	10	9.219	8.6 ± 2.7	122 ± 30	
	11	10.432	13.0 ± 2.6	49 ± 26	
	12	12.439	8.9 ± 1.9	91 ± 39	
Baksan	1	7:36:11.818(UT)	12 ± 2.4		
	2	0.435	18 ± 3.6		
	3	1.710	23.3 ± 4,7		
	4	7.687	17 ± 3		
	5	9.099	20.1 ± 4.0		
Mont Blanc	1	2:52:36.792(UT)	7		IMB−4.7 hrs.
	2	3.857	8		
	3	4.215	11		
	4	5.904	7		
	5	7.008	9		

[a]The UT times on February 23rd, 1987, are given for the first event and, for the events thereafter, the time in seconds relative to the time of the first events is given.

474

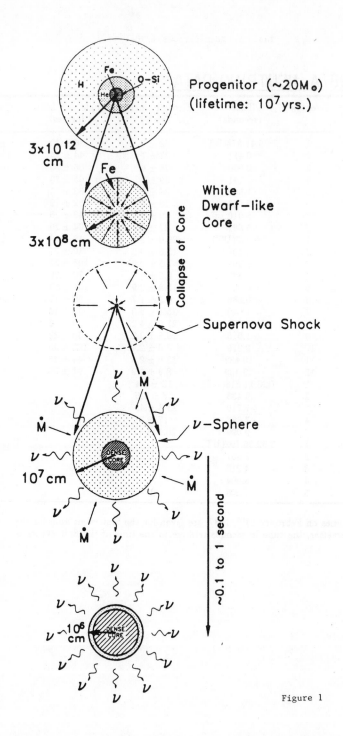

Progenitor (~20M$_\circ$)
(lifetime: 10^{7}yrs.)

3x10^{12}
cm

White
Dwarf—like
Core

3x10^{8}cm

Collapse of Core

Supernova Shock

ν—Sphere

10^{7}cm

~0.1 to 1 second

10^6
cm

Figure 1

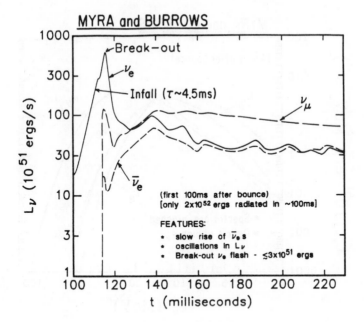

Figure 2

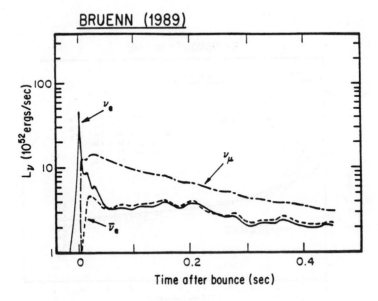

Figure 3

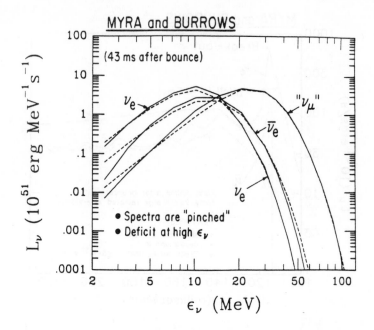

Figure 4

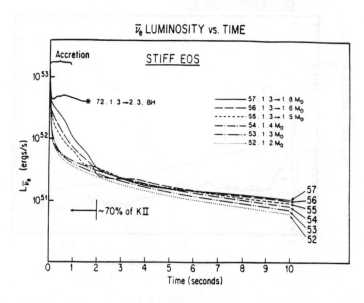

Figure 5

Preliminary Estimates of Core-Collapse Supernova Rates from the Berkeley Automated Supernova Search

Carl Pennypacker, Saul Perlmutter, Richard A. Muller, Ned Hamilton, Craig Smith, Tim Sasseen, Shawn Carlson, Heidi Marvin, Li-Ping Wang, Frank Crawford
Lawrence Berkeley Laboratory

Richard Treffers
Department of Astronomy, UC Berkeley

Sidney Bludman*
Center for Particle Astrophysics, U.C. Berkeley
and Lawrence Berkeley Laboratory

Abstract

Over the last few years, the Berkeley Automated Supernova Search has discovered 19 supernova. We present here preliminary measurements of supernova rates from a subset of these supernovæ, which were found from a sample of well-monitored galaxies. Two effects are apparent: a surprising number of the supernovæ are Type Ic supernovæ, and most (17 out of 19) of the supernovæ are in late spiral galaxies. Less luminous supernovæ may still be escaping our attention. If the Milky Way galaxy is a late spiral, the core-collapse supernova rate (i.e. for supernovae of types II, Ib, and Ic) should be at least one supernova every 44 ± 9 years (assuming H = 75 km/sec/Mpc). A crude estimate of the core collapse supernova rate in Sc and Sbc galaxies within 5 Mpc would be one supernova every 7 years. These rates have important implications for the design of supernova neutrino, gamma-ray, and gravity-wave detectors. Plans have been developed to extend our search to a better site with more sensitivity to understand the true core-collapse supernova rate in nearby galaxies.

* Currently on leave from Dept. of Physics, University of Pennsylvania

I. The Experiment

Over the last decade, based on ideas of Stirling Colgate (see Colgate, et al., 1975), we have developed the capability to search over 600 galaxies per night for supernovæ. Our limiting magnitude in the present prototype site is between 16th and 17th visual magnitude. The automated search system is described elsewhere (see Kare et. al., Pennypacker et. al, and Perlmutter et al.) and consists of an computer-controlled telescope, a Charge-Coupled-Device Camera (CCD), and computers for real-time image analysis. About thirty person-years of software development have gone into bringing this system to its present state, five miles from the Berkeley campus. This system has continuously evolved in the degree of automation since its inception, and now functions without an operator at the telescope. Candidate images are shipped over a network back to the lab, and inspected by a physicist in the morning.

One of the strengths of our system is that we can find relatively dim supernovæ, at any phase of the moon, often obscured by the background galaxy light. We have found 19 supernovæ since 1986, but in the rates discussed in this paper are only from the supernovæ found since 1988, when the automated system with a data-base came on line. We take exposures of about one minute. This exposure length is a compromise between different goals of the experiment: we want to discover under-luminous nearby events; we want to find bright, distant Type Ia supernovæ (easily seen at 7500 km/sec); and we want to come back to galaxies as quickly as possible to catch the supernova and watch its evolution. As discussed in the next section, the evolution of the spectrum of these events has made an important change in our understanding of supernova classifications. A more detailed exposition of the results of the search will be presented in a forthcoming paper (Muller et al.).

II. Background

Historically, research has identified Type II supernovæ (classified by the presence of hydrogen in the optical spectra) as core collapse supernovæ, and Type I supernovæ as thermonuclear-driven supernovæ. This simple description has been greatly complicated by the identification of sub-classes of Type I's (Ib's and Ic's), which may be core-collapses. Type Ib's and Ic's (which we shall collectively refer to as type Ibc) are characterized by forbidden oxygen and calcium lines. For example, one supernova discovered by our group (SN1987K) was observed by Alex Filippenko to transform from a Type II (with hydrogen lines) to a Type Ic. The general belief in the community is that Type Ic supernovæ are core collapse events. It has long been known (e.g, see Baade, 1941) that Sc galaxies have more supernovæ.

Supernova	galaxy	galaxy class	IAU circular	date	type	m_{CCD}
1986I	NGC 4254	Sc	4219	5/17	II	14
1986N	NGC 1667	Sc	4287	12/11	Ia	15
1986O	NGC 2227	Scd	4298	12/24	Ia	14
1987K	NGC 4651	Sc	4426	7/28	II→Ic	15
1988H	NGC 5878	Sb	4560	3/3	II	15.5
1988L	NGC 5480	Sc	4590	5/3	Ic	16.5
1989A	NGC 3687	Sbc	4721	1/19	Ia	15.3
1989B	NGC 3627	Sb	4726	1/30	Ia	12
1989L	NGC 7339	Sbc	4791	6/1	II	16
1990B	NGC 4568	Sbc	4949	1/20	Ic	16.0
1990E	NGC 1035	Sc	4965	2/15	II	16.7
1990H	NGC 3294	Sc	4992	4/9	II	16.5
1990U	NGC 7479	Sc	5063	7/28	Ic	16
1990aa	UGC 540	Sc	5087	9/4	Ic	17
1991A	UGC 6872	Sc	5153	1/2	Ic	17
1991B	NGC 5426	Sc	5163	1/12	Ia	16
1991G	NGC 4088	Sbc	5188	3/11	II	17
1991M	IC 1151	Sc	5207	3/13	Ia	16.7
1991N	NGC 3310	Sc	5227	3/30	?	15

Table I

Table caption: Supernova discovered by the Berkeley team. (SN1989B and 1991G were first found by other groups, but were discovered independently by our system.) The first column lists the supernova designation; the second column the parent galaxy, the third column the galaxy Hubble classification, the fourth column the IAU circular where the supernova discovery was announced, the fifth column the discovery date, the sixth column the type of supernova, the seventh column the observed apparent "CCD" magnitude (m_{CCD} is an uncalibrated combination of V and R photometric bands) at discovery (not peak magnitude).

III. The Discoveries

Every clear night we observe galaxies from our reference galaxy list. Although our search list includes galaxies out to recession velocity of 7500 km/sec (=100 Mpc for H=75), we begin to lose our sensitivity to Type Ic supernovæ at a distance of about 12 Mpc. The discovered supernovæ, the galaxy information, and other parameters are shown in Table I. The absolute (intrinsic) magnitude of the supernova plays an important role of how distant we would see it. For example, a Type Ia supernova in the nearby Virgo cluster of galaxies of absolute magnitude −19 yields about 12.5 apparent magnitude supernova at earth. However, we have found Type Ic supernova closer than Virgo 100 times fainter than this, which are near our magnitude limit of about 16th to 17th magnitude. We believe this indicates we are beginning to sample a previously unmeasured part of the supernova luminosity function.

IV. Preliminary Lower Limit on the Supernova Rate

The calculation of the supernova rate depends on the survey period of galaxies and the number of supernova detected. We are in the process of carefully calculating the the exact survey period for our sample of galaxies, and regard our rates as preliminary   Given our current estimate of survey period, and the fact we have seen three Type II supernovæ in the Sc/Sbc galaxies, and five Ib-c supernovæ, we find there is one core-collapse supernova per every (42 ± 9) years x h^2 per 10^{10} L_{sun}. This number is different than that in Muller et al., since Type Ia's have been subtracted from the rate. We may be   missing dimmer Type II's and Type Ic's, because of the Type Ic and Type II low luminosity. Hence this rate is a lower limit, and the true core-collapse supernova rate may be   higher.

V.   Interpretation

As initially pointed out by Becklin (1991), nearby bright infrared galaxies, produce many of the nearby, albeit cloaked in dust, supernovæ. As mentioned previously, we do not yet have sufficient data to understand the distribution in absolute luminosities of our supernovæ, but we have some evidence that there are low luminosity Type Ic events. For example, one of our events, SN1990B, had its apparent magnitude measured by Sturch, et. al, (1990), and when combined with the Tully-Fisher distance to that galaxy, is found to have an absolute magnitude of -13.3 (This supernova's magnitude was photometrically measured within a week of estimated maximum light.) This is about a factor of one-hundred times dimmer than the "standard" Type Ia supernova, and is barely detected at 12 Mpc.

Given our results on external galaxies, with an assumption of the morphological type and luminosity of the Milky Way, we can compute a Milky Way core collapse rate. The morphological type of the Milky Way galaxy is either Sbc (Hodge, 1983) or Sb (King, 1990). Assuming a luminosity of 1.7 x 10^{10} L_{sun}, (King, 1990), and assuming the Milky Way is an Sbc galaxy, we calculate one core-collapse supernova at least every 44 ± 9 years. Using luminosities for nearby galaxies from the Tully-Fisher galaxy catalog, we find for late spirals within 5 Mpc about 13 L_{10} blue luminosities. Using our rate and h = 0.75, we expect one supernova about every 7 years or so, although we emphasize that there are substantial uncertainties in this value. Most of the supernovae that we discover are near our limiting magnitude, and so there is a possibility that the rate is substantially higher. The supernova rate in these nearby galaxies is important for other detectors (neutrino and gravity wave) that have some chance of detecting supernovae. A new telescope has has been designed and partially completed by our group at Berkeley for installation at an improved site. This system can extend our sensitivity by several magnitudes, and determine whether there are additional supernovae beyond our present threshold.

Acknowledgment: This work has been supported by the U.S. Department of Energy under contract number DE-AC03-76SF00098, the Ann and Gordon Getty Foundation, and the National Science Foundation.

References

Becklin, E., 1991, in this volume.

Baade, W. A., Zwicky, F., 1941, in *Novae and White Dwarfs,* ed. A.J. Shaler, Colloque Internationale d"Astrophysique a Paris p.177.

Colgate, S.A., E.P. Moore, R. Carlson. 1975, *Pro. Astronomical Soc. Pacific,* **87,** pg. 565.

Filippenko, Alexei V., 1990, *Astronomical Journal,* **100,** 5, pg. 1575.

Hodge, Paul, 1983, *Pro. Astronomical Soc. Pacific,* **95,** pg. 721.

Kare, J.T., M.S. Burns, F.S. Crawford, P.G. Friedman, R.A. Muller, C.R. Pennypacker, S.Perlmutter, R. Williams, 1980, *Rev. of Scientific Instruments,* **59(7),** 1021-1030.

King, et. al., 1990, "The Milky Way as a Galaxy," University Science Books, Mill Valley, Ca.

Muller, R.A., Marvin, H., C.R. Pennypacker, S. Perlmutter, T. Sasseen, C. Smith, 1991, "High Rate for Type Ic Supernovæ," submitted to *Ap. J. Letters*

Pennypacker, C., F. Crawford, H. Marvin, R. Muller, S. Perlmutter, T. Sasseen, C. Smith, R. Treffers,L. Wang., 1989, in *Particle Astrophysics: Forefront Experimental Issues,* E. B. Norman editor, pp.196-197 (World Scientific).

Perlmutter, S., F. Crawford, R.A. Muller, C.R. Pennypacker, T.S. Sasseen, Smith, R. Treffers, and R. Williams, "The Status of Berkeley's Real-Time Supernova Search," *Instrumentation for Ground-Based Optical Astronomy,* ed. L.B. Robinson, New York: Springer-Verlag,p.674-680 (1988).

Sturch, T., T. Tyson, P. Guhathkurta, 1990, *IAU Circular,* # 4961, Jan 28.

van den Bergh, S., 1991, "Supernova Rates, Galactic Emission and Hubble Types", *Astronomical Journal,* in press.

van den Bergh, S., and G. Tammann1991, "Galactic and Extragalactic Supernova Rates", *Annual Reviews of Astronomy and Astrophysics.*

Solar and Supernova Neutrino Interactions

Wick C. Haxton

Department of Physics, University of Washington, Seattle, Washington 98195

Abstract

Two topics are addressed, the interactions of neutrinos during a type II supernova and the effect of current eddies on solar neutrino oscillations. The supernova discussion focuses on the nucleosynthesis that accompanies inelastic neutral current interactions of neutrinos in the mantle of a collapsing star, and on the effect of neutrino "down-scattering" and preheating on the explosion mechanism. The second half of the talk deals with the influence of solar turbulence (or density fluctuations) on the neutrino effective mass and the possibility that a time-varying neutrino flux could result. The effects of harmonic density or three-current perturbations on the oscillation probability are explored analytically and numerically.

Both the prompt [1] and the delayed [2] models of type II supernovae appear to succeed only in very special cases. This has motivated a re-examination of various aspects of the supernova model with the hope that the discovery of some missing physics might yield a more robust mechanism. As 99% of the collapse energy is radiated in neutrinos, one interesting issue is the dissipative neutrino reactions that might help drive an explosion.

In standard hydrodynamical codes various leptonic and semileptonic neutrino reactions are generally included

- $\nu_e + n \leftrightarrow p + e^-$
- $\bar{\nu}_e + p \leftrightarrow n + e^+$
- $\nu + e^- \leftrightarrow \nu + e^-$
- $e^+ e^- \leftrightarrow \nu \bar{\nu}$
- nuclear β-decay
- neutral current coherent scattering

Until very recently [3], however, inelastic neutral current reactions with nuclei were omitted

- $\nu + A \leftrightarrow \nu + A^*$
- $\nu + \bar{\nu} + A \leftrightarrow A^*$

Such reactions are intriguing because all six flavors participate. In particular, the heavy-flavor (muon and tauon) neutrino spectra are characterized by temperatures $T \sim 8$

MeV, compared to $T_{\nu_e} \sim 4$ MeV. As the cross sections for many nuclear reactions rise very steeply with T due to nuclear threshold effects on the phase space, the ν_μ's and ν_τ's may dominate many reactions. A second important factor is the rapid rise in the strength of first-forbidden transitions with three-momentum transfer. Thus such transitions begin to contribute for more energetic neutrinos. First-forbidden transitions prove to be responsible for most of the scattering by nuclei like ^{4}He, ^{12}C, ^{16}O, and ^{20}Ne that make up much of the star's mantle [3].

The sum rules governing first-forbidden transitions lead to a heating rate that is roughly independent of the nuclear species. The energy transferred by neutrino inelastic scattering in nuclear matter is [4]

$$\Delta \dot{E} \sim (90 \text{ MeV/nucleon/sec}) \left(\frac{\mathcal{L}_\nu}{10^{52} \text{ergs/sec}} \right) \left(\frac{1}{R_7} \right)^2 \left(\frac{10 \text{ MeV}}{T_\nu} \right) \tag{1}$$

where $\mathcal{L}_\nu$ is the neutrino luminosity (in all flavors), R_7 is the distance from the center of the star in units of 10^7 cm, and T_ν is the temperature of the heavy-flavor neutrinos. $R_7 \sim 1$ is a location somewhat outside of the neutrinosphere, and is thus a relevant value for heating that might influence shock wave propagation, while $\mathcal{L}_\nu \sim 10^{52}$ erg/sec is typical of type II supernovae in the first few seconds following core bounce. The resulting $\Delta \dot{E} \sim 90$ MeV/nucleon/sec is large compared to the typical binding energies of nucleons in nuclei, ~ 8 MeV, though we shall see later that this comparison is a bit misleading.

Three consequences of this heating rate have been explored. Perhaps the most interesting of these, which we call the "neutrino process", has been described in the literature [4], so I will only provide a brief summary. In neutral current inelastic scattering the energy transfer to the nucleus is typical of giant resonances, ~ 20 MeV. It follows that the scattering affects the chemistry of the star, inducing neutron, proton, or α-particle spallation from the parent nucleus, producing a new daughter nucleus. The effects are important even at great distances from the star's core. For instance at $R_7 \sim 250$, a value typical of the Ne shell in a 25 $M_\odot$ progenitor, 0.1% of all nuclei would be affected. Thus it is apparent that one has a potentially important new mode of nucleosynthesis in the case of various less abundant, odd-A isotopes.

Quantitative predictions require great care. The free nucleons produced by the spallation reactions can react with the daughter isotope to destroy the product of interest. Thus one must carefully model the nuclear chemistry that governs the processing of the coproduced nucleons. In addition the heating that accompanies passage of the shock wave through the production region can destroy fragile products.

The results of the detailed network calculations that are summarized in Ref. 4 show that many rare isotopes, including a number of the odd-A nuclei from boron through copper, owe much of their present abundance to the neutrino process. Specific nuclei due almost entirely to this production mechanism are ^{7}Li, ^{11}B, ^{19}F, ^{138}La, and ^{180}Ta. The production of ^{7}Li at ~ 0.6 of its present galactic abundance suggests that most of the contemporary ^{7}Li was synthesized along with other metals, rather than being produced in the big bang. The neutrino-induced production of ^{11}B resolves the long-standing problem that the $^{11}B/^{10}B$ ratio is inconsistent with the accepted scenario for producing these elements, spallation following high-energy cosmic ray interactions with ^{12}C. The productions of ^{10}B, ^{15}N, ^{22}Na, ^{26}Al, ^{31}P, ^{35}Cl, $^{39,40,41}K$, ^{45}Sc, $^{47,49}Ti$, $^{50,51}V$, ^{55}Mn, ^{59}Ca, and ^{63}Cu are sufficiently robust that these elements could also owe their origin primarily to the neutrino process, given the astrophysical and nuclear uncertainties in the modeling

Bruenn and Haxton [5] recently studied a second aspect of these inelastic neutrino reactions, the effect of the enhanced "downscattering" on lepton number loss during the infall stage of stellar collapse. As the cross section for coherent scattering off nuclei varies as E_ν^2, low energy neutrinos have longer mean free paths and thus larger probabilities of escaping during infall. The rate that the star radiates ν_e's determines the net lepton number that will be trapped in the core, a crucial quantity affecting the subsequent prospects for an explosion: large lepton number losses produce a smaller core, a weaker shock wave at rebound, and a thicker layer of overlying iron that the shock wave must penetrate before reaching the star's mantle [6]. (The principal energy loss occurs as the shock wave attempts to penetrate the outer iron core: ~ 8 MeV/nucleon is lost as the Fe "boils" into free nucleons and alpha particles.) As the "downscattering" permits larger lepton number losses, its occurrence reduces the chances for a successful explosion.

Bruenn [7] first observed that neutrino-electron scattering is an important neutrino energy loss mechanism during infall, serving to bring the suprathermal ν_e spectrum produced in electron capture into thermal equilibrium with the matter. This, as argued above, then decreases the trapped lepton number. The calculations by Bruenn and Haxton [5] show that, in the absence of neutrino-electron elastic scattering, inelastic neutrino-nucleus scattering has approximately the same effect on equilibrating the ν_e's to matter during infall.

Thus one might worry that very large lepton number losses would result when both mechanisms operate simultaneously, which would be disastrous in current explosion models. Fortunately, when both mechanisms are turned on, the stellar model response is highly nonlinear: the net result is only a slight increase in lepton number loss over that obtained when either $\nu_e + e \leftrightarrow \nu_e + e$ or $\nu_e + A \leftrightarrow \nu_e + A^*$ operates separately. In effect, once a robust downscattering mechanism is allowed to operate, the downscattering is limited by the rate at which thermalized ν_e's can escape. Further increasing the inelasticities has little net effect because the absence of empty, low-energy neutrino states inhibits the downscattering. This is a nice result, removing one principal concern that ν_e-A inelasticities might negatively affect the prospects for a hydrodynamic explosion.

A third aspect of ν-A interactions seems quite promising from the perspective of the explosion: neutrinos escaping at about the time of the core bounce might "preheat" iron lying ahead of the shock front, thus performing part of the work usually attributed to the shock wave.[3] In usual models about 2/3 of the shock wave losses are attributed to the dissociation of iron, and about 1/3 to neutrino losses. The notion of "preheating" suggests that this calculation is naive, overestimating the losses, since some of the neutrinos do useful work.

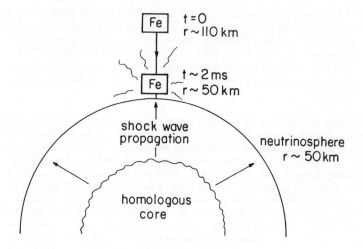

Fig. 1   Illustration of the "preheating" scenario discussed in the text.

The situation one might envision is illustrated in Fig. 1. At time t=0 sec a shock wave is produced at the edge of the homologous core and travels rapidly outward, penetrating the neutrino trapping radius (the neutrinosphere) at $r \sim 50$ km 2msec later. A volume of cold iron, initially at $r \sim 110$ km when t=0 sec, will fall inward at $v_{infall} \sim 0.6 v_{freefall} \sim 0.1c$, meeting the shock wave at the neutrinosphere 2ms later. This illustration, a qualitatively accurate representation of the early development of the shock wave, illustrates that the infalling volume of iron spends very little time at small r, where the neutrino flux is intense. Thus one concludes that luminosity, not the total fluence, is the most important factor determining the extent of the preheating.

The peak neutrino luminosity occurs at about the time the shock wave breaks through the neutrinosphere. The "neutronization burst" lasts only ~ 3 msec, but generates a peak luminosity $\sim 5 \cdot 10^{53}$ erg/sec. Thus it is beautifully timed to maximize the heating at the early times illustrated in Fig. 1. Because the neutronization burst is dominated by ν_e's, most of the heating is accomplished by charged-current reactions.

A back-of-the-envelope calculation made by Haxton [3] suggested that the preheating would be very significant. Using a Fermi-Dirac neutrino spectrum characterized by $T \sim 4$ MeV, the preheating was found to be 1.5 MeV/nucleon. Such energy deposition would quite significantly enhance the prospects for a prompt explosion. Ray and Kar [8], in a somewhat more elaborate calculation, agreed with this conclusion, finding that $0.7 \cdot 10^{51}$ ergs would be deposited in the star's mantle, roughly 25% of the total energy carried by the neutronization pulse. Unfortunately the calculation of Bruenn and Haxton [5], which embedded detailed double-differential neutrino cross sections into a full hydrodynamic code, found a heating rate only one-third as large. The origin of the discrepancy is the prediction, in the realistic calculation, that the high-energy tail of the neutrino spectrum is depleted relative to the simple Fermi-Dirac form. The highest energy neutrinos are of particular importance because of their increased phase space, including the greater portion of the nuclear Gamow-Teller distribution they can reach. If the neutrino spectrum from the full hydrodynamic calculation is accurate, the neutrino preheating, while significant, will not be a major perturbation on standard supernova models.

Thus we conclude, from the Bruenn and Haxton calculations [5], that neither the enhanced downscattering nor the preheating is as important as one might naively conclude.

The second half of this talk addresses another topic in neutrino astrophysics, the puzzling possibility that the ^{37}Cl solar neutrino counting rate varies with time [9]. Recent maximum likelihood analyses yielded a confidence level of only ~ 0.01 for a time-independent signal.[10] There have been suggestions that the counting rate is varying with a period ~ 10 years and is approximately anticorrelated with the solar cycle. A frequently discussed theoretical explanation for such an anticorrelation is the interaction of a neutrino magnetic moment with the solar magnetic field [11]. The most attractive scenario is a $\nu_{e_L} \rightarrow \nu_{\mu_R}$ oscillation, where a level crossing occurring at a special solar density allows the off-diagonal magnetic interaction to mix degenerate ν_{e_L} and ν_{μ_R} states [12].

Recently Haxton and Zhang [13] investigated another possibility that requires no new particle physics, apart from the standard Mikheyev-Smirnov-Wolfenstein (MSW) mechanism [14]. One can show that the effective mass appearing in Wolfenstein's equation depends not just on the solar density, but also on solar three-currents. While such currents presumably involve velocities $v \ll c$, and therefore do not generate large changes in the effective mass, they can have an important effect on the derivative of the effective mass. The results are particularly interesting when a series of eddy currents produces a perturbation that is approximately harmonic.

Under the assumption that the solar fluid is incompressible, $\vec{\nabla} \cdot \vec{j}(x) = 0$, one finds [13]

$$\left[E - \frac{M_D M_D^\dagger}{2E} - \sqrt{2}\, G_F(\rho(x) + j_x(x))N + i\, \frac{d}{dx} \right] |\nu_L(x) >= 0 \,. \tag{2}$$

where $j_x(x)$ is the component of the electron current along the neutrino's direction of motion. The left-handed neutrino wave function in Eq. (2) is a vector in flavor space, M_D is the mass matrix which we take to be purely Dirac, and N allows the third term to operate only on the electron component of $|\nu_L(x) >$. This is Wolfenstein's equation [14] with the solar electron density term modified by the addition of the current.

If one defines the left-handed flavor eigenstates in terms of the vacuum mass eigenstates, assuming only two flavors,

$$
\begin{aligned}
|\nu_e > &= \cos\theta_v |\nu_1 > + \sin\theta_v |\nu_2 > \\
|\nu_\mu > &= -\sin\theta_v |\nu_1 > + \cos\theta_v |\nu_2 >
\end{aligned}
\tag{3}
$$

the mass matrix in the flavor basis becomes

$$M_D = \begin{pmatrix} \overline{m} - \delta m \cos 2\theta_v & \delta m \sin 2\theta \\ \delta m \sin 2\theta & \overline{m} + \delta m \cos 2\theta_v \end{pmatrix} \tag{4}$$

where $\overline{m} = (m_1 + m_2)/2$ and $\delta m = (m_2 - m_1)/2$ where $m_1 < m_2$. Writing $|\nu(x)> = a_e(x)|\nu_e > + a_\mu(x)|\nu_\mu >$, Eq. (2) becomes, after removing an overall phase factor of $E - \frac{(m_1^2 + m_2^2)}{4E} - \frac{\sqrt{2}G_F}{2} \int_{x_i}^{x} (\rho(x') + j_x(x'))dx'$,

$$i \frac{d}{dx} \begin{pmatrix} a_e \\ a_\mu \end{pmatrix} = \begin{pmatrix} \frac{1}{\sqrt{2}} G_F(\rho + j_x) - \frac{\delta m^2}{4E} \cos 2\theta_v & \frac{\delta m^2}{4E} \sin 2\theta_v \\ \frac{\delta m^2}{4E} \sin 2\theta_v & -\frac{1}{\sqrt{2}} G_F(\rho + j_x) + \frac{\delta m^2}{4E} \cos 2\theta_v \end{pmatrix} \begin{pmatrix} a_e \\ a_\mu \end{pmatrix} \tag{5}$$

It is instructive to rewrite Eq. (4) in a basis consisting of the local mass eigenstates

$$|\nu_L(x) > = \cos\theta(x)|\nu_e > - \sin\theta(x)|\nu_\mu >$$
$$|\nu_H(x) > = \sin\theta(x)|\nu_e > + \cos\theta(x)|\nu_\mu > \tag{6}$$

where

$$\sin 2\theta(x) = \frac{\sin 2\theta_v}{[X(x)^2 + \sin^2 2\theta_v]^{1/2}}$$
$$\cos 2\theta(x) = \frac{-X(x)}{[X(x)^2 + \sin^2 \theta_v]^{1/2}} \tag{7}$$

where $X(x) = 2\sqrt{2}G_F(\rho + j_x)E/\delta m^2 - \cos 2\theta$. Thus $\theta(x)$ ranges from θ_v to $\pi/2$ as the density goes from 0 to ∞. The corresponding eigenvalues are

$$\lambda_{\substack{H \\ L}}(x) = \pm \frac{\delta m^2}{4E} [X(x)^2 + \sin^2 2\theta]^{1/2} = \pm \lambda(x)$$

If we define $|\nu(x)> = a_H(x)|\nu_H(x) > + a_L(x)|\nu_L(x) >$, the mass eigenstate amplitudes satisfy

$$i \frac{d}{dx} \begin{bmatrix} a_H \\ a_L \end{bmatrix} = \begin{bmatrix} \lambda(x) & i\alpha(x) \\ -i\alpha(x) & -\lambda(x) \end{bmatrix} \begin{bmatrix} a_H \\ a_L \end{bmatrix} \tag{8}$$

where $\alpha(x) = \sqrt{2} \frac{E}{\delta m^2} \frac{\sin 2\theta_v}{(X^2 + \sin^2 2\theta_v)} G_F \frac{d}{dx}(\rho(x) + j_x(x))$.

Consider Eq.(8) in the adiabatic limit, so that $\frac{d\rho}{dx}$, which varies smoothly over the solar radius, can be ignored, leaving $\frac{d}{dx} j_x$ to determine $\alpha(x)$. Even though one expects $|j_x| \ll |\rho|$, if j_x varies over a short distance scale, $|\frac{d}{dx} j_x|$ may be appreciable. Furthermore, we consider a pattern where the variation in j_x is sinusoidal with frequency ω; this approximates the pattern that would be seen by a neutrino passing through a region of localized eddy currents. In this case, even if $|\frac{d}{dx} j_x|$ is too small to break the nominal adiabatic condition $\lambda(x) \gg |\alpha(x)|$, nonadiabatic behavior can arise at that

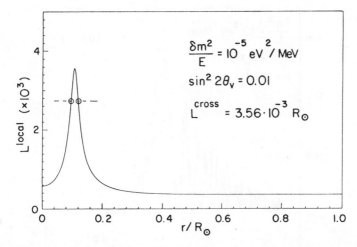

Fig. 2.   The local neutrino oscillation frequency as a function of solar radius for a neutrino produced in the sun's center with $\delta m^2/E = 10^{-5}$ eV^2/MeV and $\sin^2 2\theta_v$
$= 0.01$. The intersection with the dashed line illustrates two points where a
perturbation of fixed oscillation length could lead to a frequency matching.

point where the perturbation frequency matches the local oscillation frequency. As the
local oscillation frequency achieves its minimum value at the level crossing point, two
resonance points will generally occur for $\omega \gtrsim \lambda_c$. This is illustrated, for one example,
in Fig. 2.

To explore the consequences of such "frequency matching" we choose a current
density of the form

$$j_x(x) = h\bar{\rho}(x)\sin\omega(x - x_0) \tag{9}$$

That is, the current is proportional to an average density $\bar{\rho}(x)$, varies sinusoidally with
angular frequency ω, and has an amplitude $h \sim v/c$, where v is the current velocity.
Alternatively, one can reinterpret $j_x(x)$ as a (static) solar density variation. Although
the derivation is too complicated to reproduce here, one can treat this problem analytically in first order perturbation theory [13]. Let x_{cross} denote the level crossing
point, so that $X(x_{cross}) = 0$. As Fig. 2 illustrates, for $|\omega| \gtrsim 2\lambda(x_{cross}) = \frac{\delta m^2}{2E}\sin 2\theta_v$,
there exist two stationary phase points (i.e., points where the perturbation and local
oscillation frequencies are equal). As $|\omega| \to \frac{\delta m^2}{2E}\sin 2\theta_v$ from above, these points merge
at the crossing point. For $|\omega| < \frac{\delta m^2 \sin 2\theta_v}{2E}$, there is no stationary phase point, though
the crossing point remains the point of closest approach to frequency matching. One
can treat the physics of all three regions by exploiting the uniform approximation. If
cleverly done, the result is an expression that reduces precisely to the adiabatic and

490

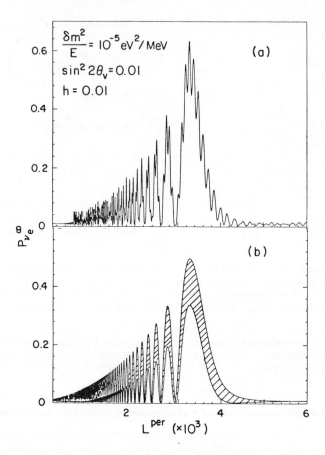

Fig. 3. Neutrino survival probabilities for a highly adiabatic transition ($\gamma_c \sim 24$) with $\sin^2 2\theta = 0.01$, $\delta m^2/E = 10^{-5}$ eV^2/MeV, and $h = 0.01$, with h the amplitude of an oscillatory perturbation (eddy currents or density fluctuations). Results are given as a function of the perturbation scale L^{per}, in units of $R_\odot$, for an exact integration (a) and for the approximate expression discussed in the text (b). The band in the approximate calculation represents the region within which P_{ν_e} will oscillate, a feature apparent in (a). The oscillation length at crossing is $3.56 \cdot 10^{-3} R_\odot$.

Landau-Zener [15] (nonadiabatic) formulae in the proper limits, but also contains the additional physics associated with a harmonic current or density perturbation [13].

Both the basic physical consequences of harmonic perturbations and the quality of the analytic result mentioned above can be illustrated through examples. Consider an electron neutrino produced at the sun's center and observed at very large distances from the sun. Figure 3a shows the result of exactly integrating Eq. (5), including the harmonic perturbation given by Eq. (9). This example corresponds to a highly adiabatic level crossing (the Landau-Zener adiabaticity parameter [15] $\gamma_c \sim 24.4$, corresponding to a crossing deep in the solar core ($x_{cross} \sim 0.1$)), with a modest mixing angle of $\sin^2 2\theta_v = 0.01$. The oscillation length at crossing L^{cross}, in the absence of any harmonic perturbation is $0.00356\ R_\odot$. The neutrino probability at long distances from the sun is given as a function of L^{per}, the oscillation length of the perturbation, which we allow to vary from 0 to $\approx 2L^{cross}$. We see, over rather broad ranges of L^{per}, substantial enhancements in P_{ν_e}, up to $P_{\nu_e} \sim 0.6$, relative to the unperturbed value $P_{\nu_e} \sim 0.008$. This is achieved for a perturbative amplitude $h = 0.01$.

Figure 3b gives our analytic description for these parameters. The envelope in Fig. 3b corresponds to an interference phase that cannot be calculated in the uniform approximation treatment. This is of no physical consequence since the envelope would disappear, leaving only the average value, for a realistic case involving either an extended solar core or a spectrum of neutrinos. The analytic formula seems to match the numerical results very well, giving the pattern of broad oscillations quite accurately. The analytic amplitude is about 80% that of the exact results, very reasonable agreement in view of the fact that the enhancements in $P_{\nu_e}^\infty$ are large and the analytic result only perturbative.

The results in Fig. 3 are remarkable in that a strong enhancement is found over a sizeable range of L^{per} for a small perturbation, $h \sim .01$. This range roughly corresponds to the height of the peak in local oscillation length illustrated in Fig. 2, which was calculated for the same parameters. The broad response in L^{per} is clearly a necessary condition if effects are to persist in less idealized cases where the perturbation is not harmonic, but instead is characterized by Fourier components distributed about an average L^{per}.

In Fig. 4 we show a large mixing angle solution, $\sin^2 2\theta_v = 0.1$, perturbed by a sinusoidal current/density fluctuation with $h = 0.005$ and a fixed $L^{per} = 0.0033$ $R_\odot$. As $E/\delta m^2$ increases one sees a sharp reduction in $P_{\nu_e}^\infty$ as a level crossing is first achieved, than an enhancement in $P_{\nu_e}^\infty$ as the effects of the perturbation come into play. The illustrated region is highly adiabatic. The peak of the first large-scale oscillation occurs at $E/\delta m^2 \sim 1.2 \cdot 10^5$ MeV/eV^2, corresponding to a local oscillation length of $1.37 \cdot 10^{-3} R_\odot$, compared to $L^{per} = 1.33 \cdot 10^{-3} R_\odot$.

The corresponding analytic result is given in the upper right-hand corner of Fig. 4. The correspondence between the exact and analytic approximations is again very close. Fig. 4 demonstrates that sizeable enhancements in $P_{\nu_e}^\infty$ can persist over an appreciable range of neutrino energies.

A qualitatively different example is given in Fig. 5a, with $\gamma_c \sim 1.55, \sin^2 2\theta_v = 0.01, L^{cross} = 0.0356 R_\odot$, and $\delta m^2/E = 10^{-6}$ eV^2/MeV. Unlike the adiabatic transitions discussed before, an interference now arises between the nonadiabatic (i.e., Landau-Zener [13]) and fluctuation contributions to the oscillation probability. Thus the net

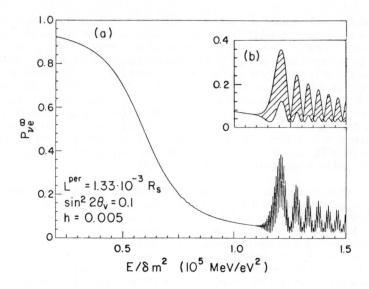

Fig. 4. As in Fig. 3, only for $\delta m^2/E = 5 \cdot 10^{-5}$ eV^2/MeV, $\sin^2 2\theta_v = 0.1$, and $h = 0.005$. The results are given as a function of neutrino energy.

effect of the perturbation is to generate an oscillation of $P_{\nu_e}^\infty$ about the Landau-Zener value, with an envelope determined by fully constructive or fully destructive addition of these two contributions. Thus no net enhancement of $P_{\nu_e}^\infty$ occurs when one averages over L^{per}, or equivalently over the neutrino energy. This is an important distinction between the adiabatic and nonadiabatic cases. One also sees that the peak of the envelope has shifted to $L^{per} \sim 0.023\ R_\odot$, significantly smaller than $L^{cross} = 0.036\ R_\odot$. The corresponding analytic result produces the envelope shown in Fig. 5b that again accurately follows the exact result, apart from some divergence in the "classically forbidden" region of large L^{per}.

These examples illustrate that rather small effects on the neutrino effective mass arising from interactions with solar current loops may significantly influence neutrino oscillation probabilities if the scale of the current loops is comparable to the local oscillation frequency. Since the velocity, location, or size of solar current eddies could plausibly be tied to solar activity, this would appear to offer a mechanism for producing fluctuations in the neutrino flux associated with the solar cycle. Such fluctuations could arise in the absence of large density fluctuations and in the absence of neutrino magnetic moments, mechanisms considered previously.

The analytic expression derived in Ref. 13 identifies $\gamma_c^{2/3}/\tan 2\theta_v$ as the parameter governing the strength of the response to currents or density fluctuations. Therefore the largest effects arise for highly adiabatic transitions with large mixing angles (γ_c

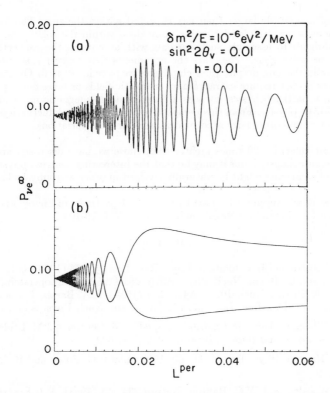

Fig. 5. As in Figs. 3, only for a nonadiabatic ($\gamma_c \sim 1.5$) transition with $\delta m^2/E = 10^{-6}$ eV^2/MeV, $\sin^2 2\theta_v = 0.01$ and $h = 0.01$.

contains a dependence on θ_v). Such transitions involve level crossings deep in the solar core. Quite generally the net effect is a substantial enhancement in $P_{\nu_e}^\infty$ over a range of neutrino energies (or, equivalently, perturbation frequencies), so that an increase in the neutrino signal would result for realistic cases involving a spectrum of neutrinos and an extended neutrino-producing solar core.

In contrast nonadiabatic transitions ($\gamma_c \sim 1$) exhibit oscillations about an average, unperturbed probability given by the Landau-Zener result. The effects, for comparable current velocities, are smaller than in the case of adiabatic transitions. Both of these observations suggest that currents (or density fluctuations) are less likely to have a substantial, net effect on spectrum- and core-averaged probabilities $P_{\nu_e}^\infty$ in this case.

For the more favorable case of highly adiabatic transitions, substantial enhancements in $P_{\nu_e}^\infty$ require amplitudes $\gtrsim 0.005$. This would correspond to velocities at least an order of magnitude larger than any expected within the sun. The location of the enhancement is near the level crossing point deep within the solar core. Thus it also becomes difficult to associate such currents with activity in the convective zone governed by the solar cycle. These are the arguments that suggest this mechanism may not operate in the sun, despite the rich physics we have illustrated. On the other hand, these effects do not require extraordinary magnetic fields or large density fluctuations, nor any new particle physics, such as a large neutrino magnetic moment, other than standard MSW physics. From this perspective current-induced fluctuations in $P_{\nu_e}^{\infty}$ may be less exotic than other suggested mechanisms for time variations.

Current velocities 20 times greater than those we have discussed are believed to arise in stellar collapse. Thus it may be that the interesting neutrino physics associated with periodic currents might be naturally realized in other astrophysical settings.

This work was supported in part by the U.S. Department of Energy and its Institute for Nuclear Theory, and by NASA under grant #ATP-90-103.

References

[1] S.A. Colgate and H.J. Johnson, Phys. Rev. Lett. 5, 573 (1960); G.E. Brown, H.A. Bethe, and G. Baym, Nucl. Phys. A375, 481 (1982); J. Cooperstein, H.A. Bethe, and G.E. Brown, Nucl. Phys. A429, 527 (1984); E.A. Baron, J. Cooperstein, and S. Kahana, Phys. Rev. Lett. 55, 126 (1985) and Nucl. Phys. A440, 744 (1985).

[2] J.R. Wilson, in *Numerical Astrophysics*, ed. J.M. Centrella, J.M. Leblanc, and R.L. Bowers (Jones and Barlett, Boston, 1985), pg. 422.

[3] W.C. Haxton, Phys. Rev. Lett. 60, 1999 (1988); G. Fuller and B. Meyer, UCSD preprint.

[4] S.E. Woosley and W.C. Haxton, Nature 334, 45 (1988); R.I. Epstein, S.A. Colgate, and W.C. Haxton, Phys. Rev. Lett. 61, 2038 (1961); S.E. Woosley, D.H. Hartmann, R.D. Hoffman, and W.C. Haxton Ap. J. 356, 272 (1990).

[5] S.W. Bruenn and W.C. Haxton, to be published in Ap. J.

[6] A. Yahil, Ap. J. 265, 1047 (1983).

[7] S.W. Bruenn, Ap. J. 340, 955 (1989); S.W. Bruenn, Ap. J. 341, 385 (1989); E.S. Myra and S.A. Bludman, Ap. J. 340, 384 (1989).

[8] A. Ray and K. Kar, Phys. Rev. Lett. 63, 2435 (1989).

[9] K. Lande and R. Davis, in Proc. 21st Int. Cosmic Ray Conf., Adelaide, 1990 (to be published); G.A. Bazilevskaya *et al.*, Yad. Fiz. 39, 856 (1984) [Sov. J. Nucl. Phys. 39, 543 (1984)].

[10] B.W. Filippone and P. Vogel, Phys. Lett. B246, 546 (1990); J.N. Bahcall and W.H. Press, Ap. J. 370, 730 (1991); J.W. Bieber *et al.*, Nature (London) 348, 407 (1990); L.M. Krauss, ibid. 348, 403 (1990).

[11] A. Cisneros, Astrophys. Space Sci. 10, 87 (1981); L.B. Okun, M.B. Voloshin, and M.I. Vysotsky, Zh. Eksp. Teor. Fiz. 91, 754 (1986) [Sov. Phys. JETP 64, 446]

and Yad. Fiz. 44, 845 (1986) [Sov. J. Nucl. Phys. 44, 544]; E. Kh. Akhmedov, Phys. Lett. B213, 64 (1988); R. Barbieri and G. Fiorentini, Nucl. Phys. B304, 909 (1988).

[12] C-S. Lim and W.J. Marciano, Phys. Rev. D 37, 1368 (1988).

[13] W.C. Haxton and W-M. Zhang, Phys. Rev. D. 43, 2484 (1991).

[14] S.P. Mikheyev and A. Yu. Smirnov, Yad. Fiz. 42, 1441 (1985) [Sov. J. Nucl. Phys. 42, 913] and Nuovo Cimento 9C, 17 (1986); L. Wolfenstein, Phys. Rev. D 17, 2369 (1978) and 20, 2634 (1979).

[15] W.C. Haxton, Phys. Rev. Lett. 57, 1271 (1986); S.J. Parke, Phys. Rev. Lett. 57, 1275 (1986).

CORE COLLAPSE SUPERNOVA RATE OUT TO 12 MEGAPARSECS

E.E. Becklin
Department of Physics and Astronomy
University of California, Los Angeles

ABSTRACT

We use the IRAS infrared fluxes from galaxies closer than 12 Mpc to estimate the core collapse supernova rate as a function of distance from the earth. The rate within 3 Mpc is dominated by the Milky Way and has a value of about 0.05 per year (one every 20 years). Between 3 Mpc and 4 Mpc the rate raises abruptly to over 1/2 per year due to the inclusion of the starburst galaxies M82 and NGC 253. This rate produces a threshold in the sense that any future detector of supernova phenomena will have a rate of almost one per year if it is sensitive to core collapse at a distance of at least 4 Mpc. Within 12 Mpc the rate is about 2 per year. The Virgo cluster of galaxies at a distance of 15 Mpc has relatively little massive star formation and contributes an additional 2 core collapse supernova per year.

I. Introduction

The connection between core collapse supernova (CCSN), massive young stars, and thermal infrared radiation (10-200μm) has been known for a number of years (e.g. Rieke et al., 1980). Core collapse supernova progenitors are massive OB type stars with M ~ 8 to 50 $M_\odot$ (e.g. Weiler and Sramek, 1988); they have high luminosity, 10^3-$10^6 L_\odot$, and temperature, T > 30,000°K. Their dominant radiation is ultraviolet (UV) (e.g. Panagia, 1973). During most, of their nuclear burning lifetime of 3 x 10^6 to 3 x 10^7 years (Maeder and Meynet, 1989) they remain in a region of space where the column density is high enough that τ(UV) > 1. Thus essentially all of the luminosity produced by these stars is absorbed by interstellar dust and re-emitted at thermal infrared wavelengths of 30-200μm. Because the massive O and B progenitors of core collapse supernova also produce the observed thermal infrared radiation, it is reasonable to expect a direct relationship between the observed supernova rate and thermal infrared emission. In equation form

$$L_{IR} \simeq L_{OB} \propto N_{OB} \propto \text{CCSN rate} \qquad \text{where} \qquad L_{IR} = \int\limits_{30\mu m}^{200\mu m} F_\lambda \, d\lambda,$$

L_{OB} is the luminosity in OB stars, N_{OB} is the number of OB stars, and CCSN rate is the core collapse supernova rate. Below we show that both theoretically and empirically such a relationship appears to exist.

Since the spectacular detection of a burst of neutrinos from the Large Magellanic Cloud Supernova 1987A, (Bionta et al., 1987, Hirata et al., 1987), many workers have asked when and where the next "detectable" supernova will most likely occur. In particular is the question of how large a neutrino detector must be built to have a reasonable chance to detect neutrinos from a number of supernova within a human lifetime. In this paper we use the relationship between thermal infrared luminosity and the core collapse supernova rate to estimate the CCSN rate within a radius of 12 Mpc. This estimate is much superior to estimates based on visual light because interstellar dust extinction obscures almost all of the UV and optical light from regions that contain massive stars in the Milky Way and more importantly in the relatively nearby starburst galaxies M82 and NGC 253.

II. M82 and NGC 253

Before proceeding with the calculation of the supernova rate it is appropriate to discuss the nearby infrared luminous starburst Galaxies M82 and NGC 253; within a volume whose radius is 4 Megaparsecs (Mpc) (1.2×10^{25}cm) these galaxies produce the largest number of supernova.

M82 is an irregular companion galaxy to the giant spiral M81 and is at a distance of about 3.5 Mpc, (e.g. Freedman and Madore, 1988). It was discovered to be a luminous infrared source by Kleinmann and Low (1970). Willner et al. (1977) made the first infrared hydrogen line measurement and showed that the ~2×10^{44} erg/sec of infrared luminosity (Telesco and Harper, 1980) was consistent with a large number (~10^7) of massive OB stars. The region where these massive stars form is about 1 Kpc in diameter at the center of the galaxy. Optical studies of the central region of the galaxy are impossible because of approximately 20 visual magnitudes of extinction (τ(dust) ~20 in the visible). Rieke et al. (1980) put together several models of stellar distributions based on assumptions about the stellar formation rates and lifetimes of massive stars and compared these results to the observations. Most important for this paper, Rieke et al. (1980) calculate a supernova rate for M82 of nearly 0.2 per year. At radio wavelengths, using the VLA, Kronberg and Sramek (1985) have shown that many compact radio sources are decaying with time; from this and their radio brightness the sources are associated with young supernova remnants. The supernova rate deduced by Kronberg and Sramek (1985) is also ~0.2 per year.

NGC 253 is an Sc galaxy at a distance of about 3 Mpc (e.g. Sandage and Tammann, 1975). It was discovered to be a strong infrared source identified with a very luminous region of massive OB stars by Becklin, Fomalont and Neugebauer (1973). It also has a total infrared luminosity of ~2 x 10^{44} ergs/sec (Telesco and Harper, 1980) which has been shown to be consistent with the amount of ionized gas resulting from massive OB stars (Wynn-Williams et al., 1979 and Rieke et al., 1980). Most regions containing the OB stars in NGC 253 can not be studied optically, because they lay behind 10-15 magnitudes of visible interstellar dust extinction (Wynn-Williams et al., 1979).

III. The Supernova Rate vs. Infrared Luminosity Relationship

a) Theory

It is beyond this paper to do a thorough theoretical study of the relationship between infrared luminosity and the CCSN rate. However, to obtain a feeling for the numbers, we have done some simple sums in three mass bins from 8 to 45 $M_\odot$. The lower limit is near the mass cutoff for CCSN (eg. Weiler and Sramek, 1988) and the upper limit is typical of the most massive stars found in the galactic H II regions. For the sums, the initial mass function of Miller and Scalo (1979), the lifetimes of Maeder and Meynet (1989) and the physical properties of Panagia (1973) were used. The total luminosity was fit to the observed value for M82. The number of Lyman continuum photons was also calculated, and agreed well within the observations error for M82. The resultant CCSN rate per century per thermal infrared luminosity of $10^{10}L_\odot$ (Q), from this simple model is: Q = 3 per century per $10^{10}L_\odot$(IR).

Several comments are in order:

   1) The infrared luminosity comes from all three mass bins approximately equally.

2) The supernova rate is dominated by stars at the low mass end. This results because the death rate must equal the birth rate (IMF) which goes as

$$\frac{dN}{dM} \propto M^{-3}$$

in this mass range (Miller and Scalo, 1979).

3) The Lyman continuum results primarily from the highest mass stars. Thus, the amount of ionized gas is not necessarily a good indicator of the supernova rate unless the IMF is uniform.

Rieke et al. (1980) made more detailed calculations using various assumptions about the IMF and starburst lifetime. From their table 2 one finds Q = 2 to 4.

b) Empirical

There are several independent ways of empirically determining the CCSN rate as a function of infrared luminosity. First, Kronberg and Sramek (1985) have determined from the decay of compact radio sources in M82 that there is one supernova every five years. Using $4 \times 10^{10} L_\odot$ for the thermal infrared luminosity (Telesco and Harper, 1980) leads to a rate Q = 5 ± 2.

Second, G.A. Tammann at the 1990 Les Houches Summer School compared the number of optical type Ib and type II supernova in nearby galaxies over the past 30 years to the infrared luminosity in these galaxies from IRAS. Correcting for supernova lost because of inclination, but not for starburst galaxies such as M82 and NGC 253, the rate is Q = 3.5 for H_o = 50 Km/sec/Mpc and Q = 8 for H_o = 75 Km/sec/Mpc.

Finally, from the Milky Way counting the number of optical supernova over the past few hundred years Tammann estimated a type Ib + type II rate of 4.3 ± 1.8 supernova per century for the Galaxy. Pulsar birth rate analysis gives a rate of 2 per century (Weiler and Sramek, 1988). Using the galactic IRAS infrared radiation from HII regions and star formation regions by Cox and Mezger (1987) as summarized by Walterbos (1988), of L_{IR} (galactic) = $5 \times 10^9 L_\odot$, we obtain the following rates, Q = 9 ± 4 from historic optical supernova and Q = 4 from pulsar birth rates.

In summary we have:

CCSN rate per century
per $10^{10}L_\Theta$

a)	Theory	3
b)	Radio Sources in M82	5 ± 2
c)	Optical Supernova in Nearby Galaxies	6 ± 2
d)	Milky Way Optical Supernova	9 ± 4
e)	Milky Way Pulsars	4

For the analysis in this paper we will use $Q = 6 ± 2$ per century per $10^{10}L_\Theta$(IR).

IV. The Core Collapse Supernova Rate as a Function of Distance.

Below we use the previously determined CCSN rate per $10^{10}L_\Theta$ of thermal infrared luminosity and the infrared luminosity from IRAS to calculate the number of CCSN per year within a volume of radius R as a function of radial distance R. The infrared fluxes come from Rice et al. (1988) for all galaxies except the Milky-Way. Because most of the galaxies are larger than the IRAS beams, Rice et al. have integrated over the surface of each galaxy to obtain the total infrared flux. A correction to the total fluxes has been made to account for flux that is not directly related to luminosity of massive stars. This correction is subjective, but following the work on our own galaxy by Cox and Mezger (1987), we assign the infrared emission not related to massive stars primarily to infrared cirrus. Infrared cirrus is dust emission associated with diffuse interstellar HI clouds. It can be identified by its characteristic flat 12 to $25\mu m$ color. For the Milky Way with a luminosity of about $10^{10}L_\Theta$ Cox and Mezger estimate that the correction factor is about 50%. A correction for diffuse infrared emission was made to all galaxies based on their infrared colors; however, errors in estimating this correction have very little effect on the overall infrared luminosity curve since most of the massive stars occur in the luminous infrared galaxies ($L_{IR} \sim 5 \times 10^{10}L_\Theta$) such as M82 and NGC 253 where the correction for cirrus is small ($\lesssim 20\%$).

Luminosities are calculated using distances given in Rice et al. except for the important starburst galaxies NGC 253, M83 and M82. Because these three galaxies dominate the CCSN rate in the local neighborhood, we use the most recent distance measurements, such as Cepheid variables, to determine their distance. The distances for these three galaxies are not determined from the Hubble flow and are thus independent of H_0. The distances used for the three important galaxies are 3.5 Mpc for M82 (Freedman and Madore, 1988), 3.0 Mpc for

NGC 253 (Sandage and Tammann, 1975) and 4.0 Mpc for M83 (de Vancouleurs, 1979). The distance for the other galaxies are based are based primarily on the Hubble flow redshift using H_o = 50 Km/sec/Mpc. If H_o = 75 Km/sec/Mpc were used instead, the basic shape of the curve would not change substantially because, although the luminosity of individual galaxies would decrease, the number of galaxies within a given volume would increase.

The resultant curve of massive star luminosity and core collapse supernova rate within a volume of radius R as a function of radial distance R out to 12 Mpc is shown in Figure 1. The curve shows that out to about 3.0 Mpc the CCSN rate is dominated by the Milky Way. Other local group galaxies such as the LMC, M31 (Walterbos, 1988) and even M33 contribute very little extra to the rate which is about 0.05 per year. In the volume between 3 Mpc and 4 Mpc the rate increases dramatically due primarily to the inclusion of the two luminous starburst galaxies M82 and NGC 253. The CCSN rate at 4 Mpc is about 0.6 per year. This means that over the next 15 years nine core collapse supernova will occur within 4 Mpc of the earth.

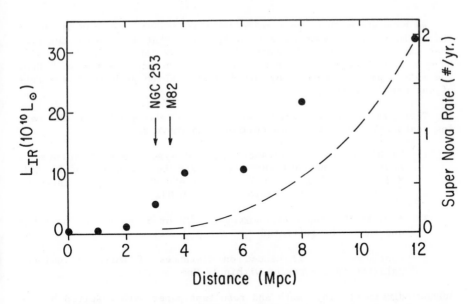

Figure 1.   The thermal infrared luminosity from massive stars M $\gtrsim$ 8$M_\odot$ and the derived core collapse supernova (CCSN) rate per year within a volume of radius R as a function of radial distance R. Solid dots are the data. Also shown with the dashed curve is the rate vs. distance if the distribution of massive stars were uniform (normalized to the observed value at 12 Mpc). The radial distance of the major contributors to the CCSN rate M82 and NGC 253 are shown.

To demonstrate the significance of the raise between 3 and 4 Mpc we have also shown the expected supernova rate if the volume out to 12 Mpc were filled uniformly. The curve has been normalized to a rate of 2 per year at 12 Mpc. The sharp raise in the supernova rate between 3 to 4 Mpc is a result of the chance placement of two intense starburst galaxies (M82 and NGC 253) relatively close to the earth. The rate out to 5 Mpc calculated here is about four times greater than estimated by Pennypacker et al. (1991) from optically identified supernova. The difference is again the optically obscured starburst galaxies. The Virgo cluster of galaxies has relatively little star formation (eg. Becklin, 1985). At a distance of about 15 Mpc it has an infrared luminosity in massive stars of $20 \times 10^{10} L_\Theta$ and a CCSN rate of about 1.5 per year. The Seyfert Galaxy NCG 1068, also at a distance of 15 Mpc, has a similar star formation luminosity and CCSN rate (Telesco et al., 1984) as the entire Virgo cluster of galaxies.

V. Conclusions and Future Work

We have shown that the core collapse supernova rate probably has a threshold at a distance of 3 to 4 Mpc such that any future detector that can detect the by-products of a supernova at this distance will have a rate of almost one per year. For distances less than 3 Mpc, the rate appears dominated by the Milky Way which probably has a rate of one per 20 years.

There is continuing work that should be attempted over the next few years related to the core collapse supernova rate.

1) With the recent advances in infrared array detectors, a dedicated supernova survey at $2.2\mu m$ is now possible in M82, M83 and NGC 253; several such surveys should be initiated to varify directly the radio results in M82.

2) A Milky Way supernova survey at 2.2 or $3.5\mu m$ is now possible and should be initiated.

3) Continued work is needed on distances of infrared luminous galaxies in the region within 12 Mpc.

Acknowledgements: This work and resultant paper were inspired by the interest, enthusiasm and encouragement of David Cline. I want to also thank Mike Jura and Jean Turner, for many important discussions and providing critical information and input.

References

Becklin, E.E., 1986. Light on Dark Matter, ed. Israel, F.P., p. 415.

Becklin, E.E., Fomalont, E.B., and Neugebauer G., 1973. Ap.J., 181, L27.

Bionta, R.M., Blewitt, G., Bratton, C.B., Casper, D., et al., 1987. Phys. Rev. Lett., 58, 1494.

Cox, P. and Mezger, P.G., 1987. NASA Conference Pub. 2466: Star Formation in Galaxies, ed. Lonsdale, C.J., p. 23.

de Vaucouleurs, G., 1979. Astronomical Journal, 84, 1270.

Freedman, W.L. and Madore, B.F., 1988. Ap.J., 332, L63.

Hirata, K., Kajita, T., Koshiba, M., Nakahata, M., Oyama, Y., et al., 1987. Phys. Rev. Lett., 58, 1490.

Kleinmann, D.E. and Low, F.J., 1970. Ap.J., 159, L165.

Kronberg, P.P. and Sramek, R.A., 1985. Science, 227, 28.

Maeder A. and Meynet G., 1989. Astron. Astrophys., 210, 155.

Miller, G.E. and Scalo, J.M., 1979. Ap.J. Supp. 41, 513.

Panagia, Nino, 1973. Astronomical Journal, 78, 929.

Pennypacker, C., et al. 1991, this volume.

Rice, W., Lonsdale, Carol J., Soifer, B.T., Neugebauer, G., Kopan, E.L., Lloyd, Lawrence A., de Jong, T., and Habing, H.J., 1988. Ap.J. Supp., 68, 91.

Rieke, G.H., Lebofsky, M.J., Thompson, R.I., Low, F.J., and Tokunaga, A.T., 1980. Ap.J., 238, 24.

Sandage, A. and Tammann, G.A., 1975. Ap.J., 196, 313.

Telesco, C.M., Becklin, E.E., Wynn-Williams, C.G., and Harper, D.A., 1984. Ap.J., 282, 427.

Telesco, C.M. and Harper, D.A., 1980. Ap.J., 235, 392.

Walterbos, R.A.M., 1988. Galactic and Extragalactic Star Formation, eds. Pudritz, R.E. and Fich, M., p. 361.

Weiler, K.W. and Sramek, R.A., 1988. Ann. Rev. Astron. Astrophys., 26, 295.

Willner, S.P., Soifer, B.T., Russel, R.W., Joyce, R.R., and Gillett, F.C., 1977. Ap.J., 217, L121.

Wynn-Williams, C.G., Becklin, E.E., Matthews, K., and Neugebauer, G., 1979. Mon. Not. R. astr. Soc., 189, 163.

Progress Report on the Berkeley/Anglo-Australian Observatory High-Redshift Supernova Search

Gerson Goldhaber, Saul Perlmutter, Carl Pennypacker, Heidi Marvin, Richard A. Muller

Lawrence Berkeley Laboratory and Center for Particle Astrophysics
University of California at Berkeley

Warrick Couch
University of New South Wales

Brian Boyle
Institute of Astronomy

I. Introduction

There are two main efforts related to supernovae in progress at Berkeley. The first is an automated supernova search for nearby supernovae, which was already discussed by Carl Pennypacker at this conference. The second is a search for distant supernovae, in the $z = 0.3$ to 0.5 region, aimed at measuring Ω. It is the latter that I want to discuss here today.

II. The Method for the Measurement

The method we intend to use to obtain a measurement of Ω is illustrated in Figure 1. It involves two steps: i) the discovery of a distant supernovae from exposures of our CCD camera at the Anglo Australian Telescope at different epochs, and ii) the rapid follow-up with spectroscopic and photometric measurements by collaborating observers. These are needed both for the identification of the supernova candidate as a type Ia supernova as well as for the measurement of the redshift. There is now evidence that the type Ia supernova is effectively a standard candle. Further work on this subject is still in progress in conjunction with the Berkeley automated nearby search. Using the combination of measured redshifts and luminosity of the discovered supernovae we will be able to plot velocity as a function of luminosity-derived distance. The curvature of such a plot is a measure of the deceleration of the Universe and hence Ω, for a cosmological constant $\Lambda = 0$.

III. What Accuracy Can We Expect?

Figure 2 shows a result from Monte Carlo calculations of the distributions of $q_0 = 1/2 \Omega$ as a function of the number of supernovae observed. The maximum light of the supernovae was assumed to follow a gaussian distribution with sigma=0.3. The lower figure shows the accuracy to which q_0 can be measured as a function of the number of supernova samples observed. Thus between 10 and 30 measured supernovae should give us a significant new measurement of Ω.

IV. How Good a Standard Candle Is a Type Ia supernova?

Figures 3 and 4 are from supernova compilations due to Barbon, Ciatti, and Rosino and Leibundgut which respectively illustrate both the photometry curve for type Ia supernovae and the redshift vs magnitude distributions for observed supernovae. Figure 5 from Miller and Branch gives the width of the magnitude distribution and shows what may be the effect of extinction due to dust in the parent galaxies. This faint tail can be statistically removed. Alternatively, photometry in the infrared, which suffers less extinction, should not show this tail.

V. The Observation of One Distant supernova By an ESO Group

What gives us the confidence that such a search is feasible is the fact that a distant supernova was discovered by an ESO (European Southern Observatory) group, including our co-worker Warrick Couch. This is illustrated in Figure 6.

VI. How Can We Find supernovae?

First I want to show one figure from the nearby search. which illustrates one of the procedures we follow in the distant search. Figure 7 illustrates the discovery of supernova 1986I. The supernova is clearly seen as a third object in the right hand figure. The discovery method is to subtract the reference image from the new image and look (automatically!) for stellar objects that appear in the subtracted image. The difference in the distant search is that a supernova candidate is not completely resolved from its host galaxy in the new image, and what we have to look for is the change in intensity between a reference and a new image.

Our procedure is to take the new image at one to 12 months after the reference image. So far we have had successful images taken at the AAT with our CCD and focal reducer illustrated in Figures 8 and 9. The epochs are: November 30, 1989; December 28, 1990; January 23, 1990; November 13 and November 18, 1990. Unfortunately runs scheduled for April, May, and August 1990 as well as January 1991 were not taken due to bad weather. Each night of observing we take about 50 images on the 1024 x 1024 pixel Thomson CCD. The length of exposure is 5 minutes. However in order to avoid cosmic rays with direct impact on the CCD, or other activation of pixels due to local radioactivity, we take two 2.5-minute exposures which we then add together in the computation stage. To accept an object as a real candidate, we demand that the 2 independent new images agree within 25%. This is summarized in Figure 8. Our search strategy is summarized in Figure 10. We expect about 1 type Ia supernova per night of observation as summarized in Figure 11.

VII. Computational Techniques

Each image gives us 1 million pixels. Thus the 2 X 50 images taken per night of observation yield about 100 million pixels, with 16 bits of information each. The problem that faced us is to analyze and compare with the 50 million pixels from the reference images within as short a time period as possible so that any supernova candidates do not have time to fade appreciably. So far due to the speed limitations of current computers at the telescope we could not perform image subtractions within this time constraint but had to develop an alternative technique of data handling. The method we developed consists of performing photometry on all the identified objects, about 1000 to 1500 per image. We then compare the photometry results between the reference images and the new images. Moreover, there was also the problem of data transfer from the images taken at the AAT to Berkeley where more extensive computing power is available to us. Using a NASA computer network, we were able to send our list of reference objects to Australia and compare the images in real time over there. Using a 32 Mbyte memory system we installed at the AAT computer, this allowed us to complete the analysis and find potential supernova candidates within 36 hours from the time the images were taken. Our procedure is summarized in Figure 12. In Figure 13 we show a typical observed magnitude distribution for 1 image.

VIII. Results from the "Engineering" Runs

The runs of November 30, 1989; December 28, 1989; and January 23, 1990 were used for taking reference images as well as "engineering" runs for developing our data handling methods. In these runs we found 5 candidates for further examination selected by the criteria in Figure11. Of these, two candidates turned out upon measurements at the NTT (European Southern Observatory's "New Technology Telescope") to be located at a galaxy. At this early stage in our program development the time taken for the analyses was too long to be able to still take a spectrum of the potential supernova. For one of these we also had further circumstantial evidence that we were dealing with a supernova by the fact that the size of the image increased between the 2 observations in November and January respectively. Also the increase in intensity we had observed, about 0.4 in magnitude, has since disappeared on later observations. Figure 14 shows the images as observed on November 30 1989 and January 23 1990 respectively. Figure 15 shows a contour plot of this candidate from November 30 while Figure 16 illustrates one of the difficulties we encountered which so far necessitates visual inspection of candidates, namely the occurrence of occasional streaks in our images These streaks are believed due to reflections of bright stars in our optical system and do reoccur in our new images. However they tend not to reoccur in precisely the same position and hence they would make it on to our candidate list of objects that have changed by more than 20% in intensity.

IX.   Results From the Most Recent Run

Since January, 1990 our only clear nights at the AAT occurred on November 13 (2 X 50 images) and November 18, 1990 (2 X 30 images).   For this run we had not expected to be able to complete our analysis in time for timely follow-up.   We surprised ourselves with the breakthrough that we were able to analyze all the data in less than 36 hours!   Figure 17 shows the distribution of "magnitude change" versus "apparent magnitude" for one image from the night of November 13, 1990.   Superimposed on this we show those events which were picked up as candidates in the 50 images.   These consist of a mixture of variable quasars,which we are following in a separate study, and possible supernovae.

At that stage we needed to use follow-up observations at another observatory,with better seeing and a finer image resolution to differentiate quasars from supernovae.   Quasars have point source image shapes, while supernovae occur in galaxies which have 'fuzzy' images.   We were able to get confirmatory measurements on five of the objects.   These turned out to be stellar, and thus presumably quasars.   The others were not followed up for lack of time.   We believe that we have now resolved this problem by developing several techniques to differentiate supernova candidates from quasars, without needing follow-up observations.   These techniques include using a data base with the time history of all our repeated measurements.   We also use an image shape analysis of our data based on the FOCAS program developed by Jarvis, Tyson, and Valdes.   With this ability to preselect we expect in future to follow-up only those candidates which are likely to be supernova.

The arrows in Figure17 indicate those objects which showed a change in magnitude from November 13 to November 18. Finally the "D" indicates the location of the ESO event as observed on the Danish telescope.   In Jan 1991, when everything was ready, the weather did not cooperate and we did not get any data.

X.   Conclusion and Future Plans

We have demonstrated that we can detect supernova of about 23rd magnitude and brighter.   In future runs, which are now scheduled at the AAT we should be able to observe and measure spectra with the help of our collaborators sufficiently rapidly to catch and identify a supernova.   Our plans now are to build and deploy a much larger CCD of four 2048 X 2048 chips. This should allow us to achieve the goals set out in this talk. While we have not yet measured Ω, we did observe the configuration shown in Figure 18.

Acknowledgements: This work has been supported by the Dept. of Energy, under contract DE-AC03-76SF00098, and the Center for Particle Astrophysics, a National Science Foundation Science and Technoloty Center, of the University of California at Berkeley, .

References

Barbon, B., F. Ciatti, and L. Rosino, 1974, in *Supernovæ and Supernova Remnants*, C. Cosmovici (ed.), D. Reidel, Dordrecht-Holland.

Leibundgut, B., 1991, in *Supernovæ*, S. Woosley, (ed.), Springer-Verlag, New York.

Miller, and D. Branch, *Astron. J.*, 1990.

Mass Density from Deceleration

$$\frac{1}{2}mv^2 = \frac{1}{2}mH_0^2R^2 = \frac{GmM}{R} = \frac{4}{3}\pi Gm\rho R^2$$

$$\rho_{\text{critical}} = \frac{3H_0^2}{8\pi G} \approx 10^{-29}\,\text{gm/cm}^3$$

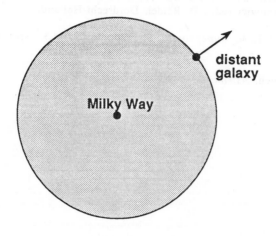

$$\ddot{R} = -\frac{GM}{R^2} = -\frac{4}{3}G\pi R\rho$$

(if $\rho = \rho_c$, at $z \sim 1$, presently decelleration $\sim 10^{-7}\text{cm/sec}^2$)

Fig. 1

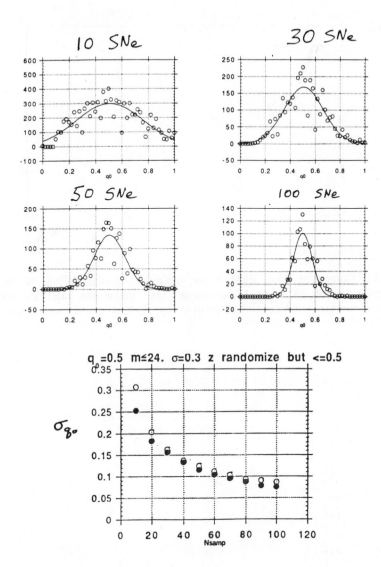

Fig.2 Monte Carlo distribution of "measured" $q_0 = \Omega/2$ for Gaussian distribution ($\sigma = 0.3$) of Type Ia maximum light. Calculations by S. Perlmutter, to be published.

512

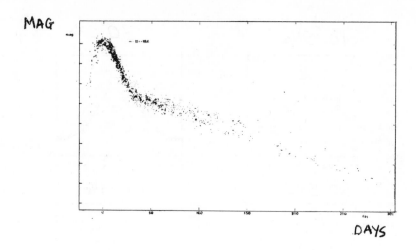

MAG

DAYS

Fig.3 Composite blue light curve obtained by the fitting of the observations of 38 Type I supernovæ. One magnitude intervals are marked on the ordinates. Compilation by Barbon, Ciatti, and Rosino (1974).

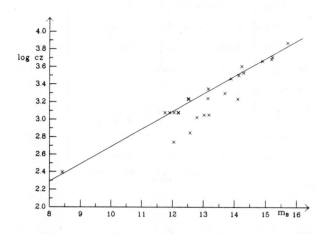

Fig. 4 The Hubble diagram of Type Ia supernovæ at maximum light. Compilation by Leibundgut (1991).

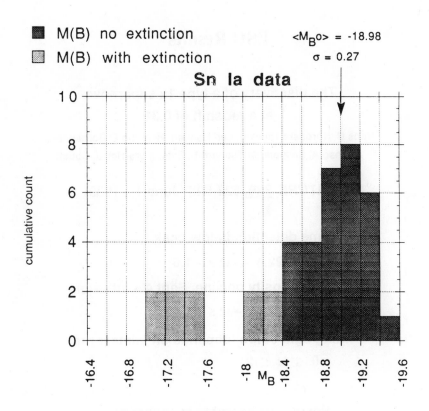

Fig. 5   Histogram of Type Ia supernovæ absolute magnitudes, based on a compilation by Miller and Branch (1990).

ESO Results

The discovery of a type 1a supernova
at a redshift of 0.31

Hans U. Norgaard-Nielsen, Leif Hansen, Henning E. Jorgensen,
Alfonso A. Salamanca, Richard S. Ellis, & Warrick J. Couch

NATURE <u>339</u>, 523-525 (June 15, 1989)

- assume detection below peak

  $B(max) \leq 22.4 \rightarrow \Omega \geq -1.6$

- from luminosity decline, Sept 4 > inflexion

  $B(max) \geq 21.5 \rightarrow \Omega \leq 3$

- fitting light curve (depends on faintest pts)

  $0 \leq \Omega \leq 2$

*Results comparable in accuracy
to galactic measurements*

Fig 6

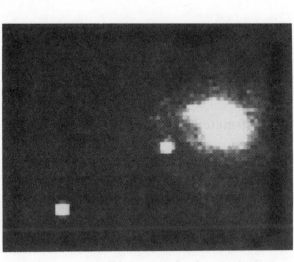

MAY 8, 1986   MAY 17, 1986

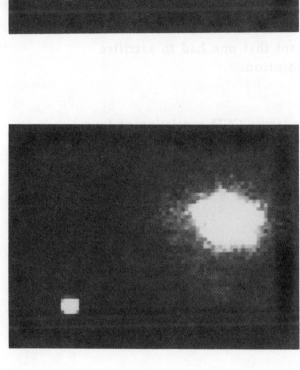

SUPERNOVA IN M99 GALAXY IN VIRGO CLUSTER

Photo by : LBL SUPERNOVA SEARCH TEAM.

Fig. 7

1989. Construction of a CCD 1024 x 1024 pixel plane with demagnifying F/1 lens for the

AAT four-meter telescope

1 pixel ↔ 1.05 arc sec.

Thus the entire CCD corresponds to a 17 x 17 (arc min)^2 area of the sky.

Here the philosophy was to maximize the area covered which meant that one had to sacrifice resolution.

A new 16x larger CCD consisting of four 2048 x 2048 pixel planes is on the drawing board. If we use this to get a factor 10 in area sampled, we will still get improved resolution.

Fig. 8

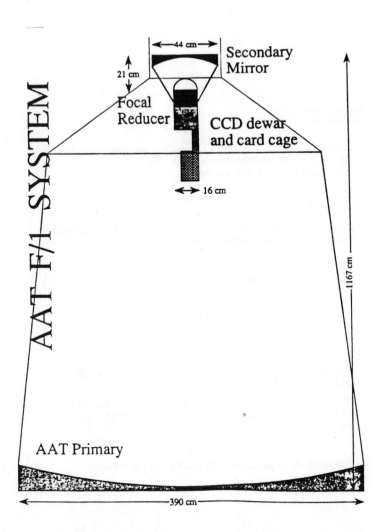

Fig 9   Schematic diagram of high redshift supernova detection hardware.   The system uses the Anglo Australian Telescope (3.9 meter primary mirror),   de-magnifying optics and 1024 x 1024 CCD at the prime focus.   The focal reducer converts the beam into a final F/# of one at the focal plane.

Search Strategy

1. Observe >50 wide fields (17'x17'), each containing >1000 objects of which >100 are galaxies at z = 0.3–0.5.
   (Exposure time: 2 x 150 seconds.)

2. Repeat observations after a month or so and look for brightness variations in these ~100,000 objects.

        Find   • Quasars

                   • Variable Stars

                   • Active Galactic Nuclei (AGN)

                   • Supernovae

3. Distinguish the supernovae (and a few AGN) from the others by high resolution imaging:

   The supernovae (and occasional AGN) appear on resolved, "fuzzy" galaxies, while the other objects all look like "sharp" point-source stars.

4. Follow up the supernovae with spectra and photometry.

Fig. 10

Supernova Discovery Rate

rate per galaxy (type Ia)	1 per 500 years
useful galaxies per image	100 galaxies
supernova visibility	1/12 year
number of images to find one supernova	60 images
observation time per supernova	6 hours

one type Ia supernova per night

Fig. 11

SEARCH FOR SNe BY INTENSITY RATIOS
(Using VISTA Program)

Reference image 5 minutes, new images 2 x 2.5 minutes, one month later.

1. Align the images with a Matching program.

2. Find location of all objects with Apparent Magnitude < 23.5.

3. Compute Integrated Intensities inside a 2.5 arc second radius I_R (Ref), I_N (New) $= I_1 + I_2$.

4. Take intensity ratios of corresponding objects $r = I_N/I_R$.

5. Cuts to accept potential candidate:

    (i) $r > 1.1$ or $r < 0.9$

    (ii) $|I_N - I_R| > 800$ counts

    (iii) inside fiducial region
            40 pixel border (15% less)

6. Cuts to accept intermediate candidate:

    (i) $|I_N - I_R| > 900$ counts

    (ii) $0.75 < I_1/I_2 < 1.25$

    (iii) Only 1 object in a 10 x 10 pixel box

7. Final candidates:

    (i) Check photometry with radii of 2.0, 2.5, 3.5, and 5 arc seconds.

    (ii) Check on distance to bright object (from contour plot).

    (iii) Compare candidates with Subtraction method.

Fig. 12

Objects found Nov '89 & Jan '90

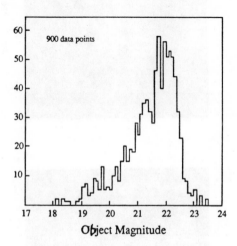

Fig. 13   Our observed object's apparent magnitude distribution for one images.   Matched objects only.

522

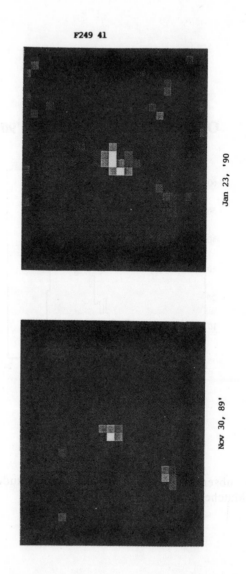

Fig. 14   The galaxy in F249-41 which has changed in magnitude and appears to have changed in shape as well.

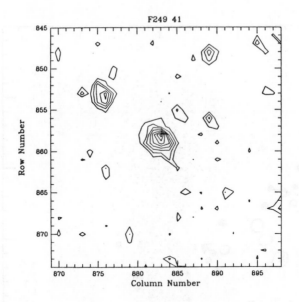

Fig. 15   Contour plot of the variable object identified as a galaxy by Jorge Melnick of the European Southern Observatory, using the NTT.

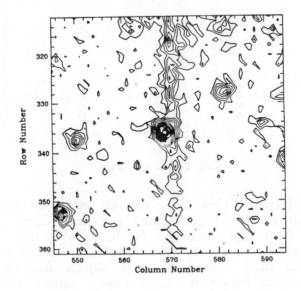

Fig. 16   Example of "Streak"

524

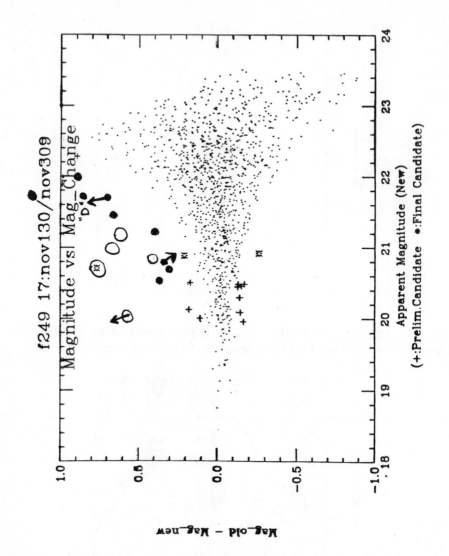

Fig. 17 Magnitude change vs. apparent magnitude for one image (solid small dots). Superimposed are (solid larger circles) all candidates from approximately 50 fields with large increase in magnitude on November 13, 1990, compared to November 30, 1989. Arrows also show change of three objects that changed from November 13 to November 18, in 1990. Circles indicate objects identified as stellar. The "D" indicates the approximate values of the Danish/ESO supernova.

f249 168, 1990:01:23, RUN 68, REGION D

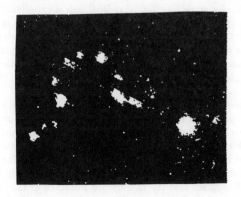

Fig. 18   A configuration observed in one of our images.

The Possibility of Radiogeochemical Limits
on Stellar Collapse Rates in the Galaxy

Calvin W. Johnson

W. K. Kellogg Radiation Laboratory
California Institute of Technology 106-38
Pasadena, CA 91125

I discuss the theoretical possibility of extracting non-trivial limits on stellar collapses in our galaxy by measuring the amount of ^{97}Tc in deeply buried molybdenum ore produced by solar and supernova neutrinos. The solar neutrino contribution is constrained by experiment and here is calculated in the framework of the non-adiabatic Mikheyev-Smirnov-Wolfenstein "solution" to the solar neutrino puzzle. A significant "poison" to a measure of the stellar collapse rate is ^{97}Tc produced by solar neutrinos on ^{97}Mo; since the nuclear structure of the latter is still unknown, a definitive answer on the practicalities of this experiment cannot yet be given.

I. INTRODUCTION

The field of neutrino astronomy has been active for more than two decades (for a review see [Ba89]), ever since Ray Davis' radiochemical experiment first measured the flux of high energy neutrinos coming from the sun [Da89]. As is evident at this Workshop, with the advent of water Čerenkov detectors neutrino astronomy is truly coming into its own. The first success of water Čerenkov experiments was the detection of neutrinos from supernova 1987A at IMB [Bi87] and Kamioka [Hi87]. Later, Kamiokande II detected solar neutrinos [Hi89,Hi90] and confirmed the solar neutrino "puzzle."

One still-outstanding problem, as discussed at this Workshop, is the rate of stellar collapses—Type II and Ib supernovae—in our galaxy. The range of rates is approximately one to ten per century, and the utility of the next generation of neutrino observatories in deducing supernova properties could hinge upon this number. Currently the rate is estimated by extrapolating from supernova rates in other galaxies, by extrapolating from historical supernova in our galaxy, and by using models of stellar populations ([Ta81,Ba83]; see also the talk by Sid Bludman at this Workshop).

In this talk I want to discuss the possibility of measuring the long-term flux of supernova neutrinos, and thus the rate of stellar collapses in our galaxy, in a radio(geo)chemical experiment. Such a measurement, or in this case probably a limit, would complement other methods for extracting the stellar collapse rate.

The Molybdenum-Technetium experiment was originally proposed [Co84] to test the stability of the solar neutrino flux on time scales of millions of years, which is the thermal Kelvin time of the solar *core*, by measuring the amout of ^{98}Tc in deeply buried molybdenum ore.

As I will show, in principle the measurement of ^{97}Tc could also measure (or, more practically, place an upper limit on) the higher-energy neutrinos from stellar collapses by using the 9 MeV threshold to neutron emission as energy filter of ^{8}B ν's. Some details are given in [Ha89]; here, for the first time, I shall include effects of neutrino mixing (specifically, the Mikheyev-Smirnov-Wolfenstein or MSW [MS86,Wo78,Wo79] mechanism) for the solar neutrinos.

Let me point out here that the low-lying nuclear structure of ^{97}Mo, which is a significant "poison" to this scheme, is unknown; therefore, although my numbers may not be as favorable as one might like, this discussion should only be taken as prototypical and not final!

2. THE MOLYBDENUM-TECHNETIUM EXPERIMENT: SOLAR NEUTRINOS

In Table 1, I list the relevant reactions and their Q-values. The lifetimes of ^{97}Tc and ^{98}Tc are 2.6 and 4.2 million years, respectively: short enough that there is no primordial technetium, but long enough to build up a sufficient integral. Of course one needs an environment relatively free from backgrounds, i.e., deeply buried (shielded from cosmic rays) ore with low uranium content. One such source is the Henderson mine in Colorado [Wo85,Ha89].

Table 1

Reaction	Threshold (MeV)
^{98}Mo$(\nu, e)^{98}$Tc	1.68
^{98}Mo$(\nu, e)^{98}$Tc$+n$	8.96
^{97}Mo$(\nu, e)^{97}$Tc	0.32

The calculation of the cross-section is straightforward [Wa75]: it is simply the integral of the phase-space times the nuclear matrix element M_{fi} times the (normalized) spectrum of the incident neutrino flux, Φ_ν:

$$\sigma = \int (\text{phase space}) \times |M_{fi}|^2 \times \Phi_\nu.$$

The nuclear matrix element is dominated by the allowed Gamow-Teller and Fermi transitions. The latter is sharply peaked to the isobar analog state (IAS), whose strength is exactly $N - Z$ and whose position is found experimentally. The former is more difficult; because the Gamow-Teller operator involves spin-flip, the strength is spread out over tens of MeV. Furthermore, although the naive sum rule gives a total strength of $3(N - Z)$, one finds experimentally that it is quenched by about two-thirds. I take the experimental $B(GT)$ strength from forward-angle (p, n) reactions on ^{98}Mo [Ra85].

If one takes the undistorted neutrino flux Φ_ν from ^{8}B one obtains a cross-section of 3×10^{-42}cm^2. In the Standard Solar Model (SSM) the total ^{8}B ν-flux is 5.6×10^6cm^{-2}sec^{-1}. Therefore, the total production of ^{98}Tc via ^{8}B in the SSM is 17×10^{-36}/sec^{-1}/atom of ^{98}Mo = 17 SNU (Solar Neutrino Units).

Of course, something appears to be seriously wrong with the Standard Solar Model. Two possibilities are that the sun is not producing as many ^{8}B neutrinos as we expect, or that those ν_e are changed into other flavors, or both.

A leading candidate is the "non-adiabatic MSW" mechanism [Pa86]. One one can parameterize the suppression [Pi87] as

$$\exp(-C_{MSW}/E_\nu)$$

The constant C_{MSW} is given by $\pi\beta^{-1}\Delta m^2 \sin^2\Theta_\nu$, where Θ_ν is the neutrino mixing angle in vacuum, Δm^2 is the mass difference between the neutrino species (only two relevant species are assumed), and $\beta = -n_e^{-1}dn_e/dr$ is the logarithmic derivative of the electron density with radius.

Values of $C_{MSW} = 10.5 \pm 5.5$ MeV are consistent with current results of all solar neutrino experiments, including the preliminary results from the gallium experiments [Ba90].

Including this suppression of the low-energy neutrinos, the cross-section falls with C_{MSW}. However, to truly match experiment one needs to constrain both C_{MSW} and the total flux $\Phi(^8\text{B})$ so that the Kamiokande result $\langle\sigma\phi\rangle = 0.46$ SSM is matched. With this constraint, one obtains 7.9 SNU if $C_{MSW} = 0$ (that is, no MSW mixing, only a low core temperature), and 5.6 ± -1.2 SNU if $C_{MSW} = 10.5 \pm 5.5$ MeV.

(Note: about half of the variation with C_{MSW} comes from the fact that the water Čerenkov detector at Kamioka can detect exotic flavors of neutrinos (ν_μ, ν_τ), albeit with a lower cross-section, into which the ν_e have been converted by the MSW mechanism, thus giving a small but non-negligible contribution to the Kamiokande results. These exotic flavors cannot contribute to induced inverse β-decay to technetium. Were it not for the ν_μ contribution to the Kamiokande results, the variation of the production of ^{98}Tc with C_{MSW} would be much flatter.)

If one could determine the ^{98}Tc content to within a fraction of a SNU, then one could then place further constraints on C_{MSW}.

3. THE MOLYBDENUM-TECHNETIUM EXPERIMENT: SUPERNOVA NEUTRINOS (IDEAL CASE)

What about other neutrino sources? Stellar collapses are prodigious sources of neutrinos: each produces about 5×10^{57} ν_e's, with an average energy of 12-15 MeV (see, e.g., section 15.2 of [Ba89]). For many years supernovae were disregarded as background in geochemical experiments, e.g., [Co84], using assumptions I shall show are inadequate.

I parameterize the neutrino spectrum as a thermal Fermi-Dirac distribution (in truth, as discussed here at the Workshop, the true spectrum should be somewhat "pinched" from a thermal distribution but I'll ignore that) and obtain

$$\sigma = \left(24.4 + 6.1\left(\bar{E} - 12\,\text{MeV}\right)\right) \times 10^{-42}\text{cm}^2.$$

This is a fairly steep function of $\bar{E}$; if $\bar{E} = 15$ MeV then the cross-section would be nearly twice that at $\bar{E} = 12$ MeV.

If there were a geologically recent supernova at 70 parsecs (much closer and life on Earth would be very uncomfortable) the total number of neutrinos passing through the Earth would be $\sim 10^{16}\,\mathrm{cm}^{-2}$. The number of ^{98}Tc atoms produced is then given by $\sigma \times N_\nu \sim 0.16$ atom of ^{98}Tc produced per mole of parent ^{98}Mo. In contrast, one calculates the equilibrium number of Tc atoms produced by solar neutrinos (in the SSM) to be $\sigma \times \Phi t_{1/2}/\ln 2 = 1900$ daughter atoms per mole of parent.

You may notice that I had to use rather clumsy units to compare the two productions. That's because I (and previous workers) considered only a single, nearby supernova event. Instead, to compare fairly, one should integrate over time and over the galaxy, to get an average flux of neutrinos from galactic supernovae, and then express the production of technetium in SNU.

The integration over the galaxy is described in [Ha89]. The two key results are that the effective distance of galactic supernovae is 4.6 kpc—that is, the average flux from over the entire galaxy is the same as if all SN were placed at a distance of 4.6 kpc from the Earth; and that the flux is

$$6.5 \times 10^3 \times \left(\frac{N_{58}}{0.5}\right) \times \left(\frac{F}{0.1\mathrm{yr}^{-1}}\right) \mathrm{cm}^{-2}\mathrm{sec}^{-1}$$

where F is the frequency of stellar collapses in our galaxy and $N_{58} \times 10^{58}$ is the number of ν_e produced per supernova. Working backwards, by assuming only one supernova, even nearby at 70 pc, the equivalent flux is only 53 $\mathrm{cm}^{-2}\mathrm{sec}^{-1}$. Thus previous work had underestimated the supernova contribution by as much as two orders of magnitude.

That's the first of two key ideas. The second is to not measure ^{98}Tc, but rather ^{97}Tc. If one restricts oneself to ^{98}Mo as a parent (and that's a catch I will address in a moment), then the 9 MeV threshold to neutron emission and ^{97}Tc acts as an energy filter, dramatically limiting contributions from the relatively lower energy solar neutrinos: between that and phase space factors, the cross-section for supernova neutrinos is at least two hundred times that for solar neutrinos!

To be specific, the cross-section for the solar contribution, without MSW-induced distortion of the spectrum, to ^{98}Mo$(\nu,e^-)^{97}$Tc$+n$ is $0.25 \times 10^{-42}\mathrm{cm}^2$, while the cross-section for the supernova contribution is at $\bar{E} = 12$ MeV is $56.6 \times 10^{-42}\mathrm{cm}^2$, rising to $128.4 \times 10^{-42}\mathrm{cm}^2$ at $\bar{E} = 15$ MeV.

The total production then of ^{97}Tc from ^{98}Mo via supernova neutrinos is then

$$0.37 \times \left(\frac{N_{58}}{0.5}\right) \times \left(\frac{F}{0.1\mathrm{yr}^{-1}}\right) \times \left(1 + \left(\frac{\bar{E} - 12\,\mathrm{MeV}}{3.6\mathrm{MeV}}\right)\right) \mathrm{SNU}.$$

This is to be compared with the solar contribution 0.70 SNU if $C_{MSW} = 0$ or $0.62^{+0.05}_{-0.08}$ SNU if $C_{MSW} = 10.5 \pm 5.5$. Thus even 2 supernova per century is a 12% effect, and more importantly, more than 3 per century exceeds the current (large) uncertainties in C_{MSW}. This seems a fascinating result! If one could measure the ^{97}Tc content in Molybdenum ore with a 10% uncertainty, then 10 SN/century would give a 6σ signal! At the least one could place stringent limits well within those now theoretically given.

This is the ideal case where only ^{98}Mo exists as a parent. Unfortunately, Nature has not been quite so kind.

4. The Molybdenum-Technetium Experiment: Supernova Neutrinos (Real Case)

The bad news is that ^{97}Mo also exists naturally and its cross-sections to produce ^{97}Tc should be comparable to those for ^{98}Mo $\rightarrow$^{98}Tc. This is somewhat mitigated by the good news that the abundance of ^{97}Mo is only 2/5 that of ^{98}Mo (9.55% compared to 24.13%).

There are to my knowledge no experimental (p, n) measurements of the Gamow-Teller strengths for ^{97}Mo. So to estimate the cross-section I use the Gamow-Teller profile from ^{98}Mo and scale by the $3(N - Z)$ sum rule. I also remove a low-lying bump in the Gamow-Teller profile that likely exists on account of the even-even nature of ^{98}Mo (Wick Haxton, private communication).

Then, the production of ^{97}Tc from ^{97}Mo via solar neutrinos is 3.4 SNU if $C_{MSW} = 0$ and 2.32 ± 0.45 SNU if $C_{MSW} = 10.5 \pm 5.5$ MeV. Here I've normalized the production of ^{97}Tc to the number of ^{98}Mo atoms. The corresponding production via supernova neutrinos is 0.08 SNU at $\bar{E} = 12$ MeV.

Putting it all together, the total production of ^{97}Tc from solar neutrinos is 4.1 SNU if $C_{MSW} = 0$ and 2.9 ± 0.5 SNU if $C_{MSW} = 10.5 \pm 5.5$MeV. That from supernova is

$$0.45 \times \left(\frac{N_{58}}{0.5}\right) \times \left(\frac{F}{0.1\text{yr}^{-1}}\right) \times \left(1 + \left(\frac{\bar{E} - 12\,\text{MeV}}{5.6\text{MeV}}\right)\right) \text{ SNU}.$$

Then 10 SN/century would be approximately a 15% effect.

5. Nuclear Structure Issues

My estimate of the contributions from ^{97}Mo are very crude. While scaling the ^{98}Mo Gamow-Teller strengths by the $3(N - Z)$ sum rule seems at first a reasonable idea, there is a serious flaw with this program. Much of the strength comes from the 'tail' of the Gamow-Teller giant resonance. The position of the centroid of the GT resonance, however, should be considerably different for neighboring even-even, odd-odd, and even-odd nuclei. In fact Petr Vogel (private communication) estimates that the ^{97}Mo GT resonance should lie about 2 MeV about that of ^{98}Mo.

The argument for this statement is simple. Because the Gamow-Teller giant resonance is a collective mode, it ought not to depend strongly on details of the nuclide, i.e., whether or not the nuclide is even-even, odd-odd, etc.. The position of the ground state, however, *will* depend strongly because of the pairing force, which lowers the ground state. Therefore, for example, the GT resonance should lie relatively higher above the ground state of an odd-even nucleus (such as ^{97}Tc) than of an odd-odd nucleus (^{98}Tc) not because the resonance shifts but because the ground state of the former is "lower". QRPA calculations of other nuclei bear out this observation (P. Vogel, private communication).

If the centroid lies higher, then the low-lying strength in the tail should be reduced. How to model this outside of a detailed calculation is not clear so I have not attempted to do so. If this is true, then the ^{97}Mo$(\nu_{\text{solar}}, e^-)$^{97}Tc production is lowered, which would be good news for limiting stellar collapse rates.

One final issue that has not been addressed before that I'd like to discuss are first-forbidden contributions. The higher energy of supernova neutrinos implies that such contributions could be non-negligible.

To estimate the first-forbidden matrix element, I take a typical dipole operator,

$$|M_{fi}|^2 = |\langle f | \tau_{\pm} \vec{q} \cdot \vec{r} | i \rangle|^2 = \frac{1}{2} q^2 |\langle f | \tau_{\pm} \vec{r} | i \rangle|^2$$

which by using the Thomas-Reiche-Kuhn sum rule [Do75]

$$\approx \frac{1}{4} \frac{q^2 A}{M_N \omega}$$

where ω is the excitation energy of the giant dipole resonance. Plugging in a typical value of $\omega \approx 17$ MeV, one gets $|M_{fi}|^2 > 0.4$. One must multiply this by the number of "modes;" in the naive Goldhaber-Teller picture there are four [Do75]. Comparing a typical value for the Gamow-Teller matrix element $|M_{fi}|^2 = N - Z \sim 10$, one estimates a ten percent correction. I have performed explicit calculations using the Goldhaber-Teller model [Do75] which agree with this estimate. This assumes, however, a $J = 0, T = 0$ initial state, which is clearly not the case here. To explore the first forbidden correction further requires a more microscopic model.

Clearly more work, both experimental and theoretical, is required before the nuclear structure issues can be resolved.

6. CONCLUSIONS

I have described a radiogeochemical method for extracting the integrated flux of neutrinos from stellar collapses in our galaxy. One expects the amount of ^{97}Tc produced by supernova neutrinos to be about 1-15% that from solar neutrinos. Because of the uncertainties inherent in measuring tiny amounts of technetium (while Kurt Wolfsberg and collaborators at Los Alamos have made heroic efforts in extracting the technetium, it is not clear that calculation of cross-sections, subtraction of backgrounds, etc., can all be done to better than 10% or even 20%), this is a tantalizing and perhaps ultimately frustrating result—an actual *measurement* of the flux and thus the rate may be just out of reach. Still, as the contributions to ^{97}Tc from supernova neutrinos appears to be of the same order as reasonable uncertainties, one could at least *place limits* on the flux. More precisely, one could rule out non-trivial regions of the parameter space of F, N_{58}, and $\bar{E}$. As models improve and if, as one hopes, we are lucky enough to have a galactic supernova in the not-too-distant future detected by the new generation of experiments, the uncertainties in such parameters will narrow and more stringent limits can be placed. Finally, a better understanding of the low-lying nuclear structure of ^{97}Mo, which at any rate is absolutely necessary to this method, would give a clearer picture of the practicality of this scheme.

This work was supported in part by NSF grants PHY90-13248 and PHY88-17296.

532

[Ba85] J.N. Bahcall and T. Piran, *Astrophys. J.* **267** (1983) L77.

[Ba89] J.N. Bahcall *Neutrino Astrophysics*, Cambridge University Press, Cambridge (1989).

[Ba90] J.N. Bahcall and H.A. Bethe, *Phys. Rev. Lett.* **65** (1990) 2233.

[Bi87] R.M. Bionta, *et al. Phys. Rev. Lett.* **58** (1987) 1494.

[Co84] G.A. Cowan, and W.C. Haxton, *Science*, **216**, (1984) 51.

[Da89] R. Davis, Jr., in *Neutrino '88*, Proceedings of the Thirteenth International Conference on Neutrino Physics and Astrophysics, edited by J. Schneps *et al.*, World Scientific, Singapore (1989) 518, and references therein.

[Do75] T.W. Donnelly, J. Dubach and W.C. Haxton, *Nucl. Phys.* **A251** (1975) 353.

[Ha88] W.C. Haxton, and C.W. Johnson, *Nature* **333** (1988) 325.

[Hi87] R.M. Hirata, *et al. Phys. Rev. Lett.* **58**, (1987) 1490.

[Hi89] R.M. Hirata, *et al. Phys. Rev. Lett.* **63**, (1989) 16.

[Hi90] R.M. Hirata, *et al. Phys. Rev. Lett.* **65**, (1990) 1297.

[MS86] S.P. Mikheyev, and A.Yu. Smirnov, *Sov. J. Nucl. Phys.* **42**, (1986) 913; *Sov. Phys.-JETP* **64**, (1986) 4; *Nuovo Cimento* **9C**, (1986) 17.

[Pa86] S.J. Parke, *Phys. Rev. Lett.* **57** (1986) 1275.

[Pi87] P. Pizzochero *Phys. Rev. D* **36**, (1987) 2293.

[Ra85] J. Rappaport *Phys. Rev. Lett.* **54** (1985) 2325.

[Ta81] G.A. Tammann, in *Supernovae: A Survey of Current Research*, M. J. Rees and R. J. Storeham, eds., D. Reidel Publishing Company, Dordrecht, Holland (1981) 371.

[Wa75] J.D. Walecka, in *Muon Physics, Vol. V*, V. W. Hughes and C. S. Wu, eds., Academic Press, New York (1975) 113.

[Wo78] L. Wolfenstein, *Phys. Rev. D* **17**, (1978) 2369.

[Wo79] L. Wolfenstein, *Phys. Rev. D* **20**, (1979) 2634.

[Wo85] K. Wolfsberg, *et al.* in *Solar Neutrinos and Neutrino Astronomy*, edited by M. L. Cherry, W. A. Fowler, and K. Lande (New York; American Institute of Physics) Conference Proceedings No. 126,(1985) p. 196.

A Large Low Energy Neutrino Detector for Oscillations and Supernovae Watch

Felix Boehm

California Institute of Technology
Pasadena, CA 91125

ABSTRACT

We describe a large, low energy, low background, Gd-loaded liquid scintillation detector with a fiducial volume of 1000 tons. Installed at a distance of 10-15 km from the San Onofre power reactors (7.5 GW), it will be capable of exploring neutrino oscillations via $\overline{\nu}_e$ disappearance down to $\Delta m^2 = 10^{-4} eV^2$ and mixing angles down to $\sin^2 2\theta \approx 0.1$. The detector will also serve as a supernova neutrino observatory. The detector reaction $\overline{\nu}_e\, p \rightarrow e^+ n$ gives rise to a correlated e^+, n signature, with the neutron, following capture in Gd, yielding an 8 MeV gamma cascade. The signal rate will be 16 neutrino events per day. Initially, a 12 ton prototype detector will be built and installed 1 km from the reactor. It will have a mass sensitivity of $10^{-3} eV^2$ and should provide access to the large mixing angle solution for Δm_{13}^2 for the MSW mechanism as currently suggested by the solar neutrino data.

1. INTRODUCTION

The goal in building this neutrino detector is twofold:

1) A 12 ton detector, to be built first, will be capable of exploring $\overline{\nu}_e$ disappearance down to $\Delta m^2 = 10^{-3} eV^2$. A 1000 ton detector will permit the exploration of neutrino oscillations down to $\Delta m^2 = 10^{-4} eV^2$, a factor 200 below the present limit provided by the Gosgen reactor experiment [1], with mixing angle limits of $\sin^2 2\theta$ of 0.1. The parameter space to be explored contain regions of possible solutions to the solar neutrino problem.

2) The 1000 ton detector will also serve as a supernova neutrino detector with low energy threshold, with a sensitivity to $\overline{\nu}_e$ neutrinos via charged current reaction and also to other neutrino flavors via neutral current electron scattering reactions.

The regions that can be explored with the 12 ton and the 1000 ton detectors are shown in Fig. 1. The Δm^2 range is a factor of 20 and 200 below that of the current best limit from the Gosgen experiment [1].

Recently, indications for the existence of neutrino oscillations have come from solar neutrino experiments. Both, the ^{37}Cl and the Kamiokande experiments show a reduction of the ^8B solar neutrino flux. This reduction may be explained by oscillations via the MSW mechanism. From the recent analyses by Bahcall and Haxton [2], by Vogel [3] and by Bludman [4], it appears that there are two solutions in the Δm^2 vs. $\sin^2 2\Theta$ parameter plane, a diagonal (or "non-adiabatic") solution with small mixing and a large mixing solution, $\sin^2 2\Theta \geq 0.4$, as illustrated in Fig. 1. For the large mixing angle solution, the range of Δm^2 required to accommodate the observations is around 5×10^{-8} - 10^{-4}eV^2, with the exclusion of a small region around 10^{-5}eV^2 ("day-night effect" [5]). The 1000 ton detector will cover the unexplored region down to and including a small portion of the large mixing angle solution.

Assuming that the mentioned Δm^2 represents Δm^2_{12}, it follows from scaling (with the square of the lepton masses) that Δm^2_{13} is expected to be in the interval of 0.003 - 0.4 eV^2. If we assume that the neutrino mixing matrix is analogous to the Kobayashi-Maskawa quark mixing matrix, in the sense that for large Θ_{12} one also expects a large Θ_{13}, it follows that the experiment will be able to cover the entire region of Δm^2_{13} suggested by the large mixing angle solution of the solar neutrino puzzle, as shown in Fig. 1. The quoted range of Δm^2 not already covered by the Gosgen experiment [1] can be best explored with a detector at $\approx 1\text{km}$ from the neutrino source. This represents the first task for the detectors described in this talk.

2. GENERAL CONSIDERATIONS FOR A LARGE LOW ENERGY DETECTOR

The $\overline{\nu}_e$ are detected with the help of the reaction $\overline{\nu}_e p \rightarrow e^+ n$. This reaction provides a specific correlated e^+,n signature. A proton rich target, such as mineral oil (H/A $\approx$ 2), to which an organic liquid scintillator is admixed is suitable. The target liquid has to have good light transmission, with a transmission length preferably greater than 10 m. As we wish to detect positrons down to 1 MeV in energy, the detector has to have good light collection. At a threshold of 1 MeV, gamma rays and electrons from the radioactive decays of radio impurities, notably Th and U, are well within the accepted spectrum. It is therefore required that their contribution be below an acceptable background rate which means that their concentrations be less than 10^{-12} g per g of material. Neutrons are detected via (n,γ) in Gd. This reaction has a very large cross section (of 0.25 Mb for the 15.7% isotope ^{157}Gd, compared to 0.3 b for (n,γ) on the proton). The capture gamma ray cascade has a summed energy of 8 MeV. This allows us to set a threshold at about 3.5 MeV, well above the gamma energy of Th and U radioactive decays. In order to reduce the hadronic component of the cosmic rays, the detector needs to be installed underground. A calibration of the detector to better than 5% is required in order to reach a mixing angle sensitivity of better than 0.1. A detailed discussion of the detector is given in ref. [6].

3. DETECTOR RESPONSE

The temporal and spatial characteristics of a neutrino reaction in a Gd-loaded scintillator are sketched in Fig. 2. The capture times were obtained experimentally, using a time to amplitude converter triggered by the 4.4 MeV gamma ray emitted from an AmBe neutron source. A pseudocumene based, 0.25% Gd loaded, liquid scintillator, NE 344a, was prepared for us by NE America. From this, several samples diluted with mineral oil (Britol 6NF HF) down to 0.01% Gd were obtained. Fig. 3 shows the capture time vs. Gd concentration. A Gd concentration of 0.05% was deemed optimal [6]. It gave a capture time of 45 μs (100 μs at 90%) and a capture ratio, R = capture in Gd / total capture = 0.77. The range of the $\approx$ 25 keV neutrons is about 7 cm and the range of the capture gamma rays is about 30 cm. This determines a radius, at 90%, of 83 cm of a "reaction sphere". Taking into account the 90% cuts in space, time, and energy, we estimate the efficiency of the detector at 90% c.l. to be about 0.65. The energy threshold for the positron event is assumed to accept 2 MeV (1 MeV positron kinetic energy + 2 mc^2) while the discriminator threshold for the 8 MeV gamma cascade will be set at 3.5 MeV, well above the 2.6 MeV gamma ray from the Th decay chain.

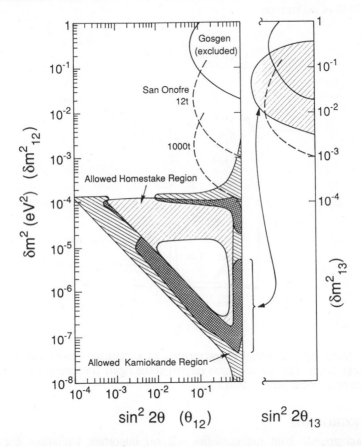

Fig.1. Two-parameter plot for neutrino oscillations. The region excluded by the Gosgen experiment is shown, as are the regions that may be explored with the proposed detectors. The allowed regions with 95% cl from the Homestake ^{37}Cl experiment and from the Kamiokande experiment are indicated by the hatched areas. Their crosshatched overlap contains a diagonal solution and a large mixing angle solution. (Based on Bludman [4] and Bahcall and Haxton [2]). A portion of the large mixing angle solution is mapped from the assumed Δm^2_{12} vs $\sin^2 2\Theta_{12}$ plane into a Δm^2_{13} vs $\sin^2 2\Theta_{13}$ plane on the right side of the figure.

The localization of an event in space was modeled with a ray-tracing Monte Carlo. Assuming a scintillation light yield of 2000 photons/MeV, a photocathode efficiency of 20%, and a light coverage of 10%, we find the 90% radius of a "resolution sphere" to be 32 cm. The "event sphere", thus, has a 90% radius of 95 cm and its volume is 3.6 m^2, 3.6×10^{-3} of the total fiducial volume for the 1000 ton detector.

Studies of the light attenuation length at 450 nm gave about 20 m for the mineral oil and 6.1 ± 1 m for the 0.05% Gd scintillator [6]. The light yield was measured to be (47 ± 2) % of anthracene.

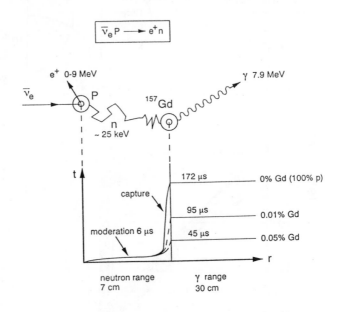

Fig.2. Temporal and spatial characteristics of a neutrino reaction in a Gd-loaded liquid scintillator. The capture times quoted in the figure for several concentrations of Gd were obtained experimentally.

4. BACKGROUNDS

Backgrounds from radioactivities will set important limitations for any low energy detector. It follows from an estimate that the contaminations from Th and U for our 1000 ton detector may not exceed 10^{-12} g Th or U per g of scintillator.

Mass spectroscopic analyses of our mineral oil samples were conducted by J. H. Chen [7] with the result that both the Th and the U concentrations were less than 0.3×10^{-12}. The Gd had impurities of Th and U of about 10^{-9} which is acceptable considering the 0.05% loading factor.

Radioactivities in the photomultipier glass, notably that from ^{40}K, must also be kept to an acceptable level. We estimate that a level of 3×10^{-8} should be acceptable considering the attenuation in the mineral oil buffer and the solid angle factor.

We have calculated, and summarized in Table 1, the uncorrelated and correlated background count rates for a 1000 ton detector, arising from the Th and U at the concentrations given in column two 2 of that table. The uncorrelated rates of 19/s and 4.1/s for a 2 MeV and 3.5 MeV threshold, respectively, give rise to an accidental coincidence rate in a 100 μs time window of 2/d. Other correlated backgrounds including that from fast neutrons following muon capture and from the ^{214}Po decay (164 μs state) in the Th chain are estimated to be about 0.1/d. In comparison, the signal rate for a 1000 ton detector, at a distance of 13 km from the 7.5 GW San Onofre reactor source will be 16/d. Cosmic ray muons will produce well recognizable and confined tracks with large energy deposits at a rate of 1/s after attenuation by an overhead shielding of 600mwe (200m rock).

Source of background	Concentration	Events s^{-1} E > 2 MeV	Events s^{-1} E > 3.5 MeV
U in scintillator	1 ppt	12	1.8
Th in scintillator	1 ppt	3	1.1
U in PMTs	100 ppb	0.4	0.04
Th in PMTs	100 ppb	0.05 ·	0.04
γ's from rocks		3	1
μ spallation		0.1	0.1
Uncorrelated background		19	4.1
Correlated background (day^{-1})		2	

Table 1. Estimates of the backgrounds for the 1000 ton detector.

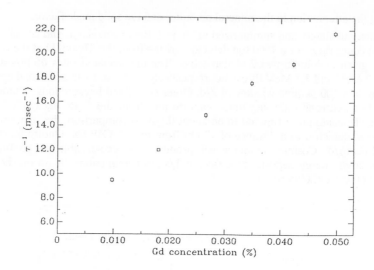

Fig.3. Inverse of the mean neutron capture time as a function of Gd concentration in the liquid scintillator (from ref.[6]).

5. THE 12 TON DETECTOR

A conceptual sketch of the 12 ton detector is shown in Fig. 4. The detector consists of an inner acrylic vessel, 2.5m high and 2.5 m diameter, filled with 12 tons of NE 344a diluted with mineral oil to give a Gd concentration of 0.05%. The central cylinder is embedded in a larger stainless steel cylinder, 3.5 m x 3.5 m, filled with pure mineral oil. This buffer region contains the 72 eight-inch photomultiplier tubes; it also provides shielding of neutrons and gamma rays. Surrounding this outer cylinder is a 5 cm thick active veto consisting of a liquid scintillator provided with a first wave length shifter. The light in the scintillator is viewed by eight narrow acrylic cylinders filled with a second shifter and topped by photomultipliers. These long cylinders are immerged in the veto liquid (with an air space to provide total reflection) and arranged equidistantly, parallel to the detector axis. The detector, placed 1 km from the San Onofre reactors, will count about 40 neutrino events per day, assuming an efficiency of about 40%. It will be shielded from the hadronic cosmic rays by 7 m of earth. The muon capture neutrons will be recognized and rejected with the help of pulse shape discrimination, a technique similar to the one used in the Gosgen experiment [1]. With a projected correlated background-event-rate of about 10/d, and a counting time of 0.5y it should be possible to reach a sensitivity for Δm^2 of $10^{-3} eV^2$, and a mixing angle limit of $sin^2 2\theta$ of 0.1. Fig. 5 gives a preliminary sketch of the underground laboratory for the 12 ton detector.

In building this detector and, especially, in developing the data acquisition system, a collaboration with LAMPF has been initiated.

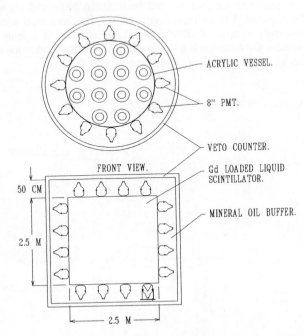

Fig.4. Sketch of the 12 ton liquid scintillator detector.

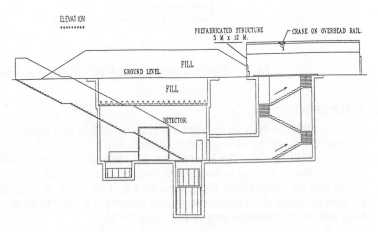

Fig.5. Sketch of the underground lab to house the 12 ton detector.

6. THE 1000 TON DETECTOR

A sketch of the 1000 ton detector is shown in Fig. 6. It is essentially an enlarged version of the 12 ton detector. The mineral oil buffer containing the 1800 photomultipliers is 1m thick. As the detector will be installed 200m underground, a muon veto will not be required. The neutrino event rate will be about 16/d, with a background of 2/d. A sensitivity to Δm^2 of $10^{-4} eV^2$ should be reached in 2 years running time. A possible site for the detector is a shallow tunnel into the mountains on the Camp Pendleton site, 13 km from the San Onofre reactors.

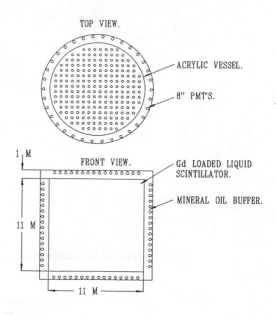

Fig. 6. Sketch of the 1000 ton liquid scintillation detector.

7. SUPERNOVAE WATCH

The 1000 ton detector installed 200 m below surface may serve, concurrently with the oscillation studies, as a supernovae watch. The event rate for a Supernova 1987A size event will be about 5/s $\overline{\nu}_e$ events > 2.8 MeV, with a correlated background of 3×10^{-5}/s. For ν_e, ν_μ, and ν_τ detection via scattering on electrons, the event rate will be 1-2/s, with an uncorrelated background of 4/s. Compared to other large detectors suitable for supernovae watch, the present detector is distinct by its specific sensitivity to $\overline{\nu}_e$ and its low neutrino energy threshold of 2.8 MeV.

8. ACKNOWLEDGMENTS

Important contributions and discussion with E. Bonvin, H.Henrikson, S. Ludtke, P. Vogel and P. Willems are gratefully acknowledged. The work is supported by the US Department of Energy.

9. REFERENCES

[1] G. Zacek et al. Phys. Rev. D 34, 2621 (1986)
[2] J.N. Bahcall and W. C. Haxton, Phys. Rev. D 40, 931 (1989)
[3] P. Vogel, Caltech Internal Report (1990)
[4] S. A Bludman, Center for Particle Astrophysics, Berkeley, Preprint TH-90-29 (1990)
[5] K. S. Hirata et al. Phys. Rev. Let. 66, 9 (1991)
[6] F. Boehm et al. NIM, in print (1990)
[7] J. N. Chen, California Institute of Technology, Geology and Planetary Sciences Division, Internal Report (1989)

THE LVD SUPERNOVA DETECTOR*

presented by E. S. HAFEN

G. BARI, M. BASILE, G. BRUNI, G. CARA ROMEO, A. CASTELVETRI, L. CIFARELLI,
A. CONTIN, C. DEL PAPA, P. GIUSTI, G. IACOBUCCI, G. MACCARRONE, T. MASSAM,
R. NANIA, V. O'SHEA, F. PALMONARI, E. PEROTTO, G. SARTORELLI, M. WILLUTZKY
University of Bologna and INFN / Bologna, Italy

J.A. CHINELLATO, C. DOBRIGKEIT CHINELLATO, A.C. FAUTH, A. TURTELLI
Department of Rajos Cosmicos, University of Campinas, Brazil

K. DE, A. DESILVA, A.M. SHAPIRO, M. WIDGOFF
Brown University, Providence, Rhode Island 02912, USA

F. ROHRBACH, A. ZICHICHI
CERN, 1211 Geneve 23, Switzerland

L. CAPUTI, G. SUSINNO
Calabria University, Cosenza and INFN / LNF, Italy

G. BARBAGLI, G. CONFORTO, G. LANDI, B. MONTELEONI, P. PELFER
INFN and University of Firenze, Italy

G. ANZIVINO, S. BIANCO, R. CASACCIA, F. CINDOLO, M. ENORINI,
F.L. FABBRI, I. LAAKSO, S. QIAN, A. RINDI, A. SPALLONE, L. VOTANO, A. ZALLO
Frascati INFN / LNF, Italy

K. LAU, J. LIU, B. MAYES, G.H. MO, L. PINSKY, J. PYRLIK, P. ROUMELIOTIS,
D. SANDERS, R. WEINSTEIN
University of Houston, Houston, TX 77004, USA

E.D. ALYEA
Indiana University, Bloomington, IN 47401, USA

G.E. KOCHAROV, V. VASILEYEV
Ioffe Physical Technical Institute, Leningrad, USSR

M. DEUTSCH, E.S. HAFEN, P. HARIDAS, B. JECKELMANN, G. JI, H.H. KUANG, T. LAINIS,
M. LI, C.S. MAO, A. PITAS, I.A. PLESS, J. TANG, S.W. WANG, X.P. WANG, , Y.R. WU,
B. XU, L. XU, Y. YUAN, G. YI, C.Z. ZHAO
Massachusetts Institute of Technology, Cambridge, MA 02139, USA

V.S. BEREZINSKY, V.L. DADYKIN, F.F. KHALCHUKOV, E.V. KOROLKOVA,
P.V. KORTCHAGUIN, V.B. KORTCHAGUIN, V.A. KUDRYAVTSEV, A.S. MALGUIN, M.A. MARKOV,
V.G. RYASSNY, O.G. RYAZHSKAYA, V.P. TALOCHKIN, V.F. YAKUSHEV, G.T. ZATSEPIN
Institute for Nuclear Research, Moscow, USSR

J. MOROMISATO, E. SALETAN, E. VON GOELER
Northeastern University, Boston, MA 02115, USA

---

* This work is supported in part by the US Department of Energy, contract # DE-ACO2-76ER03069, and the Italian INFN.

543

B. ALPAT, F. ARTEMI, C. CAPPELLETTI, P. DIODATI, P. SALVADORI
University of Perugia and INFN / Perugia, Italy

C. AGLIETTA, G. BADINO, L. BERGAMASCO, C. CASTAGNOLI, A. CASTELLINA, G. CINI,
M. DARDO, W. FULGIONE, P. GALEOTTI, C. MORELLO, G. NAVARRA, L. PERIALE,
P. PICCHI, O. SAAVEDRA, G.C. TRINCHERO, P. VALLANIA, S. VERNETTO
Institute of Cosmo-Geophysics, CNR, University of Torino, and INFN / Torino, Italy

F. GRIANTI, F. VETRANO
University of Urbino, Italy

ABSTRACT

The LVD detector and its capability for supernovae detection are reviewed. Prototype test results of its directional capability for low energy particles are presented.

1. INTRODUCTION

The LVD is a large underground detector with powerful detection capabilities for a wide variety of particle astrophysics topics.[1] The LVD detector is fully active, with tracking, timing, and energy deposition measurement interspersed throughout its volume. It is thus an especially useful tool for studying neutrinos and antineutrinos over a wide range of energies. The capabilities below 100 MeV will be discussed in this report.

2. LVD CHARACTERISTICS

The LVD consists of 190 steel portatanks, each 6 m x 2 m x 1 m, mounted in a steel support structure which fits snugly in the north part of Hall A in the Gran Sasso tunnel, as shown in Figure 1.

Each portatank contains 8 tanks of liquid scintillator surrounded by tracking chambers, as can be seen in the enlarged view of Figure 2. Each 1 m x 1 m x 1.4 m tank of liquid scintillator is viewed by three 15 cm phototubes. The pulse height and time of arrival of signals is recorded separately for each phototube. The energy deposited in a tank can be determined to $20\%/\sqrt{E(MeV)}$.[1] The energy threshold is very sensitive to the ambient radioactivity. The threshold can be operated at lower values for short periods of time. This will be done for individual tanks for 600 μsec following any trigger, so that the neutron capture signal can be detected. In addition, a lower threshold will be used for the entire detector for a few seconds following a burst of several scintillators triggering above the normal threshold. This should allow detection of lower energy neutrinos associated with supernovae.[2]

A cross section of a module, or a portatank and its surrounding tracking chambers, is shown in Figure 3. Note in Figure 2 that wide and narrow horizontal

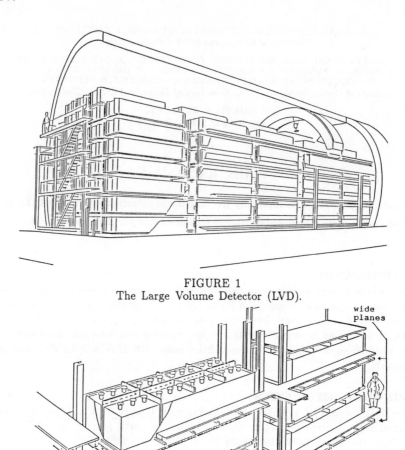

FIGURE 1
The Large Volume Detector (LVD).

wide
planes

LVD

FIGURE 2
Enlarged view of the LVD.

tracking planes alternate to allow maintenance access. The narrow horizontal planes extend only to the edge of the portatank. The tracking chambers consist of plastic extrusions of 8 gas cells, each approximately 1 cm x 1 cm x 6 m long. Hits within the gas cells are detected on 4 cm wide sense strips running parallel to the gas cells, and in 4 cm strips running perpendicular to the gas cells in L-shaped strips which bend around the corner of the module. This plane of streamer tubes plus its surrounding parallel and perpendicular sense strips is called a layer. The sense strips are staggered by about 2 cm between the two layers, for improved efficiency and angular resolution. The current design for the modules includes a 5 cm plastic spacer between the two layers. A coincidence between both layers is called a plane-pair in this document. Two modules of the type shown in Figure 3 are currently installed in the LVD structure, constituting 1% of the entire detector.

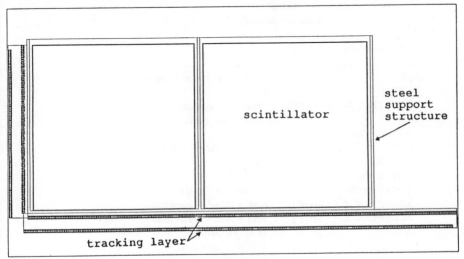

FIGURE 3
Cross section of an LVD module.

Since half of the mass is in the steel portatanks, half of the neutrino interactions from supernovae will be in the steel. Some fraction of the positrons or electrons from the interaction will be detected in the scintillator, giving a lower limit on their energy but no information about direction. Those that travel in the opposite direction will be detectable in the tracking chambers, giving no information about the energy but possibly some information about the direction.[3] Section 4 summarizes the directional capability for very short, low energy tracks using one of these modules.

The LVD properties are summarized in Table 1.

3. SUPERNOVA DETECTION IN THE LVD

The detection of neutrinos from SN1987a[4] attracted the world's attention to the astrophysics possibilities of large underground detectors.

Supernovae will be detected in the LVD scintillator via the processes shown in Table 2. The most likely process is inverse beta decay instigated by a $\bar{\nu}_e$. Such a process is followed by neutron capture. The 2.2 MeV gamma ray from this second signal will be detected 70% of the time as shown in Figure 4[1] for a ^{252}Cf source, providing a clean tag for the event. The rates given in Table 2 are the expected number detected and tagged in the scintillator within an LVD fiducial volume including 1.0 kton of scintillator for a supernova occuring 10 kiloparsecs away.

TABLE 1
Detector properties

LVD CHARACTERISTICS

SIZE	13.135 x 39.304 x 11.96 m^3
VOLUME	6174 m^3
GEOMETRIC ACCEPTANCE	7184 m^2 - steradian
MASS	3.6 kilotons (metric)
SCINTILLATOR ELEMENTS:	
SIZE	0.992 x 1.492 x .992 m^3
NUMBER	1520
TOTAL VOLUME	2232 m^3
TOTAL MASS	1.8 kilotons (metric)
ACCEPTANCE ($\geq 0 MeV$)	$6089 \pm 158 m^2$ - steradian
ACCEPTANCE ($\geq 3 MeV$)	$5983 \pm 157 m^2$ - steradian
ΔE	$20\%/\sqrt{E(MeV)}$
$E_{threshold}$	7 MeV (outer layer)
	5 MeV (inner layers)
	0.8 MeV (within gate)
TRACKING CHAMBERS:	
GAS CELL SIZE	0.9 x 614.6 x 0.97 cm^3
TOTAL CELLS	109,440
TOTAL VOLUME	58.72 m^3
SENSE STRIP WIDTH	4 cm
NUMBER	155,040
$\Delta\Theta$	1 mrad ($\geq$3 plane-pairs)
	2 mrad (2 plane-pairs)
	67 mrad (1 plane-pair)

4. RECONSTRUCTION OF SHORT TRACKS

Underground neutrino detectors face a high background of muons, some possibly associated with astrophysical sources[5]. In addition, higher energy neutrinos (greater than 112 MeV) can interact hadronically in the detector or in the surrounding rock, producing muons. Properties of some of these neutrinos will be discussed in a separate paper submitted to this conference. Thus the accurate reconstruction of muons and a thorough knowledge of the detector's acceptance for them is necessary for any neutrino study. This section explores the resolution and acceptance for short tracks.

TABLE 2

LVD SUPERNOVA RATES - 10 kiloparsecs

$(E_\nu = 20$ MeV assumed)

REACTION			DETECTION RATE / $(10^{11}/cm^2)$
$\bar{\nu}_e + p$	$\rightarrow$	$n + e^+$	800
$\nu_e + e^-$	$\rightarrow$	$\nu_e + e^-$	20
$\nu_e + {}^{12}C$	$\rightarrow$	$\nu_e + {}^{12}C^*$	10
$\nu_e + {}^{12}C$	$\rightarrow$	${}^{12}N + e^-$	6
$\bar{\nu}_e + {}^{12}C$	$\rightarrow$	${}^{12}B + e^+$	6

The most serious background for neutrinos of a few MeV is the background radioactivity of the tunnel rock and of the materials used to construct the detector. This rate sets a lower limit on the scintillator threshold which can be read and stored on tape. It also sets limits on the tracking trigger requirements. The rate measured in an earlier study[6] of the rates in the Gran Sasso tunnel was made with extruded aluminum drift chanbers. Triggering on signals detected in these aluminum chambers, the detected rate was found to be $180/m^2 - sec$. The correlated rate, or rate for double Compton scatters, was found to be $0.5/m^2 - sec$ in these chambers. These rates were also measured with the first prototype module of the design shown in Figure 3, as a part of the testing process, and will be discussed at the end of this section.

A Monte Carlo study[7] of the resolution with which the pattern recognition program[8] could find and reconstruct straight-through tracks, or muons greater than a few GeV, found that greater than 99.6% of the tracks with 3 or more plane - pairs struck could be reconstructed, with an rms of 1.0 mrad. This means, for example, that both sides of the module shown in Figure 3 were traversed, plus at least one more plane-pair in at least one other module. Initial studies[9] had

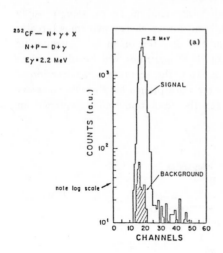
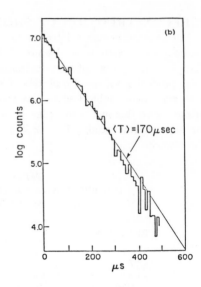

FIGURE 4

(a) Energy distribution of second signal; (b) time of arrival of second signal.

indicated that a 3-plane-pair tracking trigger was appropriate given the measured singles rate.

In the reconstruction studies,[7] 5000 events were thrown uniformly on a hemisphere containing the LVD. These events were tracked with a geometric tracking program LVD-MCGEN and also with GEANT[10]. The number (1486) which struck a scintillator with the geometric tracking program, with effectively a 0 MeV threshold in the scintillator, and the number (1458) which struck a scintillator with GEANT, with a 3 MeV threshold in the scintillator, were used to obtain the scintillator acceptances given in Table 1. This first set of studies was done for a preliminary design of the tracking chambers, with no space between the two tracking layers on each portatank.

Figure 5 shows the angular deviations between the reconstructed track and the Monte Carlo track. Figure 5a[7] shows the result when the track was generated by the straight-line Monte Carlo, LVD-MCGEN. All 996 tracks reconstructed, with an rms of 0.7 mrad due to the finite width of the sense strips and thickness of the gas cells. Figure 5b[7] shows the result when the event was tracked by GEANT without generation of secondaries, so the major differences are the scintillator energy threshold, multiple scattering, and the dead regions in the gas cells. The rms was 1.0 mrad, as expected given the amount of steel and scintillator traversed.[7] The GEANT run assumed a 100 GeV muon, so that it would be comparable to the

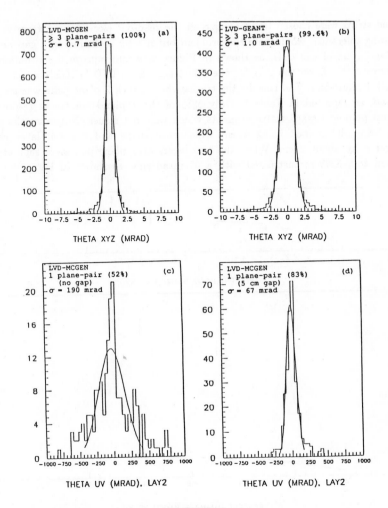

FIGURE 5
Angular deviations between Monte Carlo and reconstructed tracks.

straight line tracking program. Only 4 tracks failed to reconstruct out of 937. A track will fail when the predicted straight line trajectory intersects too many sense strips or scintillator tanks which were not struck. Closer examination of tracks not included in the initial report revealed the problem was inherent in the finite width of the L - shaped sense strips, coupled with the small separation of the gas cell layers, as discussed in a subsequent report[11] and summarized below. The L -

550

shaped strips were chosen to economize on the read-out electronics. Extensive work adjusting tolerances would help the problem but not fix it. Including a gap between the two layers of gas cells, as shown in Figure 3, would inprove the situation[11,12].

Figure 5c[11] shows the rms for the 239 tracks from LVD-MCGEN which struck exactly 1 plane-pair. The rms for the tracks which struck 2 plane-pairs was less than 2 mrad, as reported in Table 1. Only 52% of the 1 plane-pair tracks reconstruct; the rest pointed toward active elements which were not struck. Note that the angles were being determined using 4 cm wide sense strips and gas cell layer centers separated by about 2 cm. When the two layers of a plane pair were separated by a 5 cm gap, 83% reconstructed with a 67 mrad rms, as shown in Figure 5d.[11]

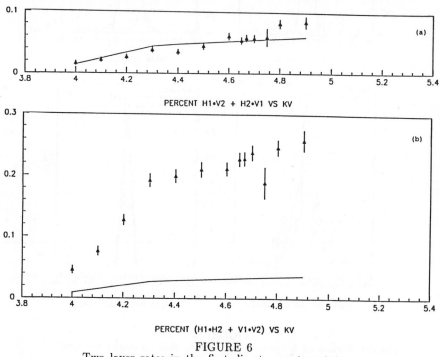

PERCENT H1•V2 + H2•V1 VS KV

PERCENT (H1•H2 + V1•V2) VS KV

FIGURE 6
Two layer rates in the first direct mount prototype.

A module of the type shown in Figure 3 was constructed[13] and tested by ramping the voltage through the plateau region and writing the resulting 1.8 million events on disk. Since a hit in any gas cell triggered the readout, this hardware performance test could also be regarded as a study of the radioactivity background. Figures 6 and 7 show the results of an offline analysis[14] of those events. The first prototype module had narrow horizontal tracking planes. The horizontal (H) layers were approximately 2 m x 6 m and the vertical layers (V) were approximately 1 m x 6 m.

Figure 6[14] shows the coincidence rate between one of the H layers and one of the V layers of gas cells, as a function of voltage and as a fraction of the triggers at that voltage. The line is the rate predicted from the singles rate, the gate width of 1 μsec, and the areas. The match between prediction (823 events) and observation (854 events) indicates that these events are random coincidences within the gate. Figure 6b[14] shows the rate for both horizontal layers or both vertical layers. The line is the predicted rate. The difference is due to double Compton scattering. This overall double Compton rate is $0.27/m^2 - sec$, half the rate measured with aluminum drift chambers. The reason appears to be that plastic cells do not provide as much material as aluminum.

To test the angular resolution of the new design, the angle the two points made with the normal to the plane was calculated for 3108 double Compton candidates. This is shown in Figure 7a.[14] Note that one expects about 4 muons from cosmic rays in this sample. Figure 7b[14] shows the angle expected for random coincidences and 7c[14] shows the angle expected for events from a uniform angular distribution of the second Compton scatter relative to the first. Note that the bin width increases as $cos\Theta$ approaches 1, or normal incidence, to be commensurate with the angular resolution for 4 cm wide strips with gas cell centers separated by 8.25 cm. Note that production modules may have a smaller separation than the prototypes, to allow room for an additional layer of resistive plate counters for timing.

The relative amounts of correlated and uncorrelated events were determined by matching the total number of candidates separated by more than 160 cm to the total number of uncorrelated events. Figure 7d[14] shows the result. Figure 7e[14] shows the result if the efficiency for obtaining the second double Compton scatter is assumed to be linear in $cos\Theta$, with 0% efficiency at normal incidence. There is some tenuous evidence discussed in Reference 14 that the remaining discrepancy between Figures 7a and 7e may provide further evidence of the energy and / or angular dependence of the source.

Note that the integrated area under the triangular curve of Figure 7a is about half that of Figure 7c, which had 100% efficiency for detecting the second Compton scatter at normal incidence. That is, the plastic limited streamer tubes are apparently half as efficient, on the average, as the aluminum extrusion drift chambers were.[6]

This exercise indicates that the tracking chambers are sensitive to low energy radiation, and when the two layers are separated by several centimeters, the angular resolution on short tracks is potentially useful.

552

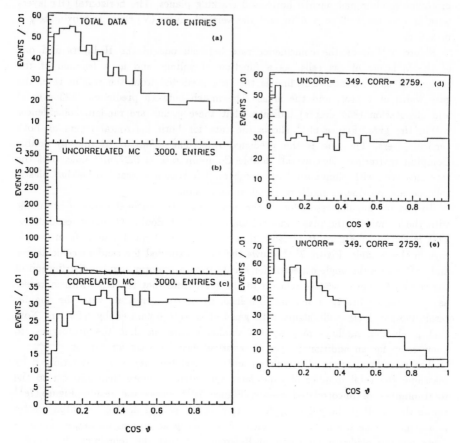

FIGURE 7

Horizontal-horizontal and vertical-vertical coincidences.

5. SUMMARY

The LVD is a large, fine-grained tracking calorimeter capable of contributing to several particle astrophysics topics. Neutrino rates from supernovae are high enough and the angular resolution from the tracking design with a spacer is good enough to potentially obtain some directional information on events occurring in the steel. Two fully-instrumented modules out of the total of 190 have now been installed in the support structure, and a full tower (20% of the total LVD) should be operational by fall 1991.

REFERENCES

1. C. Bari et al., Nucl. Instr. and Meth. A264 (1988) 5;
   C. Bari et al., Proceedings of the Workshop on Particle Astrophysics: Forefront Experimental Issues, Berkeley, December 1988, p. 320;
   C. Bari et al., Nucl. Instr. and Meth. A277 (1989) 11.

2. M. Aglietta et al., Nucl. Instr. and Meth. A277 (1989) 17.

3. L. Pinsky and B. Mayes, A Monte Carlo Evaluation of Possible Directional Information in LVD from a Stellar Collapse Event, LVD Note 48, February 1988.

4. K. Hirata et al., Phys. Rev. Lett. 58 (1987) 1490;
   R. Bionta et al., Phys. Rev. Lett. 58 (1987) 1494;
   M. Aglietta et al., Europhys. Lett. 3 (1987) 1315.

5. M. Marshak et al., Phys. Rev. Letters 54 (1985) 2079; Phys Rev. Letters 55 (1985) 1965;
   G. Battistoni et al., Phys Lett. 155B (1985) 465.

6. G. Bari et al., A Study of the Radioactivity Background at the Gran Sasso Laboratory using Proportional Tube Chambers, LVD Note 7 (87-2), January 1987.

7. E. Hafen, Muon Reconstruction in Large Underground Detectors, APC Engineering Note 89-1, May 1989;
   E. Hafen, Muon Reconstruction in Large Rectangular Detectors, Nucl. Phys. B (Proc. Suppl.) 14A (1990) 325.

8. E. Hafen, PIPR Equations, APC Engineering Note 89-2, October 1989.

9. D. Shambroom, Monte Carlo Study of a Tracking Trigger for the LVD, LVD Note 11, April 1987;
   B. Jeckelmann, Simulation of the LVD Tracking Trigger, LVD Note 13, July 1987;
   H. Kuang, A Simulation about LVD Triggering Efficiency, LVD Note 46, September 1988.

10. R. Brun et al., GEANT3 User's Guide, CERN DD/EE/84-1, May 1986.

11. E. Hafen, Comments on Track Reconstruction, LVD Note 59 (89-10), August 1989.

12. E. von Goeler, The Case for Spacers in the LVD Tracker, LVD Note 69 (89-20), November 1989.

13. P. Giusti et al., Report of the Direct Mount System at Gran Sasso, Italy, LVD Note 83 (90-11), May 1990.

14. E. D. Alyea et al., Offline Analysis of Direct Mount Test Results, LVD Note 85 (90-13), December 1990.

Expected Supernova Neutrino Signal in IMB-3 Detector [1]

Danuta Kiełczewska

The University of California, Irvine, California 92717, USA
Warsaw University, Warsaw, Poland

for the IMB Collaboration:

R.Becker-Szendy[a], C.B.Bratton[b], D.R.Cady[c], D.Casper[d], S.T.Dye[d], W.Gajewski[e],
M.Goldhaber[f], T.J.Haines[g], P.G.Halverson[e], T.W.Jones[h], W.R.Kropp[e],
J.G.Learned[a], J.M.LoSecco[c], C.McGrew[e], S.Matsuno[e], R.S.Miller[j], M.S.Mudan[h],
L.Price[e], F.Reines[e], J.Schultz[e], H.W.Sobel[e], L.R.Sulak[d], R.Svoboda[j]

[a] *The University of Hawaii, Honolulu, Hawaii 96822, USA*
[b] *Cleveland State University, Cleveland, Ohio 44115, USA*
[c] *The University of Notre Dame, Notre Dame, Indiana 46556, USA*
[d] *Boston University, Boston, Massachusetts 02215, USA*
[e] *The University of California, Irvine, California 92717, USA*
[f] *Brookhaven National Laboratory, Upton, New York 11973, USA*
[g] *The University of Maryland, College Park, Maryland, USA*
[h] *University College , London, WC1E6BT, United Kingdom*
[j] *The Louisiana State University, Baton Rouge, Louisiana, USA*

Abstract

The IMB-3 detector has been upgraded since February 1987 and is now even better prepared for detection of a gravitational collapse. As the largest existing detector, which can measure both the electron energies and angles, it would provide unique information about muon and tau neutrino properties. Thanks to the kinematics of neutrino scattering on electrons their signal could be enhanced and the direction of a source could be indicated. Observation of hundreds of electron antineutrinos would allow precise measurement of their energy spectra and the burst time evolution, which would lead to a better understanding of processes occuring in the protoneutron star.

1   INTRODUCTION

After the first observation of the neutrino signal from gravitational collapse [1,2,3], which confirmed the basic theoretical expectations, the future experiments should aim either toward a substantial increase of the observational range or toward gathering more detailed information about a particular collapse event and neutrino properties. In order to extend the observations beyond the Local Group and thus to

---

[1] *Talk presented at the UCLA SuperNova Watch Workshop, Santa Monica, CA, December 1990*

increase significantly the supernova detection rate, a new concept for experimental technique [4,5] is needed. On the other hand a gravitational collapse in our Galaxy, expected to occur once in ten years or so, would give rise to hundreds of neutrino interactions in the existing or currently built experiments [6]. To extract the most useful information from the data, the identification of the neutrino flavors is needed. To this aim it is desirable to measure not only the event energies and time of their occurence, as in scintillation tanks traditionally used for supernova searches, but also the electron direction. Imaging detectors using Čerenkov light provide this capability. The largest of them is the 8 kiloton IMB (Irvine - Michigan - Brookhaven) detector.

The importance of the direction measurement is enhanced by the fact that the angular distribution of the observed events from SN1987A exhibited in both water Čerenkov experiments the same trend of being more forward peaked than expected [7,2,8]. The paucity of the data preclude invoking new phenomena for their interpretation but the probability of a statistical fluctuation for the set of all 19 events of less than 1% is small enough to hope that new measurements would shed more light on the anomaly.

2   THE CURRENT STATUS OF THE DETECTOR

The IMB detector is located at a depth of 600 meters at the Morton Salt Mine in Cleveland, Ohio, USA. The detection of Čerenkov light emitted by relativistic particles is done in a 18m × 17m × 22.5m tank of ultra pure water surrounded on all six sides by 2048 photomultiplier tubes. The IMB-3 version of the detector operating since 1986 contains 20-cm diameter tubes mounted on 60cm × 60cm waveshifter plates for increased light collection [9]. The whole detector active volume of 6800 tons can be used for a gravitational collapse detection because of negligible background within short time interval characteristic for this phenomenon. The background comes from 2.7 Hz muons which are predominantly recorded as high energy events.

Since February 1987 the detector electronics has been upgraded [10], to make it more efficient for low energy physics and to enhance our ability to observe a supernova in the Milky Way or its neighborhood.

Thanks to better tuning of electronics, the noise has been significantly reduced and the trigger threshold could be decreased to around 10 MeV, which constitutes a substantial improvement compared to about 20 MeV in February 1987. The lower threshold increases the sensitivity to the elastic scattering on electrons, when only a half of the neutrino energy is transferred per average to electrons recorded in the detector.

In order to record bursts of high trigger rate the computer associated dead time was decreased from 55 ms to 3.5 ms. This allows bursts up to 250 Hz to be efficiently collected.

Reliable time measurements are of great importance for supernova studies. A WWV clock allows to determine absolute time of a trigger with ± 0.5 ms accuracy. In addition a 2^{18} Hz oscillator clock allows relative times of events to be measured to about $\pm 3\,\mu$s precision.

As a result of the upgrade of the data taking system all the triggers are now recorded on 8 mm tapes. Bursts are monitored on line by several algorithms that watch for anomalous high rates of low-energy triggers and notify the operator in the mine. Recently remote burst monitoring have also been installed.

The power supplies, which in February 1987 made a quarter of the detector non operational, have been replaced into more reliable. The photomultiplier tube performance has also improved, making the detector malfunctions less likely.

3   EXPECTED INTERACTION RATES

In order to estimate a detector response to a stellar collapse one needs to know the luminosity, L_{tot}, of the neutrino burst, the supernova distance, the neutrino energy spectra and the detector efficiency as a function of energy. The observation of 19 events from SN1987A provided some indications for the needed informations but the paucity of the data did not allow to resolve large uncertainties resulting from the stellar collapse models.

Models of supernova collapse provide a variety of estimates for L_{tot}, from 1×10^{53}ergs to 8×10^{53}ergs [11,12,13]. The Kamiokande collaboration estimated the energy of the $\overline{\nu_e}$ pulse to be 8×10^{52}ergs [2]. This corresponds to a total luminosity of 4.8×10^{53}ergs on the assumption of the energy equipartition among six light neutrino states. The IMB data lead to the temperature estimate of $4.2^{+1.0}_{-0.8}$ MeV, total energy carried by $\overline{\nu_e}$ of $(4.8 \pm 1.7) \times 10^{52}$ ergs and consequently L_{tot} of $(2.9 \pm 1.0) \times 10^{53}$ ergs.

For further estimates we shall use a value of 3×10^{53} ergs for the total luminosity which is conservative on the basis of the present experimental evidence and is equal to the expected binding energy of the resulting neutron star. Both, the total luminosity and the spectra change with time; in what follows we consider the time integrated L_{tot} and energy spectra.

Estimates of expected neutrino interaction rates have to be very uncertain because of the unpredictable distance to the source. For illustration purposes we will pick 10 kpc, an approximate distance to the Galaxy center.

The neutrino emission proceeds essentially in two different phases. First, during the neutronization burst lasting less than about 20 ms, the electron neutrinos carry away about 3×10^{51} ergs [11,12,13,14]. Their average energy is expected to be about 9 MeV and can cause merely ~ 1 event in the IMB-3 detector at a distance of 10 kpc.

Next, during the cooling phase lasting several seconds, the $\nu_x \, \overline{\nu_x}$ pairs are produced by thermal $e^+ e^-$ pairs through the neutral current interaction. The total energy of 3×10^{53} ergs is thus equally distributed between neutrinos and antineutrinos of three flavors .

Most of the neutrinos are emitted from a neutrinosphere , inside which they had become trapped due to high density of matter and subsequently thermalized. They then gradually diffuse out of the protoneutron star with energies distributed approximately according to the Fermi - Dirac function. The temperatures depend on the neutrino flavor. The observed SN1987A sample was generally interpreted as being due to $\overline{\nu_e}$'s with a temperature value between 3 MeV and 5 MeV. The ν_e 's are expected to have similar temperature. The muon and tau neutrinos and antineutrinos, called hereafter $\nu_{\mu|\tau}$, should be about twice as energetic, because their initial energy gets less degraded as a result of interactions in the protoneutron star.

For the Fermi - Dirac spectrum with temperature, $T(\text{MeV})$, the following formula can be used to relate the integrated neutrino flux, Φ (cm^{-2}), with the total energy, L (ergs), radiated from a source at a distance, d (kpc):

$$\Phi = \frac{1.6 \times 10^{11}}{T} \times \frac{L}{10^{52}} \times \frac{10}{d} \tag{1}$$

In a water detector the interactions take place on protons, ^{16}O nuclei and electrons, with the relative target abundances being 2, 1 and 10 correspondingly. The cross sections corrected for target abundances are plotted in Fig. 1 for all kinematically accessible reactions.

The inverse beta decay, $\overline{\nu_e} \, p \rightarrow e^+ \, n$, dominates $\overline{\nu_e}$ interactions at all energies relevant to supernova neutrinos (Fig. 1a). For ν_e the absorption on ^{16}O overcomes the elastic scattering on electrons only above 35 MeV of the neutrino energy (Fig. 1b). For $\nu_{\mu|\tau}$ (Fig. 1c) only neutral current elastic scattering is available at typical supernova energies. However in view of the fact that their energies are higher than that of $\overline{\nu_e}$ and ν_e , it is prudent not to neglect the charged current (CC) reactions that become open for $\overline{\nu_\mu}$ above 110 MeV ($\overline{\nu_\mu} \, p \rightarrow \mu^+ \, n$) and only slightly higher for ν_μ ,when the absorption on ^{16}O starts to take place. The cross section increases then by 3 orders of magnitude. This is illustrated in Fig. 2, showing the interaction rates for ν_μ and $\overline{\nu_\mu}$ with Fermi - Dirac energy spectrum and temperatures of 7 MeV to 11 MeV.

In water detectors the CC interactions of ν_μ and $\overline{\nu_\mu}$ produce muons of energies too low to emit Čerenkov light, but are detectable by their decays into electrons with energies peaking below 53 MeV. The decay electrons are more efficiently detected than the recoil electrons resulting from small momentum transfer elastic scattering of $\nu_{\mu|\tau}$.

Table 1 shows expected event rates in 6800 tons of water with trigger threshold at 10 MeV from a source of $L_{tot} = 3 \times 10^{53}$ ergs at a distance of 10 kpc.

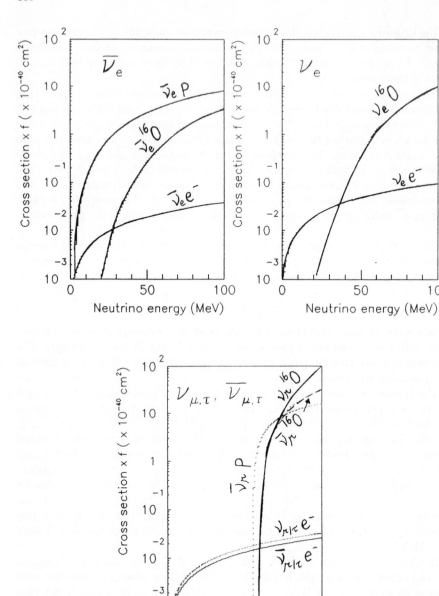

Figure 1: The cross sections for interactions with protons, ^{16}O and electrons corrected for relative target abundance. (a) for $\overline{\nu_e}$, (b) for ν_e and (c) for $\nu_{\mu|\tau}$.

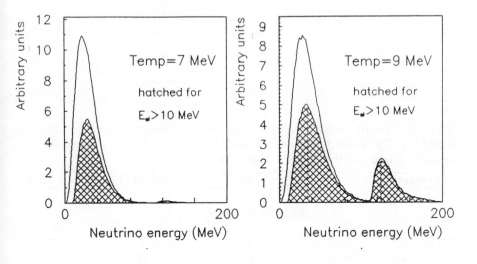

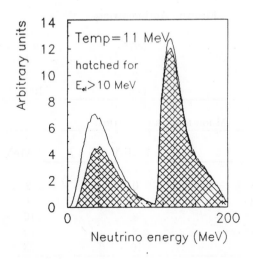

Figure 2: Interaction rates for ν_μ and $\overline{\nu_\mu}$ with Fermi - Dirac energy spectrum and temperatures of 7 MeV, 9 MeV and 11 MeV. Hatched areas correspond to events that would trigger the IMB-3 detector, i.e. the interactions with electron energy greater than 10 MeV.

Table 1: Expected neutrino signal in 6800 tons of water and the trigger threshold at 10 MeV. The source luminosity is assumed to be $L_{\nu_i} = \frac{1}{6} \times 3 \times 10^{53}$ ergs for every neutrino species and the distance is taken to be 10 kpc.

Interaction	$T_{\nu_e} = 3$ **MeV**	$T_{\nu_e} = 4$ **MeV**	$T_{\nu_e} = 5$ **MeV**
Electron Neutrinos			
$\nu_e + {}^{16}O$	0	6	25
$\nu_e + e^-$	10	17	23
Total	10	23	48
Electron Anti-neutrinos			
$\overline{\nu}_e + p$	597	998	1349
$\overline{\nu}_e + {}^{16}O$	1	6	20
$\overline{\nu}_e + e^-$	2	3	5
Total	600	1007	1374

Muon and Tau Neutrinos			
	$T_{\nu_l} = 7$ **MeV**	$T_{\nu_l} = 9$ **MeV**	$T_{\nu_l} = 11$ **MeV**
$\overline{\nu}_\mu + p$	0	2	13
$(\nu_\mu + \overline{\nu}_\mu) + {}^{16}O$	0	1	9
$(\nu_\mu + \overline{\nu}_\mu) + e^-$	8	10	11
$(\nu_\tau + \overline{\nu}_\tau) + e^-$	8	10	11
Total	16	23	44

It is seen that at all considered temperatures the signal is dominated by the $\overline{\nu_e}$ absorption on protons. The event rates are very sensitive to the assumed temperatures, especially for CC reactions, because a) the cross section is a quadratic function of energy and b) electrons below 10 MeV are not detected. For the elastic scattering reaction the dependence on the energy comes only through the threshold cut, because for the linearly rising cross section, σ, the event rate integrated over neutrino enrgy, $\int \sigma(E)\,\Phi(E)\,dE$, is proportional to the luminosity, which is fixed.

Table 1 also shows thar for the hard enough spectra muon neutrino interactions could be dominated by the charged current interactions. The estimated rates are very sensitive to the shape of the neutrino energy spectrum at high energies. Although there is no basic reason forbidding energies above 110 MeV, the stellar collapse models [6,14] indicate that high energies are depleted when compared to the Fermi-Dirac function. One should however keep in mind that the expected spectra result from the calculations of the neutrino transport through the stellar medium, which are very complicated and have to be based on various approximations [15,16,17,18]. Although some models predict that luminosities carried by different neutrino species should be equal (which was assumed for estimates reported here), a paper by Bruenn [12] shows that the luminosity of $\nu_{\mu|\tau}$ $(\overline{\nu_{\mu|\tau}})$ may also be more than twice that of ν_e $(\overline{\nu_e})$.

It is also worth noting that the kinematic properties of the SN1987A event sample were by no means well understood, particularly the angular distribution, and gave a hint that $\nu_{\mu|\tau}$ temperatures could be actually quite high [19,20].

4 LOOKING FOR MUON AND TAU NEUTRINOS.

The expected event rates are dominated by electron antineutrinos. The electrons produced by $\overline{\nu_e}$ take all the neutrino energies (less 1.8 MeV) and are emitted almost isotropically. The thresholds for the absorption on ^{16}O are at 15.4 MeV and 10.4 MeV respectively for ν_e and $\overline{\nu_e}$. The reactions with small momentum transfer are blocked by Pauli exclusion principle, i.e. electron angles are mainly large with respect to the incoming neutrinos. The electrons resulted from elastic scatterings preferentially follow the neutrino direction getting per average only half of the neutrino energy. The angular distributions should thus provide a means of selecting ν_e and $\nu_{\mu|\tau}$ events. In addition the $\nu_{\mu|\tau}$ interactions should produce higher energy electrons than ν_e.

The relative abundance of different neutrino species depend strongly on their temperatures (Table 1). For an illustration we display simulated angular distributions of a supernova event sample for an arbitrary choice of temperatures: $T_{\nu_{\mu|\tau}} = 9\,\text{MeV}$, $T_{\overline{\nu_e}} = T_{\nu_e} = 4$ MeV (Fig. 3 a,b) and $T_{\overline{\nu_e}} = T_{\nu_e} = 3$ MeV (Fig. 3 c,d). Multiple Coulomb scatterings of low energy electrons limit seriously the angular resolution and are taken into account.

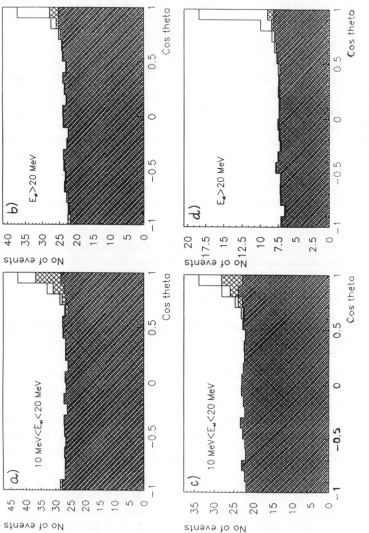

Figure 3: Expected distribution of the angles between electron and neutrino momenta for temperatures: $T_{\nu_{\mu|\tau}} = 9$ MeV, $T_{\overline{\nu_e}} = T_{\nu_e} = 4$ MeV (a,b) and $T_{\overline{\nu_e}} = T_{\nu_e} = 3$ MeV (c,d). Cross-hatched areas correspond to $\overline{\nu_e}$, hatched to ν_e and non-hatched areas to $\nu_{\mu|\tau}$ events. Electron energies are between 10 MeV and 20 MeV (a,c) or above 20 MeV (b,d).

It is clear from the figures that flavor could not be ascribed to individual events. However by selecting events with higher energy electrons (e.g. > 20 MeV), a proof of existence of $\nu_{\mu|\tau}$ interactions in the total sample would be feasible, unless the $\overline{\nu_e}$ spectrum is significantly harder, while that of $\nu_{\mu|\tau}$ softer. Again let's remind here that forward peaking in the SN1987A data suggests that the latter case is less realistic.

The angular distribution of muon decay electrons resulting from CC interactions of ν_μ and $\overline{\nu_\mu}$ is almost isotropic [20]. An extraction of these relatively highly energetic electrons could be possible only if the $\overline{\nu_e}$ spectrum was very soft. However, it will be shown that even a few events of this type could be very important, if a neutrino mass would cause observable time delays.

5 WHAT COULD BE LEARNED?

5.1 Neutrino Masses

Laboratory experiments have established the following limits on the neutrino masses [21,22,23]:

$$m(\nu_e) < 8 \text{ eV}$$
$$m(\nu_\mu) < 0.27 \text{ MeV}$$
$$m(\nu_\tau) < 35 \text{ MeV}$$

Large distances covered by supernova neutrinos offer a unique chance to detect measurable delays in arrival times resulting from non zero masses. A neutrino of mass, m (keV), and energy, E (MeV), coming from a distance, D (kpc), is delayed by t (sec), with respective to a massless particle:

$$t = 5.17 \times 10^5 \left(\frac{m}{E}\right)^2 \times \frac{D}{10} \tag{2}$$

The neutrinos that had arrived from SN1987A in Large Magellanic Cloud 50 kpc away, allowed to put limits from 20 eV to 30 eV on $m(\overline{\nu_e})$ depending on assumptions about the time profile of neutrino emission during the cooling phase. With a source a factor f closer to Earth, the masses higher by $\sqrt{f}$ could be probed. An observation of a very short neutronization burst could offer a chance for stringent $m(\nu_e)$ limits, but for a distance of 10 kpc only masses higher than $\sim$ 3 eV could be excluded, which is not an impressive improvement over laboratory limits quoted above.

A galactic stellar collapse would however offer unique possibilities for studies of $m(\nu_\mu)$ and $m(\nu_\tau)$ in a broad range of values from about 100 eV to 50 keV or more. In the estimate of the lower sensitivity limit we have taken into account that the neutrino energy is higher than that of the detected electron for an elastic scattering event.

For $m(\nu_\mu)$ or $m(\nu_\tau)$ values of hundreds eV time delays would be on the order of minutes and background constitutes no problem. Much higher masses would smear out the signal over many days and supernova events could be confounded with spallation products caused by traversing muons, natural radiation, solar or atmospheric neutrinos. However by selecting events with energy between 15 MeV and 70 MeV only the last background source remains relevant: 7 events per 10 days. In addition atmosheric neutrino background is almost isotropic, hence one could make an angular cut to select $\nu_{\mu|\tau}$ forward scatterings on electrons. With a cut on the angle θ away from supernova direction, $cos\theta > 0.8$, the background is reduced to 7 events per 100 days.

Fig. 4 shows the expected time distribution for an assumed neutrino mass of 50 keV and 80 keV and temperature of 9 MeV. The 95% c.l. upper limit on the atmospheric neutrino background is also displayed. It is seen that a pulse of $\nu_\mu, \overline{\nu_\mu}$ or $\nu_\tau, \overline{\nu_\tau}$ could be extracted from the background if masses were lower than 50 keV.

One could probe even higher muon neutrino masses in a case of hard $\nu_\mu, \overline{\nu_\mu}$ spectra. For temperatures above 10 MeV the CC muon production would dominate. Most of interacting neutrinos would have energies just above 106 MeV and therefore their arrival time would be less spread out. This is illustrated in Fig. 5 for assumed neutrino masses of higher than 100 keV. Even though the angular cut is not useful in this case, the signal would remain above the background for masses almost up to 200 keV. An observation of delayed, almost isotropic events in excess of the atmospheric neutrino background would provide a unique means of distinguishing between muon and tau neutrinos.

5.2   Neutrino Oscillations

The imaging character of Čerenkov technique and the information about particle direction would appear particularly useful for studies of neutrino oscillations. Massive neutrinos emitted in a stellar collapse can arrive to the detector in a flavor state different from the produced one. This can occur either because of resonant oscillations when traversing the high density stellar matter or because of vacuum oscillations during their long journey to Earth.

Let us assume that muon or tau neutrinos with a hard spectrum oscillate into electron neutrinos. The latter could be recognized by the characteristic backward peaking resulting from the ν_e absorptions on oxygen. This is illustrated in Fig. 6 where the distributions of angles between electron and neutrino directions are shown for two high temperatures assumed for ν_e spectrum. The detection rate of these oscillated neutrinos would be enhanced because the ν_e absorption with CC cross section takes place. Conversely, a transformation of low energy electron neutrinos into $\nu_{\mu|\tau}$ would not offer any chance for their detection.

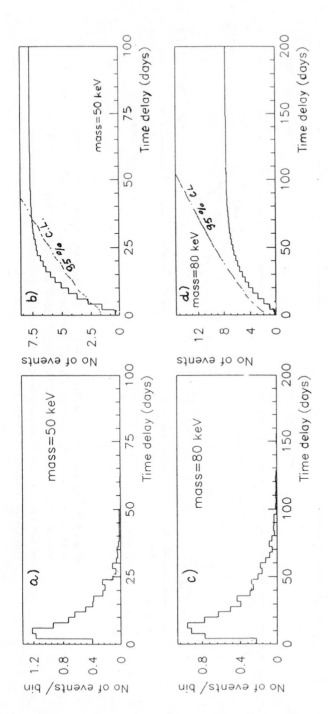

Figure 4: Simulated time distribution of $\nu_\mu, \overline{\nu_\mu}$ or $\nu_\tau, \overline{\nu_\tau}$ interactions for an assumed neutrino mass of a,b) 50 keV and c,d) 80 keV and temperature of 9 MeV. Only the events with energies between 15 MeV and 70 MeV and $\cos\theta > 0.8$ are selected. The 95% c.l. limits of the atmospheric neutrino background are also indicated in b) and d) where the cumulative number of events is shown.

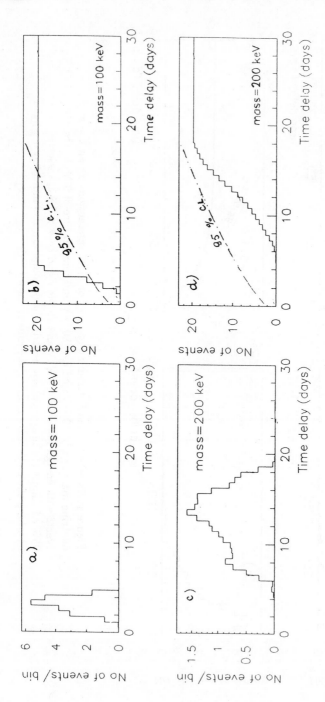

Figure 5: Simulated time distribution of $\nu_\mu, \overline{\nu}_\mu$ interactions for the assumed neutrino masses: a,b) 100 keV and c,d) 200 keV and for temperature of 11 MeV. Only the events with energies between 15 MeV and 70 MeV are selected. The 95% c.l. limits of the atmospheric neutrino background are also indicated in b) an d) where the cumulative number of events are shown.

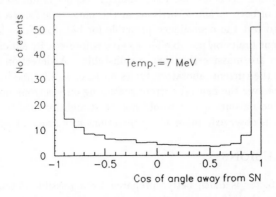

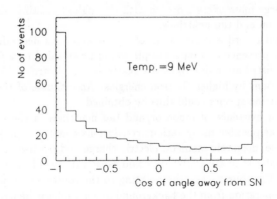

Figure 6: Diistribution of angles between electron and neutrino directions for ν_e interactions with very hard spectra: a) $T_{\nu_e} = 7$ MeV and b) $T_{\nu_e} = 9$ MeV.

5.3 Other Neutrino Properties

The SN1987A data sample gave rise to upper limits on the $\overline{\nu_e}$ mass, charge, magnetic moment and lifetime. In addition inferences have been drawn about axions and $\nu - \nu$ scattering. A possibility to see signals of various neutrino species should enable to draw many additional conclusions. For instance it has been indicated in ref. [24], that an observation of both ν_e and $\overline{\nu_e}$ arriving at the same time from a common distant source allows to test the equivalence principle for both particles and antiparticles and put stringent limits on possible force carriers foreseen by supergravity.

If neutrinos are massive they can be unstable. A detection of a $\nu_{\mu|\tau}$ signal would improve the current laboratory limits on ν_μ or ν_τ lifetime by a few orders of magnitude. However the neutral current scattering on electrons do not distinguish between neutrino flavours and it would not be straightforward to decide which of the two species is observed, unless CC interactions of ν_μ were observed.

6 CONCLUSIONS

The IMB-3 detector is much better prepared for a possible detection of a gravitational collapse than it was in February 1987. The threshold lowered to about 10 MeV and large mass offers a device with the highest existing sensitivity for the detection of muon and tau neutrinos.

IMB-3 would record a dozen or so of $\nu_{\mu|\tau}$ interactions depending on a source distance. Their presence in a data sample would be signaled by a forward peaking of the angular distribution due to elastic scattrings on electrons. They would differ from ν_e interactions by higher electron energies. An estimate of the relative fluxes of different neutrino species could thus be obtained.

The proof of existence of muon or/and tau neutrinos in the observed sample would then make possible many various conclusions about $\nu_{\mu|\tau}$ properties as limits on mass, lifetime, magnetic moment, electric charge and neutrino oscillations.

If masses grater than about 100 eV would cause a delay of heavier neutrinos with respect to the $\overline{\nu_e}, \nu_e$ burst, then the pointing to the source would provide a means for the signal separation from the background of atmospheric neutrinos. The signal of delayed neutrinos would be significant almost up to the current laboratory limit on $m(\nu_\mu)$.

The forward peaking would be of special use for a gravitational collapse in the part of Galaxy obscure for visible light. The indication of the direction of a supernova event would allow to orient detectors studying signals in various ranges of electromagnetic spectrum.

Hundreds of $\overline{\nu_e}$ would be observed. With this statistics the information about the time profile of the neutrino pulse would be greatly improved which should enhance the understanding of the process of stellar collapse. Also the $\overline{\nu_e}$ spectrum would be measured with high precision. This would allow to gain insight into the cooling phase of the explosion and state of the matter in the protoneutron star.

References

[1] R. Bionta et al., *Phys. Rev. Lett.* **58**, 1494 (1987)

[2] K. Hirata et al., *Phys. Rev. Lett.* **58**, 1490 (1987)

[3] E. N. Alekseev et al., *JETP Lett.* **45**, 589 (1987)

[4] J. Wilson, *Symposium on the Next Supernova*, Santa Monica, Feb. 1989

[5] G. Fuller et al., *Symposium on the Next Supernova*, Santa Monica, Feb. 1989

[6] A. Burrows, these proceedings

[7] C. B. Bratton et al., *Phys. Rev.* **D37**, 3361 (1988)

[8] K. Hirata et al., *Phys. Rev.* **D38**, 448 (1988)

[9] R. Claus et al.,*Nucl. Instr. and Meth.* **A261** (1987) 540.

[10] D.W. Casper, Ph. D. Thesis, 1990

[11] R. Mayle, J. R. Wilson and D. N. Schramm, *Astrophys. Journal* **318**, 288 (1987)

[12] S. Bruenn, *Phys. Rev. Lett.* **59**, 938 (1987)

[13] A. Burrows *Astrophys. Journal* **334**, 891 (1988)

[14] J. Wilson, private communication, 1990

[15] A. Burrows and J. M. Lattimer, *Astr. Journ.* **307**, 178 (1986)

[16] G. Raffelt and D. Seckel, *Phys. Rev. Lett.* **60**, 1793 (1988)

[17] W. C. Haxton, *Phys. Rev. Lett.* **60**, 1999 (1988)

[18] R. F. Sawyer, *Astrophys. Journal* **328**, 691 (1988)

[19] D. Kiełczewska, *Proceedings of the Workshop on Particle Astrophysics*, Berkeley, Dec 8-12, 1988, ed. by E. Norman (World Scientific, 1989), p. 326

[20] D. Kiełczewska, *Phys. Rev.* **D41**, 2967 (1990)

[21] E. Holzschuh, *Conference on Neutrino Physics and Astrophysics, Neutrino 90*, CERN, Geneva, 10-15 June, 1990

[22] J. Wilkerson, *Conference on Neutrino Physics and Astrophysics, Neutrino 90*, CERN, Geneva, 10-15 June, 1990

[23] Particle Data Group, *Phys. Lett.* **B239**, 1 (1990)

[24] S. Pakvasa et al., *Phys. Rev.* **D39**, 1761 (1989)

Sudbury Neutrino Observatory

G.T. Ewan, Queen's University

on behalf of the SNO Collaboration[†]

Abstract

The Sudbury Neutrino Observatory (SNO) detector is a 1000 tonne heavy water (D_2O) Čerenkov detector designed to study neutrinos from the sun and other astrophysical sources. The use of heavy water allows both electron neutrinos and all types of neutrinos to be observed by three complementary reactions. The detector will be sensitive to the electron neutrino flux and energy spectrum shape and to the total neutrino flux irrespective of neutrino type. These measurements will provide information on both vacuum neutrino oscillations and matter enhanced oscillations, the MSW effect. In the event of a supernova it will be very sensitive to muon and tau neutrinos as well as the electron neutrinos emitted in the initial burst enabling sensitive mass measurements as well as providing details of the physics of stellar collapse.

[†] The Sudbury Neutrino Collaboration: H.C. Evans, G.T. Ewan, H.W. Lee, J.R. Leslie, J.D. MacArthur, H.-B. Mak, A.B. McDonald, W. McLatchie, B.C. Robertson, B. Sur, P. Skensved (Queen's University): C.K. Hargrove, H. Mes, W.F. Davidson, D. Sinclair (Centre for Research in Particle Physics): E.D. Earle, G.M. Milton, E. Bonvin, (Chalk River Laboratories): P. Jagam, J. Law, J.-X. Wang, J.J. Simpson (University of Guelph): E.D. Hallman, R.U. Haq (Laurentian University): A.L. Carter, D. Kessler, B.R. Hollebone (Carleton University): R. Schubank, C.E. Waltham (University of British Columbia): R.T. Kouzes, M.M. Lowry, R.M. Key (Princeton University): E.W. Beier, W. Frati, M. Newcomer, R. Van Berg (University of Pennsylvania): T.J. Bowles, B.T. Cleveland, P.J. Doe, S.R. Elliott, M.M. Fowler, R.G.H. Robertson, D.J. Vieira, D.L. Wark, J.B. Wilhelmy, J.F. Wilkerson, J.M. Wouters (Los Alamos National Laboratory): E. Norman, K. Lesko, A. Smith, R. Fulton (Lawrence Berkeley Laboratory): N.W. Tanner, N. Jelley, P. Trent, J. Barton (University of Oxford).

Introduction

The Sudbury Neutrino Observatory (SNO) is based on a 1000 tonne heavy water (D_2O) detector located in a very low background laboratory 2000 m underground. The SNO experiment is made practicable by the availability, on loan, of 1000 tonnes of pure D_2O in Canada and a suitable site in the Creighton Mine owned by INCO Ltd. near Sudbury, Ontario. The objective of the SNO project is to measure the intensity, energy and direction of neutrinos from the sun and supernovae. A detailed description of the proposal is given in an earlier report (SNO-87-12)[1]. Improvements in that reference design have since been made but the same basic principles are used. The detector and laboratory are now under construction.

The use of heavy water as a detection medium enables the SNO experiment to measure both
1) the flux and energy spectrum of electron neutrinos reaching the earth and
2) the total flux of all neutrino types above an energy of 2.2 MeV.
With these two measurements, it will be possible to
1) show clearly if neutrino oscillations are occurring and
2) independently test solar models by determining the production rate of high-energy electron neutrinos in the solar core.

The use of heavy water provides an opportunity to use the sun as a calibrated source of neutrinos. Using the sun as a distant source of neutrinos would increase the sensitivity of the search for neutrino oscillations by many orders of magnitude compared to present measurements at reactors and accelerators. The detection of both electron neutrinos and all types of neutrinos from a supernova would give new information on the mechanism of stellar collapse.

As described in SNO-87-12, the proposed SNO detector makes use of three complementary neutrino reactions.

The charged current (CC) reaction

$$\text{(CC)} \quad \nu_e + d \rightarrow p + p + e^- \quad (Q = -1.44 \text{ MeV})$$

of the electron neutrino on the deuteron is unique to the SNO detector. It offers excellent spectral information, thereby providing a sensitivity to the MSW effect. The large cross section will result in more than ten events per day and make the detector some fifty times more sensitive than existing experiments. The reaction also offers some directional information and would identify the sun as the source of electron neutrinos. This reaction would also identify electron neutrinos from the initial burst in the collapse of a supernova.

The neutrino-electron elastic scattering (ES) reaction,

$$\text{(ES)} \qquad \nu_x + e^- \rightarrow \nu_x + e^-$$

the primary detection mechanism for light water detectors, is sensitive to all neutrino types, but is dominated by the electron neutrino. This reaction offers excellent directional information, but provides little information about the energy spectrum of electron neutrinos.

The third reaction occurring in the SNO detector is the neutral current (NC) disintegration of the deuteron,

$$\text{(NC)} \qquad \nu + d \rightarrow \nu + p + n \qquad (Q = -2.2 \text{ MeV})$$

observed by the detection of the gamma rays resulting from the subsequent neutron capture. This reaction is sensitive to all neutrinos equally and would be used to measure the total flux of neutrinos with a counting rate of about 10 per day above the threshold of about 2.2 MeV.

In this paper we discuss briefly the SNO detector and its role in the detection of neutrinos from supernovae.

SNO Detector

The present design of the SNO detector is shown in Figure 1. It consists of 1000 tonnes of 99.85% enriched D_2O contained in a spherical thin-walled (5 cm thick) transparent acrylic vessel which itself is immersed in 7300 tonnes of ultrapure H_2O shield. The host cavity, which is barrel-shaped and of 22 m diameter and 30 m height, is under excavation at a depth of 2070 m below surface at the (operating) Creighton Mine of INCO Ltd., near Sudbury, Canada. In the waist region the cavity is lined with low background concrete. A stainless steel liner is placed inside the cavity enclosing the light water shield. Some 9600 20-cm diameter photomultiplier tubes (PMTs) are uniformly arranged in the H_2O shield on a support structure at 2.5 m distance from the acrylic vessel. Each PMT is surrounded by a reflector to increase the light collection and the effective photocathode coverage is approximately 60%. The PMT array is sensitive to Čerenkov radiation produced by relativistic electrons and other relativistic particles in the central regions of the detector.

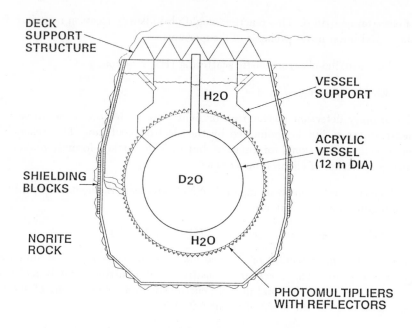

Figure 1   Line drawing of SNO detector

Neutrino interactions in the detector produce either relativistic electrons or free neutrons. The neutrons thermalize in the water and are subsequently captured generating γ-rays which produce relativistic electrons primarily through Compton scattering. The electrons from neutrino interactions or neutron captures will produce Čerenkov photons which pass through the D_2O, acrylic, and H_2O to be detected by the PMTs. The signals from the PMTs are interpreted to give the location, energy and direction of the electron based on the time of arrival of the photons at the PMTs and the locations of the particular PMTs hit.

The expected solar neutrino signal rates of about ten events per day require an extremely low background environment for the D_2O. There are two sources of background that must be guarded against. These sources are cosmic rays and radiation produced by naturally occurring radionuclides in the rock surrounding the detector and in the materials used in building the detector. The location of the detector is so deep that the background rate from cosmic-ray muons is negligible. Figure 2 shows the muon intensities at and depth underground of several laboratories.

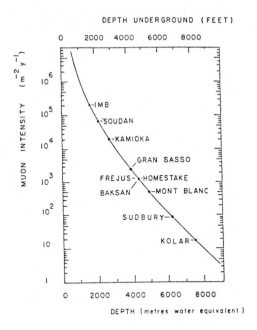

DEPTH UNDERGROUND (FEET)

Figure 2   Cosmic Ray Muon Intensity as function of depth

The purity of materials used in the detector has to be very high. The most critical components are the D_2O, H_2O, acrylic vessel and PMTs. In the D_2O and H_2O levels of U and Th in the range of 10^{-14} gram per gram of water are required. In the acrylic vessel which has less mass the tolerable level is a few times 10^{-12} gram per gram of acrylic. The PMTs are 2.5 m from the heavy water and γ-rays are attenuated in the light water shield. To keep contributions to background to a low level the PMTs are being built using glass envelopes from a special melt of low radioactivity glass being manufactured by Schott. The U and Th content of this glass is $\sim$5$-$10 times lower than normal glass.

The expected performance of the SNO detector has been calculated using an extensive series of Monte Carlo simulations of the neutrino signal processes and radioactive background processes. As an example a typical Monte Carlo calculation of the ^8B solar neutrino spectrum is shown in Figure 3. This shows the total spectrum and the components due to the ES, CC and NC reactions for D_2O with $NaC\ell$ added to increase the energy of the NC signal. Below a threshold of $\sim$5 MeV electron energy the background rises steeply and obscures the signal. The neutrinos observable from the sun are thus limited to the ^8B and hep neutrinos.

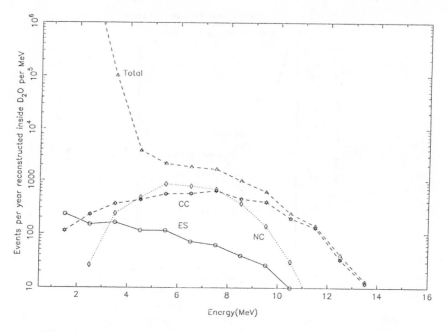

Figure 3   Monte Carlo simulation of ^{8}B spectrum in D_2O + NaCℓ

The events for the NC reaction are determined by detecting the γ-rays following the capture of the neutron released. About 2.5 tonnes of NaCℓ will be added to the heavy water. Then 83% of the neutrons will be captured by ^{35}Cℓ producing γ-rays with energies up to 8 MeV which should be observable above the background. An additional background must be considered for this reaction, corresponding to neutrons produced by the photodisintegration of deuterium by γ-rays with energies greater than 2.2 MeV. This background will arise predominantly from radioactive decay of members of the ^{232}Th and ^{238}U decay chains contained in the D_2O. This radioactive contamination can be sampled on-line to determine the background rate accurately.

The signal for the NC reaction is the production of a free neutron and alternative schemes for the detection of this neutron, such as ^{3}He counters, are being considered. One objective is to allow real-time discrimination between events produced by neutrons and other events. Extreme purity of materials is needed for such counters as they would be located in the D_2O itself.

Supernova Neutrinos

Neutrinos from Supernova 1987A were detected in the Kamioka [2] and IMB [3] experiments providing evidence that neutrinos are produced in supernovae in numbers and energy near those predicted by theory. Nearly all the events were produced by electron anti-neutrinos interacting with the H_2O.

Many calculations have been done of the production of neutrinos in a supernova. Figure 4 shows a time distribution from a paper by Burrows [4]. It is generally accepted that a short burst of ν_e will be generated by the electron capture reaction during the initial collapse. Their average energy is ~15 MeV, and the burst lasts for a few tens of milliseconds. Emission of neutrinos and anti-neutrinos in the cooling phase of the protoneutron core lasts from a few seconds to a few tens of seconds.

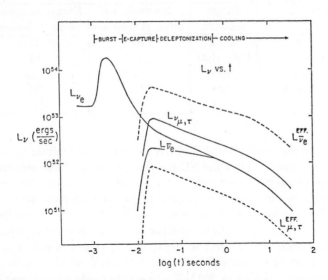

Figure 4   Neutrino luminosity versus time for the various neutrino species from paper by A. Burrows [4]

The proposed SNO detector will make a unique contribution to the study of these neutrinos because of its sensitivity to all neutrino types, because of its low-energy threshold, (~5 MeV), and because it can distinguish the various components of the neutrino flux. This detector, existing light water Čerenkov detectors and scintillator detectors all have high sensitivity for $\bar{\nu}_e$ through the charged-current reaction $\bar{\nu}_e + p \to n + e^+$. The heavy water detector has much higher sensitivity to the ν_e flux than any other detector through reaction $\nu_e + d \to p + p + e^-$. After reconstruction, the spatial distribution of events can be used to determine the ν_e and $\bar{\nu}_e$ fluxes; the majority of the charged-current events in the D_2O are from ν_e while most of the events in the H_2O are from $\bar{\nu}_e$. It is expected that the contributions from elastic scattering reactions will be less than 10% of the total event rates. In addition, the neutral-current reaction $\nu_x + d \to \nu_x + p + n$, in the heavy water can be used to measure the total neutrino intensity. Since the ν_e and $\bar{\nu}_e$ fluxes are determined independently, it is possible to deduce the total flux of other types of neutrinos. For a stellar collapse at 10 kpc, the calculated numbers of neutrino induced events in the SNO detector are listed in Table 1.

Table 1: Predicted number of events from a stellar collapse at 10 kpc

Reaction	Target medium	Events in ν_e burst /kilotonne	Events in cooling phase /kilotonne
$\nu_e + d \to p + p + e^-$	D_2O	10	33
$\nu_x + e \to \nu_x + e$	D_2O/H_2O	1	16
$\nu_x + d \to \nu_x + p + n$	D_2O	6	760
$\bar{\nu}_e + d \to n + n + e^+$	D_2O	0	20
$\bar{\nu}_e + p \to n + e^+$	H_2O	0	120

One of the important results from SN 1987A was the upper limit on the mass of the $\bar{\nu}_e$. As is clear from Table 1 the NC reaction provides a large sensitivity to all types of neutrinos and consequently it is possible that interesting mass measurements could be obtained. The present limits are 0.25 MeV for ν_μ and 70 MeV for ν_τ. For a collapse at 10 kpc, a mass of 50 eV would give a delay of 3 seconds in the ν_τ events. Although the cooling phase may last for 10 seconds, a shift of this magnitude with respect to the ν_e and $\bar{\nu}_e$ events would be easily seen. A neutrino mass in the range 50 eV< m_ν <100 keV could be measured readily. If the mass were as high as 200 keV, the events would come a year late and would be spread over half a year. This would be hard to detect against the solar neutrinos which would act as background. This sensitivity to the $\nu_{\mu,\tau}$ mass is a distinct advantage of the D_2O detector. In the event of a stellar collapse within our Galaxy it would permit the observation of finite $\nu_{\mu,\tau}$ mass with a sensitivity far better than terrestrial experiments and, perhaps, sufficient to determine whether the total mass of neutrinos is large enough to close the Universe.

Schedule

The SNO detector received funding in January 1990. Major items in the construction schedule are roughly as follows. The excavation of the cavity and the preparation of the underground laboratory will take until mid-1992, at which time the stainless steel liner and deck will be installed. Installation of the acrylic vessel, PMTs and their support structure will take place in 1993. The initial H_2O fill is scheduled to begin in the summer of 1994, with a view to completing commissioning tests by January 1995.

The expected data-taking sequence for the experiment would involve an initial filling of the acrylic tank with H_2O for a counting period of about 6 months. This would be followed by a filling with pure D_2O for a counting period of about 1 year, and finally the addition of $NaC\ell$ for subsequent operation. The H_2O fill would enable events from the ES reaction to be observed, as well as providing an initial confirmation of the detector operation. The D_2O fill would provide a clean energy spectrum from the CC reaction, to be inspected for evidence of distortions from enhanced neutrino oscillations in the Sun. Finally, the addition of $NaC\ell$ will provide a measure of the counting rate from the NC reaction, by comparing the observed spectrum with that obtained with pure D_2O.

The SNO detector is primarily designed to measure solar neutrinos, neutrino oscillations and the MSW effect. It can participate in a Supernova Watch programme as it will be sensitive to neutrinos from supernovae except during the filling stages. It can provide unique information if such an event occurs.

In the initial stage when filled with H_2O it would be mostly sensitive to electron anti-neutrinos with a sensitivity comparable to that of Kamiokande II.

During the pure D_2O fill it will be sensitive to the electron neutrinos, ν_e, produced in the initial burst as well as to $\overline{\nu}_e$. Free neutrons released by the NC reaction will be captured in deuterium and give $\sim$6.2 MeV γ-rays which will be difficult to detect above backgrounds due to radioactivity. However if an intense burst occurs in a short time such events could be detected.

When $NaC\ell$ has been added to the D_2O the NC events will be much more clearly observed and the SNO detector will measure ν_e, $\overline{\nu}_e$, ν_μ, $\overline{\nu}_\mu$ and ν_τ, $\overline{\nu}_\tau$. The present schedule predicts that this phase will start in late 1996.

References

1. G.T. Ewan et al, Sudbury Neutrino Observatory Proposal, SNO-87-12 (1987).

2. K. Hirata et al, Phys. Rev. Lett. **58**, 1490 (1987).

3. R.M. Bionta et al, Phys. Rev. Lett. **58**, 1494 (1987).

4. A. Burrows in Solar Neutrinos and Astronomy, AIP Conf. Proc. No. 126, 283 (1985).

Supernova detectors based on coherent nuclear recoil

P F Smith and J D Lewin

(Rutherford Appleton Laboratory, Chilton, Oxon., England)

1 Introduction

A number of large detectors, either planned or already in operation, are capable of observing neutrinos from a supernova in our own Galaxy or nearby, ie at distances ~10–50 kpc. However, within this radius the supernova rate is low, probably in the range 0.1–0.01/year, and to increase this rate one would like to create detectors sensitive to supernovas at distances > 1 Mpc. Within a radius ~ 4 Mpc the supernova rate is believed to approach 1/year [1]

In the case of large water detectors, the target volume required would be in the region $10^7 - 10^8$ m^3, apparently ruled out in the foreseeable future by the high cost of instrumenting such a large volume with photomultipliers (probably requiring at least 1 per 10 m^3). As an alternative approach, Cline et al.[2] have suggested the possibility of instrumenting a similar mass of natural low–activity chalk with inexpensive (boron–based) neutron counters, to observe the neutrons from nuclear excitation produced by inelastic neutrino scattering. There are two problems with this. Firstly, Monte Carlo simulations show that the neutrons random walk a distance of only about 0.5 m before being absorbed by the chalk itself, so that the detectors need to be spaced as closely as this. Secondly, even with the best natural chalk, the background neutrons produced (by alphas and fission) from U and Th contamination may provide a background count rate whose fluctuations could mimic a supernova signal.

In this paper we summarize the prospects for an alternative speculative idea, that of using coherent nuclear recoil from elastic neutrino scattering. This would allow target masses lower by a factor > 100, with more controllable background, but at the expense of more complex technology. It is of interest to ask which technique (given unlimited funding and an indefinite lifespan !) would appear the most promising for extra–Galactic supernova detection, and also whether foreseeable 21st century technological advances will offer new solutions to the technical or economic problems. No clear conclusion will emerge from this preliminary discussion, the task being clearly out of reach experimentally in the immediate future. Nevertheless a comparison of the alternative ideas is instructive in throwing further light on the problems of each, and suggesting directions for further development.

2 Coherent neutral current scattering

Coherent elastic scattering for MeV range neutrinos arises from the fact that the wavelength of the energy transfer to the nucleus exceeds the nuclear radius, so that the scattering amplitudes from the A constituent nucleons add coherently. Squaring the total amplitude results in a total cross section increasing approximately as A^2. This was first discussed by Freedman [3], and subsequently in more detail by Drukier and Stodolsky [4] after the exact form of the neutral current was established. Because the scattering amplitude from the proton is lower than from the neutron by the factor $\sim 0.5(1-4\sin^2\theta_W) \sim 0.04$, the neutrinos scatter essentially from the neutrons only, of which only half have an appropriately oriented spin. The coherence gain is thus $\sim (A-Z)/2$ for nuclear scattering. For the scattering of much lower energy neutrinos the coherence can extend over many atoms, and eventually macroscopic regions of matter, giving considerably enhanced cross sections which could be of interest in the detection of relic neutrinos. This is discussed in a separate paper [5].

Coherent nuclear scattering became of potential interest for the detection of solar and supernova neutrinos in the mid 1980s [4,6] but the first phase of development work on coherent detectors became oriented towards dark matter detection [7] in view of the importance of that problem and the much lower target masses ($\sim 1\,$kg) required. This temporarily eclipsed the application to supernova neutrino detection, but in the light of the considerable progress which has been made in the understanding of technical and background problems, it is appropriate to reassess the possibility of coherent supernova detection.

3 Coherent rates from supernova burst

Fig 1 shows the computed spectrum of nuclear recoil events from a supernova neutrino burst, assuming a simplified neutrino energy spectrum of the form $E.\exp(-E/E_0)$ with an average energy $E_{av} = 2E_0 = 16\,$MeV. The events/kg assume a total of 6.10^{57} neutrinos released at a distance of 10kpc (flux of $5.10^{11}\,$cm^{-2}). The curves show firstly the coherent gain arising from the use of heavier target nuclei, but secondly that it is necessary to be sensitive to low recoil energies (< 1–10keV) to take advantage of this.

Using these figures we list in **Table 1** the target masses for detectors based on coherent nuclear recoil, and compare these with target masses needed in the case of proposed detectors based on water or chalk targets. To provide a unified comparison we show the target masses needed to give 100 events from a supernova at either 10kpc or 1Mpc. For the chalk detector we use the cross section given by Cline et al [2]. For water detectors we use the rates calculated for H_2O and D_2O in the SNO

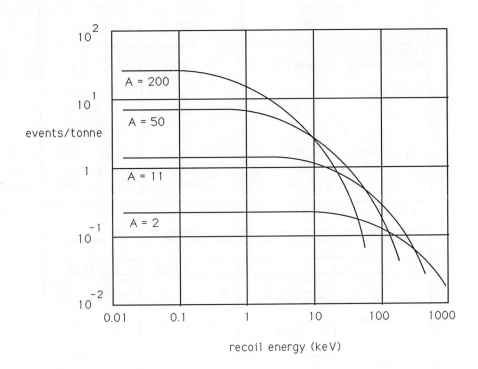

Fig 1 Nuclear recoil spectra from supernova neutrino burst, for
different target materials (note energy axis mislabelled a
factor 10 lower in ref [6]).

Table 1. Target mass (tonnes) to give 100 events from supernova at 10 kpc or 1 Mpc.

Type of target		10 kpc	1 Mpc
$CaCO_3$ [2]		1600	1.6×10^7
H_2O [8]		800	1.6×10^7
D_2O [8]		120	–
Coherent detectors. Measurement threshold E_t			
$E_t = 1\,keV$	A = 200	4	4×10^4
	A = 20	40	4×10^5
	A = 2	450	4.5×10^6
$E_t = 10\,keV$	A = 200	25	2.5×10^5
	A = 20	50	5×10^5
	A = 2	500	5×10^6

Table 2. Comparison of rms backgrounds for 10s periods

	background tonne^{-1} s^{-1}	R = 10 kpc		R = 1 Mpc	
		counts in 10s	rms ΔN	counts in 10s	rms ΔN
$CaCO_3$	10^{-2}	100	10	10^6	10^3
	10^{-4}	1	1	10^4	10^2
coherent nuclear recoil					
A = 200	10^{-2}	0.4	0.6	4000	60
	10^{-4}	0.004	0.06	40	6
A = 20	10^{-2}	4	2	40000	200
	10^{-4}	0.04	0.2	400	20
coherent with directionality					
A = 80	10^{-5}	0.0001	0.01	1	1

proposal. For the coherent detection rates we show the effect both of choice of target element and of detector energy threshold E_t.

While this comparison is of interest in highlighting the relatively low target masses potentially possible with a coherent nuclear recoil detector, it is not a correct practical comparison since, for example, the chalk detector assumes use of a pre–existing mass of natural chalk, instrumented with a relatively low mass of a boron neutron detector. To estimate the latter, we note that the neutron absorption cross section/gram for boron is about 1700 times that of chalk, so for detection of say 80% of the neutrons we would require only ~4 tonnes boron (or 24 tonnes BF_3) in 1600 tonnes of chalk.

Thus the two techniques appear in fact similar in the amount of hardware to be constructed, with the coherent recoil detectors having an apparent disadvantage in requiring more complex (eg low temperature) technology. However, the lower volume of the latter may provide a significant advantage when the background problems are taken into account, as we now discuss.

4 Background rates for chalk detector

A full assessment of background problems is complicated by the fact that the shape of the supernova neutrino burst [9] can in principle be used to identify the signal and improve the signal/background. This also depends on the extent to which the supernova signal may be spread in time by a non–zero neutrino mass [10]. For this preliminary discussion, however, we will simply estimate the constant background rate R_n and discuss whether fluctuations in the number of background events in some appropriate time interval t_s (eg 10s) could simulate a supernova burst.

In the case of a chalk detector, it is necessary first that the material is sufficiently far underground (> 3000mwe) for neutron production by muons to be negligible. The neutron background then arises from natural U and Th, giving neutrons from both fission and alpha interactions. We estimate a production ~2.10^{-6} neutrons from the chain of decays resulting from each parent U or Th decay. The measured U+Th activity is ~ 5.10^{-3} decays $g^{-1} s^{-1}$ in the case of the Gran Sasso laboratory [11], giving a neutron production rate ~ 1.10^{-8} $g^{-1} s^{-1}$. This is confirmed by our Monte Carlo simulation showing that the flux $(cm^{-2} s^{-1})$ in a cavity in the chalk should be related to the neutron production rate $(g^{-1} s^{-1})$ by the factor ~500 g cm^{-2}. From the measured total neutron flux ~$5.10^{-6} cm^{-2} s^{-1}$ in a Gran Sasso cavern [12], this again gives a production rate ~$1.10^{-8} g^{-1} s^{-1}$ in the chalk, or 10^{-2} tonne$^{-1} s^{-1}$.

Natural chalk (in dolomites [8]) exists with U and Th concentrations at least an order of magnitude lower than at Gran Sasso, and Cline et al.[2] anticipate that deposits may exist which are a further factor 10 lower. This might reduce the neutron background rate to $\sim 10^{-4}$ s^{-1} tonne^{-1}. For a 1600 tonne Galactic supernova detector there would then be a continuous background ~ 0.1 s^{-1}, with rms fluctuations ~ 1 event in, say, 10 second intervals. Thus in this case a 100 event supernova signal would always be clearly above background. However, for a 1Mpc extra–Galactic supernova detector the mass is higher by a factor 10^4, with rms fluctuations $\sim 10^2$ in 10s intervals, or ~ 30 events even in 1s intervals. Thus the background fluctuations would frequently exceed the expected 100 event signal level, unless features of the latter with time–widths $\ll 1$s can be used.

Cline et al.[2] arrive at background rates a factor ~ 50 lower than the above figures, by scaling the Gran Sasso rates to their proposed detector size. The numerical discrepancy arises principally from (a) a 20% efficiency factor mentioned in their text but omitted from their calculation and (b) the non–comparability of neutron rates measured with a detector size $\ll$ cavity size and the proposed case, detector size = cavity size. These difficulties are by–passed by our direct Monte Carlo conversion of measured flux to volume production rate, and this procedure gives the higher rates summarized above.

5 Background rates for coherent detector

There are two basic classes of coherent nuclear recoil detector under development for dark matter searches. One class consists simply of low background versions of existing ionization detectors – either semiconductors or scintillators. These present the difficulty that the ionization efficiency from nuclear recoil is typically 10–20% of that from the usual interaction with electrons, so that typical energy measurement thresholds ~ 1–10keV correspond to nuclear recoil thresholds ~ 10–50keV. There is also a material–dependent ionization threshold region which increases further the difficulty of observing nuclear recoils < 10 keV [7]. These problems can in principle be overcome by the use of a second class of detectors based on the measurement of phonons or heat pulses at low temperatures. These can consist, for example, of pure 10–100g crystals equiped with temperature sensors and operated below 100mK. Another type is based on observation of transitions in superheated superconducting granules, at temperatures usually <1K, but conceivably up to 4K. These cryogenic detectors detect essentially the whole of the nuclear recoil energy; sensitivities <1keV g^{-1} have already been demonstrated [7] with potential for improvement to <1keV in 0.1–1kg modules.

For dark matter searches, 1 kg experiments based on these ideas are clearly feasible in the foreseeable future. Scaling up to the level of a few tonnes also looks feasible since, for example, dilution refrigerators have been designed for operating multi-tonne gravitational radiation detectors below 100 mK. Scaling up to the extra–Galactic supernova level of 10^4–10^5 tonnes looks a more formidable task, possibly unrealistic for temperatures < 100 mK but still just conceivable for temperatures in the 1–4 K region.

Assuming technical feasibility, the background problem can be assessed from the knowledge already gained from studies of dark matter experiments. Again, natural radioactivity is the main problem, this time the continuous spectrum of photons and electrons from the U and Th decay chains. A background objective of 1 event day^{-1} kg^{-1} is certainly regarded as feasible, corresponding to ~10^{-2} tonne^{-1} s^{-1}, and by the development of techniques allowing discrimination of nuclear recoil events one can envisage reducing this by a further factor 100.

We conclude that, for similar levels of optimism, *the background rates/tonne achievable in a low temperature detector are in fact comparable to those in a natural chalk detector, but the target mass itself is a factor 40–400 lower.* Thus the absolute background rates are also lower by this factor, giving rms fluctuations in a 10 s interval < 1 for a 10 kpc detector and < 20 for a 1 Mpc detector. Though considerably better than the chalk background, this could still simulate a 100 event signal at a rate ~1/year. However, prospects also exist for obtaining some directional sensitivity in nuclear recoil detectors, which (since the supernova events would all lie in the same recoil hemisphere) would reduce the probability of a simulated signal to a negligible level. Illustrative background estimates for both types of detector are summarized in **Table 2**.

6 Conclusions

6.1 Nearby supernova.

It can be seen from the above that coherent nuclear recoil detectors would not necessarily offer any advantage for the detection of supernova neutrinos within our own Galaxy. Conventional water detectors are already known to be feasible on the 1000 tonne scale, and on this scale both these and the proposed chalk detector have sufficiently low background for the observation of a 100 event supernova neutrino burst. A low temperature coherent detector with the same sensitivity would be remarkably small (~10 tonnes) but to achieve the necessary energy threshold < 1–10 keV requires difficult low temperature detector technology, currently being developed only for target masses in the region 0.1–1 kg.

6.2 *Extra–Galactic supernova.*

The incentive for reaching distances > 1Mpc is firstly to obtain a higher supernova rate, and secondly to obtain information on neutrino mass from the time spread of events. The necessary scale up of target mass, by a factor 10^4–10^5 for the same signal size, brings with it firstly a problem of prohibitive cost, and secondly a problem of background – since the latter increases as (target mass)$^{0.5}$. In principle, of course, this is soluble by increasing the target mass further, for a given supernova distance (for example increasing the signal by a factor 100 while the background fluctuations increase by a factor 10), but this again is likely to be precluded by overall size and cost considerations.

It is in this situation that the coherent nuclear recoil detector could in principle prove advantageous, since with its factor >100 gain in volume and better control of the activity of the target material, the background levels could be made much lower, and in addition various techniques could be developed (in particular nuclear recoil discrimination and directionality) to provide an unambiguous signature for the supernova burst. Against this would appear to be the serious difficulty of implementing the difficult low temperature technology on the scale of 10^4–10^5 tonnes.

6.3 *21st century technology*

Experience of the past has, however, shown that it is misleading to apply the current standards of difficulty and economic feasibility to projects which are at least one generation in the future. The production and detection of neutrino beams has become possible on a scale which would have appeared unrealistic 25 years previously, and totally unforeseen 50 years previously when the neutrino was first postulated. Vacuum–tube based computers filling a room in 1960 can now be etched on the surface of a small piece of silicon, and moreover at a small fraction of the cost.

Further major advances are foreseen in the next 30–50 years, based on nanotechnology, or the ability to construct molecular machines which would themselves be capable of assembling macroscopic objects from their atomic constituents with hitherto unprecedented precision and economy [13]. The ability to lift and reposition single atoms and portions of molecules using the scanning tunneling microscope has already been demonstrated, and rapid progress is expected towards the first purpose–built molecular machines. This technology will have several important effects on the future of large scale projects in particle astrophysics. Key points relevant to the above discussion are:

(a) It is envisaged that extraction of basic materials will become possible at a fraction of present–day costs, and that macroscopic detector structures could be replicated *in–situ* from the basic materials. Thus the large scale instrumentation of a 10^8 tonne (a 300 m cube) volume of chalk might be achievable in the future at a realistic cost. It is similarly conceivable that new types of photon detector (equivalent to photomultipliers) could be replicated in the same way, allowing the economic instrumentation of very large volumes of natural water. New techniques for large scale removal of residual uranium and thorium contamination of the chalk or water target material might also be envisaged.

(b) Specialised composite material structures will be "grown" at low cost from their atomic constituents, with high precision and negligible radioactive contamination. This would allow the components of low temperature nuclear recoil detectors to be produced in quantities which could not be considered at the present time.

It is therefore not out of the question that an extra–Galactic supernova detector, apparently prohibitively costly with today's technology, could nevertheless become a realistic proposition in the future, and this prospect encourages and justifies continuing development work on the basic detector ideas.

References

1 See paper by E Becklin, this workshop.

2. D B Cline et al., Nucl.Phys.B (Proc. Supp.) 14A (1990) 348; Astro. Lett. and Communications 27 (1990) 403.

3 D Z Freedman, Phys.Rev. D9 (1974) 1389.

4 A Drukier, L Stodolsky D30 (1984) 2295.

   L Stodolsky, Comments Nucl. Part. Phys. 18 (1988) 157.

5 P F Smith, these proceedings (astroparticle physics conference). For a summary of coherent cross section formulae see P F Smith, Nuovo Cimento 83A (1984) 263 (note misprint in eq 4.2 of that paper, for which the square brackets should be positioned as in eq 3.2 of the same paper).

6 P F Smith, Zeits.Phys. 31 (1986) 265.

7 For a recent review, see P F Smith, J D Lewin, Phys. Rep. 187 (1990) 203.

8 Sudbury Neutrino Observatory Proposal SNO–87–12 (1987).

9 See, for example, A Burrows, this workshop and Ann.Rev.Nucl.Part.Sci. 40 (1990) 181.

10 J Wilson, this workshop.

11 E Bellotti et al., report INFN/TC–85/19.

12 A Rindi et al., Nucl.Instrum.Meth.Phys.Res.A272 (1988) 871.

13 See, for example, review by C Schneiker in "Artificial Life" , Proc Interdisciplinary Workshop on Synthesis of Living Systems (C Langton, ed., Los Alamos, 1987) 443.

EXTRA GALACTIC SUPERNOVA DETECTOR,
NEUTRINO MASS AND THE SUPERNOVA WATCH

DAVID B. CLINE

Departments of Physics & Astronomy, University of California at Los Angeles
405 Hilgard Avenue, Los Angeles, California 90024

ABSTRACT

The detection of finite neutrino mass above a few eV would have profound consequences on cosmology and particle physics. Supernova neutrino bursts provide the most reliable technique to detect finite neutrino mass. We report on techniques to detect extra-galactic supernova bursts that would be sensitive to $\sim$ eV neutrino mass and galactic supernova detection (SuperNova Watch) that is sensitive to the $20-50$ eV mass range. A brief remark is made on a real-time supernova network being discussed.

1 THE SUPERNOVA WATCH

The detection of neutrinos from SN 1987 A by the Kamiokande II, IMB and other detectors demonstrated that the final stages of a type II SN was more or less as had been expected by theory and modeling studies. However, there were many questions left open about the details of the collapse and the properties of the neutrinos that are emitted. In the future the detection of the neutrino burst by several terrestial detectors would have a profound effect on may diverse fields of physics as elementary particles (neutrino mass), nucleon physics (equation of state) and so forth. Recently a workshop was held at UCLA to discuss the SuperNova Watch and to discuss a possible Real Time Network that could link the different detectors. In Table 1 we list some of the SuperNova Detectors operating, being constructed or being planned around the world. This is an impressive array of detectors. The basic idea of a SuperNova Watch Network is shown schematically in Figure 1. A Real Time SuperNova Watch discussion group has been formed following the UCLA meeting and progress in the development of this concept is expected during the next year or so.

2   SUPERNOVA NEUTRINO BURSTS

Figure 2 shows the expected luminosity function and mean energy of the ν_e, $\bar{\nu}_e$, $\overset{(-)}{\nu_\mu}$, $\overset{(-)}{\nu_\tau}$ neutrinos from the collapse. The important times are*:

(i)	Prompt ν_e Burst	$\sim (3-6)\,\mathrm{ms}$
(ii)	Rise time and time internal to acretion pulse	$\sim (100-200)\,\mathrm{ms}$
(iii)	Width of the acretion pulse	$\sim 400\,\mathrm{ms}$
(iv)	Explosion starts	$\sim (300-600)\,\mathrm{ms}$
(v)	Start of neutrino cooling	$\sim (300-600)\,\mathrm{ms}$
(vi)	Full width of cool-down	$\sim (10-20)\,\mathrm{sec}$

These characteristic times of the SuperNova neutrino emission process are set by scale of the SuperNova collapse dynamics[1] . In turn the detailed study of these characteristic time structures in a future SuperNova event would provide real detailed information about the dynamics of the SuperNova process. In addition, these time structures are important if we are to use the SuperNova neutrinos flight time to determine SuperNova mass (see section 3 for more details).

---

* I have been coached in this subject by Jim Wilson

These times also set the scale of the techniques that may be used to detect a finite neutrino mass by the time of flight method. In Figure 3 we show the difference between a "time of flight" mass measurement at an accelerator and using a supernova.

3   DETERMINATION OF NEUTRINO MASS OF COSMOLOGICAL SIGNIFICANCE

One or more stable neutrinos, with a mass in the region of 30 eV, could supply the missing or dark matter. The mass relationship is $\Omega_\nu = 1$ for $M_{\nu_x} = 92\ h^2 eV$, where H is the Hubble Constant (h = 1 for H = 100 km/s^{-1}MPC^{-1}; for h = 1/2, $M_{\nu_x} = 23$ eV to give closure of the Universe and for h = 0.6, $M_{\nu_x} = 33$ eV). We thus consider the neutrino mass range of 10 – 40 eV to be of cosmological significance. There are no known laboratory techniques to uniquely detect such a mass directly. It is possible that some form of neutrino oscillation experiment could be used to infer a mass in this range, however, this will depend on the uncertain level of neutrino mixing. The only technique that is known to provide a unique mass measurement is to use the difference in flight times for neutrinos from a distant supernova that goes as

$$\Delta t = 51.4\,R_{\mathrm{MPC}}\left[\left(\frac{M_{\nu_x}}{E_{\nu_x}}\right)^2 - \left(\frac{M_{\nu_e}}{E_{\nu_e}}\right)^2\right]\ \mathrm{sec}$$

where M_{ν_x} and M_{ν_e} are measured in electro volts, E_{ν_x}, E_{ν_e} are measured in MeV and R (the distance to the supernova) is measured in Mega Parsecs. The μ and τ neutrinos are expected to have higher average energies since they escape from deep inside the supernova core[1] .

We assume an instantaneous source of ν_x neutrinos with a distribution of the form[2]

$$E_{\nu_x}^2\,e^{-E_{\nu_x}/T}$$

and an assumed detection efficiency that scales as $E_{\nu_x}^3$ (i.e. the cross section scales like $E_{\nu_x}^2$ and detection of secondary products like E_{ν_x}) gives

$$\delta t = 51.4 R_{\mathrm{MPC}}\frac{\int E_{\nu_x}^5\left(\frac{M_{\nu_x}}{E_{\nu_x}}\right)^2 e^{-E_{\nu_x}/T}\,dE_{\nu_x}}{\int E_{\nu_x}^2\,e^{E_{\nu_x}/T}\,dE_\nu}$$

and for the case of μ, τ neutrinos we expect[2] [5] $T \simeq 25/3$, giving

$$\delta t = 0.037 \, M_{\nu_x}^2 \, R_{MPC} \text{ sec}$$

For a galactic supernova $R_{MPC} = 0.01$MPC and for $M_{\nu_x} = 30$ eV we find

$$\delta t = 330 \text{ms}$$

Note that the mean time separation and shape of the time distribution are altered in a characteristic manner by the different neutrino masses. It is this characteristic that must be used to extract a cosmologically significant mass from a Galactic Supernova. The detailed time distribution for a galactic supernova is shown in Figure 4. Table 2 lists the reactions that can be used for supernova burst detection and the experimental techniques that can be employed. Note from Table 2 that no detector will provide information on all the possible channels. In this note we are mainly concerned with the prospects for extracting a neutrino mass of the μ and τ neutrino if the mass value is in the 10 - 40 eV mass range of cosmological significance. Note also that the mean time distribution for these cases are

$$\delta t = 35 \text{ ms} \qquad M_{\nu_x} = 10 \text{ eV}$$

$$\delta t = 590 \text{ ms} \qquad M_{\nu_x} = 40 \text{ eV}$$

In a sense the galaxy is simply too small to obtain large time differences for cosmologically interesting neutrino mass. From Fig. 1 it is clear that the shape of the time pulse changes with M_{ν_x} and that the mean width of the initial pulse is

$$\delta t \simeq 400 \text{ ms} \quad \text{for} \quad M_{\nu_x} \simeq 10 \text{ eV}$$

$$\delta t \simeq 1 \text{ sec} \quad \text{for} \quad M_{\nu_x} \simeq 40 \text{ eV}$$

and thus

$$(\delta t)_{\text{pulse width}} > (\delta t)_{\text{time difference } (-) \text{ between } \overset{(\tau)}{\nu}_x \text{ and } \overset{(-)}{\nu'}_x \text{ neutrino arrival}}$$

It is clear that a very large number of events and very good time resolution is required to resolve the effects in the lower neutrino mass range due to the shape of the time pulse near the origin. We propose to use the derivative of the pulse to give a zero crossing estimate of the arrival times to obtain the required accuracy.

We now consider the expected event ratio for various channels for planned or proposed detectors for supernova detection in the 1990's and beyond. Table 3 lists the approximate event rates for several detectors in the construction or planning stage (ICARUS, SNO, LVD, MACRO) and for two newly proposed detectors[2] :

i) Super Kamiokande

ii) SNBO (Supernova Neutrino Burst Observatory) as well as the existing IMB and Kamiokande II detectors (other detectors are likely too small to give additional information).

We refer to Ref. [3] for discussions of the proposed Super Kamiokande detector. The SNBO detector would have the active mass of 100,000 tons of $CaCO_3$ and would be instrumented with a large number of neutron detectors. The detector concept has been described in Refs. [2] and [4]. This detector is mainly sensitive to ν_μ and ν_τ neutrinos due to the dynamics of the neutral current process which strongly discriminates against lower energy ν_e and $\bar{\nu}_e$ events, thus, it is a ν_μ, ν_τ detector. Therefore, this zero crossing technique could be applied to obtain the desired time resolution.

4 A SUPERNOVA NEUTRINO BURST OBSERVATORY

A new concept in supernova neutrino burst detection has recently been proposed. The detector uses high energy neutrinos from stellar collapse to drive neutral-current inelastic-scattering excitation of nuclei in the detector. The excited nuclei can decay by emitting neutrons. These neutrons would be detected by inexpensive counters. There are four important points regarding this proposed detector:

(1.) large, relatively pure deposits of SNBO detector material exist in nature in well-shielded sites. Preparation of the detector material should therefore be minimal. In addition, detectors are relatively simple, inexpensive BF_3 neutron counters. The technology of these detectors has been available since the 1950's and is well-understood. The SNBO potentially is an inexpensive, easily-prepared, and easily-maintained neutrino-burst detector that would run for decades with little mainte-nance. This is important because the stellar collapse rate in our galaxy is only one event every ten to one hundred years so that neutrino detectors will probably have to run for several decades before the next galactic supernova is seen.

(2.) Because the basic design of the SNBO is to place neutron counters into holes drilled into the detector medium, the mass of the SNBO is easily increased by drilling new holes and adding new counters. The advantage of the SNBO over water detectors is clear. The feasibility of the detector may be tested by placing a neutron counter into a chunk of detector medium which could then be transported to a neutrino source (such as the Los Alamos Meson Physics Facility) and tested. Should the detector indeed prove feasible, a small-scale SNBO could then be started with only a few neutron counters in place. As experience with the detector grows, more holes could be drilled and more counters added. This is the only proposed cosmic neutrino detector which could be scaled up to very large size. The ultimate mass of the SNBO could reach 10^5 tons or more, which would give tens of thousands of counts for a galactic supernova. This large number of counts could yield detailed information on the supernova mechanism as well as provide a good limit on the μ and τ neutrino masses.

(3.) Because the detector medium produces neutrons almost exclusively by neutral cur-rent reactions, the SNBO would be able to count neutrinos of all three flavors coming from a supernova. In addition, the threshold for the detector, which is set by the neutron separation energy of the nuclei of the detector medium, is high ($\gtrsim 12\,\mathrm{MeV}$) so that only high energy neutrinos would be counted. In the case that there are no oscillations between neutrino flavors, the average energy of electron neutrinos from a supernova is roughly $15\,\mathrm{MeV}$, while that of the μ and τ neutrino are more like 20-25 MeV. As a consequence, the SNBO would essentially count only μ and τ neutrinos.

Neutrinos with a mass in the range 10-50 eV are again becoming attractive as candidates for the dark matter of the universe. The upper limit on the mass of the electron neutrino is currently 18 eV. Improvement in ^{3}H endpoint experiments could reduce this upper limit below the cosmologically interesting range. Any MSW resonant neutrino oscillation explanation for the solar neutrino puzzle would suggest that, at best, $\Delta m^2 \approx 10^{-6} - 10^{-4}\,\mathrm{eV}^2$ so that m_{ν_e} would be very small but conceivably $m_{\nu_\mu} \simeq 10^{-2}\,\mathrm{eV}$ and "see-saw" schemes for neutrino masses would then suggest $m_{\nu_\tau} \sim 10\,\mathrm{eV} - 40\,\mathrm{eV}$. This would leave the τ neutrino as the potential closure neutrino. The only way to measure the mass of μ and τ neutrinos in this mass range is, at present, by time-of-flight measurements from supernovae, a task for which the SNBO is ideally suited because of its ability to operate for a long time, its large number of counts and its particular sensitivity to μ and τ neutrinos.

(4.) If vacuum mixing angles and mass difference between ν_e and ν_μ or ν_τ are as expected in any of the MSW oscillation solutions to the solar neutrino problem, then neutrino flavors might be mixed on their way down the density gradient in the supernova. This would complicate the clear identification of high energy neutrinos as corresponding to ν_μ and ν_τ. We point out, however, that in a mass hierarchy in which $m_{\nu_\tau} > m_{\nu_\mu} > m_{\nu_e}$ the anti-neutrinos have no resonant transitions so that a $\bar{\nu}_e$ signal in LVD or a water detector serves to give a "fiducial mark" to calibrate the $\nu_e\,(\bar{\nu}_e)$ flux. This, coupled with a detailed time history of events from SNBO, could give interesting constraints on neutrino masses and mixing angles in a region of parameter space potentially very different from solar neutrino experiments. We note that since our experiment involves neutral currents we can still identify high energy peaks from $\bar{\nu}_\mu$ and $\bar{\nu}_\tau$ and place mass limits as discussed in point.

Note that the SNBO detector could also provide a higher mass limit on one of the neutrinos if the rate of events is considerably below the prediction. This would imply that one of the neutrinos has a mass of > 100 eV or is unstable. The two newly proposed detectors provide additional information that could be used to uniquely detect a ν_μ or ν_τ mass even in the 10 eV range. This is due to the fact that the Super Kamiokande detector gives adequate numbers of $\nu_x e \to \nu_x e$ events to possibly make the separation in the high mass case and the SNBO detector (when combined with the other detector results) provides a very large number of pure ν_μ and ν_τ events. With 5000 ν_μ and 5000 ν_τ events it should be possible to make a unique separation of low mass neutrinos (i.e. $M_{\nu_\mu} \simeq 0$; $M_{\nu_\tau} \sim 30$ eV) and mixed cases such as ($M_{\nu_\mu} \simeq 10$ eV, $M_{\nu_\tau} \sim 30$ eV). (Note that the zero crossing technique is difficult to use in the case of a mixed, overlapping ν_e, ν_x sample.)

The construction and long term operation of such detectors in a self triggered "Supernova Watch" detection mode is essential to determine if the μ and τ (and possibly ν_e) neutrinos have mass values that are important for the cosmology of the universe. At this time there appears to be no other viable proposal of techniques to carry out the important measurement. In addition to the neutrino mass determination the detection of thousands of $\bar{\nu}_e$, ν_e and ν_μ/ν_τ events from a future supernova will provide crucial information about other properties of neutrinos (such as a magnetic moment in the range of $10^{-11} - 10^{-14} \mu_B$) and exotic $\nu\nu$ interactions as well as the dynamics of stellar collapse and explosion.

Detailed response calculations have been carried out at LLNL for the configuration shown in Figure 5[2] . The SNBO could be constructed in a limestone deposit in places like Arizona. Figure 6 shows a conceptual design of such a detector. Prototype neutron detectors are being developed by the collaboration (see Figure 7)[2] .

5   POSSIBILITY OF EXTRA GALACTIC SUPERNOVA DETECTION

The search for SuperNova neutrino bursts poses a serious problem: the rate of Galactic SuperNova is expected to be small ($\sim 1/50$ years) and the distance to the next large galaxy is large. At the SuperNova Watch meeting a lively discussion of the distance to a galaxy was held (see Figure 8)[5] . It was concluded that at a distance of $\sim (1 - 2)$ Mpc the SN II rate would be ~ 1 per year. In order to detect such SuperNova, a very massive SNBO-type detector is required ($\sim 10^7 - 10^8$ tons). This might be accomplished if a technique can be found to construct such a detector for the cost of ~ 5 dollars per ton.

Now we can compare the prospects for detecting a finite mass neutrino from the use of galactic and extra galactic supernova as given in Table 4. While the lowest mass detector is one using very low temperature sensors (such as 300 mK superconducting grams) the most easily constructed is the SNBO[6] .

6   ACKNOWLEDGEMENT

I wish to thank George Fuller, Jim Wilson, Peter Smith and Adam Burrows as well as other members of the SNBO Collaboration for the help they have given me.

REFERENCES

[1] J. Wilson and R. Mayle, Livermore Preprint (1988) and to appear in Ap.J.

[2] R. Boyd, et al., Proposal to Study a New Type of Neutrino Burst Detector: The SuperNova Neutrino Burst Observatory (SNBO), UCLA, 1991.

[3] R. Cowsik and J. McClelland, Phys.Rev.Lett., 29, 660 (1972).

[4] D. Cline, et.al., A New Method for Detection of Distant Supernova Neutrino Bursts, Astro.Phys.Letts. and Comm. (1989).

[5] Private Communication, P.F. Smith.

[6] See the proceeding of the UCLA SuperNova Watch Workshop to be published as a UCLA preprint, 1991.

[7] T. Piran and J. Wilson (unpublished) and D. Cline, et.al., Proposal to Study a New Type of Neutrino Burst Detector: The Supernova Neutrino Burst Observatory (SNBO), UCLA preprint 1990.

[8] D. Cline, Neutrino Astronomy, the proceedings of the 14th Texas Symposium on Relativistic Astrophysics, Dallas, Texas, (1988), to be published by the New York Academy of Sciences.

[9] A. Burrows, M. Turner and R.E. Brickman, Phys.Rev.D (1989).

[10] The event rate estimates for the SNO detector come from P. Doe, "The SNO Detector", published in Observational Neutrino Astronomy, editor: D. Cline, publisher: World Scientific, 1988, p.92.

Real Time
SuperNova Watch
Network

CONCEPT:

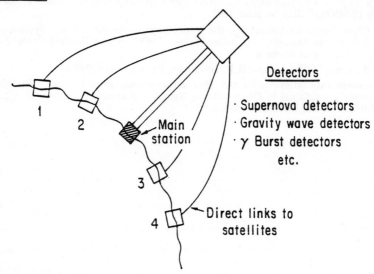

Detectors

· Supernova detectors
· Gravity wave detectors
· γ Burst detectors
    etc.

Direct links to
satellites

Main
station

PURPOSE OF NETWORK:

1) To Provide a <u>Real Time Detection</u> of a <u>Neutrino Burst</u> –

   possible to provide real time information to various detectors

2) To Detect <u>Very Faint Signals</u> that are <u>Not Observable in a Single Detector</u>

   (or between gravity wave and SN, etc.)

   Figure 1 A schematic of a Real Time SuperNova Watch
   Network connecting several detectors.

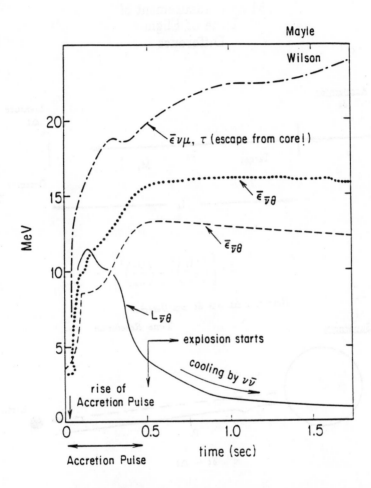

Figure 2 The neutrino luminosity function and average energies of the various neutrinos vs. time.

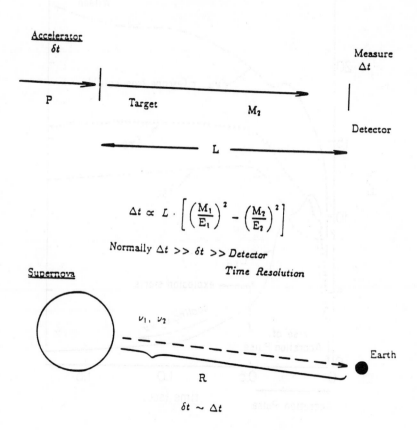

Figure 3 Schematic of the technique to measure neutrino mass using time of flight from a SuperNova explosion compared to a similar accelerator experiment.

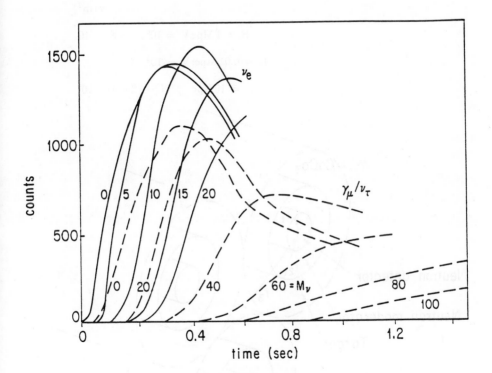

Figure 4 Time distribution for the neutrino interactions from a Galactic SuperNova as a function of the
neutrino mass. The solid curves refer to ν_e and the
dashed to ν_μ / ν_τ.

Scaling Rule:

$$N_{events} \simeq \frac{(20 - 40)}{R^2} V \qquad R\,(Kpc)$$

$$v\,(m^3)$$

$$R = 1\,Mpc\,V = 10^6, \qquad N \sim 40$$

$$R = 0.01\,Mpc; V = 10^5$$

$$= 10\,Kpc \qquad N \sim (2 - 4) \times 10^4$$

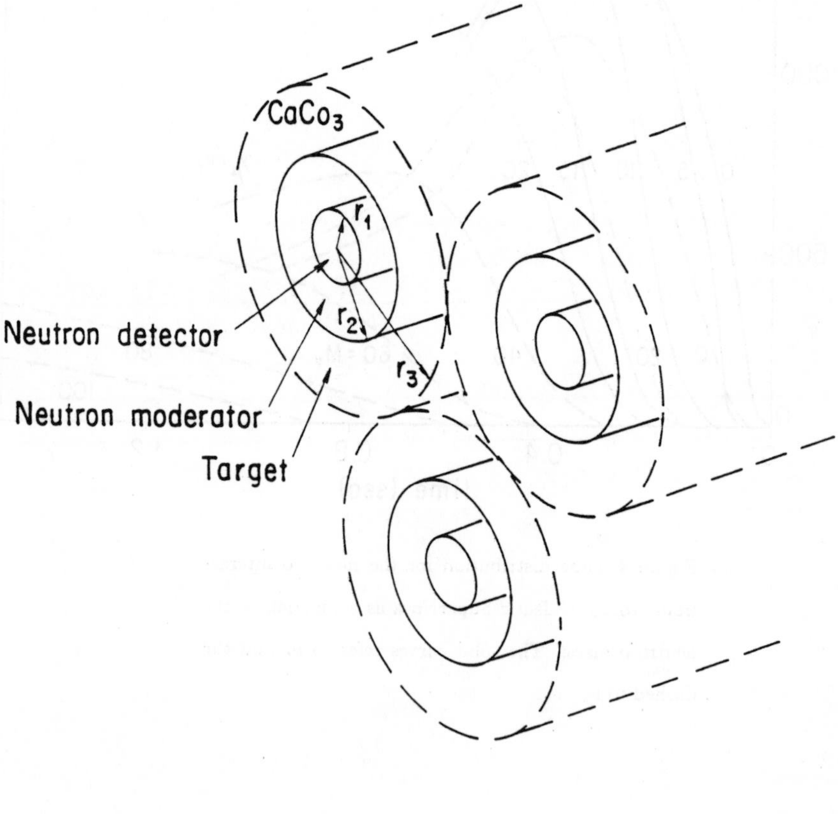

Figure 5   Concept of a distant supernova neutrino burst detector.

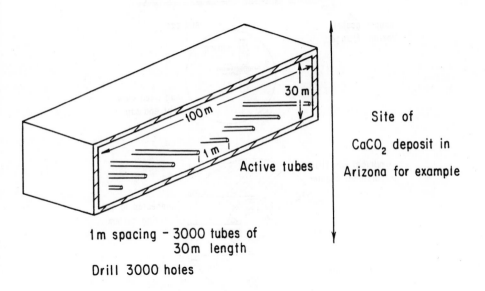

Conceptual Detector Location

(10^5 Tons)

30m x 30m x 100m

30 m

100 m

1 m

Active tubes

1m spacing – 3000 tubes of 30m length

Drill 3000 holes

Site of

$CaCO_2$ deposit in

Arizona for example

Figure 6 A schematic layout of the SNBO Detector that uses Neutron Detectors.

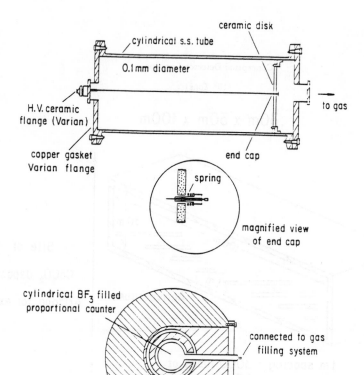

Figure 7 Schematic of a BF_3 prototype detector for SNBO.

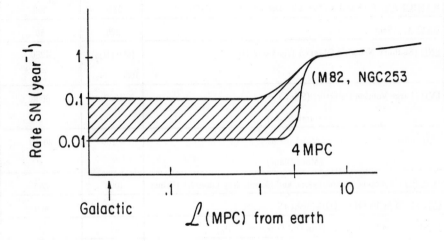

Figure 8 Estimated SN II rate as a function of distance
from Earth from the recent UCLA workshop.

TABLE 1

SUPERNOVA DETECTORS IN OPERATION, CONSTRUCION OR DESIGN		
FOR SUPERNOVA WATCH	MASS	RATE FOR GSN
IMB (UCIrvine-Michigan-Brookhaven) (H_2O)	6800	1300
KAMIOKANDE (Kamioka Neutrino Detector Exp.) (H_2O)	2140	500
BAKSAN (Scint.)	200	50
SNO (Sudbury Neutrino Observ.) ($H_2O + D_2O$)	1600 (H_{20}) 1000 (CO_{20})	1200
LVD (Large Volume Detector) (Scint.) (Gran Sasso)	1800	6000
LSD (Large Scintillation Detector) (Scint.) (Mont Blanc)	~ 100	30–50
MACRO (Monopole, Astrophysics, and Cosmic Ray Observ.) (Scint)	1000	350
CALTECH SCINTILLATOR (Scint.)	1000	400
LSND (Liquid Scintillating Neutrino Detector) (CH_4)	200	90
HOMESTAKE ($^{37}C\ell$, ^{137}I)	1000	420
ICARUS (Imaging of Cosmic And Rare Underground) (^{137}I)	1000–3000	60–180
BAKSAN ($^{37}C\ell$) Signals (^{40}Ar)	3000	130
SNBO (SuperNova Burst Observ.) ($CaCO_2$)	100,000	10,000
JULIA (Joint Underwater Lab. and Inst. for Astrophysics) (H_2O)	40,000	10,000
LOW TEMPERATURE DETECORS (?) (?)	~ 4	~ 150
≥ 15 OPERATING/PLANNED SN DETECTORS IN WORLD		

TABLE 2

SUPERNOVA NEUTRINO DETECTION IN THE 1990'S

REACTIONS:	$\bar{\nu}_e p \to e^+ n$	$\nu_x e \to \nu_x e$	$\nu_x N \to \nu_x N$	$\nu_x N \to \nu_x N^*$ ↳ n
PARAMETERS:				
CROSS SECTION	LARGE (KII, SK, IMB, LVD)	SMALL ~E_ν^2 (ICARUS)	LARGE For Coherent Process	LARGE At High E_{ν_x} SNO/SNBO
NEUTRINO ENERGY ESTIMATE	YES ~E_e	PARTIAL $E_{\nu_e} \sim f(E_e)$	NO	NO But a threshold may set E_{ν_x}
ν DIRECTION	NO	YES	NO	NO
TIME INFORMATION	YES	YES	YES	YES
DOWN TIME (guess)	≳ 10%	~ 30%	?	(could be small)
MAXIMUM DETECTOR SIZE	2×10^5 Tons (H$_2$O) LENA ≲ 10^4 Tons Liq. Scint. (LVD)	$\sim 2 \times 10^5$ Tons (H$_2$O) or ~ 10^3 Tons Cryogenic (ICARUS)	Kilograms No detector proposed so far	$\sim 10^5 - 10^7$ Tons of CaCO$_3$ (SNBO) or ~ 10^3 Tons D$_2$O(SNO)
BACKGROUNDS	SMALL If e^+ and n capture detected - OK for H$_2$O Galactic Signal	SMALL If directionality used to reject background	?	DEPENDS on Radioactivity of material

TABLE 3

COMPARISION OF FUTURE SUPERNOVA ν DETECTORS

PROCESS: DETECTORS:	$\bar{\nu}_e p \to e^+ n$ $\bar{\nu}_e d \to ppe^-$	$\bar{\nu}_e e \to \bar{\nu}_e e$	$\bar{\nu}_x e \to \bar{\nu}_x e$ $x = \mu, \tau$	$\nu_e N \to N^* \nu_e$ $\hookrightarrow n$	$\nu_\mu N \to N^* \nu_x$ $\hookrightarrow n$	ν_e Prompt
ICARUS (3kT)	–	~ 140	25	–	–	4^*
SNO (1kT)	~ 500	60	20	~ 200	~ 400	5^*
($D_2O + H_2O$) (H$_2$O Shield + D$_2$O)						$5 - 20^*$
LVD/MACRO (3kT) scent	~ 1000	–	–	–	–	–
Kam.II/IMB	(~ 480)	(~ 60)	(~ 20)	–	–	–
SUPER Kam. (30 kT) H$_2$O	~ 4000	~ 600	200	–	–	$\sim 5^*$
SNBO (100kT)	$(100\text{'s})?$			$\sim 100\text{'s}$	10,000	–
COMMENTS:	measure t_ν, $E_\nu \sim E_e$ No Direction	$t\nu$ E_ν estimated from E_e Θ_ν Measured		$t\nu$ only No E_ν ! No Θ_ν		$\Delta t \simeq 10$ ms

* Depends on Energy Spectrum of Prompt ν_e and Detector Threshold

TABLE 4

	Mass	Rate(ν_x)	Distance	M_ν
	To Measure $M_\nu \sim 2\,eV$			
	Must Use Extragalactic Supernova			
H_2O	3×10^4 T (Super Kam.)	~ 200	10 Kpc	20 eV
	(10^8) T	~ 200	0.6 Mpc	2 eV
$CaCO_2$	10^5 T (SNBO)	2×10^4	10 Kpc	20 eV
	(10^7) T	300	0.6 Mpc	2 eV
Cryogenic	10 T*	30*	10 Kpc*	20 eV*
(T < 300 MK)	(10^4) T*	30*	0.6 Mpc*	2 eV*

LIST OF PARTICIPANTS

David Alsop
Department of Physics
LeConte Hall
University of California Berkeley
Berkeley, CA 94720
tel: (415) 643-5105
fax: (415) 643-5204
e-mail: alsop%Lange.Berkeley.Edu@
    jade.berkeley.edu

Katsushi Arisaka
University of California, Los Angeles
405 Hilgard Avenue
Los Angeles, CA 90024-1547
tel: (213) 825-4925

M. Atac
MS 223 Fermilab
P.O. Box 500
Batavia, IL 60510
tel: (708) 840-3960
fax: (708) 840-4343

Barry Barish
Caltech
Div. of Math & Astronomy
256-48
Pasadena, CA 91125
tel: (818) 356-6684

Steven W. Barwick
Physics Dept.
UCI
Irvine, CA 92717
e-mail: barwick%price.hepnet@Lbl

Eric Becklin
Departments of Physics & Astronomy
University of California, Los Angeles
405 Hilgard Avenue
Los Angeles, CA 90024
tel: (213) 206-0280

David Bennett
Lawrence Livermore National Laboratory
IGPP, L4-13
Livermore, CA 94550
tel: (415) 423-0660
fax: (415) 423-0238
e-mail: gov%"bennett@sunlight.llnl.gov"

Lars Bergström
Department of Physics
Stockholm University
Vanadisvagen 9
S-113 46 Stockholm
Sweden
tel: [46-8] 16-46-12
fax: [46-8] 34-78-17
e-mail: jnet%"lbe@sesuf51"

Andrew Berkin
Department of Physics
Waseda University
Okubo 3-4-1
Shinjuku-ku, Tokyo 169
Japan
tel: (03) 203-4141 ext-732443
fax: (03) 200-2567
e-mail: berkin@jpnwas00

V. S. Berezinsky
Institute of Nuclear Research
Profsouuznaya 7A
Moscow 117312
USSR

Karl van Bibber
L-288
Lawrence Livermore National Laboratory
P.o. Box 808
Livermore, CA 94550
tel: (415) 423-8949
e-mail: pegasys@slacasp

Robert W. Birge
50-256
Lawrence Berkeley Laboratory
Physics
1 Cyclotron Road
Berkeley, CA 94720
tel: (415) 486-4421
fax: (415) 486-5401

Sidney Bludman
Center for Particle Astrophysics
301 Le Conte Hall
University of California, Berkeley
Berkeley, CA 94720
tel: (415) 642-9190
or (415) 486-7188
e-mail: bludman@lbl.gov

Furio Bobisut
Universita Degli Studi Di Padova
Dipartimento Di Fisica
"Galileo Galilei"
Via F. Marzolo, 8
35131 Padova
Italy
tel: (049) 844-111
fax: (049) 844-245
e-mail: bobisut@padova.infn.it

Felix Boehm
Norman Bridge Laboratory of Physics 161-33
Caltech
Pasadena, CA 91125
Tel: (818) 356-4281
fax: (818) 568-8263
e-mail: boehm@caltech

J. R. Bond
CITA
Department of Physics
University of Toronto
Toronto, ONT M5S 1A7
Canada

Alessandro Bottino
Dipartimento di Fisica Teorica
Universitá di Torino
Via P. Giuria 1
10125 Torino
Italy
tel: 39-11 6527202
fax: (39-11) 6699579
e-mail: VAXTO::Bottino
or Bottino@VAXTO.INFN.IT

Peter Bosetti
III. Physikalisches Institute
Lehrstuhl B fur Experimentalphysik
der RWTH AACHEN - Physikzentrum
Sommerfeldstrasse
D - 1500 Aachen
GERMANY

Richard N. Boyd
Department of Physics
Ohio State university
174 W. 18th Ave.
Columbus, OH 43210
tel: (6tel: (614) 292-2875
fax: (614) 292-7557
e-mail: boyd@ohstpy

David Brahm
Caltech 452-48
Pasadena, CA 91125
tel: (818) 356-6644
e-mail: brahm@theory3.caltech.edu

Kerry N. Buckland
Physics Department
University of California San Diego
9500 Gilman Drive
La Jolla, CA 92093-0319
tel: (619) 534-4940

Robert Burman
MP-Division MS H846
Los Alamos National Laboratory
Los Alamos, NM 87545
tel: (505) 667-5574
fax: (505) 665-1712
e-mail: burman@lampf

Adam Burrows
University of Arizona, Tucson
1118 E. 4th St.
Tucson, AZ 85721
tel: (602) 621-6820

Eric Carlson
Lyman Laboratory of Physics
Harvanrd University
Cambridge, MA 02138
tel: (617) 495-2868
fax: (617) 495-0416
e-mail: carlson@huhepl

Per Carlson
Mann Siegbahn Institute of Physics
Frescativagen 24
10405 Stockholm
Sweden
tel: 46 8 161044
fax: 46 8 158674
e-mail: pc@sesuf51

Michael Cherry
Louisiana State University
Department of Physics & Astronomy
202 Nicholson Hall
Baton rouge, LA 70803-4001
tel: (504) 388-8591
e-mail: cherry%phepds.span@star.stanford.edu

David B. Cline
Departments of Physics & Astronomy
University of California, Los Angeles
405 Hilgard Avenue
Los Angeles, CA 90024-1547
tel: (213) 825-1673
fax: (213) 206-1091
e-mail: laraneta@uclahep
or maryanne@uclahep

James M. Cline
Physics Department
Ohio State University
174 West 18th Street
Columbus, OH 43210
tel: (614) 292-6957
fax: (614) 292-8261
e-mail: jcline@ohstpy.bitnet

Thomas L. Cline
NASA/Goddard Space Flight Center
Laboratory for High Energy Astrophysics
Code 661 Greenbelt, MD 20771
tel: (301) 286-8375
fax: (301) 286-3391

John Cornwall
Physics Department
University of California Los Angeles
405 Hilgard Avenue
Los Angeles, CA 90024-1547
tel: (213) 825-3165

Donald Coyne
Physics Department/Physics Board
Natural Sciences 2
University of California Santa Cruz
Santa Cruz, CA 95064
tel: (408) 459-4754
fax: (408) 459-3043
e-mail: coyne@slactbf

R. L. Davis
Institute for Advanced Study
School of Natural Science
Princeton, NJ 08540
e-mail: bitnet%"davis@iassns.bitnet"

David S.P. Dearborn
Lawrence Livermore National Laboratory
Physics
P.O. Box 808 L-23
Livermore, CA 94550
tel: (415) 422-7219

Brian L. Dougherty
Varian Building Room 220
Physics Department
Stanford University
Stanford, CA 94305
tel: (415) 723-1880
fax: (415) 725-6544

Alan M. Dressler
Cargnegie Institution of Washington
Mt. Wilson & Las Campanas Observatories
813 Santa Barbara Street
Pasadena, CA 91101-1292
tel: (818) 304-0245
e-mail: dressler%ociw@caltech.edu

R.W.P. Drever
Caltech
130-33
Pasadena, CA 91125

Ruth Durrer
Princeton University
Physics Department
Joseph Henry Laboratory
Princeton, NJ 08544
tel: (609) 258-4371
fax: (609) 258-6360
e-mail: rd@pupthy.princeton.edu

George Efstathiou
Department of Physics
University of Oxford
Nuclear and Astrophysics Building
Keble Road
Oxford OX1 3RH
England

George T. Ewan
Queens University
Physics Department
Kingston
Ontario, K7L 3N6
Canada
tel: (613) 545-2698
fax: (613) 545-6463
e-mail: ewan@mips2.phy.queensu.ca

Hume A. Feldman
Canadian Institute for Theoretical Astorphysics
McLennan Laboratory
University of Toronto
60 St. Greorge Street
Toronto, ON M5S 1A1
Canada
tel: (416) 978-1777
fax: (416) 978-3921
e-mail: feldman@orca.cita.toronto.edu

Ervin Fenyves
Physics Program
University of Texas
P.O. Box 830688
Richardson, TX 75083
tel: (214) 690-2971
fax: (214) 690-2848

W. R. Frazer
Senior Vice President
Academic Affairs
Office of the President
University of California
300 Lakeside Drive 22nd Floor
Oakland, CA 94612-3550
tel: (415) 987-9020
fax: (415) 987-9209

Joshua A. Frieman
Theoretical Astrophysics MS-209
Fermilab
P.O. Box 500
Batavia, IL 60510
tel: (708) 840-2226
e-mail: frieman@fnal

George Fuller
University of California, San Diego
Department of Physics
Center for Astrophysics
San Diego, CA 92093
tel: (619) 534-6329
fax; (619) 534-0173
e-mail: gfuller@sdph1.ucsd.edu

Mary K. Gaillard
University of California Berkeley
Department of Physics
50A-3115
Berkeley, CA 94720
tel: (415) 642-3561

Kenneth S. Ganezer
EATON
Pacific-Sierra Research Cororpation
Santa Monica Boulevard
Los Angeles, CA 90025
tel: (213) 820-2200

Graciela Gelmini
Physics Department
University of California, Los Angeles
405 Hilgard Avenue
Los Angeles, CA 90024-1547
tel: (213) 825-4293
e-mail: gelmini@uclahep

Gilles Gerbier
CEN SACLAY
DPhPE/SEPh
F-9119 GIF-sur Yvette CEDEX
France
tel: 33-1 69 08 35 20
fax: 33-1 69 08 76 36
e-mail: dphcls::gerbier
or gerbier@frsac53

Giovanni Gervasio
INFN – Sezione di Milano
Via Celoria 16
20133 Milano
Italy
tel: 39-2-2392329
fax: 39-2-366583
e-mail: gervasio@vaxmi.infn.it

Gian Giudice
Theory Group
Department of Physics
University of Texas at Austin
Austin, TX 78712
tel: (512) 471-4073
fax: (512) 471-9637
e-mail: giudice@utaphy

Norman K. Glendenning
Nuclear Science Division – Theory Group
Unviersity of California Berkeley
MS 70A-3307
Lawrence Berkeley Laboratory
1 Cyclotron Road
Berkeley, CA 94720
tel: (415) 486-5420
fax: (415) 486-5401
e-mail: nkg@lbl.bitnet

Gerson Goldhaber
2160 50A
Lawrence Berkeley Laboratory
Berkeley, CA 94720
tel: (415) 486-6210
fax: (415) 486-5101
e-mail: gerson@slacvm

W. Glassley
Lawrence Livermore National Laboratroy
Livermore, CA 94550

Kim Griest
Department of Astronomy
University of California Berkeley
Berkeley, CA 94549
tel: (415) 642-4960
e-mail: bitnet%"griest@lbl.bitnet"

P. Gondolo
Physics Department
University of California Los Angeles
405 Hilgard Avenue
Los Angeles, CA 90024-1547
tel: (213) 825-8527

John Mace Grunsfeld
Caltech 220-47
Pasadena, CA 91125
tel: (818) 356-8400
e-mail: jmg@citsrl.caltech.edu

Elizabeth S. Hafen
Laboratory for Nuclear Studies
MIT
Cambridge, MA 02139
tel: (617) 253-6084
or (617) 253-6081
fax: (617) 565-8316
e-mail: chas@mitlns

Christian A. Hagmann
Department of Physics
Unviersity of California Berkeley
Berkeley, CA 94720
tel: (415) 642-8923
fax: (415) 642-8497

Todd Haines
Dept. of Physics & Astronomy
University of Maryland
College Park, MD 20742

Mark Halpern
Department of Physics
University of British Columbia
Vancouver, BC V6T 2A6
Canada
tel: (604) 228-2673
fax: (604) 228-5324
e-mail: halpern@ubcmtsg

David S. Hanna
Department of Physics
McGill University
3600 Unviersity Street
Montreal, P.Q. H3A 2Y8
Canada
tel: (514) 398-6510
fax: (514) 398-3733
e-mail: hanna@physics.mcgill.ca

Jay Hauser
Physics Department
University of California Los Angeles
Los Angeles, CA (0024-1547
fax: (213) 206-1091
e-mail: hauser@uclahep

Michael Hauser
Laboratory for Astronomy & Solar Physics
Code 685
NASA/Goddard Space Flight Center
Greenbelt, MD 20771
tel: (301) 286-8701
fax: (301) 286-8709

Wick Haxton
Physics Department
University of Washington
Seattle, WA
e-mail: haxton%gamow@phast.phys.washington.edu

Yudong He
Department of Physics
University of California Berkeley
Berkeley, CA 94720
e-mail: yudong@underdog.berkeley.edu

W.P. Hong
Physics Department
University of California, Los Angeles
405 Hilgard Avenue
Los Angeles, CA 90024-1547

Calvin Johnson
Div. of Physics, Math & Astronomy
Caltech
Pasadena, CA 91125
tel: (818) 356-4241

Guiru Jing
Department of Physics
University of California Berkeley
Berkeley, CA 94720

Mark Kamionkowski
Enrico Fermi Institute
5640 South Ellis Avenue
Chicago, IL 60637
tel: (312) 702-6041
e-mail: edu%"marck@oddjob.uchicago.edu"

Andrzej Kotucki
Department of Physics
University of British Columbia
Vancouver, B.C. V6T 2A6
Canada
tel: (604) 228-3645
fax: (604) 228-5324
e-mail: userbrit@ubcmtsg

D. M. Kielczewska
Department of Physics
University of California Irvine
Irvine, CA 92717

Ernst Kreysa
Max-Planck-Institut für Radioastronomie
Aut Dem Hügel
D-5300 Bonn1
Germany
fax: (02 28) 525-229

Arthur Kosowsky
Unviersity of Chicago
Physics Department
5640 South Ellis Avenue
AAC 137
Chicago. IL 60637
tel: (312) 702-8203
fax: (312) 702-5863
e-mail: arthur@oddjob.uchicago.edu

William Kropp
Physics Department
University of California Irvine
Irvine, CA 92717

George R. Lake
Department of Astronomy FM-2
University of Wahsington
Seattle, WA 98195
tel: (206) 543-71-6
fax: (206) 685-0403
e-mail: lake@uwaphast.bitnet

Kenneth Lande
Astronomy & Astrophysics Dept.
University of Pennsylvania
Philadelphia, PA 19104

Andrew Lange
Department of Physics
366 LeConte Hall
Unviersity of California Berkeley
Berkeley, CA 94720
tel: (415) 642-6577
fax: (415) 643-5204
e-mail: Lange.Berkeley.Edu@lilac.berkeley.edu

Mark Le Gros
University of British Columbia
Department of Physics
6224 Agricultural Road
Vancouver, B.C. V6T 2A6
Canada
tel: (604) 228-4334
faxL (604) 228-5324
e-mail: userbrit@ubcmtsg.bitnet

Steven Levin
JPL
MS 506 Building 169
4800 Oakwood Drive
Pasadena, CA 91109
tel: (818) 354-1917

Gang Liu
Department of Physics
University of California Irvine
Irvine, CA 92717

Philip Lubin
Physics Department
University of California Santa Barbara
Santa Barbara, CA 93106
tel: (805) 893-8432
e-mail: lubin@voodoo

Jane MacGibbon
Code 665
NASA/Goddard Space Flight Center
Greenbelt, MD 20771
tel: (301) 286-3966
fax: (301) 286-3391
e-mail: postmaster@dftbit

Bengt Magnusson
Department of Physics
University of California Santa Barbara
Santa Barbara, CA 93106
tel :(805) 893-2058
fax: (805) 893-8597
e-mail: bengt@voodoo
e-mail: voodoo::bengt

William Mahoney
JPL
4800 Oak Grove Dr.
Bldg 180, Rm 904
Pasadena, CA 91109
tel: (818) 354-3405

Robert A. Malaney
L-413
IGPP
Lawrence Livermore National Laboratroy
Livermore, CA 94550
tel: (415) 423-8129
fax: (415) 423-0238
e-mail: ram@jake.llnl.gov

Richard Markeloff
Univ. of Wisconsin-Madison
High Energy Physics
MS 223
Fermilab
Batavia, IL 60510
e-mail: fnald::markeloff

John Mather
NASA/GSFC
Code 685 Greenbelt, MD 20771
tel: 301-286-8720
email: 6959::mather

W. McLatchie
Dean of Graduate Studies & Research
Queen's University
Kingston
Ontario K7L 3N6
Canada
tel: (6130 545-6079
fax: (613) 545-6300

Ian McLean
Astronomy Department
University of California, Los Angeles
405 Hilgard Avenue
Los Angeles, CA 90024
tel: (213) 825-1140

Brad Meyer
L-413
LLNL
Livermore, CA 94550
e-mail: gov%"meyer%sunlight.llnl.gov@lll-lc.llnl.gov

Stephan S. Meyers
Massachusetts Institute of Technology
Center for Space Research
Room 20B-145
18 Vassar Street
Cambridge, MA 02139
tel: (617) 253-8153 (office)
tel: (617) 253-4824 (lab)
fax: (617) 253-7014
e-mail: meyer@ir.mit.edu

Agnieszka Mlinowska
University of California Berkeley
Berkeley, CA 94720

Emil Mottola
Theoretical Division T-8
Los Alamos National Laboratory
M.S. B285
Los Alamos, NM 87545
tel: (505) 667-7646
fax: (505) 665-3003
e-mail: emil%pion@lanl.gov

Brian Newport
University of Chicago
Enrico Fermi Institute
c/o Physics Department
University of Utah
Salt Lake City, UT 84112
tel: (801) 581-4296
fax: (801) 581-4801
e-mail: brainor edu%"brian.newport@
   physics.utah.edu"

Angela Olinto
University of Chicago
5640 South Ellis Avenue
Chicago, IL 60637
tel: (317) 702-8206

James Pantaleone
Physics Department
University of California
Riverside, CA 92521
tel: (714) 787-5602
e-mail: pantaleone@ucrphys
or: 28018::pantaleone

Jun Park
Physics Department
University of California Los Angeles
405 Hilgard Avenue
Los Angeles, CA 90024-1547
tel: (213) 825-2673
fax: (213) 206-1091
e-mail: park@uclahep

Roberto Peccei
Chairman
Physics Department
University of California Los Angeles
405 Hilgard Avenue
Los Angeles, CA 90024-1547
tel: (213) 825-4935
e-mail: peccei@uclahep

622

P. James E. Peebles
Princeton University
Department of Physics
Jadwin Hall
P.O. Box 708
Princeton, NJ 08544
tel: (609) 452-4386
or (609) 258-4386

Carl Pennypacker
University of California, Berkeley
Space Sciences Lab.
Berkeley, CA 94720
e-mail: 42084::Pennypacker

Leroy Price
Physics Department
University of California Irvine
Irvine, CA 92717
tel: (714) 856-6995

Stuart Raby
Physics Department
Ohio State University
Columbus, OH 43210
e-mail: raby@ohstpy

Subhash Rajpoot
Physics & Astronomy
CSU Long Beach
1250 Bellflower Blvd.
Long Beach, CA 90840
tel: (213) 985-4943
e-mail: jfkg0001@beach1.csulb.edu

Bharat Ratra
Theoretical Astrophysics 130-33
Caltech
Pasadena, CA 91125
tel: (818) 356-8307
or (818) 356-4597
e-mail: bvr@tapir.Caltech.EDU

Frederick Reines
Department of Physics
University of California, Irvine
Irvine, CA 92717

Ted Ressell
University of Chicago
Department of Astronomy & Astrophysics
5640 S. Ellis Ave
Chicago, IL 60637
e-mail: ressell@oddjob.uchicago.edu

Jeffrey Robbins
Physics Editor
Oxford University Press
200 Madison Avenue
New York, NY 10016
tel: (212) 679-7300 x-7269
fax: (212) 725-2972

Sun Hong Rhie
Lawrence Livermore National Laboratory
IGPP, L-413
700 East Avenue
Livermore, CA 94550
tel: (415) 294-5805
e-mail: sunhong@sunlight.llnl.gov

Leslie J. Rosenberg
Enrico Fermi Institute
University of Chicago
Chicago, IL 60637
tel: (312) 702-7486
fax: (312) 702-1914
e-mail: ljr@fnal

Leszek Roszkowski
CERN
1211 Geneva 23
Switzerland
tel: (4122) 767-4411
fax: (4122) 782-3914
e-mail: LeszekR@cernvm

Hector Rubinstein
Physics Department
University of Stockholm
Vanadisvagen 9
S-113 46 Stockholm
Sweden
e-mail: rug@sesuf51.bitnet

Bernard Sadoulet !
Director
Center for Particle Astophysics
NSF Science and Technology Center
University of California, Berkeley
301 LeConte Hall
Berkeley, CA 94720
tel: (415) 642-4705
fax: (415) 642-1756

Michael Salamon
Physics Department
201 JFB
University of Utah
Salt Lake City, UT 84112
tel: (801) 581-4785
fax: (801) 581-4801
e-mail: salamon@physics.utah.edu
or salamon@utahcca.bitnet

Martin John Savage
Serin Physics Laboratory
Freylinghuysen Road
Busch Campus
Theoretical Physics
Rutgers University
Piscataway, NJ 08855-0849
tel: (201) 932-4374
e-mail: savage@ruthep

David Seckel
Bartol Research Institute
University of Delaware
Newark, DE 19716
tel: (302) 451-1846
fax: (302) 451-1843
e-mail: seckel%bartol.span@star.stanford.edu

E. Shaya
University of Maryland
College Park, MD 20742
e-mail: shaya@umaip.span

Dennis Silverman
Physics Department
University of California Irvine
Irvine, CA 92717
tel: (714) 856-5149
e-mail: djsilver@uci

Michael Smith
W.W. Kellogg Radiation Laboratory
Caltech
Pasadena, CA 91125
tel: (818) 356-3745
fax: (818) 564-8708
e-mail: msmith@caltech

Peter F. Smith
Ruthorford Labs
Shilton
Didcot
Oxford OX11 0QX
England
fax: 44-235-446-733

George R. Smoot
Lawrence Berkeley Laboratory
Physics Division
50-232
1 Cyclotron Rd.
Berkeley, CA 94720
tel: (415) 486-5237

Dan Snowden-Ifft
Dept. of Physics
LeConte Hall
Univeristy of California, Berkeley
Berkeley, CA 94720

Henry W. Sobel
Physics Department
University of California, Irvine
Irvine, CA 92717
e-mail: hsobel@ucivmsa

Floyd Stecker
NASA Goddard Space Flight Center
Code 662
Greenbelt, MD 20771
tel: (301) 286-6057
fax: (301) 286-3391
e-mail: stecker@lheavx.nasa.gov

Chris Stoughton
Fermilab
P.O. Box 500
Batavia, IL 60510
tel: (708) 840-2440
fax: (708) 840-3867
e-mail: stoughto@fnal

Nikolaos Tetradis
Physics Department
Stanford University
Stanford, CA 94305
tel: (415) 723-4232
e-mail: visth@slacvm

R. Tuluie
Center for Relativity
University of Texas at Austin
Austin, TX 78712
e-mail: jnet%"tuluie@utaphy"

Michael S. Turner
University of Chicago
Department of Astronomy & Astrophysics
5640 S. Ellis Avenue
Chicago, IL 60637
tel: (312) 702-7974

Neil Turok
Department of Astronomy
Princeton University
Princeton, NJ 08544

Brian G. Turrell
Head of Physics
Department of Physics
University of British Columbia
6224 Agricultural Rd.
Vancouver, B.C. V6T 2A6
Canada
tel: (604) 228-3150
fax: (604) 228-5324
e-mail: userbrit@ubcmtsg

Roger Ulrich
Astronomy Department
University of California Los Angeles
405 Hilgard Avenue
Los Angeles, CA 90024
tel: (213) 825-4270

Tanmay Vachaspati
Dept. of Physics & Astronomy
Tufts University
Medford, MA 02155 e-mail: tvachasp@tufts

Jin Wang
University of Illinois
Astronomy & Astrophysics
1002 West Green Street
Urbana, IL 61801
tel: (217) 244-5469
fax: (217) 244-7638
e-mail: jinwang@rigel.astro.uiuc.edu

Richard Watkins, Jr.
Astronomy and Astrophysics Center
University of Chicago
5640 South Ellis Avenue
Chicago, IL 60637
tel: (312) 702-9752
e-mail: watkins@oddjob.uchicago.edu

James R. Wilson
LLNL
L-035
P.O. Box 808
Livermore, CA 94550
tel: (415) 422-1659
fax: (415) 422-3389

Ed Wishnow
Department of Physics
University of British Columbia
Vancouver
Canada
tel: (604) 228-2673

Karl Y. Yee
Physics Department
University of California, Irvine
Irvine, CA 92715
tel: (714) 856-6870